Signal Processing for Security Technologies

Series Editor

M. Emre Celebi
Baton Rouge, Louisiana, USA

More information about this series at http://www.springer.com/series/13765

Richard Jiang • Somaya Al-maadeed
Ahmed Bouridane • Danny Crookes
Azeddine Beghdadi
Editors

Biometric Security and Privacy

Opportunities & Challenges in The Big Data Era

Springer

Editors
Richard Jiang
Department of Computer
and Information Science
Northumbria University
Newcastle upon Tyne
United Kingdom

Ahmed Bouridane
Department of Computer
and Information Science
Northumbria University
Newcastle upon Tyne
United Kingdom

Azeddine Beghdadi
Institut Galilée
Université Paris 13
Paris, France

Somaya Al-maadeed
Department of Computer Science
and Engineering
Qatar University
Doha, Qatar

Danny Crookes
School of Electronics, Electrical Engineering
and Computer Science
ECIT Institute, Queen's University Belfast
Belfast, Antrim, UK

Signal Processing for Security Technologies
ISBN 978-3-319-47300-0 ISBN 978-3-319-47301-7 (eBook)
DOI 10.1007/978-3-319-47301-7

Library of Congress Control Number: 2016958827

Printed on acid-free paper

This Springer imprint is published by Springer Nature
The registered company is Springer International Publishing AG
The registered company address is: Gewerbestrasse 11, 6330 Cham, Switzerland

Preface

Biometrics in modern computer science is defined as the automated use of biological properties to identify individuals. The early use of biometrics can be dated back to nearly 4000 years ago when the Babylon Empire legislated the use of fingerprints to protect a legal contract against forgery and falsification by having the fingerprints impressed into the clay tablet on which the contract had been written. Nowadays, the wide use of the Internet and mobile devices has brought out the booming of the biometric applications, and research on biometrics has been drastically expanded into many new domains.

The research trends in biometric research may be categorized into three directions. The first direction is toward the broader Internet and mobile applications. This brings out a number of new topics to utilize biometrics in mobile banking, health care, medical archiving, cybersecurity, and privacy as a service, etc. These new applications have created a huge market of billion dollars for biometric technologies and the industry needs comes back to push the research further and vigorously. The second direction is towards algorithmic development, which includes the investigation of many new AI techniques in biometrics, such as fuzzy approaches, ensemble learning, and deep learning. These new approaches can often help improve the accuracy of automated recognition, making many new applications available for business. Especially, with the vast amount of data coming from billions of users on internet/mobile, biometrics now becomes a new Big Data challenge in its streaming, processing, classification and storage. The third research direction aims at discovering more types of biometrics for various uses. Besides the conventional fingerprints and signatures, other types of biometrics (such as iris, vein pattern, gait, and touch dynamics) have been investigated in recent biometric research. Their combination as multimodal biometrics is another popular way to exploit these types of biometrics in research.

This book includes 16 chapters highlighting recent research advances in biometric security. Chapters 1–3 present new research developments using various biometric modalities including Fingerprints, Vein Patterns and Palmprints. New tools and techniques such as Deep Learning are investigated and presented. Chapter 4 reports a new biometric recognition approach based on the acoustic features of human ears. Chapters 5–9 discuss new research works relating to a number of dynamic behavioural biometric traits. Chapters 10–13 focus on face recognition, which is the most popular topic in biometrics. Chapter 14 carries out a survey of biometric template protection, a very important topic in biometric privacy and security. Chapter 15 investigates the use of biometrics for better security in cloud computing and Internet of Things. Chapter 16 reports the new EU legislation on biometrics, which should help technology developers be aware of the legal aspects of biometric technologies.

The target audience for this book includes graduate students, engineers, researchers, scholars, forensic scientists, police force, criminal solicitors, IT practitioners and developers who are interested in security and privacy related issues on biometrics. The editors would like to express their sincere gratitude to all distinguished contributors who have made this book possible, and the group of reviewers who have offered insightful comments to improve the quality of each chapter. A dedicated team at Springer Publishing has offered professional assistances to the editors from inception to final production of the book. We thank them for their painstaking efforts at all stages of production.

Richard Jiang
Newcastle upon Tyne, UK

Contents

Chapter 1
Fingerprint Quality Assessment: Matching Performance and Image Quality

Zhigang Yao, Jean-Marie Le Bars, Christophe Charrier, and Christophe Rosenberger

1.1 Introduction

The disadvantage of biometric recognition systems is chiefly attributed to the imperfect matching in contrast with traditional alphanumeric system. Because of this, sample quality is more important for image-based biometric systems, and so is fingerprint image used for the Automatic Fingerprint Identification System (AFIS). Matching of fingerprint images is generally divided into three classes: correlation-based, image-based, and minutiae matching, among which the last one is acknowledged as the primary solution so far [10]. In this case, good quality sample is basically a prerequisite for extracting reliable and sufficient minutia points, and is hence the essential factor for the overall matching performance. The effect of sample quality to the matching performance is defined as the utility of a biometric sample [12]. Therefore, most of the Fingerprint Quality Assessment (FQA) approaches (or fingerprint quality metrics) rely on two aspects: subjective assessment criteria of the pattern [8] and sample utility. In addition, most of the quality metrics are also evaluated in terms of the utility. [1]. However, this property is limited by matching configurations, i.e., sample utility varies as the matching algorithm changes because no matching approach proposed so far is perfect or robust enough in dealing with different image settings though their resolution is similar to each other (normal application requires gray-level images of 500-dpi according to the ISO).

This chapter compares the existing solutions of the FQA in terms of a methodological categorization [4]. Such a comparison analyzes whether those quality metrics based on multi-feature are really able to take the advantages of the employed

Z. Yao • J.-M. Le Bars • C. Charrier • C. Rosenberger (✉)
Normandie Univ, UNICAEN, ENSICAEN, CNRS, GREYC, 14000 Caen, France
e-mail: zhigang.yao@ensicaen.fr; jean-marie.lebars@unicaen.fr; christophe.charrier@unicaen.fr; christophe.rosenberger@ensicaen.fr

R. Jiang et al. (eds.), *Biometric Security and Privacy*, Signal Processing for Security Technologies, DOI 10.1007/978-3-319-47301-7_1

features. Similarly, quality assessment approaches rely on a prior-knowledge of matching performance still need discussion, especially the prediction to the matching performance. Our work gives a study of these potential problems in an experimental manner. Each of the selected quality metrics in this chapter represents a typical solution in the existing studies.

This chapter is organized as follows: Sect. 1.2 presents a brief review of the categorization of the existing FQA solutions. In Sect. 1.3, the description of trial fingerprint quality metrics is given. Experimental results are given in Sect. 1.4. Section 1.5 concludes the paper.

1.2 Background

Yao et al. [4] categorize prior work in FQA into several classes in terms how this problem is solved. Typical FQA solutions can be summarized as:

1. Single feature-based approaches: these could be further divided into solutions rely on the feature itself or a regularity [18] observed from the employed feature. For instance, standard deviation [13] at block-wise is a brief factor which somehow measures the clarity and differentiates the foreground block of fingerprint. Some studies also obtain relatively good result by using a single feature, such as the Pet's hat wavelet (CWT) coefficients [16] and the regularity of fingerprint Discrete Fourier Transform (DFT) [6], and Gabor feature [17]. These features also represent the solution of FQA in different domain. In addition, the "relatively good result" here means that those solutions perform well in reducing the overall matching performance because we believe that the evaluation of a quality metric is basically a biometric test which involves both genuine matching and impostor matching errors.
2. FQA via segmentation-like operations: these kinds of solutions are divided into two vast classes at first, including global-level and local-level approaches. Mostly, local-level approaches estimate a quality measure to a fingerprint block in terms of one or several features or indexes, such as directional information and clarity [3, 9, 13, 15]. Some other local-level approaches choose to determine whether a block is a foreground at first [23], and then give a global quality measure to the fingerprint image. This type of solutions implemented globally are further divided as non-image quality assessment and image-based approach. Yao et al. [4] propose one FQA approach by using only minutiae coordinates, meaning that no real image information is used in assessing fingerprint quality. Image-based solutions are basically achieved by performing a segmentation at first, and then estimate the quality of the foreground area according to one or more measurements [4].
3. FQA approaches by using multi-feature: these could be carried out by using either fusion or classification. For example, some studies combine several quality features or indexes together via a linear (or weighted) fusion [5, 7, 15, 25]. The linear fusion is basically used for a specific scenario because coefficient is a constraint of this kind of solution. Similarly, fusion of multiple features or

experts outputs could also be achieved via other more sophisticated approaches such as Bayesian statistics [20] and Dezert-Smarandache (DS) theory [26]. The effectiveness of the fusion algorithm itself and differences between multiple experts outputs impact on the fused result. For instance, it is quite difficult to look out an appropriate way to fuse results generated by two different metrics, where one gives continuous output and another generalize a few discrete numbers. This chapter considers only FQA problem of the AFIS rather than any multi-modal, score/cluster-level fusion, and some fusion related issues.

FQA via multi-feature classification [14, 15] basically employs one (or more) classifier(s) to classify fingerprint image into different quality levels. Obviously, this kind of solution depends on the classifier itself. In addition, the robustness and the reliability of the prior-knowledge used by learning-based classification or fusion also impacts on the effectiveness of the quality metric, particularly when generalizing a common solution such as the state-of-the-art (SoA) approach [24].

In addition, some studies propose to use knowledge-based feature by training a multi-layer neural network [18]. However, it is essentially an observed regularity of the learnt feature and external factors such as classifier and tremendous training data set are also required.

According to the discussion above, one can note that fingerprint quality is still an open issue. Existing studies are mostly limited in these kinds of solutions, where learning-based approaches are chiefly associated with the prior-knowledge of matching performance which is debatable for a cross-use. Grother and Tabassi [10] have introduced that quality is not linearly predictive to the matching performance. This chapter gives an experimental study to analyze this problem by comparing FQA approaches selected from each of the categorized solutions.

1.3 Trial Measures

In order to observe the relationship between the quality and the matching performance, several metrics carried out by using each of the categorized solution are employed in this study, given as follows.

1.3.1 Metrics with Single Feature

As mentioned in Sect. 1.2, we first choose one quality metric generalized by using a single feature. The selected metric is implemented via the Pet's Hat continuous wavelet, which is denoted as the CWT as mentioned in Sect. 1.2. The CWT in a window of W is formulated as

$$\text{CWT} = \sqrt{\frac{\sum_W |c_i|}{W}} \tag{1.1}$$

where c_i is wavelet coefficient and the windows size depends on the image size, for example, 16 pixels for gray scale images of 512 dpi. In our study, the CWT is implemented with two default parameters, a scale of 2 and angle of 0. We choose this quality metric because it outperforms the SoA approach in reducing the overall error rate for some different image settings. **Note** that the resolution of fingerprint image is about 500-dpi, which is the minimum requirement of the AFIS [19].

1.3.2 Segmentation-Based Metrics

Fingerprint segmentation is one way to separate the foreground area (ridge-valley pattern) from the background (vacuum area) formed by input sensor(s). This operation is in some measure equivalent to the quality assessment of a fingerprint image because the matching (or comparison) is mainly dependent on the foreground area. It is reasonable that a fingerprint image with relatively clear and large foreground area can generate a higher genuine matching score than those characterized in an opposite way. In this case, many studies use segmentation-based solutions to perform quality assessment. This section gives two metrics based on segmentation-like operations to show how foreground area is important to quality assessment. The first one is an image-independent quality metric and the second is dependent on the image pixel (Fig. 1.1).

1.3.2.1 FQA via Informative Region

The image-independent approach employed in this chapter is known as the MQF [29] which uses only the coordinates information of the minutiae template of the associated fingerprint image. Figure 1.2 gives a general diagram of this quality metric.

As depicted by the diagram (Fig. 1.2), the convex-hull and Delaunay triangulation are used at first for modeling the informative region of a fingerprint image in

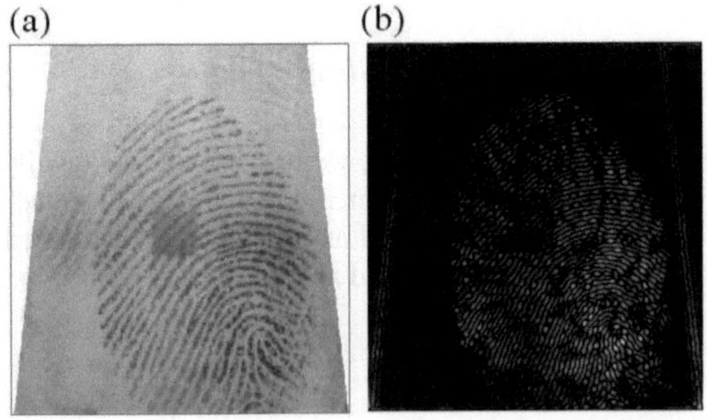

Fig. 1.1 Example of CWT (**b**) of a fingerprint image (**a**)

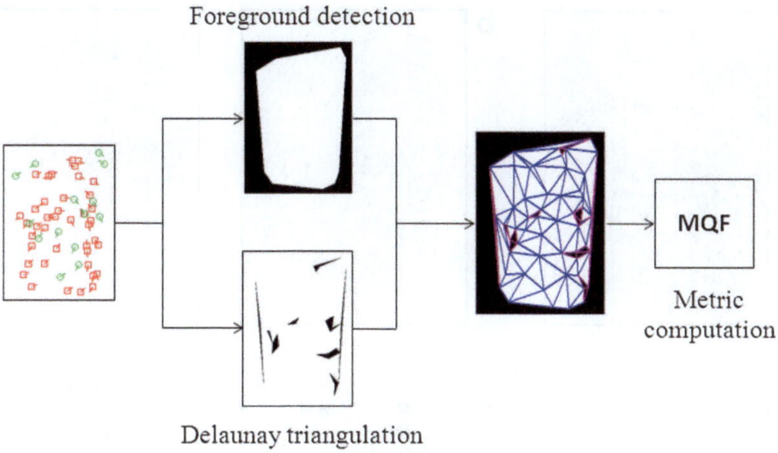

Fig. 1.2 Diagram of the framework of the MQF

terms of the detected minutiae points. Next, some unreasonable-looking triangular areas marked by pink are removed from the informative region. The remaining area of the informative region hence represents the quality of the associated fingerprint or the minutiae template [29].

This quality metric is chosen because it is a new solution of the FQA and it outperforms the SoA approach in some cases though only minutiae coordinates are used. The details of this metric can be found in the reference article and are not given here.

1.3.2.2 FQA via Pixel-Pruning

Another segmentation-based quality metric is denoted as MSEG [4] which performs a two-step operation to a fingerprint image, including a coarse segmentation and a pixel-pruning operation. The pixel-pruning is implemented via categorizing fingerprint quality into two general cases: desired image and non-desired image. Figure 1.3 illustrates such a categorization.

Obviously, an AFIS basically prefers keeping images like Fig. 1.3a because it is more probably to give reliable and sufficient feature. Figure 1.3b shows two images that are not desired subjectively because the left one has some tiny quality problems and the right one is relatively small and both may lead to low genuine matching or high impostor matching. In this case, a better quality assessment can be done if one can make a clearer difference between the desired image and the non-desired image. The MSEG employs a gradient measure of image pixel to prune pixels of non-desired image as much as possible. Figure 1.3c, d illustrates the result of pixel-pruning operation of two kinds of images.

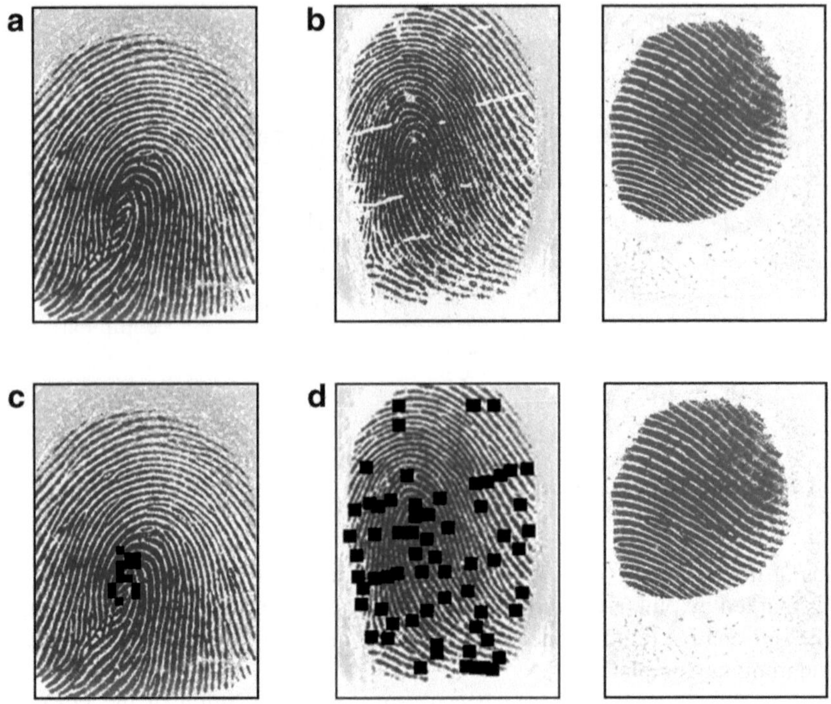

Fig. 1.3 Demonstration of pixel-pruning approach. (a) Desired. (b) Non-desired. (c) Desired after pixel-pruning. (d) Non-desired after pixel-pruning

1.3.3 FQA via Multi-feature

Similarly, we also choose two quality metrics that rely on multi-feature and both are implemented via a prior-knowledge of matching performance. By using this kind of solutions, an experimental comparison can be made between different approaches, especially one can find that solutions based on multiple features do not really take the advantages of the employed features because of the effect generated by the variation of image specifications, so is the employed prior-knowledge generated form the big data [21]. The first one is classification-based approach which is the SoA solution known as NFIQ [24]. The NFIQ estimates a normalized matching score of a fingerprint sample by sending a set of quality features (11 features) to a neural network model. The NFIQ algorithm remapped the estimated matching score into five classes denoted by integers from 1 to 5 where 1 indicates the best quality level.

On the other hand, we choose quality metric based on multi-feature fusion which is actually a No-reference Image Quality Assessment (NR-IQA) [22] solution used for FQA by integrating multiple features with a set of weighted coefficients. The selected approach is denoted as Qabed [7], which is basically defined as

$$Q = \sum_{i=1}^{N} \alpha_i F_i \,, \tag{1.2}$$

where N is the number of quality features F_i $(i = 1, \ldots, N)$, α_i are the weighted coefficients obtained by optimizing a fitness function of a genetic algorithm. The fitness function is defined as a correlation between linearly combined quality value and genuine matching score [11]. Maximizing such a linear relation is somehow equivalent to the concept that quality predicts matching performance. The weighted coefficient is dependent on a training set of fingerprint samples. We choose this approach because it performs well in predicting the matching performance in comparison with the SoA quality metric.

1.4 Experimental Results

Some existing studies propose to calculate correlation between different metrics [2] for comparing the behavior of them. However, this is not completely observable, because there is no explicit linear relation among every group of variant quality metrics. Generally, this kind of measure is to observe the similarity between two different variables such as wavelet coefficients. In this case, to compare the studied metrics, we simply provide experiment results of two evaluation approaches, one is a validation approach relied on Enrollment Selection (ES) [30] and another is an evaluation method with multiple bins of sorted biometric samples [6].

1.4.1 Software

In the experiment, we use two matching systems where one is the OpenSource NBIS [27] and another is a commercial fingerprint SDK known as "id3". The NIST software contains several modules, among which the MINDTCT is used for generating INCITS 378-2004 standard minutiae template and the matching scores are calculated via Bozorth3. The commercial SDK has six options of the existing minutiae template standards and the minutiae templates of ISO/IEC 19794-2:2005 standard [19] have been extracted in the experiment. Similarly, a corresponding matcher has also been implemented with the SDK. By using these two sets of programs, the comparative study is accomplished via an interoperate analysis of the selected quality metrics.

Table 1.1 Dataset specification

DB	Sensor	Dim.	Resolution (dpi)
00DB2A	Low-cost capacitive	256×364	500
02DB2A	Optical	296×560	569
04DB1A	Optical	640×480	500
04DB2A	Optical	328×364	500
04DB3A	Thermal	300×480	512
CASL2	Optical	328×356	512
CASR2	Optical	328×356	512

1.4.2 Database et Protocol

In the experiment, one dataset of the 2000 Fingerprint Verification Competition (FVC) test, one of FVC2002, three of FVC2004, and two CASIA[1] datasets are employed. Each of the FVC datasets includes 800 images of 100 individuals, 8 samples per individual. The CASIA database contains fingerprint images of 4 fingers of each hand of 500 subjects, where each finger has 5 samples. In this study, we create the two re-organized databases by using samples of the second finger of each hand, and they are, respectively, denoted as CASL2 and CASR2. Therefore, each sub-database has 2500 images of 500 individual (5 samples per individual) (Table 1.1).

The image size of each dataset is different from one another and the resolution is over 500-dpi. A glance of the datasets is given by several samples in Fig. 1.4. In this study, the experiment includes two parts, one is utility-based evaluation and another is quality-based evaluation. The evaluation approach employed in the experiment is based on the Enrollment Selection (ES) [28].

1.4.3 Results

1.4.3.1 ES with Quality

The evaluation task is a comparison between variant frameworks of fingerprint quality metric. We use each group of quality values and two types of matching scores to perform enrollment selection for each dataset. The global Equal Error Rate (EER) values obtained by the selected quality metrics are given in Table 1.2.

One can found that the quality metrics providing the lowest global EER are not always ones based on multi-feature, even for an associated vendor such as NBIS matching software of the NFIQ. The quality metric based on a single feature

[1]http://biometrics.idealtest.org/detailsDatabase.do?id=3.

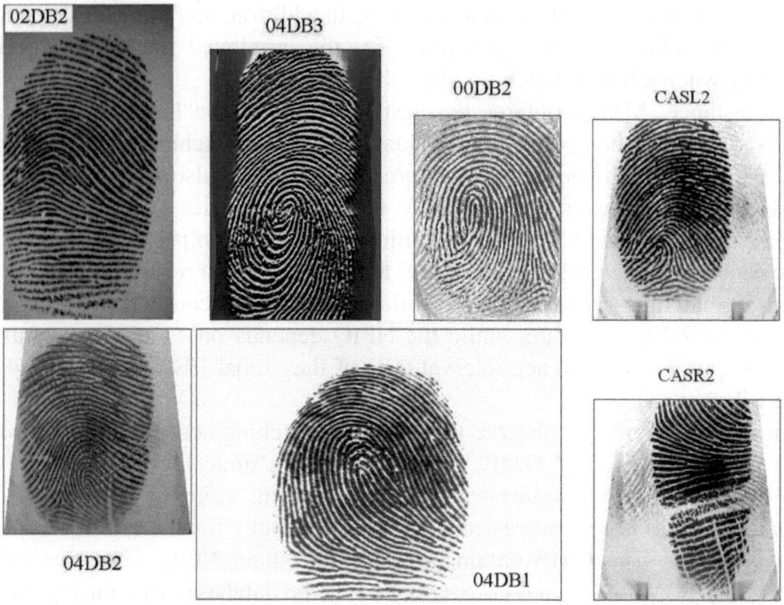

Fig. 1.4 Illustration of dataset samples

Table 1.2 Global EERs obtained via ES with quality metrics

	QM				
DB	NFIQ (%)	QMF (%)	MQF (%)	CWT (%)	MSEG (%)
00DB2A (N.)	4.97	6.57	5.03	4.93	**4.50**
02DB2A (N.)	13.33	11.11	11.18	11.11	**10.79**
04DB1A (N.)	15.37	**14.72**	14.98	17.53	16.54
04DB2A (N.)	**13.32**	16.64	15.02	14.16	14.05
04DB3A (N.)	7.47	7.36	**6.87**	7.00	7.18
CASL2 (N.)	43.09	40.64	40.48	**40.09**	42.30
CASR2 (N.)	43.51	41.39	40.62	**40.45**	43.20
00DB2A (S.)	0.22	0.40	0.76	**0.09**	0.10
02DB2A (S.)	0.11	0.30	0.12	**0.10**	0.20
04DB1A (S.)	2.66	1.74	**1.73**	1.91	1.93
04DB2A (S.)	3.86	3.94	3.43	3.33	**3.24**
04DB3A (S.)	1.89	1.66	1.51	1.59	**1.51**
CASL2 (S.)	40.92	42.72	42.19	42.35	**38.61**
CASR2 (S.)	38.20	41.26	40.94	39.70	**35.97**

"NBIS" and "SDK" are two sets of matching scores

Note: NFIQ and QMF rely on multi-feature and prior-knowledge of GMS; MQF and MSEG are based on segmentation, CWT is a single feature-based metrics

(CWT) also performs well on many datasets. In addition, both the CWT and MSEG demonstrate relatively good generality for the employed matching algorithms, especially when a better matching algorithm is involved.

For instance, MSEG obtains the best results from the last four of the seven employed dataset when performing evaluation with the matching scores of the SDK, while the results obtained from other three databases are also not bad. Particularly, MSEG decreases the error rates more than other metrics for the two difficult databases: CASL2 and CASR2. In addition, the CWT also performs well for most of the databases. The QMF and NFIQ do not give dominant results, especially when the NBIS matching scores are used in the experiment because QMF relies on the GMS of the NBIS software, while the NFIQ depends on 11 quality features (or real metrics). The confidence interval (CI) of the global EER values are given in Table 1.3.

Furthermore, one can observe the effect of matching scores to the knowledge-based metrics: NFIQ and QMF. The NFIQ obtains quite high (bad) EER values from the two CASIA datasets when NBIS matching scores are employed in the evaluation, while it generalizes relatively better results for the two datasets when using the SDK. The QMF obtains better results than NFIQ from five (02DB2, 04DB1, 04DB3, CASL2, and CASR2) of the seven databases when using the NBIS matching scores because its training is independently performed for each dataset via the NBIS matching scores, meaning it is appropriate to a specific scenario. However, in comparison with the knowledge-free metrics, both the two metrics do not show a higher performance though they employ different sets of features. Meanwhile, the MQF is a no-image quality metric but the performance is not bad in comparison with the NFIQ and QMF, especially when using the NBIS matching scores because it relies on the minutiae extractor associated with the NBIS software. In this case, one can observe that a good matching algorithm and a relatively good dataset (such as 00DB2, 02DB2 and 04DB3) may blurs the effect of a quality metric, i.e., it is easier to approach to a relatively better performance if the matcher is relatively robust. Thus, it is really necessary to perform an offline biometric test via "bad" datasets. In addition, it is possible to consider that the implementation of a metric should be independent from the matching performance if we emphasize its "generality." The effect of matching performance to quality metrics is further discussed in Sect. 1.4.4.

1.4.3.2 Isometric Bins

The ES with sample's quality reveals the best of quality metrics' capability in reducing the error rate. In this section, another evaluation is performed by using an approach based on isometric bins of the samples that had been sorted in terms of quality [6]. We don't assert that quality metric is fully able to predict matching performance due to the diversity of matching algorithms. In this case, this kind of evaluation is somehow to demonstrate the linearity between a quality metric and the performance of a matcher. The NFIQ is used as a reference, while the QMF, MQF, and CWT represent metrics based on multi-feature fusion, segmentation,

Table 1.3 The 95 % confidence interval of EER of each quality metric

DB	QM				
	NFIQ	QMF	MQF	CWT	MSEG
00DB2A (N.)	[0.0492 0.0502]	[0.0651 0.0663]	[0.0497 0.0509]	[0.0488 0.0499]	**[0.0450 0.0461]**
02DB2A (N.)	[0.1326 0.1340]	[0.1104 0.1118]	[0.1109 0.1128]	[0.1103 0.1119]	**[0.1068 0.1084]**
04DB1A (N.)	[0.1529 0.1545]	**[0.1464 0.1480]**	[0.1491 0.1506]	[0.1744 0.1762]	[0.1645 0.1662]
04DB2A (N.)	**[0.1321 0.1344]**	[0.1651 0.1676]	[0.1489 0.1515]	[0.1407 0.1425]	[0.1396 0.1413]
04DB3A (N.)	[0.0741 0.0752]	[0.0730 0.0742]	**[0.0681 0.0693]**	[0.0694 0.0706]	[0.0712 0.0723]
CASL2 (N.)	[0.4296 0.4322]	[0.4059 0.4070]	[0.4043 0.4054]	**[0.4004 0.4015]**	[0.4213 0.4247]
CASR2 (N.)	[0.4337 0.4364]	[0.4134 0.4145]	[0.4057 0.4068]	**[0.4039 0.4050]**	[0.4307 0.4332]
00DB2A (S.)	[0.0021 0.0023]	[0.0040 0.0043]	[0.0074 0.0078]	**[0.0008 0.0009]**	[0.0009 0.0011]
02DB2A (S.)	[0.0011 0.0013]	[0.0029 0.0032]	[0.0011 0.0013]	**[0.0010 0.0011]**	[0.0013 0.0016]
04DB1A (S.)	[0.0268 0.0276]	[0.0172 0.0178]	**[0.0171 0.0177]**	[0.0188 0.0194]	[0.0189 0.0195]
04DB2A (S.)	[0.0390 0.0402]	[0.0378 0.0389]	[0.0338 0.0349]	[0.0327 0.0338]	**[0.0318 0.0328]**
04DB3A (S.)	[0.0190 0.0195]	[0.0162 0.0167]	[0.0148 0.0154]	[0.0159 0.0164]	**[0.0117 0.0122]**
CASL2 (S.)	[0.4087 0.4097]	[0.4266 0.4278]	[0.4213 0.4225]	[0.4229 0.4241]	**[0.3856 0.3866]**
CASR2 (S.)	[0.3815 0.3825]	[0.4119 0.4132]	[0.4087 0.4102]	[0.3963 0.3977]	**[0.3592 0.3603]**

and single feature, respectively. We do not use all databases and metrics because these results are enough to show what the quality predicting matching performance is. The results obtained by using two types of matching scores (NBIS and SDK) are given by global EERs' plots in Figs. 1.5 and 1.6, respectively. One can found that the EER values of the bins obtained by some of the quality metrics are monotonically decreasing, which assert the purpose of proving the validity of a quality metric. Loosely speaking, this kind of property demonstrates the so-called quality predicting matching performance. On the other hand, it shows the similarity or linear relationship between the quality scores and GMS. This could be observed with correlation coefficients between the two measurements.

In the experiment, the maximum GMS for each sample is calculated to demonstrate such an observation, see Table 1.4. For instance, when MSBoz is used, the Pearson correlation coefficients of NFIQ for 00DB2A and QMF for 02DB2 with respect to the maximum GMS are −0.4541 and 0.5127. Similarly, this kind of correlation also could be found for the monotonically decreased cases when MSSDK is employed. Here, we simply gives the result of some opposite cases, where the Pearson coefficients of CWT for 04DB1A, NFIQ for 02DB2A, and MQF for 04DB1A with respect to the maximum GMS of MSSDK are 0.0444, −0.2596, and 0.0585, respectively. These non-correlated values or some negative correlated cases such as the CWT in Fig. 1.5c are mostly caused by outliers of either the metric or the matching algorithm. Meanwhile, with the results in Table 1.2, Figs. 1.5 and 1.6 together, it reveals that quality predicting matching performance is not always reached linearly, such as the CWT for 04DB2A shown by the three sets of results. The global EERs in Table 1.2 demonstrate that the two metrics perform relatively better for determining the best cases of sample quality, while no linear relationship were found between them and both employed matching algorithms according to Figs. 1.5d and 1.6d, so is learning-based metric such as Figs. 1.5d and 1.6b.

Table 1.4 Pearson correlation between metrics and maximum GMS

DB	QM			
	NFIQ	QMF	TMQ	CWT
00DB2A (N.)	−0.4541	−0.0014	−0.0268	0.2885
02DB2A (N.)	−0.3308	0.5217	0.3940	0.2626
04DB1A (N.)	−0.1579	0.2601	0.0027	0.0122
04DB2A (N.)	−0.3937	−0.0177	0.1450	0.1684
04DB3A (N.)	−0.3063	0.5922	0.3132	0.4604
00DB2A (S.)	−0.4379	−0.0021	0.0402	0.3246
02DB2A (S.)	−0.2596	0.3254	0.3732	0.3230
04DB1A (S.)	−0.1970	0.3734	0.0585	0.0444
04DB2A (S.)	−0.5843	0.0615	0.1309	0.1961
04DB3A (S.)	−0.4131	0.4142	0.4371	0.6121

"NBIS" and "SDK" are two sets of matching scores

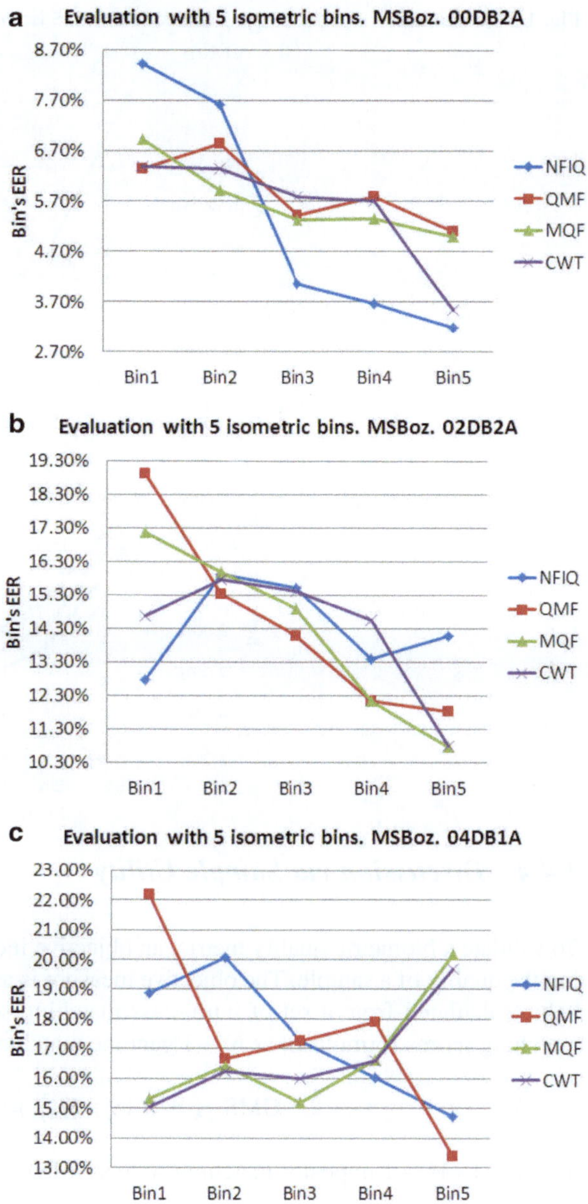

Fig. 1.5 Enrollment selection with quality metrics. (**a**) 00DB2A, (**b**) 02DB2A, (**c**) 04DB1A, (**d**) 04DB2A, (**e**) 04DB3A

Fig. 1.5 (continued)

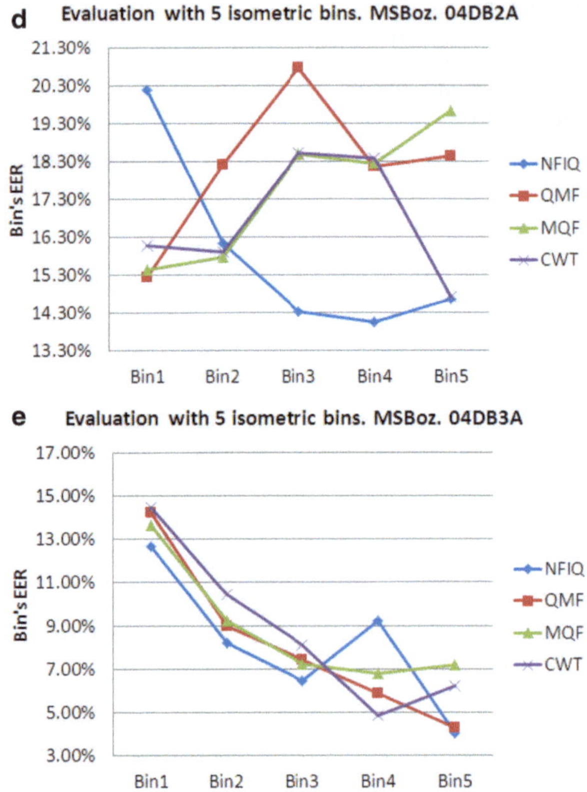

1.4.4 Discussion via Sample Utility

To validate a biometric quality metric, an objective index [30] is used for representing the quality of a sample. The objective measure is an offline sample EER (SEER) value calculated from a set of intra-class matching scores and a set of inter-class matching scores formulated as $N - 1$ genuine matching scores (GMS)

$$\mathrm{GMS}_{i,j,k} = R\left(S_{i,j}, S_{i,k}\right)\ j{\neq}k \tag{1.3}$$

and $N - 1{\times}M - 1$ impostor matching scores (IMS)

$$\mathrm{IMS}_{i,j,l,k} = R\left(S_{i,j}, S_{l,k}\right)\ i{\neq}l \text{ and } j{\neq}k, \tag{1.4}$$

where N and M denote sample number and individual number of a trial dataset, R is a matcher, and $S_{i,j}$ indicates the jth sample of the ith individual ($S_{l,k}$ is similar).

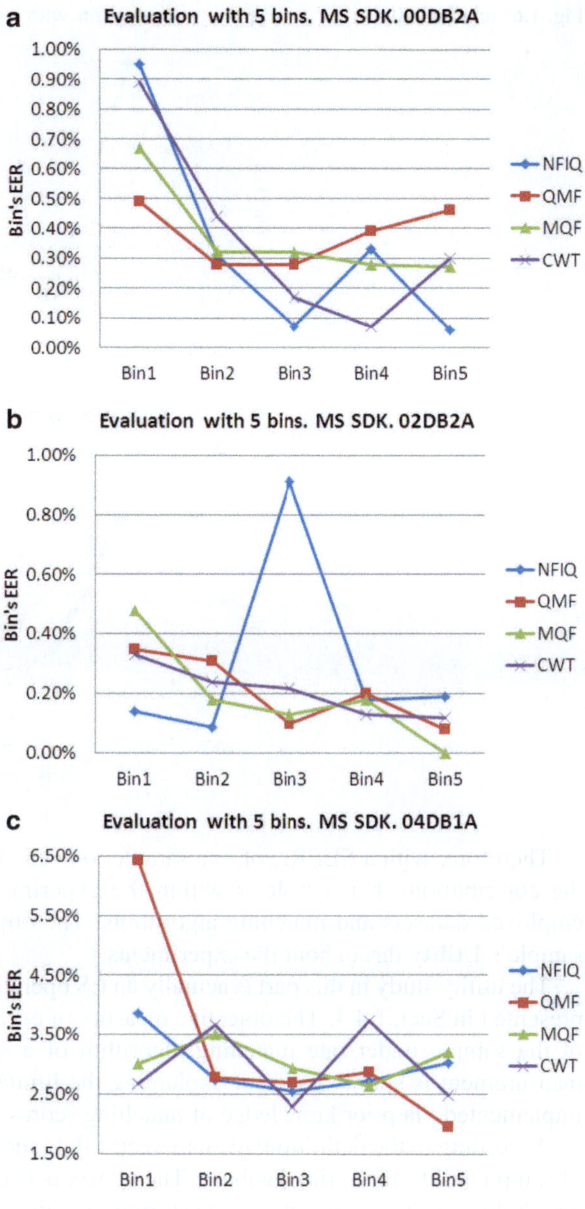

Fig. 1.6 Enrollment selection with quality metrics. (**a**) 00DB2A, (**b**) 02DB2A, (**c**) 04DB1A, (**d**) 04DB2A, (**e**) 04DB3A

Fig. 1.6 (continued)

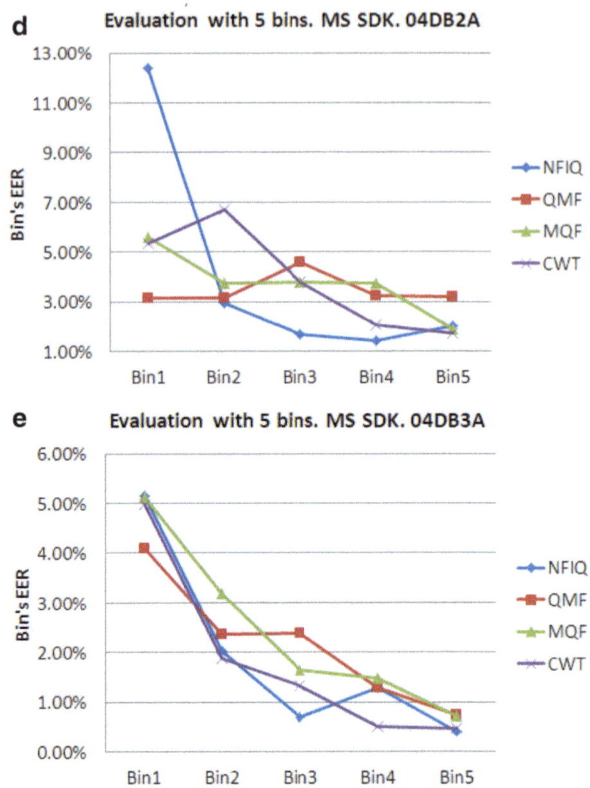

Therefore, with a $SEER_{i,j}$ of one sample, one can have a measure of how much the contribution of a sample is within the experimental framework consisted of employed datasets and matching algorithms. The objective measure is denoted as sample's **Utility** throughout the experiments.

The utility study in this part is actually an ES operation with the objective indexes presented in Sect. 1.4.4. The objective measure of each sample reflects the behavior of the sample under one matching algorithm of a specific vendor. This kind of measurement is simply used for explaining the limitation of those quality metrics implemented via prior knowledge of matching scores.

According to the definition given in Sect. 1.4.4, one can obtain an M-by-N matrix of sample utility for a trial database. The matrix is hence used as a quality result by which the enrollment selection is performed via interoperate matching algorithms, see graphical results in Fig. 1.7.

Figure 1.7 gives the plots of global EER values obtained by using ES with sample utility values, where Fig. 1.7a is the result based on NBIS matching scores (MSBoz) and Fig. 1.7b is generated from the SDK's matching scores (MSSDK). In the experiment, first of all, the utility value of each sample ($SEER_{i,j}$) with

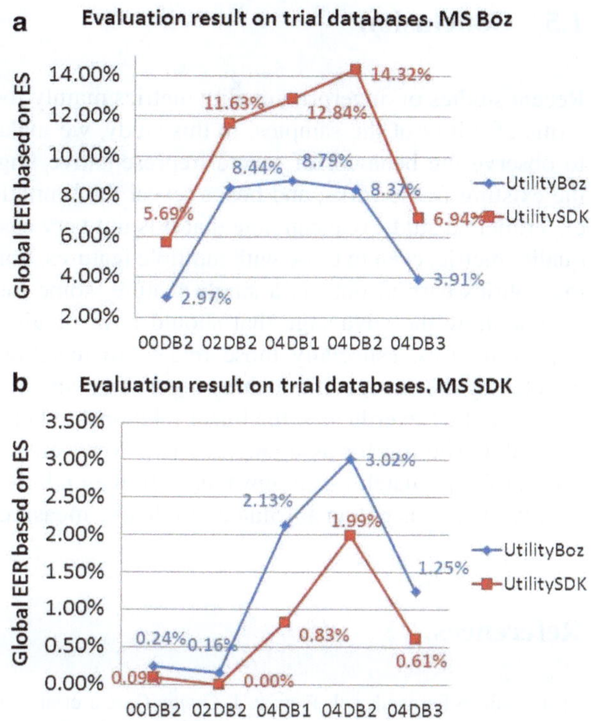

Fig. 1.7 Enrollment selection with objective measures. (**a**) Matcher of NBIS. (**b**) Matcher of SDK

respect to each matcher is calculated, respectively. In this case, two matrices of the sample utility were figured out and then used for enrollment selection. The utility values correspond to NBIS software and the SDK are denoted as "UtilityBoz" and "UtilitySDK," by which the global EER values calculated with ES are plotted in the figure, indicating by blue and red points, respectively.

The enrollment selection task chooses the best sample of one individual as the enrollment in terms of their utility values. In this case, the best performance of a matching algorithm obtained from a trial dataset cannot go over the global EER value. Apparently, the utility value is mostly dependent on the performance of the matching algorithm which is illustrated by two set of plots. In addition, according to the results, we believe that a quality metric based on a prior knowledge of matching score is not fully able to predict the matching performance in a cross-use. In fact, one can consider that whether two genuine samples should produce high GMS when one of them is not able to give reliable and sufficient features [29]. Besides, it is not clear that how much the prior knowledge is close to the ground-truth of sample quality.

1.5 Conclusion

Recent studies of fingerprint quality metrics mainly focus on reducing error rates in terms of utility of the samples. In this study, we make an interoperability analysis to observe the behavior of several representative fingerprint quality metrics from the existing frameworks, and hence reveal the limitations of this issue. Among the experimental study, one can note that it is not very easy to achieve a common good quality metric, even to those with multiple features. For instance, by comparing with the metrics carried out via a single feature, some metrics based on multi-feature do not show the advantage that should have obtained after fusion. Utility-based quality metrics, especially those related to matching scores are more probably affected by the change of matching algorithm, which is clearly brought out with the experiments. Nevertheless, the linear relationship between GMS and quality values is a valid criterion for assessing quality. However, it is not absolutely appropriate for a different matching circumstance. To the end, the offline trials also reveal that quality metrics is not an absolutely predictive measure for matching performance.

References

1. F. Alonso-Fernandez, J. Fierrez, J. Ortega-Garcia et al., A comparative study of fingerprint image-quality estimation methods. IEEE Trans. Inf. Forensics Secur. **2**(4), 734–743 (2007)
2. S. Bharadwaj, M. Vatsa, R. Singh, Biometric quality: a review of fingerprint, iris, and face. EURASIP J. Image Video Process. **2014**(1), 1–28 (2014)
3. R.M. Bolle, S.U. Pankanti, Y. Yao, System and method for determining the quality of fingerprint images, US Patent 5,963,656, 5 Oct 1999
4. C. Charrier, C. Rosenberger, Z. Yao, J.-M. Le Bars, Fingerprint quality assessment with multiple segmentation, in *IEEE International Conference on Cyberworlds (CW)*, Gotland, Oct 2015
5. T.P. Chen, X. Jiang, W.Y. Yau, Fingerprint image quality analysis, in *2004 International Conference on Image Processing, 2004. ICIP '04*, vol. 2 (2004), pp. 1253–1256
6. Y. Chen, S.C. Dass, A.K. Jain, Fingerprint quality indices for predicting authentication performance, in *Audio-and Video-Based Biometric Person Authentication* (Springer, Berlin, 2005), pp. 160–170
7. M. El Abed, A. Ninassi, C. Charrier, C. Rosenberger, Fingerprint quality assessment using a no-reference image quality metric, in *European Signal Processing Conference (EUSIPCO)* (2013), p. 6
8. J. Fierrez-Aguilar, J. Ortega-Garcia et al., Kernel-based multimodal biometric verification using quality signals, in *Defense and Security* (International Society for Optics and Photonics, Bellingham, 2004), pp. 544–554
9. H. Fronthaler, K. Kollreider, J. Bigun, Automatic image quality assessment with application in biometrics, in *Conference on Computer Vision and Pattern Recognition Workshop. CVPRW'06* (IEEE, New York, 2006), p. 30
10. P. Grother, E. Tabassi, Performance of biometric quality measures. IEEE Trans. Pattern Anal. Mach. Intell. **29**(4), 531–543 (2007)
11. R.-L.V. Hsu, J. Shah, B. Martin, Quality assessment of facial images, in *Biometric Consortium Conference, 2006 Biometrics Symposium: Special Session on Research at the* (IEEE, New York, 2006), pp. 1–6

12. ISO/IEC 29794-1:2009. Information technology? Biometric sample quality? Part 1: Framework. August 2009
13. B. Lee, J. Moon, H. Kim, A novel measure of fingerprint image quality using the Fourier spectrum, in *Society of Photo-Optical Instrumentation Engineers (SPIE) Conference*, ed. by A.K. Jain, N.K. Ratha, vol. 5779 (2005), pp. 105–112
14. G. Li, B. Yang, C. Busch, Autocorrelation and dct based quality metrics for fingerprint samples generated by smartphones, in *2013 18th International Conference on Digital Signal Processing (DSP)* (IEEE, New York, 2013), pp. 1–5
15. E. Lim, X. Jiang, W. Yau, Fingerprint quality and validity analysis, in *Proceedings. 2002 International Conference on Image Processing. 2002*, vol. 1 (2002), pp. I-469–I-472
16. L. Nanni, A. Lumini, A hybrid wavelet-based fingerprint matcher. Pattern Recognit. **40**(11), 3146–3151 (2007)
17. M.A. Olsen, H. Xu, C. Busch, Gabor filters as candidate quality measure for nfiq 2.0, in *2012 5th IAPR International Conference on Biometrics (ICB)* (IEEE, New York, 2012), pp. 158–163
18. M.A. Olsen, E. Tabassi, A. Makarov, C. Busch, Self-organizing maps for fingerprint image quality assessment, in *2013 IEEE Conference on Computer Vision and Pattern Recognition Workshops (CVPRW)*, June 2013, pp. 138–145
19. Organization for Standardization. Iso/iec 19794-2:2005: information technology-biometric data interchange formats-part 2: finger minutiae data (2005)
20. N. Poh, J. Kittler, A unified framework for biometric expert fusion incorporating quality measures. IEEE Trans. Pattern Anal. Mach. Intell. **34**(1), 3–18 (2011)
21. N.K. Ratha, J.H. Connell, S. Pankanti, Big data approach to biometric-based identity analytics. IBM J. Res. Dev. **59**(2/3), 4:1–4:11 (2015)
22. M. Saad, A.C. Bovik, C. Charrier, Blind image quality assessment: a natural scene statistics approach in the DCT domain. IEEE Trans. Image Process. **21**(8), 3339–3352 (2012)
23. L. Shen, A. Kot, W. Koo, Quality measures of fingerprint images, in *Proceedings of AVBPA*. LNCS, vol. 2091 (Springer, Berlin, 2001), pp. 266–271
24. E. Tabassi, C. Wilson, C. Watson, Nist fingerprint image quality. NIST Res. Rep. NISTIR7151 (2004)
25. X. Tao, X. Yang, Y. Zang, X. Jia, J. Tian, A novel measure of fingerprint image quality using principal component analysis (PCA), in *2012 5th IAPR International Conference on Biometrics (ICB)*, March 2012, pp. 170–175
26. M. Vatsa, R. Singh, A. Noore, M.M. Houck, Quality-augmented fusion of level-2 and level-3 fingerprint information using DSm theory. Int. J. Approx. Reason. **50**(1), 51–61 (2009)
27. C.I. Watson, M.D. Garris, E. Tabassi, C.L. Wilson, R.M. Mccabe, S. Janet, K. Ko, *User's Guide to NIST Biometric Image Software (NBIS)*. NIST Interagency/Internal Report (NISTIR) - 7392 (2007)
28. Z. Yao, J.-M. LeBars, C. Charrier, C. Rosenberger, Fingerprint quality assessment combining blind image quality, texture and minutiae features, in *International Conference on Information Systems Security and Privacy*, Feb 2015
29. Z. Yao, J.-M. LeBars, C. Charrier, C. Rosenberger, Quality assessment of fingerprints with minutiae Delaunay triangulation, in *International Conference on Information Systems Security and Privacy*, Feb 2015
30. Z. Yao, J.-M. Le Bars, C. Charrier, C. Rosenberger, A literature review of fingerprint quality assessment and its evaluation. IET J. Biometrics (2016)

[illegible]

[illegible] Z. [illegible] literature review of [illegible] assessment and evaluation. BMC [illegible] (2016)

Chapter 2
A Novel Perspective on Hand Vein Patterns for Biometric Recognition: Problems, Challenges, and Implementations

Septimiu Crisan

2.1 Introduction

In biometric applications, a relatively new technology is emerging, namely the optical scanning of superficial vein patterns. In order to be viable, a biometric parameter has to be easily identifiable but hidden from view so that it cannot be reproduced or simulated. It can be observed that the veins of the human body do not leave external marks like fingerprints, are not easily falsifiable like the voice, cannot be disguised like face traits, and are extremely hard to covertly extract during and after the lifetime of an individual in order to be reused by an impostor. In the same time, the technology used to acquire the vein pattern has reduced costs and is not invasive, requires minimal cooperation from a person, and is largely a noncontact procedure that allows it to be used where hygienic concerns are an issue [1].

Some of the most important requirements for a biometric system are the uniqueness and permanence of the biometric parameter used for recognition. Even in the case of complete uniqueness, a biometric system should be sensitive enough to be able to accurately discriminate between samples acquired from different individuals.

A review of the scientific literature shows that the visual structure of the veins is a unique property of an individual both in the retina [2, 3] and in the hand [1, 4–7]. Furthermore, it is often assumed that the localization of arteries, veins, and capillaries is specific to each person [7, 8]. Due to the novelty of the technology, the scientific studies related to the uniqueness of the vein model are rather scarce.

S. Crisan (✉)
Department of Electrotechnics and Measurements, Faculty of Electrical Engineering, Technical University of Cluj-Napoca, Cluj-Napoca, Romania
e-mail: septimiu.crisan@ethm.utcluj.ro

R. Jiang et al. (eds.), *Biometric Security and Privacy*, Signal Processing for Security Technologies, DOI 10.1007/978-3-319-47301-7_2

From a medical point of view, the cardiovascular system is formed first in the human body. The exact reason for the actual shape and path of veins, arteries, and capillaries is not completely known but, until now, from the study of the scientific literature, the probability of finding two individuals with the same vein pattern is very low. In vitro studies of the cells' spatial distribution show the automatic forming of blood vessels and the migration of cells in order to create a connected vascular network. The migration process and the dynamic aggregation result in a fractal-like behavior at both a small and a large scale [9]. Taking this premise into account, while it is impossible to predict the future blood network arrangement, a realistic vein model simulation has to take into account different aspects such as:

- The local anatomy,
- The blood irrigation requirements, and
- Other case-specific hemodynamic constraints—veins anastomose frequently, redundant vein paths.

In this manner, while there is a comfortable variation degree for a discrimination detection system, the veins are not randomly formed. Thus, in order to guarantee the uniqueness parameter, designing and implementing a vein pattern recognition system is not a trivial task.

A possible vein network arrangement belonging to a person's hand can be observed in Fig. 2.1.

The second property mentioned in this chapter is the permanence of the vein pattern. A biometric recognition system is only useful if an individual can be identified after subsequent scans on different timeframes. For blood vessels, there are three processes that can modify partially or totally their network:

- Natural changes of the vascular system over the course of a healthy individual's life
- Changes in the vascular network due to traumas or diseases
- Changes of the blood vessels due to surgical interventions

From the genesis of the blood vessels during gestation, most differences in the pattern as an individual grows up are related to the overall size and position of the network. Veins will get thicker or thinner or exhibit irregularities but the general path will remain mostly unchanged. Taking into account the fact that this model is unaffected by superficial wounds or lacerations of the skin, it is a viable biometric parameter for scans taken at large intervals of time from each other [3]. In extreme cases, such as surgery that can modify—through sectioning, rerouting, grafts, etc.—the vein model, the biometric device can reenroll the individual or compensate the modifications between two successive scans by using automated algorithms.

Condensing the three presented processes that can modify the vein pattern, several concerning factors are:

- The degree of pigmentation or discoloration of the skin. Color changes triggered by sun exposure, pigmentation due to old age, or even the native color of the skin do not interfere significantly with the vein scanning process as validated in [11]

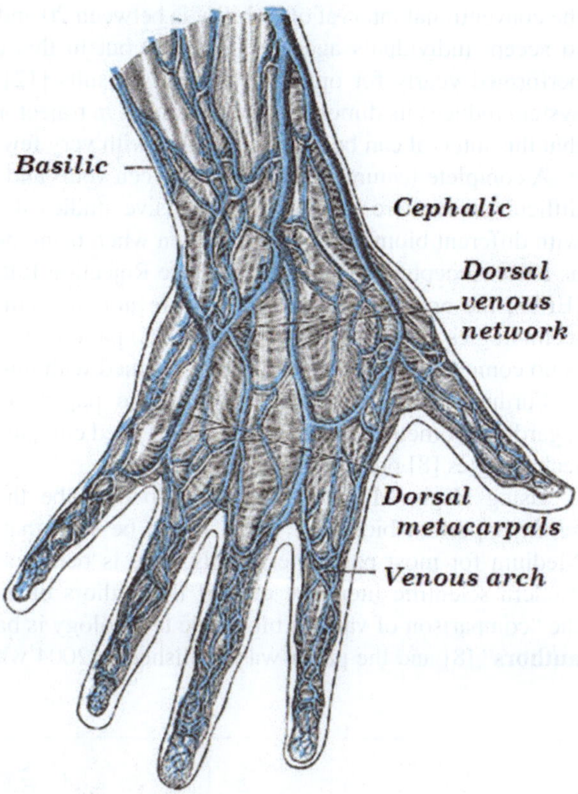

Fig. 2.1 Possible vein network in the back of the hand [10]

- Blood loss—a relevant factor since a lower quantity of blood could diminish the absorption rate at the vein level
- Medical conditions that are known to cause blood vessel constriction or dilatation
- Reduced number of blood cells, anemia, or other diseases that may modify the normal amount of deoxidized hemoglobin
- Deep skin cuts or surgical procedures that may potentially modify the vein model (although common skin problems should not interfere with the actual detection of the vein pattern)
- Environmental factors such as differences in altitude, prolonged change in hand orientation, physical stress, etc.

From a permanence point of view, using the vein pattern as a biometric feature is correct because it is a parameter with predictable modifications during the lifetime of an individual and the types of surgery or diseases that can completely modify the model in the hand region are rare and can be compensated through reenrollment. Nevertheless, in order to minimize the complexity of the scanning algorithms, vein pattern detection should be performed on individuals close to adulthood for a less drastic modification of the blood vessel network from one scan to the next. In [12], it is also observed that, generally, no major growth happens during the adult life and the conventional interval of stability is between 20 and 50 years. It is also suggested to accept individuals aged less than 20 but in this case, reenrollment should be performed yearly for optimum scanning results [12]. At a later age, the vascular system reduces its dimensions and changes in trajectories. Experiments have shown that this interval can be safely extended with very few exceptions [1].

A complete feature comparison between veins and other biometric parameters is difficult since there are no comprehensive studies showing correlated experiments with different biometric methods. Even when using pure technical parameters such as False Acceptance Rate (FAR), False Rejection Rate (FRR), or Equal Error Rate (EER), the environment conditions are not the same between scans of different technologies; The Failure to Enroll (FTE) parameter is often undescribed and there is no common dataset of individuals scanned with multimodal biometric devices.

Furthermore, a highly cited scientific paper containing very valuable data regarding biometrics has one of the most used comparison charts between biometric technologies [8] presented in Fig. 2.2.

Using **H**igh, **M**edium, and **L**ow to describe the fulfillment of each of the seven important biometric traits, it can be seen that hand veins are classified as Medium for most parameters. This table is being consistently reused throughout modern scientific literature even if the authors of the original paper declare that the "comparison of various biometric technology is based on the **perception of the authors**" [8] and the paper was published in 2004 when vein biometrics was in its

Biometric Identifier	Universality	Distinctiveness	Permanence	Collectability	Performance	Acceptability	Circumvention
DNA	H	H	H	L	H	L	L
Ear	M	M	H	M	M	H	M
Face	H	L	M	H	L	H	H
Facial thermogram	H	H	L	H	M	H	L
Fingerprint	M	H	H	M	H	M	M
Gait	M	L	L	H	L	H	M
Hand geometry	M	M	M	H	M	M	M
Hand vein	M	M	M	M	M	M	L
Iris	H	H	H	M	H	L	L
Keystroke	L	L	L	M	L	M	M
Odor	H	H	H	L	L	M	L
Palmprint	M	H	H	M	H	M	M
Retina	H	H	M	L	H	L	L
Signature	L	L	L	H	L	H	H
Voice	M	L	L	M	L	H	H

Fig. 2.2 Biometric technology comparison according to [8]

Fig. 2.3 Biometric parameters of various technologies according to [5]

False Acceptance Rate (FAR)
& False Rejection Rate Comparison (FRR)

Authentication Method	FAR (%) =	If FRR (%) =
Face recognition	~ 1.3	~ 2.6
Voice pattern	~ 0.01	~ 0.3
Fingerprint	~ 0.001	~ 0.1
Finger vein	~ 0.0001	~ 0.01
Iris/Retina	~ 0.0001	~ 0.01
Fujitsu Palm vein	**< 0.00008**	~ 0.01

infancy. On the other side of the spectrum, new synthetic data from Fujitsu [5] place palm vein recognition over fingerprints, face, voice, or iris with accuracy parameters on the same level or higher than retinal scans as seen in Fig. 2.3.

It can be seen that, due to the lack of scientific studies regarding actual performance experiments between different biometric technologies, any comparison is inherently biased.

Various research concerning vein patterns points to the viability of this parameter as a strong biometric trait when required scanning conditions are met. Fingerprints and iris scanning have the advantage of more complex patterns and can perform adequately even under less than perfect conditions. On the other hand, veins are intricate but the total model has less extractable features and requires perfect scanned images in order to have a high discrimination rate.

The main advantage of vein patterns as a biometric feature lies in the sum of its parts. Most biometric features are consistent in the case of vein models without major drawbacks allowing the technology to potentially substitute other traditional methods.

2.2 Vein Pattern Scanning Using Optical Methods

While most superficial veins are good candidates for biometric recognition, the veins in upper limb extremities are preferred. Finger or hand veins have intricate structures but they reside very close to the surface of the skin and can be easily acquired. In addition, hands and fingers are directly observable with reduced pilosity and sufficient mobility and they create minimal acceptance issues from the individuals being scanned by a biometric system.

This section of the chapter will reveal the optical background for vein scanning and propose a modular structure for an accurate vein scanning hardware device based on previous research.

2.2.1 Vein Pattern Visualization

While almost invisible under normal lighting conditions, vein patterns can be visualized if the blood vessels are exposed to infrared radiation. Due to the different absorption rates of infrared radiation in various types of tissue, a vein scanning device is able to pinpoint the location of veins while ignoring arteries and the surrounding tissue.

To achieve this effect, lighting should be performed under a tight optical window, namely 760–870 nm which is consistent with the near infrared portion of the electromagnetic radiation spectrum. This radiation is strongly absorbed by the deoxidized hemoglobin (Hb) present in the vein vessels and it is slightly less absorbed—near the top of the window—by the oxidized hemoglobin (HbO) in the arteries as seen in Fig. 2.4.

It is also worth noted that, as the diameters of arteries are as small as approximately 1/3 of those of targeted veins in the finger or hand, it is reasonable to assume that most of the visualized blood vessels are veins [12]. In addition, water, very commonly found in tissues, has a very low absorption rate at this specific radiation domain.

Hemoglobin is the main component of the red cells found in the blood stream that carries oxygen from the lungs through arteries and helps in the transport of carbon dioxide from tissues through veins back to the lungs. The high level of absorption is due to the fact that a single red cell contains about 280 million hemoglobin molecules [14].

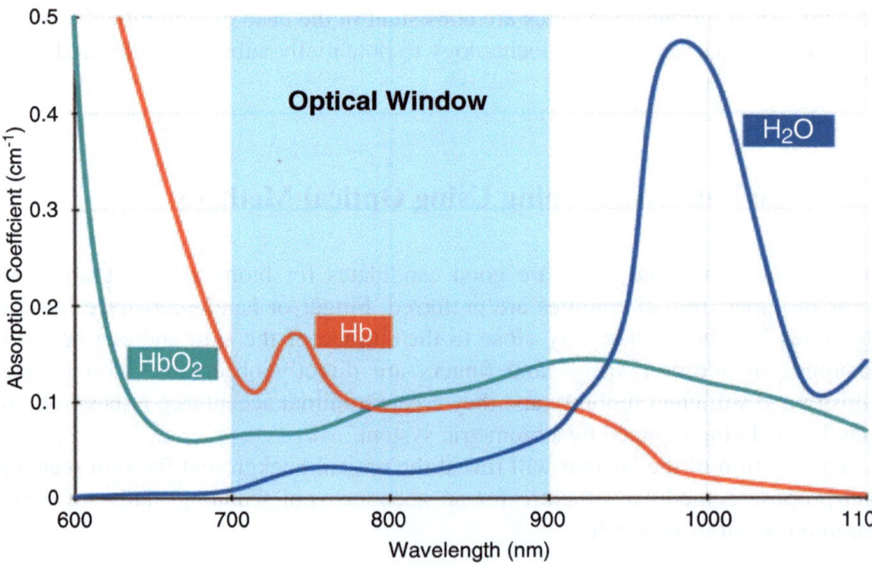

Fig. 2.4 Near infrared optical radiation window for a vein pattern recognition device [13]

The nature of the scanning method also helps secure the validity of the biometric parameter, since the presence of blood in the veins acts as a liveness proof; if blood ceases to flow through the blood vessels, the absorption method yields inconclusive results.

Using this optical window, a vein recognition system can be implemented but several factors have to be taken into account such as:

- Tissue optical diffusion,
- Depth of the scanned veins,
- Strong radiation filtering due to the water in the tissues or the tissues themselves, and
- Diffuse and specular reflections from the outer layers of the skin—specific lighting configurations have to be devised.

As mentioned earlier in the chapter, the veins of the upper extremity are divided into two sets, superficial and deep. There are many connections between these two sets of veins but due to the optical constraints of the scanning method and the relative depth of each of these two sets, the system is only able to detect superficial veins—since they are placed immediately beneath the integument between the two layers of superficial fascia [15]. The actual measured range of the optical penetration is in the range of 0.1–3 mm [11]. Two optical coefficients determine the total acquisition distance, an absorption coefficient $\propto_a$ and a scattering coefficient $\propto_s$.

The resulting image of a vein pattern under near infrared radiation can be seen in Fig. 2.5.

2.2.2 Structure of a Hand Vein Recognition Device

Most hand vein pattern recognition devices used for research and algorithm testing follow the same recipe and usually contain the same hardware modules as observed in [16–19].

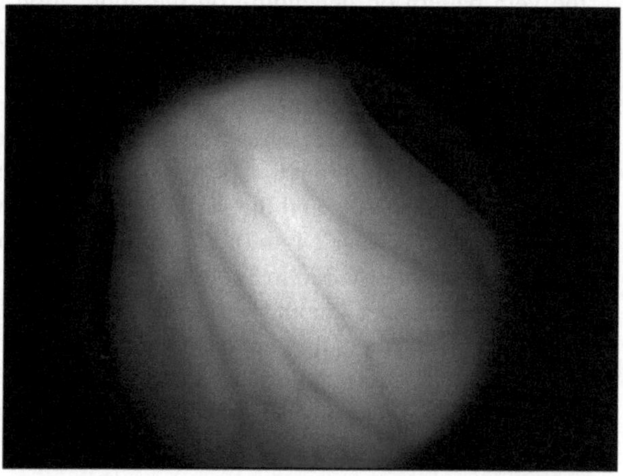

Fig. 2.5 Low resolution vein scan using the NIR optical window

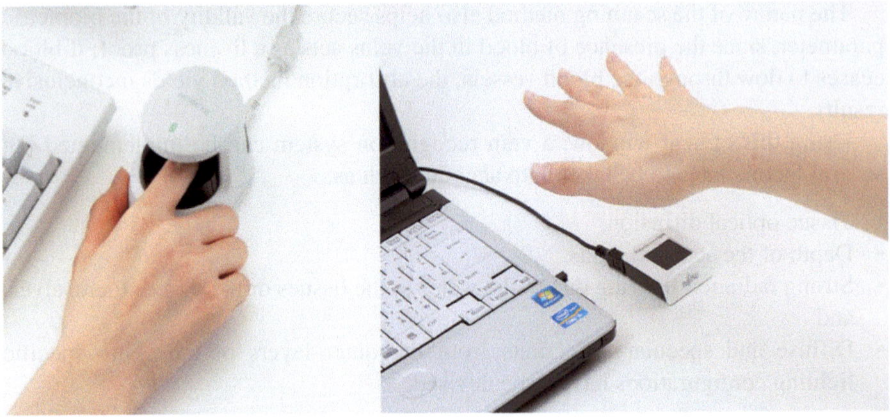

Fig. 2.6 *Left*: Hitachi finger vein scanner. *Right*: Fujitsu palm vein scanner [5, 6]

Commercial applications have a strong emphasis on finger veins and palm veins [5, 6] all research being coordinated by two major companies, Hitachi and Fujitsu. Figure 2.6 depicts working vein pattern scanners from the entities mentioned above.

Outside of the realm of commercial applications, a review of the scientific literature has failed to observe a complete proposal for an accurate vein pattern recognition system.

Since vein capturing is an optical process and through several years of iterations and experiments [11, 20–23], a complete optic-electrical structure for a hand vein pattern detection device has been devised and is being presented in this chapter. The scanning modules are optimized for the palm veins and dorsal hand veins but they can be extended for finger or forearm veins with minimal modifications.

The complete structure of a hand vein scanner involves the use of several components and modules:

- A CCD or CMOS camera with high sensitivity to the sub-spectrum of infrared radiation used. For biometric purposes, the camera has to take a snapshot of the vein pattern or, depending on the application requirements, offer real-time image processing with the help of progressive scan algorithms. The subject of the cameras will be revisited later in this chapter since several important parameters must be modified in order for the system to capture accurate vein patterns.
- An illumination source, either single or multispectral, capable of providing constant radiation without hotspots, variable intensity and achieve a high contrast between blood vessels and the surrounding tissue, without illumination artifacts [11]. The central wavelength of the emitted radiation must be a part of the tissue optical window described in Sect. 2.1 and it has to be arranged in a configuration that diminishes specular and—to some extent—diffuse skin reflections.
- A set of optical filters that increase the quality of the raw pictures taken. Taking into account the fact that the skin is a highly reflective medium that interferes with the acquisition process, previous research has documented the use of polarizing

filters, light guides, and foil diffusers [11, 20, 21]. In addition, an infrared band-pass filter matching the spectral signature of the radiation source has to be employed in order to reduce environmental influences. The filter characteristics also need to take into account possible red-shift or blue-shift from the angle of the lens and radiation emitters' position.

- Depending on the setup, a mechanical constraint system has been employed by several researchers [16–19] in order to force the hand position under the scanner. In this manner, the resulting vein scans are captured from the same position in space thus simplifying the processing algorithms. In the same time, a constrained system diminishes one of the advantages of a vein pattern recognition device—the possibility of a full no-contact and hygienic procedure.
- A sample position detection with rotation and translation extraction. If a constraint system is not used, the scanning device has to gather all relevant data regarding the spatial representation of the user sample. It is a dual module since the hand presence must first be detected using optical, ultrasound, or microwave sensors and then the orientation of the hand has to be inferred using different technologies—mono or stereo cameras, hand motion capture, photogrammetry, or structured lighting [24].
- A liveness detection mechanism. While veins offer significant native spoof protection due to the nature of the acquisition process that requires flowing blood, there are possible fraud techniques that employ materials with similar absorption and transmission characteristics as real blood vessels or living human tissues. In constrained systems, capacitive arrays and additional optical sensors can be employed in a multimodal liveness proof system that reduce the spoofing attempt success. In a free-hand position device, the use of laser grids, stereo cameras, and complex software algorithms [24] can mitigate the identification risks.

While position invariance is a difficult task, all modern scientific approaches presented in this chapter only use the sensing element—infrared sensitive camera—for determining vein pattern trajectories. As mentioned in the last paragraph, using one presented solution by adding a structured light scanner and using several photogrammetry algorithms, the relative position and orientation of the hand to the camera can be inferred.

Software vein processing algorithms can then remap the vein model on a "flat" surface by compensating the tilt angle. This effectively solves the pitch and roll problem for a significant angular range—simulations and experiments have shown $\pm 30°$. Yaw solving is purely a software rotation algorithm based on hemodynamic constraints or—after software thinning—a bifurcation/ending point count and is a relatively known method.

A representation of a hand vein capturing hardware device and its modules can be visualized in Fig. 2.7.

One of the roles of a hardware biometric device is to provide sufficient accurate data to lower the computational resources needed by the software algorithms used for processing the vein pattern. The resulting data should be as noiseless as possible and provide a good contrast between the veins and the surrounding tissue [25].

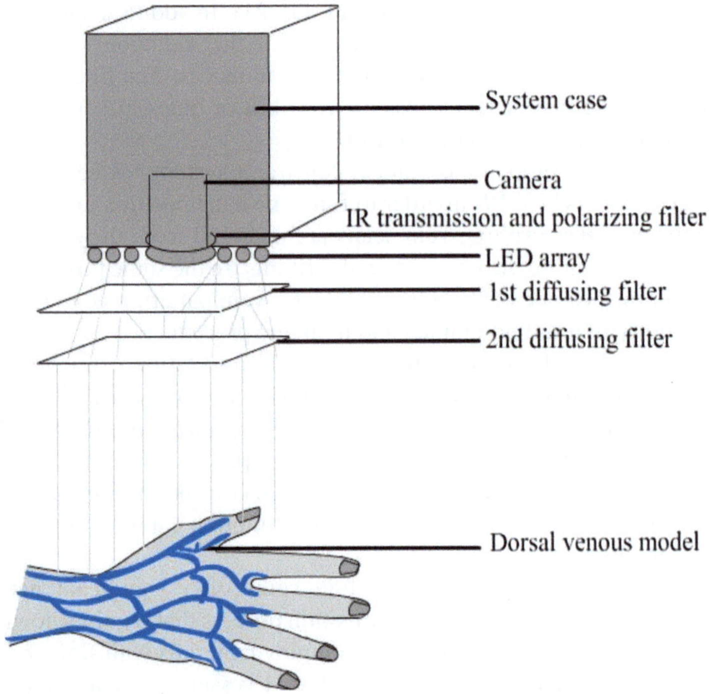

Fig. 2.7 Possible structure and modules for a hand vein scanning device [11]

By using all the filters described earlier, many of these concerns are eliminated since the CCD camera will only record the relevant data and the price of the hardware device remains low. Adequate filtering also permits the use of higher or lower wavelength infrared sources thus increasing the flexibility of the system and increasing the number of applications it can be used for. Depending on the type of application, the result of the scanning process is either a single image or a sequence that must be processed in real time.

Results of the efficiency of these hardware modules have been partially demonstrated and quantified in [11, 23, 24] and current research is focused on finalizing a large database of accurate raw scans of the vein pattern on the dorsal part of the hand.

2.3 Problems and Challenges in Vein Pattern Applications

In comparison with other traditional biometric parameter—fingerprints, voice, signature, or the iris pattern—the vein model as a biometric trait is not completely studied. There are several scientific questions still unanswered regarding hardware and software approaches to the correct feature extraction of vein patterns. There are no known studies on the possible multispectral character of the radiation sources or the influence of chromophores, melanin, or adipose tissues on vein scanning accuracy [26]. The influence of the environmental factors is not quantified [27] and in some commercial systems and scientific literature there are few systems with robust algorithms that support 1:n applications—recognition and not solely verification [24, 27].

On another vein, the low-cost entry point for a makeshift vein hardware scanner has created many opportunities for scientific researchers to implement algorithms and feature extractors. However, the quality of the raw images acquired from these systems is often very low [24, 28, 29] and the effect has been opposite: there are inconsistencies between different hardware devices to the point that raw images differ substantially for cross-reference algorithm verification. Furthermore, the samples used by different researchers are often confined to their own research and are unavailable to the general public.

This determines **a first set of important problems**: *the acquisition technology is not standardized, the algorithms are often optimized for a small number of samples, and the methods and techniques proposed in various scientific papers cannot be replicated due to the lack of access to the original image set.* These problems and limitations are easy to spot in reference papers such as [30] that determines vein parameters using image acquisition under multispectral illumination techniques [31], variable sliding kernels [32], using Gabor filters or other examples such as [33–35] where the acquired images are suffering from the uneven illumination or the algorithms cannot be reproduced by other authors in their detection systems. These problems are difficult to solve because the domain has not reached a level of maturity specific to other biometric methods.

The proposed hardware structure in this chapter aims to help in solving these inconsistencies by providing a modular acquisition system and using it as a potential standard proposal, thus accelerating the development of algorithms and methods that can benefit from correctly acquired samples using a robust system.

A second set of problems is determined by the *lack of a consistent vein pattern image database that allows for the study and implementation of applications for biometric feature extraction.* These two sets of problems are interconnected since a proper hardware device helps create accurate databases of vein patterns. In the same time, an increased number of correct databases reduce the need for a perfect hardware device for *all* researchers.

As mentioned, generating images of the vein model implies the development of an experimental system—with relatively modest costs—but with a noticeable complexity regarding the construction and the detection modules. In this case, the lack of image sets that can be used as a study base substantially reduces the quantity and quality of analysis methods and algorithms that can be implemented in the scientific world. In other biometric domains, these databases exist and allowed for the rapid advancement of the respective biometric processing technologies. Regarding the vein pattern, there are a select few recent attempts for the finger veins [36–40], palm veins (MSP and CASIA) [41–44], or wrist veins (PUT) [45, 46]. One of the most important areas for vein pattern image acquisition, the veins in the back of the hand—the main focus of the research by the author of this chapter—is not represented properly. Databases of this kind are both low in numbers and in the number of downloadable samples, at the time of writing this chapter the Bosphorus Database has 1575 collected images from 100 subjects—mostly from the left hand [47], while GPDS Database offers just 1020 images collected using geometric constrains [48, 49]. One of the larger vein pattern databases with 2040 samples—NCUT (North China University of Technology; hand-dorsa vein dataset)—has been unavailable to researchers outside of China.

It can also be observed that, in order to supplement these real samples, **most biometric methods benefit from synthetic databases**. The most obvious example is the SFINGE fingerprint database [50] but other parameters can also be generated such as the iris pattern [51], palm print [52], or finger veins [53]. In the palm or dorsal veins, these synthetic databases are not existent and a section of this chapter will present the advances and refinements from the research initiated in [54] in order to complete the real images database with a system that can generate synthetic veins in the back of the hand. Even if the morphogenesis of the vein patterns is still a relatively unknown process [14], there is still sufficient data in order to create realistic models of the vein structure. Simulated "raw" images have been obtained by recreating the pattern starting from influence and crossing points while taking into account the anatomic, hemodynamic constraints and the way in which superficial veins often anastomose.

Together with the obvious biometric applications, the localization, acquisition, and visualization of the vein pattern has **important implications in medicine**. Needle insertion for intravenous access is a common procedure with an incidence of 80 % of the patients found in hospitals [55, 56]. Although a peripheral vein can be accessed on the first try, for a significant number of patients, the medical staff can need from 2 up to 10 tries for successful needle insertion [56, 57]. The causes of multiple tries are determined by: lack of venipuncture skills, lack of appropriate medical care [58, 59], or one of the medical situations commonly defined as difficult peripheral venous access [60]. In all these cases, visualizing the vein pattern by using a contrasting technique to separate them from the surrounding tissue can improve the success rate of the venipuncture [61, 62]. In the medical field, there are a number of commercial implementations created to solve this problem, for example: VeinViewer [63], AccuVein [64], Veinsite [65], or VascuLuminator [66] but *commercial systems are often restrictive, with high acquisition costs and a proprietary interface that does not allow for adjusting acquisition values* [56, 61].

Using modern methods for visualization, there is a basis for experiments regarding the education of the medical personnel and the development of researches into assisted venipuncture by augmenting the vein pattern visibility in the intravenous site.

2.4 Modern Perspectives on Vein Structure Recognition

Since the first articles regarding the use of veins for biometric recognition, the technology used has employed low-cost, low-quality hardware devices for the vein scanning process. In addition, there are very few studies regarding vein patterns characteristics and virtually no information in terms of ergonomics, reliability, and performance of vein pattern identification devices. As mentioned in the beginning of the chapter, the use of synthetic databases can reduce the need for high-performance hardware devices and software detection algorithms can be tested against multiple cases. These include resilience to different hand poses, device placements, and orientation of the biometric parameter under the scanning device in both constrained and free-hand scenarios.

2.4.1 Ergonomics and Hand Pose Assessment in Vein pattern Identification

A review of the literature has shown that there is no known data regarding ergonomics and hand poses in a vein pattern biometric system. Individuals will have medical or personal preferences and different hand orientations in an unconstrained biometric device. Furthermore, even in the case of a geometric constraint for the system, the angle of attack for the gripping mechanism and the relative height of the forearm in relation with the dorsal or palm part of the hand are different from scan to scan. Automated algorithms can compensate to a certain degree but if the variation is over a predetermined threshold, the acquisition will suffer.

Using an inertial motion capture glove—part of the Perception Neuron full body suit [67]—172 individuals were scanned with 12 positions for each hand. The first six hand poses were unforced and each individual was asked to place their closed fist with the dorsal part of the hand pointing upwards towards the sensing system. A visual guiding system comprised of a pair of triangulation based proximity sensors and two visible lasers was used as an indication for a relative placement of the sampled hand. The last six poses were constrained using a fixed cylinder grip underneath the sensing system. Using the Perception Neuron available sensor fusion and inverse kinematics algorithms, the position and orientation of the forearm, hand, and fingers was calculated for each scan. Figure 2.8 presents the inertial glove used and the test system. In this setup, no actual veins were scanned, all the experiments were directed at determining average user hand positions or natural state poses for the hands under the scanner.

In an unconstrained scenario, after analyzing the position and angles of each subsequent scan for a person it can be observed that the average deviation from the first to last sample is 12–27° in the horizontal X axis and 4–16° in the horizontal Y axis. In addition, the last three samples exhibit a lower angle deviation due to the adaptation of the user to the scanning system. For each individual, four subsequent

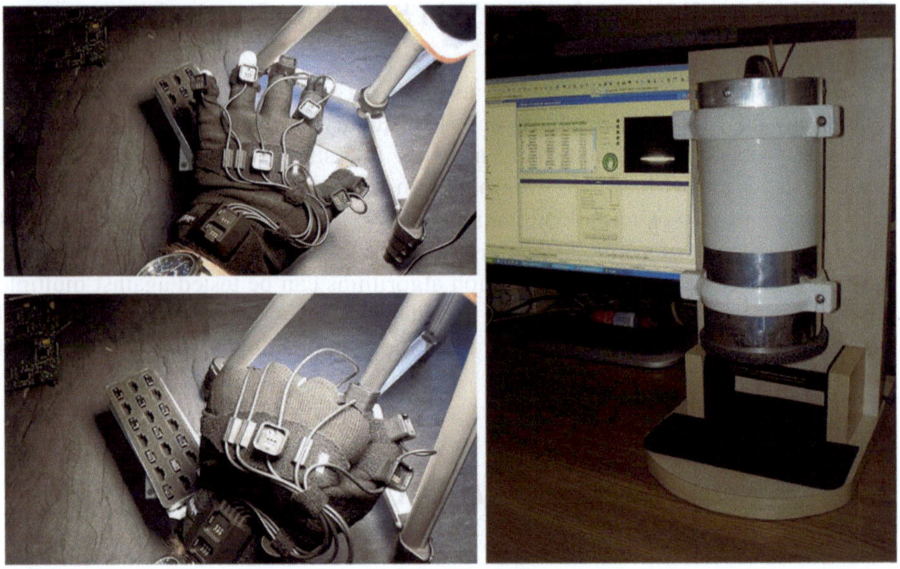

Fig. 2.8 Ergonomics test system and inertial glove hand poses

scans taken at 1 h intervals were acquired. Sets three and four exhibited the lowest angle deviation in the same set but there was no visible correlation to the angle of the first two sets. Further research is required to estimate the importance of muscle memory or acquired learning but the results show that untrained users will converge towards a static position and orientation given enough tries. Coupled with the fact that horizontal angle differences are easily compensated through software algorithms, an unconstrained system is a viable option for the veins in the back of the hand.

The difficulty arises in the accurate determination of all vertical, *Z* axis poses' deviation and for 18 individuals, the difference from the vertical axis was between 7 and 19°. Vein scanning becomes impossible after a vertical orientation threshold, experimentally determined to be 10–15° in the vertical plane so that veins are not visually lost due to hand occlusion. In order to solve this problem, a "suggestion" mechanism was used. Each user has been allowed to freely position the hand under the scanner but a visual cue system comprised of the two visible lasers was devised. Each laser blinks with different speeds until the user achieves a correct range of the vertical axis orientation underneath the sensor and the lasers.

The experimental results have helped in tweaking the angles and position of the system for user ergonomics and correct scanning. However, for the proposed hardware setup described in Sect. 2.2, it is cumbersome and difficult to use a contact based position sensing device. Using the research results obtained in [24], a module

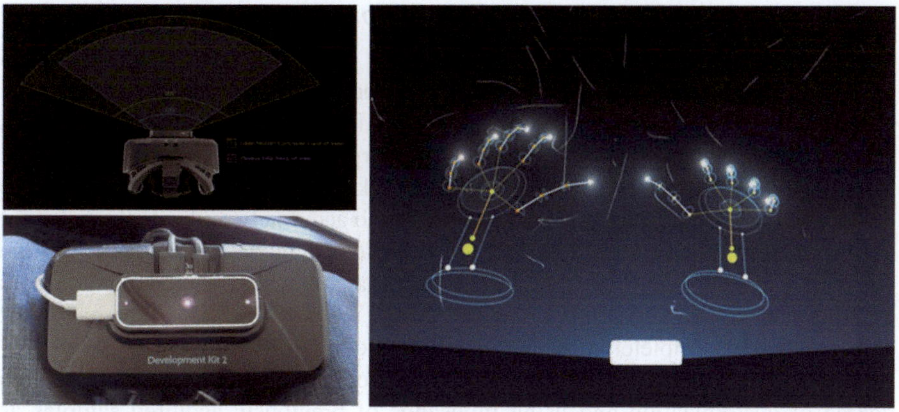

Fig. 2.9 Leap Motion camera attached to an Oculus Rift virtual reality headset

containing a stereo infrared camera [68] is introduced in the system case. After calibrating the orientation and matching the results with the data obtained from the inertial system, the vein pattern recognition device is able to determine position and orientation of the user's hand in both fist closed and open fingers position.

Several edge cases have been tested where the users' hand poses have been constrained by their body position and by surrounding objects. In order to accurately simulate real-life scenarios, a virtual reality system has been employed. Using an Oculus Rift DK2 model coupled with the stereo Leap Motion camera used for the hand pose study, users are being presented with various environments and scenarios. The hardware components and the test data from the Leap Motion camera are shown in Fig. 2.9.

In order to test the correct placement (both height and orientation) of a fixed vein pattern recognition system, each user was presented with a virtual vein scanner at different height, distance, and orientation towards the user. For each case, the user needs to insert the hand under the scanner unaided. Relative position and angles of the hand to the simulated vein scanner have been recorded using Leap Motion raw data correlated to the coordinate system of the Oculus Rift external camera.

Minimum hand angle deviations on all axes have been recorded for a relative position of the simulated scanner between the shoulder level and ½ distance between shoulder and complete downward pointing hand. By analyzing the distribution of the user heights across the test study—1.51–1.93 m—the correct placement for a vein scanner—in normal conditions—is 1.21–1.44 m from the ground level—satisfying 92 % of the user dataset.

Using virtual reality environments, biometric and medical visualization data can be exposed in a rich, collaborative manner while creating user scenarios difficult to achieve using real constraints.

2.4.2 Synthetic Vein Pattern Generation

As mentioned in Sect. 2.3, vein patterns do not benefit from synthetic databases or generation platforms. The apparent chaos in the forming of blood vessels presents several challenges as opposed to synthetic fingerprint or iris generation where the rules have clearer outlines. Since veins have less key points and extractable features than other traditional parameters, it is important to accurately replicate the behavior of the vessels down to the level of local angles or model direction inside the hand. In addition, since—to some extent—local anatomy and hemodynamic needs together with several signaling molecules dictate the overall shape of the pattern, it is impossible to completely predict the exact structure of the total vein network.

The goal of a synthetic database is to provide plausible samples with a high degree of customization and using thousands of previously acquired samples as reference, it is a possible endeavor.

In the creation of the proposed vein simulator application, the software workflow involves using Embarcadero Delphi as the simulation programming software, Autodesk 3D Studio Max for creating hand masters or blanks, and Epic Unreal Engine for accurate rendering of final hands and hand poses. Blanks are created for both hands and the correct angle of the model is taken into account when the simulated blood vessels are introduced into the hand.

One of the important rules in vein creation is related to the preservation of connectivity—veins cannot be unconnected—and the development of the model branches has to follow a statistical distribution that should be efficient in irrigating the entire hand tissue.

In addition, simulated hands should not be perfect, veins do not possess the same thickness, their depth greatly varies underneath the skin and the hand may be covered with hair, have a distinct curvature or there could be significant environmental influences.

External factors should also be taken into account. Variations in illumination can completely modify the accuracy of the scanned pattern as can be seen in Fig. 2.10. Camera performance and noise, its sensitivity to the desired spectrum, position of the hand in relationship with the scanner, or the lack of uniformity of the lighting system are all important error-generating situations.

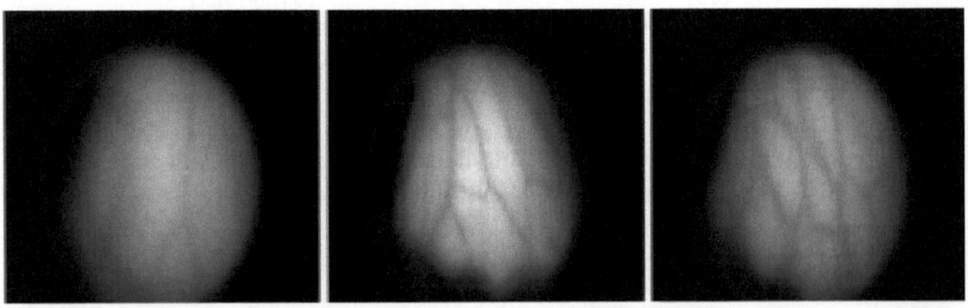

Fig. 2.10 Scanned hand vein pattern images under different illumination and acquisition scenarios [54]

The proposed hardware system in Sect. 2.2 of the chapter solves several of these issues by providing a high-quality raw scan of a hand vein pattern. Nevertheless, any synthetic image should be able to contain as many error-generating situations as possible. As mentioned before, vein pattern algorithms are often optimized for small datasets or perfect scanning conditions and a setup that allows for the creation of "fake" vein models should establish the real accuracy of an algorithm.

The procedure used to extract features from real vein images is based on the algorithm created by the authors in [11, 20]. Parameter extraction in a normal vein recognition algorithm starts with a local adaptive threshold and a thinning operation performed on the vein model including optimizations of the branches such as pruning and elimination of unconnected segments [11, 22]. Synthetic models start from a known network of nodes, terminations, and intersections—using an updated version of the crossing number [69]—and create the connecting segments based on several rules—segments have a slight curvature, more than one segment can connect to an intersection but not to a node, longer segments are created first, etc. Starting from these influence points, hand dimension constraints are applied so that vein points do not fall outside the hand model and veins occupy at least 80 % of the entire hand surface. In this way, for each influence point, the software will extrapolate the branches in between, while obeying the general flow and direction of the model.

The model is optimized to resemble a near infrared scan but—based on known behavior of radiation inside the human tissue—other wavelengths can be tested and simulated.

The simulation algorithm creates a structure having a series of intersecting curves with a width of 1 pixel. Creating the desired thickness is performed using a dilation algorithm with automatic or manual constraints—total dimensions of the model determine overall thickness, longer and major veins can have a larger surface, etc.

A snapshot of the actual dilation process involving automated vein thickness is shown in Fig. 2.11.

The vein pattern is embedded in the generated blank hand and the whole model is scaled based on a lookup table containing statistical data regarding the average

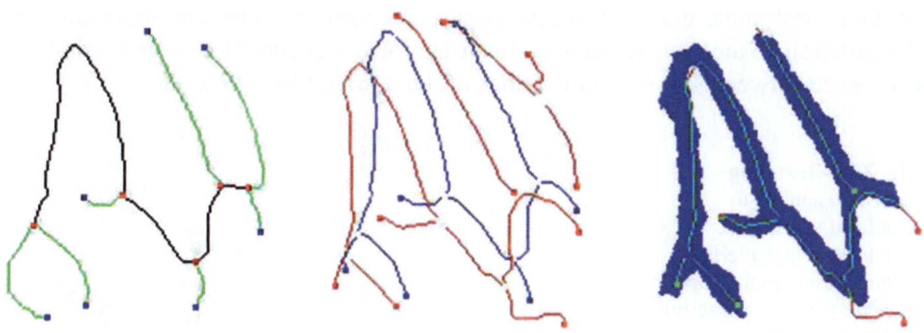

Fig. 2.11 Reconstruction of the vein pattern at full width. *Left*: simulated vein centerline with key points, *middle*: trajectories for possible vein thickness, and *right*: complete vein model after dilation algorithm [54]

correlation between age, sex, and hand dimensions. Subsequently, a gradient texture is applied on both the vein model surroundings and the outer surface of the skin. The texture values are calculated using a contrast variation coefficient C_v based on the following formula [54]:

$$C_v = \frac{\sum_{y=1}^{m}\sum_{x=1}^{n}\left(P_{x,y} - P_{x+1,y}\right)}{xy} + \left[\max\left(P_{x,y} - P_{x+1,y}\right) - \min\left(P_{x,y} - P_{x+1,y}\right)\right]/10 \tag{2.1}$$

where:

$P_{x,y}$ represents the intensity value of a pixel at coordinates x, y in the image

$P_{x+1,y}$ represents the intensity value of a pixel at coordinates $x + 1, y$ in the image

m, n are the width and the height of the area of interest in pixels

max, min are the maximum and minimum values of the differences between adjacent pixels

A representation of the outcome of the texturing algorithm can be observed in Fig. 2.12. The simulated vein pattern is also pruned and optimized according to the hand and vein geometry constraints presented earlier.

A robust vein simulation platform allows researchers to create edge-case scenarios for recognition algorithms. In the same time, by correlating the results of the ergonomics and position case study presented in Sect. 2.4.1, simulated vein patterns can be mapped in the desired position and orientation to better mimic real-life cases.

2.4.3 Vein Biometrics in a Connected World

Traditionally, for biometric systems, the main concern is the identification of the individual and the storage of the template in a simplified form in order to serve as a comparison based system. Due to tight computational requirements and preservation of data constraints, many of the parameters relevant to a medical application are discarded since the system has to compensate for vessel constriction or dilation, age difference between scans, modifications of the normal blood flow, etc.

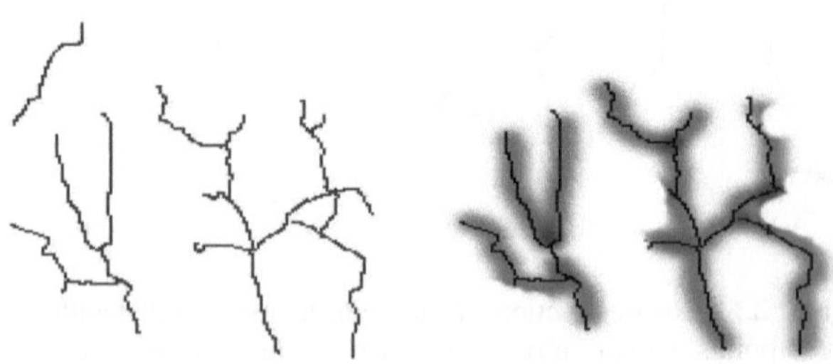

Fig. 2.12 Texturing algorithm applied to simulated veins. *Left*: simulated vein centerline, *right*: texture gradient applied to vein model—for insertion into simulated hand

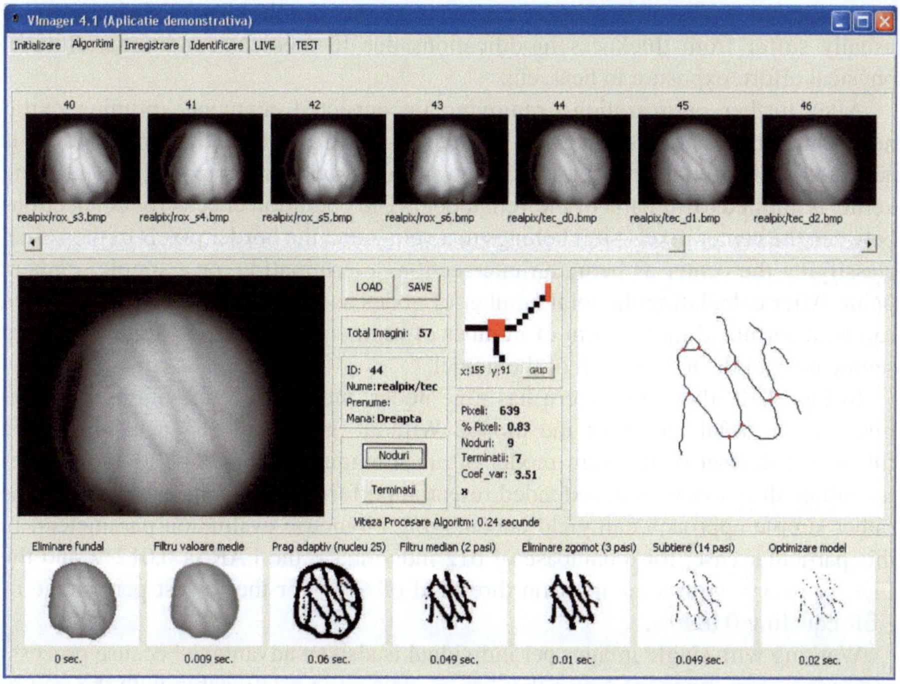

Fig. 2.13 Vein image processing and feature extraction algorithms in the VEINSIM application [11]

The algorithms are built in such a way that they isolate, enhance, and then reduce the vein model to a series of lines, usually one pixel wide as shown in Fig. 2.13. In this particular case, the dataset storage is performed by emphasizing and extracting the unique features for any individual and converting them to an array of values representing, for example, the number of segments, endings, and bifurcations in the pattern, the total segment length, relative angles between lines, intervein distances, etc.

To exemplify the process depicted in Fig. 2.13 and briefly mentioned in Sect. 2.4.2, a particular set of algorithms used in previous research for vein feature extraction involves several stages of image processing. After the background is successfully removed, a median filter is applied to the dorsal hand image. Using a locally adaptive threshold with a 9–25-pixel nucleus, the image is binarized with a strong emphasis on the vein pattern. The artifacts created by the thresholding process have a low degree of importance for vein scanning because even if the veins become artificially thicker, the last step of the process is a thinning algorithm. Using a custom thinning operation that has strong rules preserving connectivity between diagonal and vertical lines—following the natural flow of the blood model—the process ensures that all the veins are reduced to a 1-pixel width set of lines.

Even without a destructive operation such as the thresholding algorithm, veins usually suffer from thickness modifications due to medical conditions, altitude, physical effort, exposure to heat, etc.

After further optimization—removing unconnected segments, pruning extraneous branches, and restoring lost segments—the feature recognition module is activated. Using the feature extraction and storage model presented in [11], a sliding kernel is applied to all parts of the vein model. The operation checks for connections between the center pixel—if it belongs to a vein—and the border pixels of the kernel classifying the center as being an intersection, a termination, or a simple segment point. After calculating the total number of nodes and terminations, additional data can be determined in the form of number of segments, all relative angles between segments, and the total length of the model.

In Fig. 2.13, the vein pattern has nine nodes and seven terminations, and 639 pixels is the total length of the model. Without any compression techniques, a full stored dataset of the vein model of an individual occupies bellow 400 bytes including all relevant data. Extended research performed in [1] has shown that this rather simple approach can yield encouraging biometric evaluation parameters. In this particular case, for a database of 612 individuals, the FAR is 0.012 % and the FRR is 1.03 % with a comparison threshold of 67 % for the lowest percentage of EER equaling 0.092 %.

Working with single images per individual is also an advantage because processing speed is an important factor that has to scale up in relationship with the size of the sample database.

All these factors define the workflow of a vein detection device as a standalone machine with limited processing and storage capabilities that only serves the identification/verification function. While this is the main task of a vein recognition system, the ability to have devices permanently connected creates unique opportunities. As it has been mentioned in Sect. 2.3 of this chapter, there are very few studies performed on vein patterns, their permanence and uniqueness are viable but not fully tested. There are no researched correlations between the age of the individual and the state of their vein model, the skin color/pigmentation influence, or the difficulty of enrollment at different timeframes.

As biometric technologies mature and the number of civilian security application increases, the amount of collected data will require a shift in the processing and storing model. Using the Big Data paradigm, it can be observed that vein pattern recognition follows closely the three main features of a large dataset such as [70]:

- Volume: vein patterns and the collateral acquired data already exhibit large volumes in current applications; the size will continue to increase at a disproportionate rate.
- Variety: there are many different types of data associated with biometric parameters, as text, extracted features, sensor data, raw scans, individual data, and more.
- Velocity: data is arriving continuously as streams of data, and the goal is to obtain useful information from it in real time.

In addition to the classic 3V model, current Big Data representations include two more Vs:

- Variability: there are potential changes in the structure of the data and how the data can be interpreted.
- Value: the intrinsic value of the information, in the case of biometric parameters the value resides in their own ability to provide security and in the total asset cost insured by a biometric security scan.

The first working models for large datasets were based mainly on the homogenous and structure behavior of global data. Due to the unprecedented rise in collected data "often dispersed across independent systems that are difficult to access, fuse and mine due to disparate nature and granularity" [71], most modern Big Data approaches treat large volumes of information as they are created—unstructured and heterogeneous.

There are clear advantages of non-homogenous data collection in an emerging biometric parameter such as the hand veins. Big Data frameworks provide versatility and adaptability to increasing datasets and offer unprecedented insights in statistical correlation of multimodal features of the individual together with the vein biometry.

Using a modern Big Data framework and workflow, Fig. 2.14 presents a working model for a series of permanently connected vein recognition devices of different types.

Such an infrastructure is capable of acquiring raw images and delivering them online to be stored, processed, and analyzed without the constraints of a real-time verification system. While individual privacy is a very important parameter when it comes to sensitive biometric data, vein images can be anonymously collected without any tracing to the owner of the vein model. The communication channel can also be encrypted using a dual-key pair and the relevant data is stored only for future comparison and analysis.

A dual—Hadoop Map-Reduce and Apache Spark—system for data processing has been chosen for their different capabilities. Both are processing technologies which are able to handle large volumes of data by parallelizing the operations. The biggest difference between them is that Map-Reduce stores all the intermediary results generated during processing on the hard drive; Spark on the other hand uses only the random access memory and occasionally the hard drive—when it is necessary. Depending on the allocation of resources, both systems can function in parallel. Currently Spark is faster and there are hints that it will replace Map-Reduce in the foreseeable future.

There are a number of storing systems for the input data (raw data) that would allow storing large amounts of unstructured data. The ones that are worth mentioning in the context of biometric datasets are the following:

- Plain files (csv, binary files, etc.)
  - In this case, there will be only a limited number of file formats and all these formats need to be supported by the component that will be responsible for the parsing of the input data

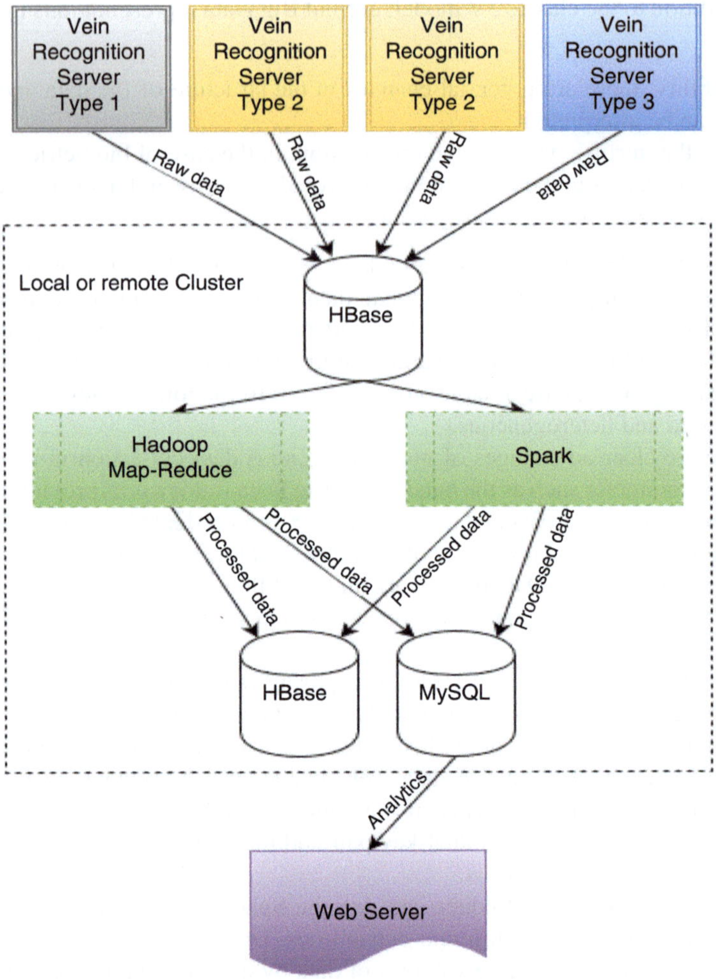

Fig. 2.14 Proposed infrastructure and workflow for a network of vein recognition devices

  - This approach is a good match with the distributed system proposed in Fig. 2.10 allowing for parallel processing but the main drawback is that it is extremely difficult to extract certain types of data (for example, cross-correlated field information, e.g., young male subjects from a certain day, or the data for which the processing returned an FTE error, etc.)
- MongoDB
  - Complete NoSQL database that allows for the storage of large amounts of data—preferred in many Big Data implementations.

  - In the context of large processing batches for biometric raw data and associated information, MongoDB is not optimized for computational-heavy parallel operations being fast to write and slow to read when accessed from Hadoop or Spark.

- HBase
  - Apache HBase is colloquially called the Hadoop database because it is a NoSQL database running over Hadoop. As mentioned in [72], "it combines the scalability of Hadoop by running on the Hadoop Distributed File System (HDFS), with real-time data access as a key/value store and deep analytic capabilities of Map Reduce." Balancing the distributed nature of the HDFS with the need to access relevant information, HBase is both capable of querying individual records and offering complex correlations for reports on very large datasets.
  - It is implemented to work with Map-Reduce as well as Spark—being built on HDFS.
  - As it is designed to support queries on massive datasets, HBase is optimized for read performance. When writing data, HBase maintains consistency at the expense of slower write operations [73].
  - Major advantages: random access, single-row lookups and updates, and processing of adjacent key ranges.

Hadoop and its underlying structure are built for managing large sets of data, so a columnar store is a natural complement. Databases normally store information in rows and are therefore optimized for accessing one record at a time. Columnar storage systems serialize and store data by column, optimizing searches and reads across massive datasets [74]. In the context of fast access to information, new technologies such as Apache Parquet—containing per-column data compression—can become the norm in developing specialized storage and data analysis networks in a connected world.

The use of an additional SQL database provides a fast framework for statistics and analytics, processed data is used to populate different SQL tables, and data can be displayed according to a random number of query parameters on a web server or locally.

Such a system is also scalable, for a Proof of Concept processing, local clusters can be inexpensively built. There are use cases where there is just the need to test the software code that processes some specific data and assesses how the results are in relation with the expected result. For this case, there is no need to have access to a large cluster, with a lot of powerful machines. A smaller cluster could be built on a local machine by using three different Docker containers which are configured to communicate between each other.

For the production environment, a cluster of commodity servers could be used that will have to cover the following aspects:

- Data availability—a replication factor ≥ 2 is required and has to be set on the cluster that will assure no data is lost

- Low latency—gigabyte connection between machines in the cluster
- Scale up as the volume of data increases

The model presented in Fig. 2.14 takes into account the future proliferation of vein scanning devices of different types—ranging from mobile devices to fully fledged vein detection systems. Local behavior remains the same; the system will extract relevant biometric parameters for comparison or enrollment. The difference is that, using this proposed framework, local devices can send raw images anonymously to the local or cloud cluster for long-time storing. Using the powerful computational resources of such a cluster, several important scenarios are created:

- Global performance of local vein pattern recognition systems can be assessed based on their type, and matching scores and error rates can be determined from the raw images and the result reported by the local system.
- Algorithm testing can be performed on the collected raw images, for each enrolled individual an untraceable image is stored in the cluster and is available for further analysis.
- Biometric features are just one parameter of an individual scanned hand. Various amounts of data can be collected for correlation purposes. Features such as age, sex, hand shape, vein visibility, degree of pigmentation, hair density, specular or diffuse reflection coefficients, etc., can be stored along with the raw images.
- Collecting large amounts of data offers—as mentioned—unique opportunities for large-scale studies. How is the real permanence of the vein pattern as the individual grows older? What modifications occur in the model during the lifespan of an individual? How much of an influence does light or dark skin have on the robustness of the vein scanner? What is the discrimination rate between individuals? As more raw data is collected, algorithms will have to increase in their complexity since identical features between different people will probably be revealed.
- Statistics for all collected data allows for complex case studies over long timeframes and enables long-term monitoring of individual or global biometric parameters.

While the concept of Big Data is extremely new and biometric applications supporting the paradigm are virtually nonexistent, it is worth mentioning that all technologies and protocols described in this workflow and the subsequent discussion are open-source and are driven entirely by a large user community. While proprietary technologies will probably gather large groups of followers in the foreseeable future, using adaptive open-source software in the first steps of a domain can yield faster and more relevant results in creating a workable standardized pipeline.

It is clear that there are growth opportunities for biometric parameters and especially for emerging technologies such as the vein pattern recognition in the age of the Internet of Things and Big Data paradigms.

2.5 Conclusions

This chapter has presented the current state of vein patterns used as biometric parameters as well as the important challenges and problems that are inherent to the technology. The current degree of acceptance and applicability of the technology raises three sets of important problems described in this work. Several solutions and proposals have been devised in order to mitigate the issues. Using modern methods and approaches to the scanning technology, a complete hardware setup for the extraction of accurate vein patterns has been presented together with user case studies regarding ergonomics, hand placement, and orientation in both constrained and unconstrained setups.

Through the use of extremely new virtual, augmented, and mixed reality devices, ease of use and level of acceptance for hand vein biometric recognition can be quantified. In addition, these technologies will help in the creation of more accurate hand simulations and user scenarios impossible to replicate in real life. Experimental results are shown from the usage of inertial hand motion capture suits for gathering statistics on hand position, user ergonomics, and the correlation and storage of images taken with their position and rotation angle. It also foreshadows the advantages of the immersive data visualization for this technology including extraction of the superficial vein patterns for virtual teaching and medical demonstration.

Equally important, the simulation of vein patterns for synthetic database generation—as a response to the lack of real/simulated hand vein images—is described in the chapter. Research data has been presented with encouraging results in the creation of realistic hands and hand vein models with a high degree of customization for detection algorithm testing, adding to previous research.

The chapter also analyzes the possible inclusion of modern paradigms such as Internet of Things and Big Data into the normal workflow of a connected biometric network. As the quantity of biometric data increases and can be stored for subsequent analysis, several crucial experiments can be performed on datasets impossible to obtain until now. In addition, cross-correlations using additional user data can be inferred and all gathered data can also improve the creation and implementation of vein pattern recognition algorithms.

It is the author's opinion that future research has to involve all challenges presented in this chapter. It is equally important to standardize vein pattern acquisition as it is to create real and synthetic databases for algorithm testing and implementation. It is also vital to perform more studies on each of the main seven biometric features as they relate to the use of vein patterns. As the industry embraces the use of Big Data, biometrics can also greatly benefit from the storage and analysis of unstructured and heterogeneous data and answer important questions regarding the viability of all biometric technologies.

Acknowledgments This work was supported by a grant of the Romanian National Authority for Scientific Research and Innovation, CNCS-UEFISCDI, project number PN-II-RU-TE-2014-4-2196.

References

1. S. Crisan, Researches concerning the development of biometric applications using infrared radiation. Ph.D. Thesis, Technical University of Cluj-Napoca, 2008
2. A. Jain, R.M. Bolle, S. Pankati, *Biometrics: Personal Identification in a Networked Society* (Kluwer Academic Publishers, Dordrecht, 1999)
3. J.L. Wayman, Technical testing and evaluation of biometric identification devices, in *Biometrics: Personal Identification in Networked Society* (Kluwer Academic, Dordrecht, 1998)
4. Biometric identification using subcutaneous vein patterns (2005), Luminetx. http://www.biotechweblog.com/50226711/biometric_identification_
5. Fujitsu Develops Technology for World's First Contactless Palm Vein Pattern Biometric Authentication System (2015), Fujitsu Laboratories Limited, Tokyo Japan, http://www.fujitsu.com/global/news/pr/archives/month/2003/20030331-05.html
6. Introducing Hitachi Finger Vein Authentication (2005), Hitachi Corporation Japan. http://www.hitachi.co.jp/Prod/comp/fingervein/global/
7. P.J. Phillips, A. Martin, C.L. Wilson et al., An introduction to evaluating biometric systems. IEEE Comput. **33**(2), 56–63 (2000)
8. A.K. Jain, A. Ross, S. Prabhakar, An introduction to biometric recognition, invited paper. IEEE Trans. Circuits Syst. Video Technol. **14**(1), 4–20, (2004)
9. A. Gamba, D. Ambrosi, A. Coniglio et al., Percolation, morphogenesis, and burgers dynamics in blood vessels formation. Phys. Rev. Lett. **90**(11), 118101 (2003)
10. H. Gray, *Anatomy of the Human Body* (Lea & Febiger, Philadelphia, 2000), 1918; Online Edition Bartleby.com, fig 574
11. S. Crisan, I.G. Tarnovan, T.E. Crisan, Radiation optimization and image processing algorithms in the identification of hand vein patterns. Comput. Stand. Interfaces **32**(3), 130 (2010)
12. S.Z. Li, A.K. Jain (eds.), *Encyclopedia of Biometrics*, 2nd edn. (Springer Science + Business Media, New York, 2015), p. 565, p. 380
13. T.G. Phan, A. Bullen, Practical intravital two-photon microscopy for immunological research: faster, brighter, deeper. Immunol. Cell Biol. **88**, 438–444 (2010)
14. A. Nadort, The hand vein pattern used as a biometric feature. Master Thesis, Nederlands Forensisch Instituut, 2007
15. V. Paquit, J. Price, Near-infrared imaging and structured light ranging for automatic catheter insertion. *Medical Imaging* (Oak Ridge National Laboratory, Oak Ridge, TN, 2006)
16. D. Huang, Y. Tang, Y. Wang, L. Chen, Y. Wang, Hand vein recognition based on oriented gradient maps and local feature matching, in *Computer Vision–ACCV 2012* (Springer, Berlin/Heidelberg, 2012), pp. 430–444
17. S. Zhao, Y. Wang, Y. Wang, Extracting hand vein patterns from low-quality images: a new biometric technique using low-cost devices, in *IEEE International Conference on Image and Graphics, Chengdu, Sichuan, China* (2007), pp. 667–671
18. Y. Wang, K. Li, J. Cui, L. Shark, M. Varley, Study of hand-dorsa vein recognition, in *Advanced Intelligent Computing Theories and Applications.* Lecture Notes in Computer Science, vol. 6215 (Springer, 2010), pp. 490–498
19. S. Benziane, A. Merouane, A. Benyettou, A low-cost IR imaging system for hand vein biometrics parameter's extraction. Adv. Comput. Sci. Eng. **11**(2), 51 (2013)
20. S. Crisan, I.G. Târnovan, T.E. Crisan, Vein pattern recognition. Image enhancement and feature extraction algorithms, in *15th IMEKO TC4 Symposium*, Iasi, Romania, 2007
21. S. Crisan, I.G. Tarnovan, T.E. Crisan, A low cost vein detection system using near infrared radiation, in *IEEE SAS*, San Diego, USA (2007)
22. S. Crisan, I.G. Tarnovan, B. Tebrean, T.E. Crisan, Correlation of near and far infrared vein recognition for unified processing and simulation, in *XIX IMEKO World Congress Fundamental and Applied Metrology*, Lisbon, Portugal (2009)
23. S. Crisan, I.G. Tarnovan, B. Tebrean, T.E. Crisan, Hand vein biometric authentication in optical multi-touch systems, in *International Conference on Advancements of Medicine and Health Care through Technology* (Springer, Berlin/Heidelberg, 2011), pp. 124–127

24. S. Crisan, B. Tebrean, S. Emerich, *14th IMEKO TC10 Workshop Technical Diagnostics New Perspectives in Measurements, Tools and Techniques for System's Reliability, Maintainability and Safety*, Milan, Italy, June 27–28 (2016)
25. S. Crisan, T.E. Crisan, C. Curta, Near infrared vein pattern recognition for medical applications. Qualitative aspects and implementations, in *1st International Conference on Advancements of Medicine and Health Care through Technology* (2007)
26. D. Zang, Z. Guo, Y. Gong, *Multispectral Biometrics, Systems and Applications* (Springer International Publishing, Switzerland, 2016)
27. C. Wilson, *Vein Pattern Recognition: A Privacy-Enhancing Biometric* (Taylor & Francis, Boca Raton, FL, 2010). ISBN 978-1-4398-2137-4
28. R. Sheetal, G. Parkash, G. Kanwal, Image processing in hand vein pattern recognition system. Int. J. Adv. Res. Comput. Sci. Softw. Eng. **4**(6), 427–430 (2014)
29. R.D. Prasanna, P. Neelamegam, S. Sriram, N. Raju, Enhancement of vein patterns in hand image for biometric and biomedical application using various image enhancement techniques. Procedia Eng. **38**, 1174–1185 (2012)
30. M.V.N.K. Prasad, I. Kavati, K. Ravindra, Hand vein authentication system using dynamic ROI, in *Security in Computing and Communications* (Springer, Berlin/Heidelberg, 2013), pp. 203–212
31. J.-C. Lee, C.-H. Lee, C.-B. Hsu, P.-Y. Kuei, K.-C. Chang, Dorsal hand vein recognition based on 2D Gabor filters. Imaging Sci. J. **62**(3), 127–138 (2014)
32. C.B. Hsu, J.C. Lee, S.J. Chuang, P.Y. Kuei, Gaussian directional pattern for dorsal hand vein recognition. Imaging Sci. J., **63**(1), 56–62 (2015)
33. J.-C. Lee, C.-P. Chang, W.-K. Chen, Palm vein recognition based on directional empirical mode decomposition. Opt. Eng. **53**(4), 043102 (2014)
34. Y. Wang, K. Zhang, L.-K. Shark, Personal identification based on multiple keypoint sets of dorsal hand vein images. IET Biom. **3**(4), 234–245 (2014)
35. B. Xu, Y Liu, L.K. Mestha, R. Loce, Vascular pattern localization via temporal signature, in *Engineering in Medicine and Biology Society (EMBC), 2013 35th Annual International Conference of the IEEE* (IEEE, 2013), pp. 3710–3713
36. A. Kumar, Y. Zhou, Human identification using finger images. IEEE Trans. Image Process. **21**(4), 2228–2244 (2012)
37. Y. Yin, L. Liu, X. Sun, SDUMLA-HMT: a multimodel biometric database, in *Biometric Recognition* (Springer, Berlin, 2011), pp. 260–268
38. B.T. Ton, R.N.J. Veldhuis, A high quality finger vascular pattern dataset collected using a custom designed capturing device, in *2013 International Conference on Biometrics (ICB)* (IEEE, 2013), pp. 1–5
39. C. Zhang, X. Li, Z. Liu, Q. Zhao, H. Xu, F. Su, The CFVD reflection type finger vein image database with evaluation baseline, in *Biometric Recognition* (Springer, Berlin, 2013), pp. 282–287
40. M. Shahrimie, M. Asaari, S.A. Suandi, B.A. Rosdi, Fusion of band limited phase only correlation and width centroid contour distance for finger based biometrics. Expert Syst. Appl. **41**(7), 3367–3382 (2014)
41. D. Zhang, Z. Guo, G. Lu, L. Zhang, W. Zuo, An online system of multi-spectral palmprint verification. IEEE Trans. Instrum. Meas. **59**(2), 480–490 (2010)
42. D. Zhang, Z. Guo, G. Lu, MSP PalmPrint Multimodal Database. http://www4.comp.polyu.edu.hk/~biometrics/MultispectralPalmprint/MSP.htm
43. Y. Hao, Z. Sun, T. Tan, C. Ren, Multi-spectral palm image fusion for accurate contact-free palmprint recognition, in *Proceedings of IEEE International Conference on Image Processing*, IEEE, USA (2008), pp. 281–284
44. Y. Hao, Z. Sun, T. Tan, C. Ren, CASIA PalmPrint Multimodal Database. http://www.idealtest.org/dbDetailForUser.do?id=6
45. R. Kabacinski, M. Kowalski, Vein pattern database and benchmark results. Electron. Lett. **47**(20), 1127 (2011)

46. R. Kabacinski, M. Kowalski, CIE Biometrics. PUT Vein Database. http://biometrics.put.poznan.pl/vein-dataset/Descriptionpalm+wrist
47. B. Sankur, Bosphorus hand geometry and back of the hand vein database. http://bosphorus.ee.boun.edu.tr/hand/Home.aspx
48. M.A. Ferrer, A. Morales, L. Ortega, Infrared hand dorsum images for identification. Electron. Lett. **45**(6), 306–308 (2009)
49. M.A. Ferrer, A. Morales, L. Ortega, GPDS veins database. http://www.gpds.ulpgc.es/download/index.htm
50. R. Cappelli, SFinGe: an approach to synthetic fingerprint generation, in *International Workshop on Biometric Technologies (BT2004)* (2004), pp. 147–154
51. J. Zuo, N.A. Schmid, X. Chen, On generation and analysis of synthetic iris images. IEEE Trans. Inf. Forensics Secur. **2**(1), 77–90 (2007)
52. Z. Wei, Y. Han, Z. Sun, T. Tan, Palmprint image synthesis: a preliminary study, in *2008 15th IEEE International Conference on Image Processing (ICIP 2008)* (IEEE, 2008), pp. 285–288
53. F. Hillerström, A. Kumar, On generation and analysis of synthetic finger-vein images for biometrics identification (vol. 17). Technical Report No. COMP-K (2014)
54. S. Crisan, I.G. Târnovan, T.E. Crisan, A hand vein structure simulation platform for algorithm testing and biometric identification. *16th IMEKO TC4 Symposium*, Florence, Italy, 2008
55. A. Sabri, J. Szalas, K.S. Holmes, L. Labib, T. Mussivand, Failed attempts and improvement strategies in peripheral intravenous catheterization. Biomed. Mater. Eng. **23**, 93–108 (2013). doi:10.3233/BME-120735
56. S. Juric, B. Zalik, An innovative approach to near-infrared spectroscopy using a standard mobile device and its clinical application in the real-time visualization of peripheral veins. BMC Med. Inform. Decis. Mak. **14**(1), 100 (2014)
57. L.L. Kuensting, S. DeBoer, R. Holleran, B.L. Shultz, R.A. Steinmann, J. Venella, Difficult venous access in children: taking control. J. Emerg. Nurs. **35**(5), 419–424 (2009)
58. P. Larsen, D. Eldridge, J. Brinkley, D. Newton, D. Goff, T. Hartzog, N.D. Saad, R. Perkin, Pediatric peripheral intravenous access: does nursing experience and competence really make a difference? J. Infus. Nurs. **33**(4), 226–235 (2010)
59. L.C. Hadaway, D.A. Mill, On the road to successful IV starts. Nursing **35**(Suppl O), 1–14 (2005)
60. K. Yen, A. Riegert, M.H. Gorelick, Derivation of the DIVA score: a clinical prediction rule for the identification of children with difficult intravenous access. Pediatr. Emerg. Care **24**(3), 143–147 (2008)
61. S. Juric, V. Flis, M. Debevc, A. Holzinger, B. Zalik, Towards a low-cost mobile subcutaneous vein detection solution using near-infrared spectroscopy. Sci. World J. **2014**, Article ID 365902, 1–15 (2014)
62. A. Shahzad, M.N.M. Saad, N. Walter, A.S. Malik, F. Meriaudeau, A review on subcutaneous veins localization using imaging techniques. Curr. Med. Imaging Rev. **10**(2), 125–133 (2014)
63. L.L. Chapman, B. Sullivan, A.L. Pacheco, C.P. Draleau, B.M. Becker, VeinViewer-assisted intravenous catheter placement in a pediatric emergency department. Acad. Emerg. Med. **18**(9), 966–971 (2011)
64. E. Hemricourt (2011) AccuVein AV300: arrivée d'un nouvel outil médical de prime abord révolutionnaire, AccuVein primary Website: http://learn.accuvein.com/2010/08/accuvein-av300-praised-as-revolutionary-device-in-news-sante-europe/
65. F.B. Chiao, F. Resta-Flarer, J. Lesser, J. Ng, A. Ganz, D. Pino-Luey, H. Bennett, C. Perkins Jr., B. Witek, Vein visualization: patient characteristic factors and efficacy of a new infrared vein finder technology. Br J Anaesth **110**(6), 966–971 (2013)
66. N.J. Cuper, R.M. Verdaasdonk, R. de Roode, K.M. de Vooght, M.A. Viergever, C.J. Kalkman, J.C. deGraaff, Visualizing veins with near infrared light to facilitate blood withdrawal in children. Clin. Pediatr. (Phila) **50**(6), 508–512 (2011)
67. W. Mason, Perception neuron review: in-depth with the $1,500 motion capture suit. http://uploadvr.com/perception-neuron-review/

68. M. Gibbs, Leap Motion 3D Controller (2015), NetworkWorld, http://www.networkworld.com/article/2225036/software/leap-motion-3d-controller-is-fantastic.html
69. G.T. Diefenderfer, Thesis on "Fingerprint Recognition" at Naval Postgraduate School, Monterey, California, 2006
70. Jagdale AR, Sonawane KV, Khan SS (2014) Data mining and data pre-processing for Big Data. Int. J. Sci. Eng. Res. **5**(7), 1156 (2014). ISSN 2229-5518
71. D. Galar, Big data analytics for diagnosis: a new era for measurements, Invited Lecture in IMEKO TC-10 Workshop on New Perspectives in Measurements, Tools and Techniques for System's Reliability, Maintainability and Safety, Milan, Italy, Conference Proceedings – ISBN: 978-92-990073-9-6 (2016)
72. S. Haines, Introduction to HBase, the NoSQL Database for Hadoop (2014). Pearson InformIT, http://www.informit.com/articles/article.aspx?p=2253412
73. R. Grehan, Review: HBase is massively scalable—and hugely complex (2014). InfoWorld, http://www.infoworld.com/article/2610709/database/review--hbase-is-massively-scalable----and-hugely-complex.html
74. S. Yegulalp, Apache Parquet paves the way for better Hadoop data storage (2015). InfoWorld Tech Watch, http://www.infoworld.com/article/2915565/big-data/apache-parquet-paves-the-way-towards-better-hadoop-data-storage.html

Chapter 3
Improving Biometric Identification Performance Using PCANet Deep Learning and Multispectral Palmprint

Abdallah Meraoumia, Farid Kadri, Hakim Bendjenna, Salim Chitroub, and Ahmed Bouridane

3.1 Introduction

Security technology is an important aspect of any system and in a wide variety of applications such as public security, access control, and banking. So, ensuring this security is becoming an increasingly important problem as modern technology is integrated into the majority of these applications. To ensure this security, one of the best solutions that have proven their effectiveness and robustness in this field is the personal identity recognition [1]. Traditionally, two person identification methods are used, which employ knowledge based or token-based method. Unfortunately, these are time-consuming, inefficient, and expensive. Thus, to overcome these limitations we should use something that really characterizes the given person such as their behavioral or physiological traits. These are called biometric characteristics and the technology of the identification based on these characteristics is called biometrics [2]. This technology offers some significant advantages over traditional

A. Meraoumia (✉) • F. Kadri
Electrical Engineering Laboratory (LAGE), University of Ouargla, Ouargla, Algeria
e-mail: ameraoumia@gmail.com; ge.fkadri@gmail.com

H. Bendjenna
LAboratory of Mathematics, Informatics and Systems (LAMIS), University of Tebessa, Tebessa, Algeria
e-mail: hbendjenna@gmail.com

S. Chitroub
Laboratory of Intelligent and Communication Systems Engineering (LISIC), USTHB, Bab Ezzouar, Algeria
e-mail: s_chitroub@hotmail.com

A. Bouridane
Northumbria University, Newcastle, UK
e-mail: ahmed.bouridane@northumbria.ac.uk

R. Jiang et al. (eds.), *Biometric Security and Privacy*, Signal Processing for Security Technologies, DOI 10.1007/978-3-319-47301-7_3

methods; a card-based access system can control the access of authorized card, but not the one who is in possession of the card. Also, systems using PINs require that an individual only knows a specific number to gain entry. Therefore, biometric emerges as a stronger recognition solution that eliminates hassles, cuts cost, and most of all, provides the highest level of safety and convenience [3]. Thus, there are many biometric systems for person recognition in the literature, among them the palmprint, which we will investigate in this paper, has been widely used in biometric systems owing to its properties of uniqueness, reliability, and stability. On the other hand, compared with other biometrics, the palmprint has several advantages such as low-resolution; low-cost capture devices and it is difficult to fake a palmprint [4].

A typical biometric system is composed of four main modules: acquisition, features extraction, features matching, and decision. Generally, improving the biometric system performance requires to improve one of these modules. Therefore, several works focus on both acquisition and feature extraction modules in order to improve the system performance. An important key factor of the acquisition module is to provide the user a friendly and simple interface to interact with the system. Also, the quality and amount of the acquired data affect significantly the system's accuracy. Thus, several studies for palmprint-based personal identification systems have focused on improving the performance of palmprint images captured under visible light (grayscale or color image). However, during the past few years, some researchers have considered multispectral images to improve the efficiency of these systems [5]. This technique can give different information from the same palmprint modality using an acquisition device to capture, for example, the palmprint images under visible and infrared light resulting into several spectrum images.

On the other hand, compared with the acquisition module, one of the foremost advantages offered by the feature extraction module is flexibility. The reason for that is the possibility to use a wide range of algorithms to extract some discriminant information of the biometric trait. Consequently, researchers can focus their efforts on improving the performance of the system rather than on designing new acquisition device, which can be a very expensive and time-consuming process. Hence, a lot of classification methods rely on feature extraction method. These methods have great limitations because there is usually a large amount of intra-class variability. Recently, a new area of research in machine learning called deep learning has emerged [6]. The idea of these techniques is to discover multiple levels of representation with the hope that higher level features can represent effectively the discriminant characteristics of the biometric modality. In this chapter, we try to enhance biometric system performance by using the multispectral imaging and a new and simple deep learning method called PCANet [7]. Using multispectral images which not only increases the effectiveness of the palmprint recognition but also helps the person to trust on the security systems based on biometric technologies. Moreover, the PCANet deep learning extracts only the palmprint features which can increase the inter-class variability and decrease the intra-class variability, as a result, a very important improvement in the identification system accuracy can be obtained.

3.2 Image Features

One of the most critical problems in the pattern recognition is the selection of suitable characteristics for a compact and optimal representation of the image. So, an image contains various features, including geometrical features, lines, singular points, and texture. For that, many approaches for the feature extraction process have been suggested in the literature [4]. In general the proposed techniques can be divided into three different categories: line-based technique (e.g., edge detectors or Gabor filter with thresholding), appearance-based technique (e.g., Linear Discriminant Analysis (LDA) or Independent Component Analysis (ICA)), and texture-based technique (e.g., Discrete Cosine Transform (DCT) or Gabor filter). Feature vector extracted from the original image is based on one of the above-mentioned techniques or a combination of different techniques.

In order to obtain a better image representation that contains the most discriminatory information, many researchers have used the texture as a source of information. With over a half century of digital image processing, texture is still a very fertile field of research attracting many mathematical algorithms and techniques developed in order to obtain a better image characterization that characterizes the spatial variations within it and then facilitates image processing task. Recently, a new texture based technique (PCANet) using the deep learning approaches is developed. This technique is potential and simple for texture classification and object recognition.

3.3 Proposed Methodology

As mentioned above, our objective is to design a biometric identification system using PCANet deep learning and multispectral palmprint. As all biometric systems, our system works in two phases: enrollment and identification. During enrollment phase, the feature vector of each palmprint modality is represented by a feature vector extracted by the PCANet deep learning method (see next section). After that, all feature vectors (extracted from all enrolled persons) are used in order to train one of the four classifiers (SVM, RBF, RFT, and KNN) and so constructing the biometric identification system. As for identification phase, the same feature vector is extracted from the test spectral band of palmprint modality and then is used as an input vector to the classifier in order to decide whether to accept or reject this person. The multimodal biometric identification system architecture is based on the fusion at matching score level where the normalized scores of two or more unimodal systems (each one is based on one spectral band of palmprint image) are fused.

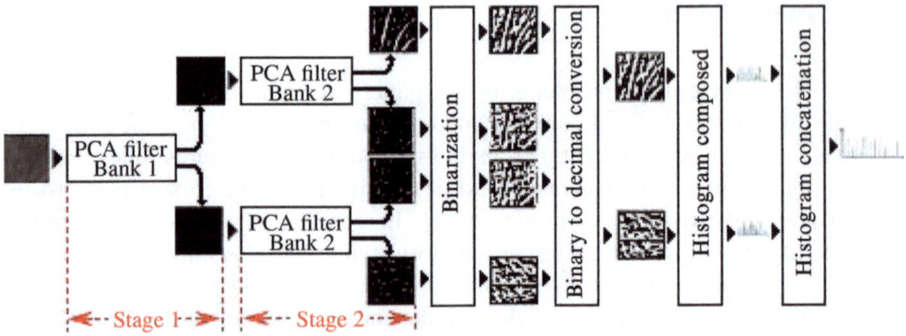

Fig. 3.1 Block diagram of the PCANet deep learning based feature extraction using two stages

3.3.1 Feature Extraction

Feature extraction is a crucial module in any pattern recognition application since good classification results are directly related to the uniqueness and variability of the extracted features used to distinguish between different patterns [8]. In our scheme, the PCANet deep learning is used to extract the feature vector of each spectral band.

3.3.1.1 PCANet Deep Learning

PCANet is a simple deep learning baseline for image classification which comprises only the very basic data processing components: (1) cascaded Principal Component Analysis (PCA), (2) binary hashing, and (3) histograms. Thus, the block diagram of PCANet algorithm presented in Fig. 3.1 can be summarized as follows [7, 9]:

(1) **PCA Filter bank:** As shown in Fig. 3.1, the PCA filter bank comprises two stages of filter bank convolutions. In the first stage, the filter banks are estimated by performing principal components (PCA) over a set of vectors where each vector represents the $k_1 \times k_2$ points around each pixel.
For each vector we take the mean of the entries and then subtract the mean from each entry of it. Then, we perform PCA over these vectors and retain the principal components W (size of $k_1 \cdot k_2 \times L_{S_1}$) where L_{S_1} is the primary eigen vectors. Next, each principal component (column of W) is a filter and may be converted to $k_1 \times k_2$ kernel which is convolved with the input image. So,

$$I_\ell(x, y) = h_\ell(x, y) * I(x, y) \quad \text{where} \quad 1 \leq \ell \leq L_{S_1} \tag{3.1}$$

where $*$ denotes the discrete convolution and I_ℓ is the resulting filtered image using the h_ℓ filter. So, using the L_{S_1} columns of W we take each input palmprint image I and convert it into L_{S_1} output images.

The second stage is constructed by iterating the algorithm from the first stage (Filter bank convulsions) over each of the output image. For each output images I_ℓ we take the vector (points around each pixel), the mean of the entries and then subtract the mean from each entry of the vector is computed. The vectors produced are then concatenated together and we estimate another PCA filter bank (with L_{S_2} filters). Finally, each obtained filter is convolved with I_ℓ to produce a new image.

$$I_{\ell,m}(x,y) = h_m(x,y) * I_\ell(x,y), \qquad 1 \leq m \leq L_{S_2} \text{ and } 1 \leq \ell \leq L_{S_1} \tag{3.2}$$

Hence, using all output images of the first stage and by repeating convulsion process for each filter, $L_{S_1} \cdot L_{S_2}$ output images can be produced.

(2) Binary hashing: In this step, the $L_{S_1} \cdot L_{S_2}$ output images obtained in second layer are converted into binary format by using a Heaviside step function whose their value is one for positive entries and zero otherwise.

$$I^B_{\ell,m}(i,j) = \begin{cases} 1 & \text{if } I_{\ell,m}(i,j) \geq 0 \\ 0 & \text{otherwise} \end{cases} \tag{3.3}$$

where $I^B_{\ell,m}$ is a binary image. In addition, around each pixel, we view the vector of L_{S_2} binary bits as a decimal number. This converts the L_{S_2} outputs into a single integer-valued (image).

$$I^D_\ell(i,j) = \sum_{m=1}^{L_{S_2}} 2^{m-1} I^B_{\ell,m}(i,j) \tag{3.4}$$

where I^D_ℓ is an image whose every pixel is an integer in the range $[0, \ 2^{L_{S_2}-1}]$.

(3) Histogram composition: Each hashed images (I^D_ℓ) is partitioned into N_B blocks and a histogram of the decimal values in each block (B) is computed. These blocks can be non-overlapping (disjoint) or they can be overlapping (depending on the application). Thus, the feature of I^D_ℓ is obtained by concatenating all B histograms such as:

$$\upsilon^{\text{hist}}_\ell = [B^{\text{hist}}_1, B^{\text{hist}}_2, \ldots B^{\text{hist}}_{N_B}] \tag{3.5}$$

Finally, after this encoding process, the feature vector of the input image I is then defined as:

$$\upsilon_I = [\upsilon^{\text{hist}}_1, \upsilon^{\text{hist}}_2, \ldots \upsilon^{\text{hist}}_{L_{S_1}}]^T \tag{3.6}$$

In conclusion, the hyper-parameters of the PCANet include the filter size (k_1, k_2), the number of filters in each stage (L_{S_i}), the number of stages (N_s), and the block size for local histograms in the output layer (B).

3.3.1.2 Flexibility Property

If we compare PCANet deep learning with several feature extraction methods, it is clear that PCANet deep learning can capture textural features in an adaptive manner. Therefore, it can extract the relevant image features with an effective flexibility. This flexibility aspect of this method is ensured thanks to the two parameters: number of stages (N_s) and number of filters in each stage (L_{S_i}), which are variable and adaptive depending on the intended application. The size of the extracted feature vectors is also variable depending on the user choice.

Thus, in our case (two stages), the outputs number of the first stage is equal to the number of selected filters in this stage (L_{S_1}). Moreover, in the second stage, the outputs number is equal to the number of selected filters in this stage (L_{S_2}) multiplied by the number of selected filters in the previous stage (S_1). In general, repeating this process for N_s stages produces γ outputs, where

$$\gamma = \prod_{i=1}^{N_s} L_{S_i} \tag{3.7}$$

As mentioned above, the feature vector is represented by the histograms formed from output images (γ images). This histogram is a function that counts the number of observations that fall into each of the bins, for that, the feature vector length of each L_{S_2} output images is equal to the total number of bins ($\eta = 2^{L_{S_2}}$ bins). Furthermore, the feature vector that characterizes the texture of a given palmprint image is generated by concatenating of L_{S_1} histograms and their length (ξ) is given as follows:

$$\xi = L_{S_1} \cdot \eta = L_{S_1} \cdot 2^{L_{S_2}} \tag{3.8}$$

For example, in the case of two stages ($N_s = 2$) with 3 and 4 filters in the first and second stage ($L_{S_1} = 3$ and $L_{S_2} = 4$), the size of the extracted features vector is equal to $\xi = 3 \times 2^4 = 48$. It is also interesting to note that the length of the output feature vectors is independent of input image size.

3.3.2 *Classification*

In our scheme, the palmprint is typically analyzed by PCANet deep learning technique. After applying this technique to the palmprint images, the produced histogram is selected to construct feature vector. Furthermore, all histograms (for all enrolled persons) are concatenated and an appropriate classifier is trained in order to construct the system database. In the original work which uses PCANet deep learning (in face recognition), a multi-class SVM classifier is used. Thus, in our scheme, three additional classifiers are tested which are Radial Basis Function (RBF) [10], Random Forest Transform (RFT) [11], and K-Nearest Neighbor (KNN) [12].

3.4 Experimental Results and Discussion

The proposed scheme is tested and validated using two popular and publicly available multispectral palmprint databases; The CASIA database [13], from the Institute of Automation of the Chinese Academy of Sciences and the PolyU database [14] from the Hong Kong Polytechnic University. These databases are largely used in academic domain to understand the challenges. Also, they are the only available multispectral datasets for evaluating the techniques developed by the researchers.

3.4.1 Experimental Databases

Multispectral palmprint image database from CASIA has a total of 7200 images obtained from 100 persons. In this database, the images of each hand are collected in two separate sessions. The time interval between the two sessions is more than 1 month. In each session, there are three samples. Each sample contains six palm images which are captured at the same time with six different electromagnetic spectrums. Wavelengths of the illuminator corresponding to the six spectrums are 460, 630, 700, 850, 940 nm and white light, respectively. Between two samples, certain degrees of variations of hand postures are allowed. All palm images are low resolution (<150 dpi) stored as 8-bit gray-level images per band.

Multispectral palmprint image database from PolyU contains images captured with visible and infrared light. Four palmprint images for each person, including *Red*, *Green*, *Blue*, and *near-infrared* spectral band, are collected. Six thousand multispectral palmprint images were collected from 500 persons. These images were collected in two separate sessions. In each session, the person provides 6 images for each palm, so there are 12 images for each person. Therefore, 48 spectrum images of all illumination from 2 palms were collected from each person. The average time interval between the first and the second sessions was about 9 days. Also, all palm images are low resolution (<150 dpi) stored as 8-bit gray-level images per band.

3.4.2 Identification Test Results

To evaluate the efficiency of our proposed method, four palmprints (spectral band images) of each person were randomly selected for enrollment, and the rest, eight palmprint images, are used as test images for identification, respectively. Thus, for CASIA database, there are totally 800 genuine comparisons and 39,600 impostor comparisons that are generated (a database contains 100 persons). Furthermore, for PolyU database, 3200 genuine comparisons and 638,400 impostor comparisons were generated (a database contains 400 persons).

The identification test results are divided into two parts. In the first part, we present the performance of the unimodal systems using our proposed scheme to choose the system (spectral band) yielding the best performance. This part includes a section for selecting the best parameters for the PCAnet deep learning. The test results in the second part are devoted to evaluate the performance of the multimodal identification systems. It is important to note that a biometric identification system can work into two modes: open-set identification and closed-set identification. In open-set mode, the system indicates that the person presenting the acquired biometric data is an enrolled person or not in the system database. In the other case (closed-set mode), the system selects the identity of the person whose reference has the highest degree of similarity with the acquired biometric data. In our study, the proposed method was tested through the two modes.

Finally, it might be of interest to note that the original image (each spectral band of palmprint) must be preprocessed to obtain only the area information of the palmprint (Region Of Interest (ROI) sub-image). In our work, we use the preprocessing method described in [15].

3.4.2.1 Performance of the Unimodal Systems

The objective of the first part was to evaluate the system performance when we use information from each modality (each spectral band), but before this, it is very necessary to choose the best configuration of PCANet. Thus, a series of experiments were carried out using the 460 nm spectral band (from CASIA database) in order to select the best parameters of PCANet. Their objective is to study the impact of each parameter (number of filters, filter size, and block overlap ratio) on the system accuracy. Note that, in our scheme a PCANet with two stages (two-stage networks : PCANet^{S_2}) is used, however, in the first stage we examine all the parameters, whereas, in the second stage and due to the greater number of tests, we limited this study to only the number of filters. Also, always due to the greater number of the tests, in the one-stage networks (PCANet^{S_1}), for each case we fix two parameters and test the third. Therefore, our empirical test results can be divided into two subparts: the first subpart includes the preliminary results obtained from PCANet^{S_1}. In the second subpart, the presented results were obtained using PCANet^{S_2}.

The goal of the first part was to evaluate the system performance when we use PCANet^{S_1} with varying each time one parameter. For this, we found the performance, for each parameter, in form of Genuine Acceptance Rate (GAR) (performance of open-set identification system). Thus, concerning the impact of the number of filters, the filter size of the PCANet^{S_1} is fixed to 7×7 with non-overlapping blocks. We vary the number of filters from 1 to 13 and compute each time the GAR. In Fig. 3.2a we show the obtained results. In this figure, we plot the system performance as a function of the number of filters. The reason that this figure was generated was to show how the number of filters might have an effect on the performance of our system. We observe that the identification accuracy becomes very high at certain number of filters ([5 ... 11]), where it actually exceeds 99.70 %

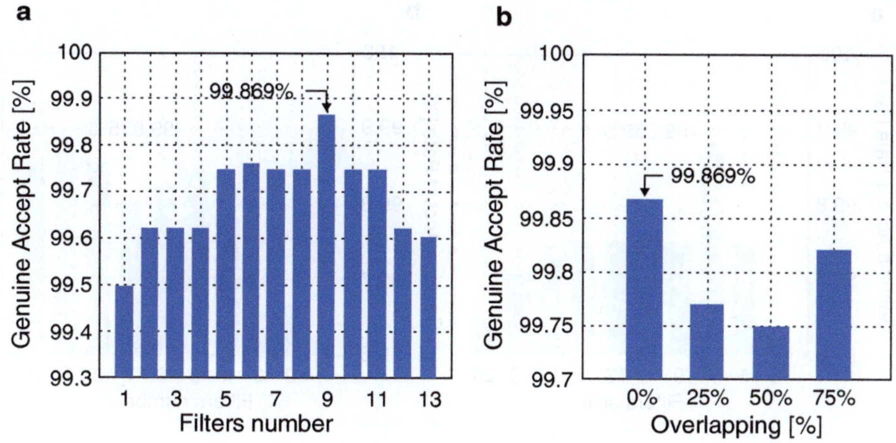

Fig. 3.2 Performance of the unimodal system as function of filters number and overlap between blocks in the first stage. **(a)** Variation of the GAR as function of filters number and **(b)** Variation of the GAR as function of block overlap ratio

and a slight decrease in identification accuracy as we go to lower or higher numbers of filters. In addition, the experimental results indicate that nine filters improve better the system performance in terms of GAR, in this case the system achieves a higher accuracy with GAR = 99.869 % at a threshold T_o equal to 0.6841.

Now, after adopting the best number of filters (nine filters), we conduct several experiments to test the effectiveness of this scheme under several block overlap ratio. So, the system performance under 0, 25, 50, and 75 % overlap was evaluated and Fig. 3.2b shows the open-set identification system performances resulting from the four ratio. It showed that the best results were obtained with non-overlapping blocks (0 % ratio). Thus, the best GAR is about 99.869 % (best Equal Error Rate (EER) equal to 0.1311 %) at the threshold T_o equal to 0.6841. As for the other results, if the 25 % ratio is used, our identification system can achieve an EER = 0.2286 % at the threshold $T_o = 0.5965$. In the case of using the 50 % ratio, EER is equal 0.250 % at the threshold $T_o = 0.5790$. Finally, using 75 % ratio, the minimum EER was 0.1785 % at the threshold $T_o = 0.6073$.

Finally, we examine the impact of the filter size ($n \times n$) on the identification system accuracy. Thus, in this experiment, we compare the obtained system accuracies, by different filter sizes. Ten filters, with $n \in [3, 5, 7, \ldots, 21]$, are used and the experimental results are plotted in Fig. 3.3a. This figure reports the system performance for the different filter sizes, with their corresponding GAR, achieved on the test set. Comparing the achieved GAR, it is clear that applying filter size of $7 \times 7 (n = 7)$ improves the performance with a height GAR equal to 99.869 %. It is important to see that almost all sizes give an excellent performance (GAR > 99.700 %).

In the second subpart, our system (PCANet^{S_2}) contains two stages. Thus, by adopting the previous best parameters for the first stage, the presented results (see

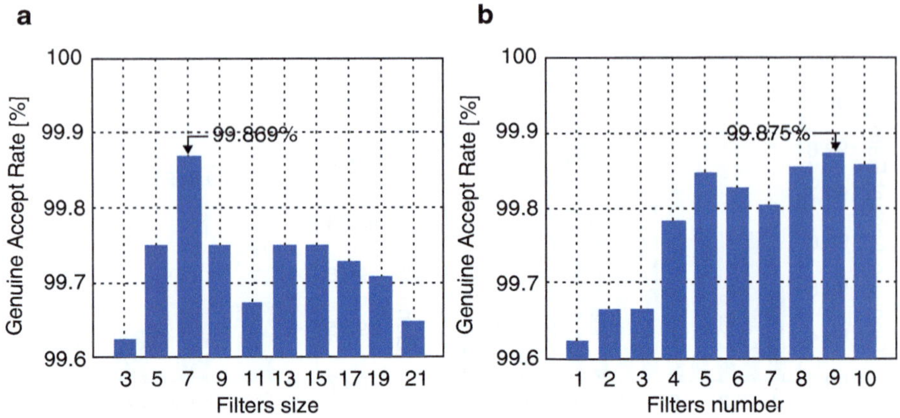

Fig. 3.3 Performance of the unimodal system as function of filter sizes (first stage) and filter number (second stage). **(a)** Variation of the GAR as function of filter sizes (first stage) and **(b)** Variation of the GAR as function of filters number (second stage)

Fig. 3.3b) were obtained by changing only the number of filters in second stage. In this stage, the filter size and block overlap ratio is fixed as those in the first stage (7×7 and 0 %). As shown in Fig. 3.3b, always in this stage, a perfect performance (>99.700 %) was given by several filter numbers ($\geq$4). These results suggest that nine filters achieve high accuracy comparing with the other numbers (GAR = 99.875 % and EER = 0.125 % at $T_o = 0.6052$).

- **CASIA database:** the goal of the last section of the first part was to evaluate the system performance when we use information from each modality (spectral band). For this, we found the performance under different modalities (460, 630, 700, 850, 940 nm, and white light). However for all spectral bands, a PCANet^{S_2} with the previously obtained best parameters is used. Thus, in order to see the performance of the identification system, we usually present, in Table 3.1, the average open/closed-set identification results as a function of the spectral band. Thus, in the case of open-set identification mode, this table shows that the 460 nm spectral band offers better results in terms of the EER. In this case, the identification system can achieve an EER of 0.125 % at a threshold $T_o = 0.6052$. Thus, as shown in the table, a minimum EER above to 0.500 % can be obtained with all bands, except the 850 and 940 nm bands which give, respectively, 0.949 % ($T_o = 0.5928$) and 1.250 % ($T_o = 0.5905$). Furthermore, the effectiveness of all spectral bands is shown in Fig. 3.4a, where the Receiver Operating Characteristic (ROC) curves (False Rejected Rate (FRR) against False Accept Rate (FAR)) and EER obtained from experimental results are illustrated. In the case of a closed-set identification, a series of experiments were carried out to select the best spectral band, this has been done by comparing all bands and finding the band that gives the best identification rate. Always, Table 3.1 presents the experiments results obtained for all spectral bands. From this table, the best

Table 3.1 Unimodal open/closed-set identification system performance (CASIA database)

	Open-set identification		Closed-set identification	
Spectral bands	T_o	EER	ROR	RPR
460	0.6052	0.125	99.500	40
630	0.6936	0.148	99.250	43
700	0.6704	0.182	98.875	21
850	0.5928	0.949	96.375	42
940	0.5905	1.250	96.625	55
WHT	0.5502	0.300	98.875	22

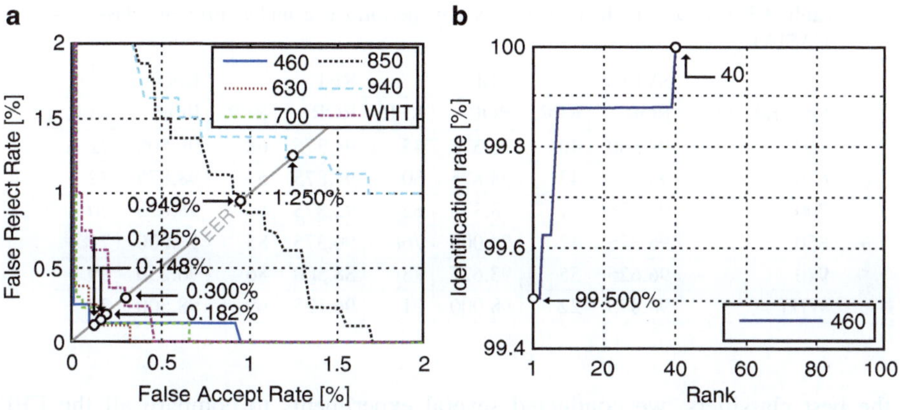

Fig. 3.4 Unimodal open/closed-set identification test results under different spectral bands. **(a)** *ROC* ROC curves (FRR against FAR) under different spectral bands and **(b)** under CMC curve for the best case (460 nm spectral band)

results of Rank-One Recognition (ROR) produce 99.500 % with lowest Rank of Perfect Recognition (RPR) of 40 in the case of 460 nm spectral band. A similar result is obtained by the 630 nm spectral band (ROR = 99.250 % with RPR = 43). Poor results are obtained by 850 and 940 nm bands (ROR < 97.000 %). Finally, Fig. 3.4b plots the Cumulative Match Characteristics (CMC) curve for the best case (460 nm spectral band). The results (unimodal identification system) presented in this part clearly indicate that our proposed systems can achieve comparable performance to the best of the state-of-the-art systems like those in [16, 17].

Finally, in order to obtain the performance of the different classifiers (SVM, RBF, RTF or KNN), the feature vector of an unknown palmprint (unknown spectral band) is presented to the system and the system will try to match this modality with the desired classifier. Thus, our goal is to evaluate the identification system performance when using information from each classifier. So, after applying the PCANet deep learning technique on each spectral band and in order to select

Table 3.2 Open-set identification system performance under different classifiers (CASIA)

	SVM		RBF		RFT		KNN	
Spectral bands	T_o	EER	T_o	EER	T_o	EER	T_o	EER
460	0.6052	0.125	0.6052	0.625	0.6324	1.000	0.2682	0.250
630	0.6936	0.148	0.7776	0.125	0.6187	1.250	0.2505	0.187
700	0.6704	0.182	0.6834	0.375	0.6450	1.376	0.2398	0.257
850	0.5928	0.949	0.6456	1.652	0.5671	3.500	0.3466	1.184
940	0.5905	1.250	0.6315	2.250	0.6041	3.125	0.3316	1.057
WHT	0.5502	0.299	0.6080	0.750	0.6229	0.954	0.2680	0.335

Table 3.3 Closed-set identification system performance under different classifiers (CASIA)

	SVM		RBF		RFT		KNN	
Spectral bands	ROR	RPR	ROR	RPR	ROR	RPR	ROR	RPR
460	99.500	40	98.375	45	94.375	68	98.500	22
630	99.249	43	98.875	50	94.875	59	98.875	32
700	98.875	21	98.375	62	92.875	64	98.625	30
850	96.375	42	94.000	76	90.375	85	96.000	71
940	96.625	55	93.625	78	88.500	86	96.250	99
WHT	98.875	22	98.000	91	96.125	69	98.500	37

the best classifiers, we conducted several experiments to compare all the EER and ROR metrics for several classifiers to find the classifier that gives the best identification rate. Also, for the open-set identification mode, Tables 3.2 and 3.3 show the system performances resulting from the four classifiers for the open-set and closed-set identification modes. These tables showed that the best results were obtained with SVM and RBF classifier in the open-set and closed-set identification modes. Thus, the RFT classifier achieved a poor result; it can achieve an EER equal to 0.954 % at a threshold $T_o = 0.6229$ and an ROR equal to 96.125 % with a lowest RPR = 69 for, respectively, open-set and closed-set identification modes. So, if we take into consideration these accuracies, we believe that this technology (PCANet deep learning technique) may be sufficient for representing the majority of biometric characteristics within biometric modality.

- **PolyU database:** In the experimental results presented in this part, the feature extraction method (PCANet deep learning which uses the same parameters chosen above) is tested and evaluated on the PolyU database with 400 persons in which we performed a comparison study between the four classifiers. Similarly, the goal of the first subpart was to evaluate the system performance when using PCANet deep learning based feature extraction and the information from the four spectral bands (BLUE, GREEN, RED, and Near-InfraRed (NIR)) modalities. Thus, the effectiveness of this method using all classifiers can be shown in Tables 3.4 and 3.5 for the two modes of identification, where the EER, the

Table 3.4 Open-set identification system performance under different classifiers (PolyU)

	SVM		RBF		RFT		KNN	
Spectral bands	T_o	EER	T_o	EER	T_o	EER	T_o	EER
Blue	0.4620	0.000	0.6239	0.063	0.5921	0.600	0.5100	0.000
Green	0.5190	0.000	0.7000	0.010	0.6201	0.421	0.6630	0.000
Red	0.7261	0.000	0.8670	0.000	0.8111	0.105	0.7311	0.000
NIR	0.4919	0.000	0.8042	0.002	0.5336	0.112	0.5381	0.000

Table 3.5 Closed-set identification system performance under different classifiers (PolyU)

	SVM		RBF		RFT		KNN	
Spectral bands	ROR	RPR	ROR	RPR	ROR	RPR	ROR	RPR
Blue	100.00	1	99.563	2	97.211	16	100.00	1
Green	100.00	1	99.250	2	96.666	23	100.00	1
Red	100.00	1	99.969	2	98.151	7	100.00	1
NIR	100.00	1	99.719	2	98.705	11	100.00	1

ROR, and the RPR obtained from experimental results are illustrated. From these tables, we can observe that the open-set identification accuracy becomes very high at certain classifiers for all spectral bands, where it reaches 100 % for, respectively, the open-set and closed-set identification mode. In these results we can remark that SVM and KNN provide similar precisions for identification for all spectral bands. But, regarding the time processing, SVM classifier is very efficient compared with KNN classifier (0.8 s instead 3.0 s). Also, RBF classifier offers best results for all spectral bands (EER > 99.250 %). For this database and compared with to the best of the state-of-the-art systems, the results presented in this part clearly indicate that the PCANet deep learning based feature extraction technique can achieve comparable performance to the best of the state-of-the-art techniques.

3.4.2.2 Performance of the Multimodal Systems

Recently, research in the field of biometrics for identification purposes has increasingly investigated the use of multiple biometric modalities (multimodal biometrics). Multimodal biometrics refers to the use of more than one biometric modality for person identification. Research into multimodal biometric systems has eventuated because unimodal biometric systems may suffer from several limitations such as: noise in data, intra-class variations, and inter-class similarity [18]. Thus, the growing interest in the use of multiple modalities in biometrics is due to its potential capabilities for overcoming certain important limitations of unimodal systems. Researchers believe that the use of multimodal biometrics will provide a more reliable and robust system. However, the goal of this experiment was to investigate the systems performance when we fuse information from some spectral bands. In fact, at such a case the system works as a kind of multimodal system with a single biometric (palmprint) but multiple instances. Therefore, biometric information presented by different modalities (spectral bands) are fused to make the system efficient.

The multimodal system combined the different information based on data fusion. In fact, in biometric systems, four possible levels of fusion can be used for integrating data from two or more modalities [19]. These levels are the sensor level, the feature level, the matching score level, and the decision level. In our study, the fusion at matching score level is used. Currently, this level appears to be the most useful fusion level because of its good performance and simplicity. In their former approach, a scalar fused score is obtained by normalizing the input matching scores into the same range and then combining such normalized scores. Normalization is done by assigning the lowest available value the new value 0, changing the highest available value to 1, and updating the values in between accordingly (*min–max* normalization [20]). During the system design, we experiment four fusion rules {sum-score (SUM), min-score (MIN), max-score (MAX), and mul-score (MUL)}.

The CASIA database contains six spectral bands. By using these bands, several combinations can be performed. In our work, we limited the test for only two combinations which are (460 nm–630 nm–700 nm) and (850 nm–940 nm–white). Thus, in the case of open-set identification (Table 3.6), it is clear that the first combination can reduce the EER to zero (100 % improvement) when the SUM, MIN, and MUL rules are used. Compared with these rules, poor results are obtained by MAX rule, but, even this rule improves the unimodal identification results by 88 %. For the second combination, it is very clear that the fusion does not improve the system performance (0 % improvement) for all fusion rules. The non-improvement of the system performance is justified by the fact that the two unimodal systems based, respectively, on 850 and 940 nm have higher errors. For a further demonstration of the efficiency of our scheme, we have also performed a closed-set identification mode by applying all fusion rules on the matching scores obtained from the unimodal system and calculated the ROR and RPR values (see Table 3.7). Also, in the first combination, a very higher improvement (100 %) is remarked compared with the best case in the unimodal system. The best result of ROR produces an accuracy of 100.00 % with lowest RPR of 1 in the case of SUM and MUL rules. In addition, the system can operate with an ROR equal to 99.875 % (RPR = 2) and 98.628 % (RPR = 3) for, respectively, MIN and MAX rules. In the

Table 3.6 Multimodal *open-set* identification system performance (CASIA)

	SUM		MIN		MAX		MUL	
Combinations	T_o	EER	T_o	EER	T_o	EER	T_o	EER
460–630–700	0.8449	0.000	0.7813	0.000	0.9999	0.015	0.6960	0.000
850–940–WHT	0.5886	0.125	0.5175	0.125	0.9255	0.125	0.1835	0.125

Table 3.7 Multimodal *closed-set* identification system performance (CASIA)

	SUM		MIN		MAX		MUL	
Combinations	ROR	RPR	ROR	RPR	ROR	RPR	ROR	RPR
460–630–700	100.00	1	99.875	2	98.628	3	100.00	1
850–940–WHT	99.750	15	99.750	29	95.875	30	99.750	14

second combination, the best result of ROR is obtained by the SUM, MIN, and MUL rules and it is equal to 99.750 % with RPR equal to 15, 29, and 14 for, respectively, SUM, MIN, and MUL rules.

Similarly, the second database (PolyU database) was tested using the Multimodality concept. Thus, the two classifiers SVM and KNN do not need to apply a fusion process. But for the other classifiers, it is necessary to fuse some spectral bands to improve the system performance. So, by using the four spectral bands (BLUE, GREEN, RED, and NIR), several combinations can be performed. Thus, in our work, we limited the test to only for two combinations which are RGB (color image) and the fusion of all spectral bands (named RGBN) for multispectral image. In the case of open-set identification, the two combinations (RGB and RGBN) can reduce the EER to zero (100 % improvement) when the SUM, MIN, and MUL fusion based rules are used (SVM classifier), which, in this case, is efficient than several previous works obtained using PolyU database. Errors equal to 0.003 % ($T_o = 0.9990$) and 0.004 % ($T_o = 0.9990$) were obtained when using the MAX rule based fusion. In the case of RFT classifier, the two combinations achieve a better result. Thus, the best result (best EER), given by RGBN combination and MUL rule, is 0.057 % at $T_o = 0.7223$. Finally, for the closed-set identification mode, also, a series of experiments was carried out to select the best combinations and fusion rule. The obtained results are also improved by the fusion. From the obtained results, we remark that in general the performances of the system are significantly improved by using the tested combinations. However, the ROR is exceeded to 99.553 % when using the SUM, MIN, and MUL fusion rules for the two classifiers (equal to 100 % for RBF classifier). Poor results (ROR = 98.850 % with RPR = 17) are obtained with MAX rule.

3.4.3 *Comparison Study*

In order to show the effectiveness of the PCANet deep learning via the other methods, a comparison study with some works in the literature is performed. Thus, in this work, PCANet as a feature extraction method is used. This method can capture the information from the texture in an image and provide a very high robustness which efficiently describes the image characteristics. The proposed scheme was tested using both PolyU and CASIA databases as a popular and available multispectral palmprint database. The extensive experiments presented in this work validate the robustness and the effectiveness of the deep learning over the basic feature extraction techniques. However, a very high identification accuracy (EER = 0.000 % and ROR = 100.00 %) is obtained. Generally, EER is reported in literature to prove the worth of the techniques. For that, a fair comparison of some related and existing techniques must be made on both databases in the same experimental setup.

The performance of the proposed method is compared with Local Line Directional Pattern (LLDP) [21], Log-Gabor Feature extraction with dimension reduction using Kernel Discriminant Analysis and classification by Sparse Representation Classifier (LG-KDA-SRC) [22], log-Gabor with hamming distance (LG-Hm) [23], Neighboring Direction Indicator (NDI) [24], Double Orientation Code (DOC) [25], Block Dominant Orientation Code (BDOC) [26], BDOC and Block-based Histogram of Oriented Gradient (BDOC-BHOG) [27], Wavelet Fusion with ACO [28], Local Micro-structure Tetra Pattern (LMTrP) [29], RootSIFT-based feature fusion [30], Contour Code (CC) [31], Double Half-Orientation (DHO) [32], Block-wise Gaussian Derivative Phase Pattern Histogram with reducing dimension through Kernel Discriminant Analysis (BGDPPH-KDA) [33].

Table 3.8 summarizes the main works which use the CASIA and PolyU multispectral palmprint databases. From this table, which represents only the open-set identification, it is observed that the proposed PCANet method achieves the best identification performance among other methods listed in the same table. Particularly, the EER of two-stage networks method is 0.000 %, which is actually quite the best result as compared to other related works. Most of the techniques in literature are tested only on PolyU multispectral database even in recent research works [24, 29, 33]. There are really few existing works on CASIA multispectral database. We think that this work gives a new and recent technique which can achieve a best identification performance in all bands without a need of fusion for PolyU Multispectral database and a very satisfactory performance in unimodal system (EER = 0.125 %) and best performance with fusion in multimodal system (EER = 0.000 %) for CASIA Multispectral database reputed for an EER much greater than 1.000 % in recent and existing methods of state-of-the-art given in Table 3.8. Furthermore, our robust palmprint recognition system can be improved by modifying the PCANet based feature extraction process like increasing the PCANet number of stages to effectively eliminate unnecessary features, or performing a hamming distance instead of the histogram in the binary hashing phase.

3.5 Conclusion and Further Work

In this paper, a palmprint identification system is presented. Thus, we have investigated the feasibility and effectiveness of using the deep learning architecture for palm identification. In this study, a simple deep learning architecture for feature extraction, called PCANet, is used. This method can capture the information from the texture in an image and provide a very high robustness which efficiently

Table 3.8 Performances comparison of our scheme with the state-of-the-art

Papers	Methods	Database	Size	Accuracy (EER%)
Luo et al. [21]	LLDP	PolyU	250	0.026 (Blue)
Raghavendra and Busch [22]	LG-KDA-SRC	CASIA	100	7.950 (one band)
				1.640 (fusion of all bands)
Bounneche et al. [23]	LG-Hm	PolyU	250	0.025 (best Green)
				0.012 (best fusion: Green, Red)
Fei et al. [24]	NDI	PolyU	250	0.013 (best Red)
				0.009 (best fusion : Blue, Red, NIR)
Fei et al. [25]	DOC	PolyU	250	0.012 (best Red)
Chen [26]	BDOC	PolyU	250	0.016 (best Red)
				0.007 (best fusion: Green, Red)
Hong et al. [27]	BDOC-BHOG	PolyU	250	0.016 (best Red)
				0.007 (best fusion : Green, Red)
Kisku et al .[28]	Wavelet Fusion with ACO	CASIA	100	3.125
Li and Kim [29]	LMTrP	PolyU	250	0.000 (best only for Red and Blue)
Yan et al. [30]	RootSIFT-based feature fusion	CASIA	100	0.160
Khan et al. [31]	Contour Code (CC)	PolyU	250	0.024 (best Red)
				0.006 (fusion of bands)
		CASIA	100	0.615 (White)
				0.300 (fusion of bands)
Fei et al. [32]	DHO	PolyU	250	0.013 (best Red)
Tamrakar and Khanna [33]	BGDPPH-KDA	PolyU	250	0.000 (best : Red and NIR)
		CASIA	100	0.150 (one band)
				0.040 (best three bands)
Our	Two-stage PCANet^{S_2}	PolyU	400	0.000 (best: four bands)
				No need to fusion
		CASIA	100	0.125 (best 640 nm)
				0.000 (best: fusion of three bands)

describes the image characteristics. The proposed scheme was tested using two popular and available multispectral palmprint databases. The extensive experiments presented in this work validate the robustness and the effectiveness of the deep learning over the basic feature extraction techniques. However, a very high identification accuracy (EER = 0.000 % and ROR = 100.00 %) is obtained. For further improvement, we project in our future work to investigate other deep learning architectures like Convolutional Neural Networks (CNN), Restricted Boltzman Machines (RBM), and conventional encoder-decoder model (RNNencdec). Also, in the multimodal system technique, we focalized on other fusion levels like image level, feature extraction level, and decision level to improve their effectiveness.

Acknowledgements First of all, I have been deeply indebted to CHERGUI MARIA for proofreading this chapter. So, a special thanks should be addressed to her. In addition, I am very grateful to the electrical engineering laboratory (LAGE), Ouargla university, Algeria as well as the laboratory of mathematics, informatics and systems (LAMIS), Tebessa university, Algeria.

References

1. R. Shyam, Y.N. Singh, Identifying individuals using multimodal face recognition techniques. Proc. Comput. Sci. **48**, 666–672 (2015)
2. A.K. Jain, A. Ross, S. Prabhakar, An introduction to biometric recognition. IEEE Trans. Circuits Syst. Video Technol. **14**(1), 4–20 (2004)
3. A. Meraoumia, S. Chitroub, A. Bouridane, An efficient palmprint identification system using multispectral and hyperspectral imaging, in *Modeling Approaches and Algorithms for Advanced Computer Applications*. Studies in Computational Intelligence, vol. 488 (Springer, Berlin, 2013), pp. 155–164
4. D. Tamrakar, P. Khanna, Noise and rotation invariant RDF descriptor for palmprint identification. Multimed. Tools Appl. **75**(10), 5777–5794 (2016)
5. X. Xu, L. Lu, X. Zhang, H. Lu, W. Deng, Multispectral palmprint recognition using multiclass projection extreme learning machine and digital shearlet transform. Neural Comput. Appl. **27**(1), 143–153 (2016)
6. K. Usha, M. Ezhilarasan, Fusion of geometric and texture features for finger knuckle surface recognition. Alexandria Eng. J. **55**(1), 683–697 (2016)
7. T.-H. Chan, K. Jia, S. Gao, J. Lu, Z. Zeng, Y. Ma, PCANet: a simple deep learning baseline for image classification? IEEE Trans. Image Process. **24**(12), 5017–5032 (2015)
8. W. Zhao, S. Du, Spectral-spatial feature extraction for hyperspectral image classification: a dimension reduction and deep learning approach. IEEE Trans. Geosci. Remote Sens. **54**(8), 4544–4554 (2016)
9. W. Xinshao, C. Cheng, Weed seeds classification based on PCANet deep learning baseline, in *Asia-Pacific Signal and Information Processing Association Annual Summit and Conference (APSIPA)*, Hong Kong (2015), pp. 408–415
10. S. Pino-Povedano, C. Bousoño-Calzón, F.J. González-Serrano, Radial basis function interpolation for signal-model-independent localization. IEEE Sensors J. **16**(7), 2028–2035 (2016)
11. Z. Geler, V. Kurbalija, M. Radovanović, M. Ivanović, Comparison of different weighting schemes for the kNN classifier on time-series data. Knowl. Inf. Syst. **48**(2), 331–378 (2016)
12. A. Seal, D. Bhattacharjee, M. Nasipuri, Human face recognition using random forest based fusion of à-trous wavelet transform coefficients from thermal and visible images. Int. J. Electron. Commun. **70**(8), 1041–1049 (2016)

13. CASIA database. Available at http://biometrics.idealtest.org/
14. PolyU. database. Available at http://www4.comp.polyu.edu.hk/~biometrics/
15. D. Zhang, Z. Guo, G. Lu, L. Zhang, W. Zuo, An online system of multispectral palmprint verification. IEEE Trans. Instrum. Meas. **59**(2), 480–490 (2010)
16. D.R. Kisku, P. Gupta, J.K. Sing, C. Hwang, Multispectral palm image fusion for person authentication using ant colony optimization, in *Proceedings of International Workshop on Emerging Techniques and Challenges for Hand-Based Biometrics (ETCHB)*, Istanbul (2010), pp. 1–7
17. Y. Hao, Z. Sun, T. Tan, C. Ren, Multispectral palm image fusion for accurate contact-free palmprint recognition, in *Proceedings of International Conference on Image Processing*, San Diego, CA (2008) pp. 281–284
18. M.-H. Lim, S. Verma, G. Mai, P.C. Yuen, Learning discriminability-preserving histogram representation from unordered features for multibiometric feature-fused-template protection. Pattern Recogn. **60**, 706–719 (2016)
19. A. Al Abdulwahid, N. Clarke, I. Stengel, S. Furnell, C. Reich, Continuous and transparent multimodal authentication: reviewing the state of the art. Clust. Comput. **19**(1), 455–474 (2016)
20. A. Jain, K. Nandakumar, A. Ross, Score normalization in multimodal biometric systems. Pattern Recogn. **38**(12), 2270–2285 (2005)
21. Y.-T. Luo, L.-Y. Zhao, B. Zhang, W. Jia, F. Xue, J.-T. Lu, Y.-H. Zhu, B.-Q. Xu, Local line directional pattern for palmprint recognition. Pattern Recogn. **50**, 26–44 (2016)
22. R. Raghavendra, C. Busch, Novel image fusion scheme based on dependency measure for robust multispectral palmprint recognition. Pattern Recogn. **47**(6), 2205–2221 (2014)
23. M.D. Bounneche, L. Boubchir, A. Bouridane, B. Nekhoul, A. Ali-Cherif, Multi-spectral palmprint recognition based on oriented multiscale log-Gabor filters. Neurocomputing **205**(12), 274–286 (2016)
24. L. Fei, B. Zhang, Y. Xu, L. Yan, Palmprint recognition using neighboring direction indicator. IEEE Trans. Hum.-Mach. Syst. **99**, 1–12 (2016)
25. L. Fei, Y. Xu, W. Tang, D. Zhang, Double-orientation code and nonlinear matching scheme for palmprint recognition. Pattern Recogn. **49**, 89–101 (2016)
26. H. Chen, An efficient palmprint recognition method based on block dominant orientation code. Int. J. Light Electron Optics **126**(21), 2869–2875 (2015)
27. D. Hong, W. Liu, J. Su, Z. Pan, G. Wang, A novel hierarchical approach for multispectral palmprint recognition. Neurocomputing **151**(Part 1), 511–521 (2015)
28. D.R. Kisku, P. Gupta, J.K. Sing, C. Hwang, Multispectral palm image fusion for person authentication using ant colony optimization, in *Proceedings of International Workshop on Emerging Techniques and Challenges for Hand-Based Biometrics (ETCHB)* (IEEE, New York, 2010), pp. 1–7
29. G. Li, J. Kim, Palmprint recognition with local micro-structure tetra pattern. Pattern Recogn. **61**, 29–46 (2017)
30. X. Yan, W. Kang, F. Deng, Q. Wu, Palm vein recognition based on multi-sampling and feature-level fusion. Neurocomputing **151**(Part 2), 798–807 (2015)
31. Z. Khan, A. Mian, Y. Hu, Contour code: robust and efficient multispectral palmprint encoding for human recognition, in *Proceedings of the IEEE International Conference on Computer Vision* (2011), pp. 1935–1942
32. L. Fei, Y. Xu, D. Zhang, Half-orientation extraction of palmprint features. Pattern Recogn. Lett. **69**, 35–41 (2016)
33. D. Tamrakar, P. Khanna, Kernel discriminant, analysis of block-wise gaussian derivative phase pattern histogram for palmprint recognition. J. Vis. Commun. Image Represent. **40**, Part B, 432–448 (2016)

Chapter 4
Biometric Acoustic Ear Recognition

Mohammad Derawi, Patrick Bours, and Ray Chen

4.1 Introduction

Today's consumer electronic devices concerning sound reproduction, such as headphones and earphones, are widely used as peripherals to PCs, mobile phones, MP3 players, and in general various sound systems. These are very common and domesticated products; however, with their popular manifestations, security issues arise. Imagine headphones, besides serving as a listening device, also served as a scanning device for personal authentication. This peripheral, which is versatile because it can connect to anything with an audio port (typically in terms of a jack or mini-jack interface), but also very user-friendly, because of its intuitive and simple usage, could make a great innovation by incorporating the act of being a security measure.

The idea of acoustic ear recognition in this chapter is about using the human ear, more precisely the acoustic properties of the pinna and ear canal (defined as the outer ear), as a "fingerprint." It means utilizing the ear as a personal feature, which eventually can be captured and registered by a system for person recognition. The measuring device should then be a pair of modified headphones (which has a microphone build in its cavity as the sensor). This concept constitutes what is called a biometric system, where the ear is the biometric modality and the measuring device is used to capture a biometric sample for recognition.

M. Derawi (✉) • R. Chen
Smart Wireless Systems, Norwegian University of Science and Technology, Teknologiveien 11, 2815 Gjøvik, Norway
e-mail: mohammad.derawi@ntnu.no

P. Bours
Norwegian Information Security Laboratory, Teknologiveien 11, 2815 Gjøvik, Norway

R. Jiang et al. (eds.), *Biometric Security and Privacy*, Signal Processing for Security Technologies, DOI 10.1007/978-3-319-47301-7_4

Utilizing the ear as a biometric modality, commonly image based, is relatively still a newcomer in the fields of biometric recognition techniques. There is not as much research ongoing within this topic, and no commercial application has yet become available. Acoustical based ear recognition is even more uncommon than image based [1, 37]. Only few several papers by Philips researchers have been reported to deal with acoustic ear recognition, where the performance results have been found to be of EERs from 0.8 % to 18.0 % [1]. However this could change, because ears have certain advantages over the more established biometrics, as they have a rich and stable structure that changes little with age [6]. The ear does not suffer from changes in facial expression, and is firmly fixed in the middle of the side of the head so that the immediate background is predictable, whereas face recognition usually requires the face to be captured against a controlled background. Further on unlike fingerprints, "ear-fingerprints" are less likely to be left behind on glossy surfaces, and even if it happens, it will not leave any acoustical characteristics behind. In this respect acoustic ear recognition may lead to a more secure biometric. This book-chapter will deal with the holistic development of a biometric system based on acoustic ear recognition. Its main purpose is to map and investigate the many aspects within acoustic ear recognition, and based on the most influential ones, make attempts on deducing well performing recognition systems.

4.2 Ear Biometrics and Acoustics

Acoustic ear recognition addresses a large number of knowledge domains, where the two major topics are acoustics and biometrics. Acoustics is a discipline which has existed for a very long time. In the sixth century BC, the Greek philosopher Pythagoras observed that some musical intervals sounded more beautiful than others. The answers were found in terms of numerical ratios representing the harmonic overtone series on a string [9]. Biometrics in the other hand, starting with fingerprints more than a hundred years ago, first became an independent field for itself for about a decade ago [8]. This is due to the fact that each biometric-related topics grew out of different disciplines. For example, the study of fingerprints came from forensics and pattern recognition, voice recognition evolved from signal processing, and privacy concerns arose from the public policy arena.

The following two sections will deal with the acoustics of the outer ear and when a headphone is coupled to it.

4.2.1 *The Ear as a Biometric*

Using the ear as a biometric modality is a newcomer in the fields of biometric recognition techniques. There are relatively not many ongoing research within this topic, in which most of them deals with investigating unused methods in order to improve the performance. Therefore there is yet no well-established fully automated ear recognition system. Before looking at the different known methods, a review of why at all bother using the ear as a biometric will be examined. In order to make a biometric characteristic practical, the following seven properties described by [19] must be valid to some degree:

1. *Universality*: Every person should have the biometric characteristic.
2. *Uniqueness*: No two persons should be the same in terms of the biometric characteristic.
3. *Permanence*: The biometric characteristic should be invariant over time.
4. *Collectability*: The biometric characteristic should be measurable with some (practical) sensing device.
5. *Performance*: The technology used should have a certain accuracy, speed, and robustness.
6. *Acceptability*: The particular user population and the public in general should have no (strong) objections to the measuring/collection of the biometric.
7. *Circumvention*: The technology should be ease of use of a substitute.

In case of the ear, there are no questions about its universality. Every normal human being is born with a pair of ears.

But what about its uniqueness? There is no direct way to prove that each of us has a unique pair of ears, but one can assert that this is probable and as a matter of fact, studies within forensics have shown evidence of so. Ears have played a significant role in forensic science for many years and one of the most famous works among ear identification was made by Alfred Iannareli [17]. His first study compared over 10,000 ears drawn from a randomly selected sample in California, and the second study examined fraternal and identical twins, in which physiological features are known to be similar. The evidence from these studies suggested the ears of the individuals to be unique, though identical twins were found to have similar, but not identical, ear structures especially in the Concha and lobe areas [10]. Further on a paper has been written by the Netherlands Forensic Institute, discussing the ability to identify persons by their ear from surveillance videotapes [15]. Thus forensic research supports the validity of the ear's uniqueness. Now, does the ear hold true for permanence, in other words, is it comparable over time? It is obvious that the structure of the ear does not change radically over time. The French police officer and biometrics researcher Alphonse Bertillon, who created anthropometry-measurement of the human individual for the purposes of understanding human physical variation, wrote the following in his seminal work on biometrics [6]:

> The ear, thanks to these multiple small valleys and hills which furrow across it, is the most significant factor from the point of view of identification. Immutable in its form since birth, resistant to the influences of environment and education, this organ remains, during the entire life, like the intangible legacy of heredity and of the intrauterine life

The clinical literature reports that although the ear canal and pinna assumes adult form at the time of birth (the most critical period for the development of the outer ear is during the first 3 months of pregnancy), these two structures continue to grow until age 7–9 years, at which time adult shape and size is attained [4]. Other studies suggest also that the ears grow throughout your lifetime on average by 0.22 mm a year [14]. This seems neglectable for biometrics systems operating on a daily basis, and one could imagine such systems to undergo annual/milestone maintenance routines, where renewals for such could occur.

With regard to collectability, it can be said that since the ears are an outer body organ and thus visually obtainable, any camera/video device from some distance, which, for instance, could constitute in a passive recognition system, is able to capture an image of the ear, much like in face recognition [28]. Though sometimes headgear or hair may cover the ears, then it will require a more active recognition system. That could prompt the use of more specific-made measuring devices, where either the user need to lift the device to their ear(s), or lean their head to line their ear(s) with the respective measuring device. For example, the Phillips research department in the Netherlands have used a mobile phone, a pair of earphones, and a pair of headphones, modified with an integrated microphone, as measurement devices for acoustic ear recognition, resulting in great authentication performance [1, 2] (this will thoroughly be examined in later sections). The devices presented are all easy accessible; however, the performance and circumvention would depend upon the specific measuring device itself. The wondering now would be if the measuring device is practically appropriate. That would depend on the usage situation and the user population, which leads to the acceptability property, perhaps less inherently connected with a particular biometric.

Unlike fingerprint recognition where people often relate/label establishing their identity through a print of their finger with the feeling of being criminalized [39], ear recognition is yet not labelled by any means by the public. Face recognition can be regarded as a privacy violation by the public [7, 18], because people in general do not like pictures taken of their faces by strangers. Also perhaps if the measure device, in this case a camera, is visible to the subjects, they might feel too exposed and surveillanced, which can prompt negative feelings towards the biometrics system. This obtrusive nature during the picture capturing for face recognition would probably also be present for ear recognition. But maybe lesser objections would be raised with regard to privacy violation, simply because an image of the ear is not the same as an image of the face. Most people can easily describe an unknown face and intuitively recognize the unknown person if their face is revealed, but the same thing cannot be said about an ear, thus making it unidentifiable for the normal trained eyes.

It has been clarified that the human ear inherit the seven properties described by [19]. Universality, uniqueness and permanence has been proved through forensic and clinical studies, collectability, performance, circumvention, and acceptability are assessed to be valid of some degree based on successful usage of easy accessible measure devices and the anonymity of the ear to the naked eyes.

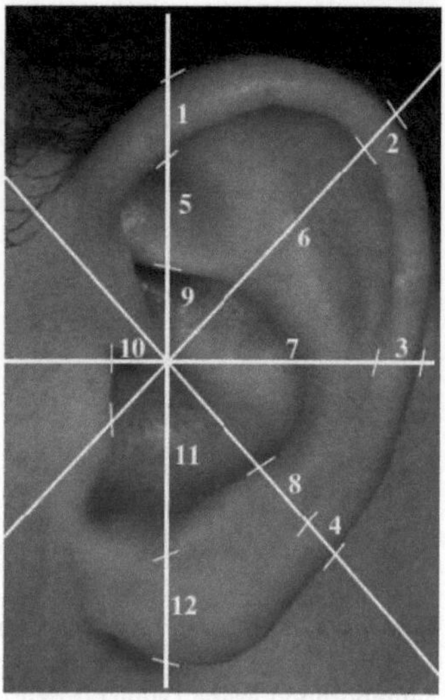

Fig. 4.1 Iannarelli's manual ear measuring system. Each number refers to a measuring point

4.2.2 Image Based Ear Recognition

The common approach to ear recognition is by processing captured images of the ear. The first thorough work was done by Iannareli [17], who developed an ear classification system based on manual measurements. It is more or less based on the geometry of the ear, the locations of the anthropometric measurements can be seen in Fig. 4.1. Burge and Burger [10] were the first to investigate the ear as a biometric in the context of machine vision.

Inspired by the work of Iannarelli, they conducted a proof of concept study by practically implementing a computer vision based system. Each subject's ear was modeled as an adjacency graph built from the Voroni diagram of its Canny extracted curve segments. They devised a novel graph matching algorithm for authentication which took into account the erroneous curve segments which could occur in the ear image due to changes such as lighting, shadowing, and occlusion. Figure 4.2 shows the stages in building the ear biometric graph model. They found that the human ear was ideal for passive identification because the features were robust and could be reliably extracted from a distance. Though they identified the problem of occlusion by hair as a major obstacle and proposed the use of thermal imagery to overcome this obstacle.

The work of Iannarelli and Burge and Burger have shown the availability of ear biometrics and laid the first foundations in which it can be built upon. Many have followed and done studies and proposed biometric systems based on different techniques of image processing of the ear. Moreno et al. used feature points of outer ear contour and information obtained from ear shape and wrinkles for ear recognition [5]. Hurley et al. treated the ear image as an array of mutually attracting particles that act as the source of Gaussian force field. The original image is

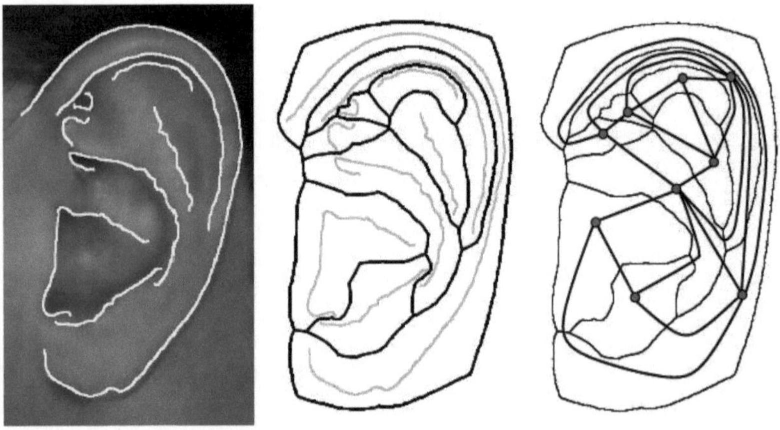

Fig. 4.2 Graph model by Burge and Burger: Stages in building the ear biometric graph model. A generalized Voroni diagram (*center*) of the Canny extracted edge curves (*left*) is built and a neighborhood graph (*right*) is extracted

described by a set of potential channels and positions of potential wells [16]. Mu presented a long axis based shape and structural feature extraction method, the shape feature is consisted of the curve-fitting parameters of the outer ear contour, the structural feature is composed of ratios of between the length of key sections and the length of the long axis [25]. Statistical holistic analysis, especially principal component analysis (PCA), has proved to be one of the most popular approaches to ear recognition. Victor et al. [40] applied PCA to both face and ear recognition and concluded that the face yields a better performance than the ear. However, Chang et al. [11] conducted a similar experiment and reached a different conclusion, namely no significant difference was observed between face and ear biometrics when using PCA. For ear recognition in 3D, Hui Chen et al. [12] proposed an Iterative Closest Point (ICP) based 3D ear recognition. Ping Yan et al. [41] proposed an automatic 3D ear recognition system, which includes automatic ear extraction, and ICP based ear recognition. More recently Ali et al. [3] used wavelet transform for feature extraction of manually cropped ear images. Matching was carried out using Euclidean distance, achieving results of 94.3 % correct recognition rate. Li Yuan et al. [42] addressed two major issues regarding 2D ear recognition, namely automating ear extraction from 2D images using improved active shape model, and the study of acceptable head rotation range for ear recognition.

4.2.3 Acoustic Based Ear Recognition

So far only two papers written by researchers at Philips Research Laboratories have addressed acoustic ear recognition, and in addition the Sandia Corporation has claimed the first US patent on acoustic ear recognition. P. Tuyls et al. in [37] evaluated acoustic ear recognition, aiming at a special approach to feature extraction in order to be combined with biometric template protection systems [24, 36, 38]. The acoustic measuring device was a pair of full-size headphones with integrated microphones plus small tubes fitted onto the microphones such that measurements close to the ear-canal entrance could be made. Two experiments were carried out, where the Headphone-to-ear-canal Transfer Functions were measured as the response of a short noise burst, played through the headphones. The first experiment was performed on 45 individuals, eight measurements per person. The headphones were kept fixed during these measurements, which entailed only background noise, as an error factor, was taken into account. The biometric template was represented as a real vector of size 512. The first 256 components corresponded to the left ear and the remaining to the right ear. The second experiment was performed on 65 individuals, eight measurements per person. In contrast to the first experiment, after each measurement the headphones were taken off and put on again. Hence not only background noise, but also noise caused by repositioning errors of the headphone was taken into account. The influence of repositioning errors can be viewed between the plots in Fig. 4.3.

The data for each person were afterwards split randomly into two sets: six measurements for training and two measurements for validation. Principal component (PC) and random orthogonal (RO) transformations were used to reduce the number of features and also increase the validation performance. Further on combinations of PC with Quantized Key Extraction, and Fuzzy Commitment based schemes were made to further increase the performance. It was proved that the performance of the method based on the RO transformations was superior to the method based on the PC, but the first mentioned had major security vulnerabilities. The repositioning errors in the second experiment resulted in a decrease in performance of roughly five times the first experiment (the results are shown in Tables 4.1, 4.2, and 4.3). Akkermans et al. in [1] did a practical proof of concept, somehow similar to the second experiment in [37] described above. Instead of only using headphones, a mobile phone and a pair of earphones with integrated microphones were also used as test platforms (see Fig. 4.4 for pictures of the test platforms).

The features were extracted from the impulse response of a noise signal, first transformed into its frequency domain, then divided with the original input signal (assuming the signal was expressed in amplitude, possibly as direct raw voltage output from the microphone) in order to get a transfer function. The influence of the phase was eliminated, although they could contain significant discriminating information about an individual, they may also lead to larger intra-class variations, i.e., variations amongst measured transfer functions for the same subject. Robustness was expected to be gained when using amplitude information only. Experiments were performed on 31 subjects for both the earphones and headphones. Eight ear transfer functions were measured per person. For the mobile phone 17 persons were enrolled again with eight measurements per person. The obtained number of the

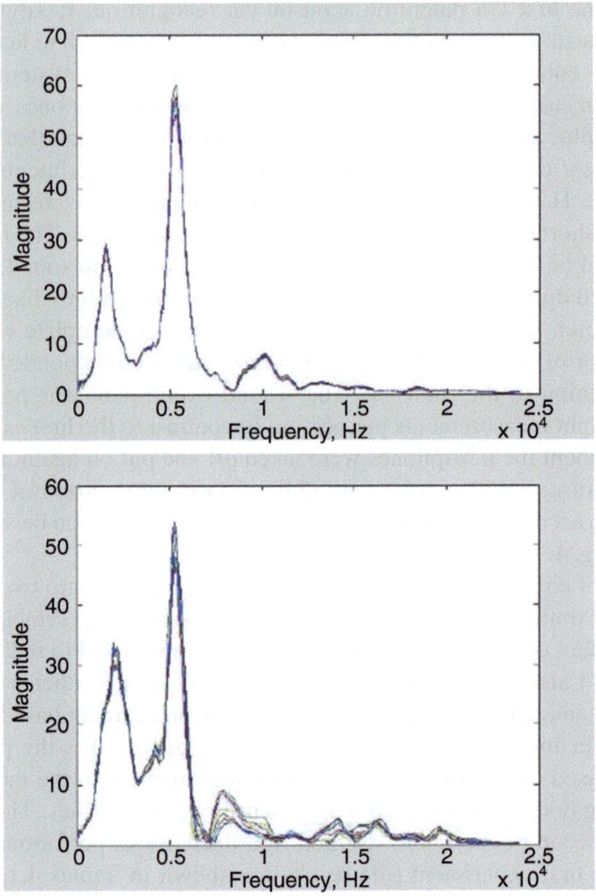

Fig. 4.3 *Left curve*: Data obtained from left-ear measurements for one individual without repositioning errors. *Right curve*: Data obtained from left-ear measurements for one individual with repositioning errors [37]

Table 4.1 False rejection rate (FRR) and false acceptance rate (FAR), minimum key size (Min KS) in bits for various values of δ using the PC and RO transformations for the first experiment [37]

	PC transformation				RO transformation			
δ	Min KS	Mean KS	FRR (%)	FAR (%)	Min KS	Mean KS	FRR (%)	FAR (%)
0.1	17	18.8	13.4	0.0	400	423	17.2	0.0
0.5	15	17.2	5.2	0.0	390	405	5.4	0.0
1.0	14	15.1	3.1	0.0	369	388	2.8	0.002
1.5	12	13.5	2.7	0.005	340	370	2.9	0.01
2.0	10	12.3	2.7	0.008	327	343	0.6	0.05
3.0	8	9.1	2.8	0.09	281	295	0.8	0.7

Table 4.2 False rejection rate (FRR) and false acceptance rate (FAR), minimum key size (Min KS) in bits for various values of δ using the PC and RO transformations for the second experiment [37]

	PC transformation				RO transformation			
δ	Min KS	Mean KS	FRR (%)	FAR (%)	Min KS	Mean KS	FRR (%)	FAR (%)
0.1	10	11.7	51.4	0.02	405	427	66.5	0.002
0.5	7	8.6	29.5	0.2	406	416	45.1	0.009
1.0	5	6.7	20.8	0.9	382	399	24.7	0.09
1.5	4	4.9	11.1	3.2	368	380	14.92	0.3
2.0	2	4.0	9.5	7.4	356	365	7.8	1.1
3.0	2	2.9	5.3	13.3	321	330	2.4	4.4

Table 4.3 Comparison of the performance of the PCA based scheme with the PCA+QKE based scheme [37]

	PC transformation			PC transformation + QKE		
Experiment	Key length	FRR (%)	FAR (%)	Key length	FRR (%)	FAR (%)
First experiment	18	13.5	0.0	100	3.89	0.0
Second experiment	5	11.15	3.2	100	2.76	1.68

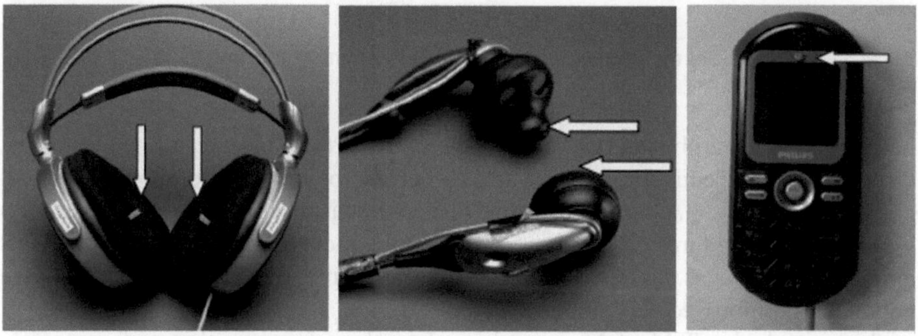

Fig. 4.4 *Left*: Headphone with integrated microphones. The white tubes as the *arrows* are pointing at, are mounted at the microphones. *Middle*: Earphone pieces with integrated microphones. The *arrows* mark the placement of the microphones. *Right*: Mobile phone. The *arrow* again marks the placement of the microphone

features was 256 frequency components per microphone, making the earphones and headphones each have 512 components and the mobile phone 256 components. An example with three subjects' data can be viewed in Fig. 4.5.

The average correlation between the measurements of the right and left ear of each person was determined. It was found that using both ears only gave marginally better discrimination capabilities since the acoustic left and right ear responses were quite similar and differed only 10 % in terms of correlation. Recognition performance was tested by investigating the FAR and FRR using the impostor and genuine distributions for the correlation measure. It was observed that the

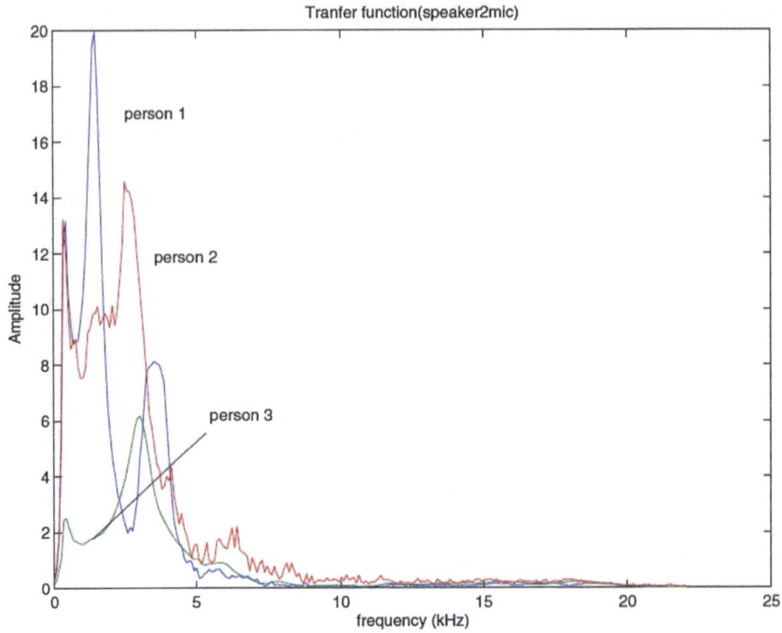

Fig. 4.5 An example of the frequency response of the ear transfer function for three persons [1]

headphones and earphones gave roughly the same performance, resulting in an equal error rate of, respectively, 7 % and 6 %, whereas for the mobile phone the equal error rate was 15 %. As an experiment, Fishers linear discriminant analysis (LDA) was applied to the three ear databases to select the most discriminating components amongst the subjects. The performance in terms of receiver operating curves were significantly improved, as can be viewed in Fig. 4.6.

As a conclusion Table 4.4 was made to show the ear recognition performance in terms of EER as a function of frequency range of excitation signal, with the effect of applying LDA.

The US Patent [26] has claimed the invention which teaches systems and methods for verifying or recognizing a person's identity based on measurements of the acoustic response of the individual's ear canal. The way the system is described are the following:

> The system comprises an acoustic emission device, which emits an acoustic source signal s(t), designated by a computer, into the ear canal of an individual, and an acoustic response detection device, which detects the acoustic response signal f(t). A computer digitizes the response (detected) signal f(t) and stores the data. Computer-implemented algorithms analyze the response signal f(t) to produce ear-canal feature data. The ear-canal feature data obtained during enrollment is stored on the computer, or some other recording medium, to compare the enrollment

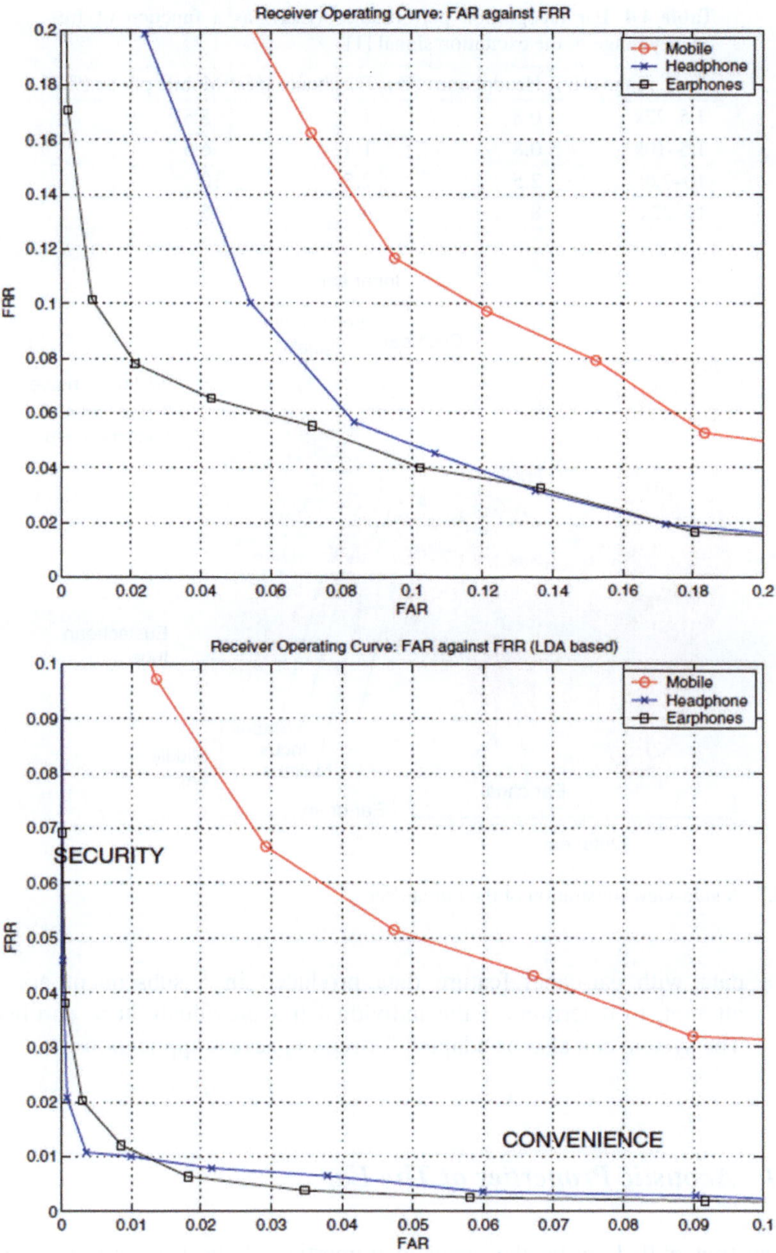

Fig. 4.6 *Left*: Receiver operating curves without Fisher LDA transformation. *Right*: Receiver operating curves using Fisher LDA transformation

Table 4.4 Ear recognition performance (EER) as a function of frequency range of the excitation signal [1]

Freq. range (Hz)	Headphones (%)	Earphones (%)	Mobile phone (%)
1.5–22k	0.8	1	5.5
1.5–10k	0.8	1.4	6.5
10–22k	2.5	2.5	10
16–22k	8	6	18

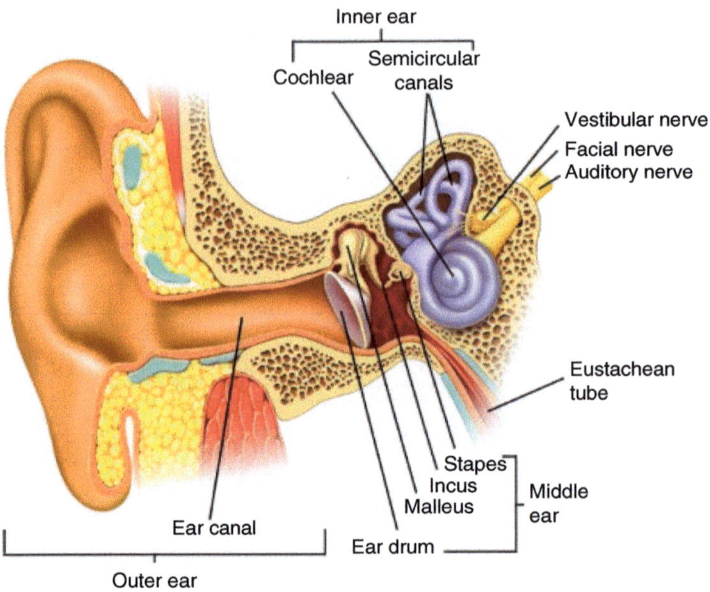

Fig. 4.7 A side-view illustration of the human ear

data with ear-canal feature data produced in a subsequent access attempt, to determine if the individual has previously been enrolled. The system can also be adapted for remote access applications.

4.2.4 Acoustic Properties of The Ear

This section will describe the acoustic properties of the ear, or more precisely, the outer ear, which consist of the pinna (also called auricle) and the ear canal. Illustrations of the ear can be seen in Figs. 4.7 and 4.8.

The first illustration shows a side view of the whole ear, which comprise of the outer ear, middle ear, and inner ear. The outer ear functions as an acoustic funnel, where the external ear (the visible part of the ear) mainly contributes by the effect of

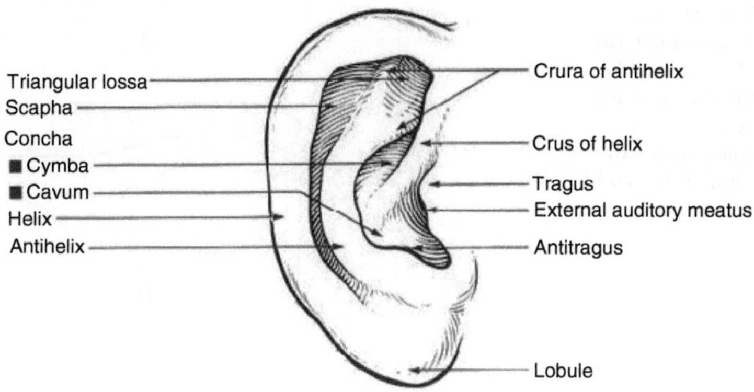

Fig. 4.8 A front-view illustration of the external ear

pinna diffraction and concha wave motion, which filter and characterize the sound for each individual, and the ear canal contributes by resonating certain frequencies in order to optimize an individual's dynamic hearing [27]. The middle ear transfers the air pressure variations into motion by the eardrum (also called tympanic membrane) and through the three ossicles (bones), Malleus, Stapes, and Incus, which make up a mechanical lever system, the motions are further transformed into fluid flow at the oval window of the cochlea (inner ear). In other words, the primary function of the middle ear is to efficiently transfer acoustic energy from compression waves in air to fluid-membrane waves within the cochlea. The first steps of auditory processing happen in the cochlea, where the so-called inner and outer hair cells pick up the fluid waves and due to chemical processes fire neural spikes through the auditory nerve, which at the end are interpreted by the brain and a sound will then be heard.

Much research has been done of the outer ear with regard to evaluation of the auditory system. The purpose of such could, for instance, in clinics be to diagnose patients with ear diseases [4], in which case a treatment might be recommended. In case of impaired hearing, knowledge of the dimensions and acoustic properties of the ear canal are needed in order to fit and adjust a hearing aid to the patient. For the industry it is common to extract and model head related transfer functions (HRTF), which is a response that characterizes how ones ear receives a sound from a point in space. When pairing HRTFs for two ears, it can be used to synthesize a binaural sound that for an individual seems to come from a particular point in space. In other words, HRTFs are able to make sounds more authentic in sound reproduction systems, for example, in headphones or 2.0 speaker systems to have the effect of surround sound.

According to Shaw [31], the factors that govern the sound transformation from free-field to the ear drum can be divided into two major areas:

1. The head, torso, and pinna flange acting as diffracting bodies.
2. The Concha and the ear canal acting as resonators.

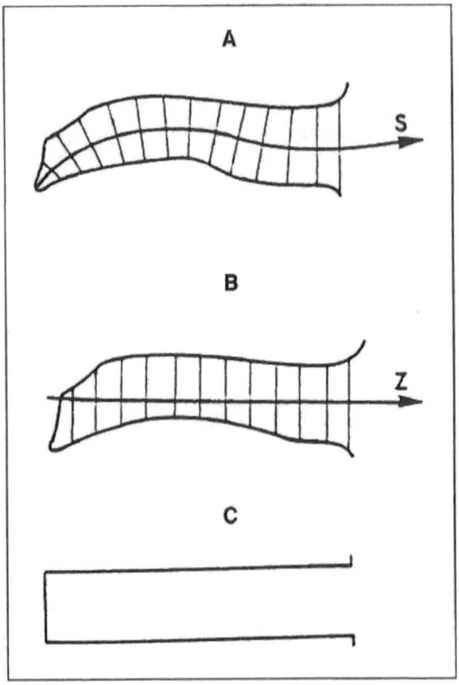

Fig. 4.9 The ear-canal dimensions indicating: (**a**) The curved axis along the length of the ear canal and slices of varying cross-sectional areas. (**b**) Uniform parallel slices with a straight axis. (**c**) Uniform tube [33]

When the outer ear is excited by a broad-band signal (e.g., broad-band noise) from the free-field, certain frequencies are emphasized and others are not. The magnitude of difference in the amount of sound pressure at the ear drum to that in the free-field is due to the pressure gain achieved by the outer ear.

The ear canal in reality has a complex slightly S-shaped form (see Figs. 4.9 and 4.10).

Its dimensions have been extracted by various methods [4], the average of those are listed in Table 4.5.

In order to analyze the ear canal as a resonator, it will be treated as a straight tube of uniform cross-section that is open at one end (concha region) and closed at the other (tympanic membrane). It contains air which will resonate in response to sounds of various frequencies. The frequency that matches the natural resonant frequency of the ear canal is amplified more than other frequencies. The wavelength of the fundamental resonance of this one-open, one-close ended tube is four times its length. That means, only one-quarter of the wave can fit into the tube at any one pass. Thus, the ear canal is called a quarterwave resonator. The resonant frequency of the ear canal can be calculated by the formula:

$$f = \frac{c}{4l} \tag{4.1}$$

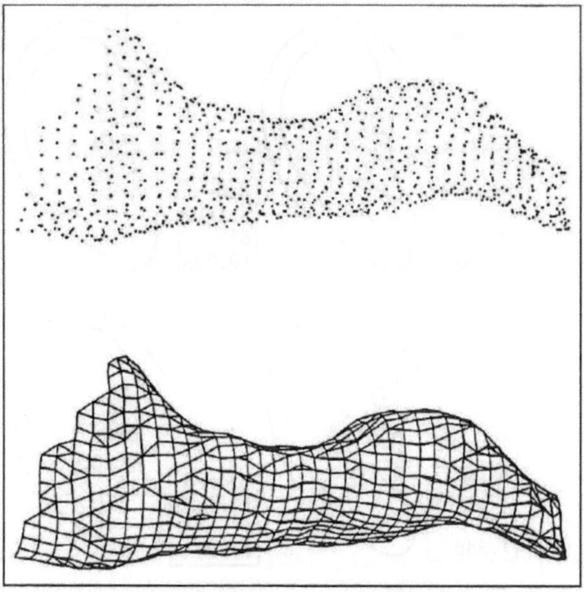

Fig. 4.10 Three-dimensional display of the geometry of the ear canal from 978 coordinate points measured from the earmold impression. The longitudinal cross-section reflects the parallel slices at 1 mm apart [33]

Table 4.5 Average of ear-canal geometry [4]

Parameters	Dimensions
Length of the ear canal	25(±2.5) mm
Longest diameter	9.4(±1.5) mm
Shortest diameter	4.8(±0.5) mm
Ear-canal volume	1014.0(±15.4) mm^3

where f is the resonant frequency, c is the velocity of sound in air (given as 343 m/s), and l is the length of the ear canal (25 mm). Inserting the known variables into 4.1 yields the following:

$$f = \frac{343}{4 \cdot 25 \cdot 10^{-3}} \tag{4.2}$$

Therefore, the resonant frequency of an ear canal 25 mm in length is 3430 Hz. However it is important to note that any variation in the length of the canal can alter the resonant frequency.

The Concha, with a volume of approximately 4 cm^3, is a broad shallow cavity that is partially divided by the crus helias. The lower part, the cavum, is tightly coupled to the canal whereas the upper part, the cymba, is connected to the fossa. These structures clearly have specific acoustical attributes, which are discussed later, whereas the structures extending from the concha, such as the helix, the anti-helix, and the lobule, seem to function collectively as a simple flange [35].

Studies of the concha have been made by Shaw [31, 32] where the first six modes of the human concha under blocked meatus conditions (blocked ear canal) were identified and measured. Figure 4.11 illustrates the essential characteristics of these modes.

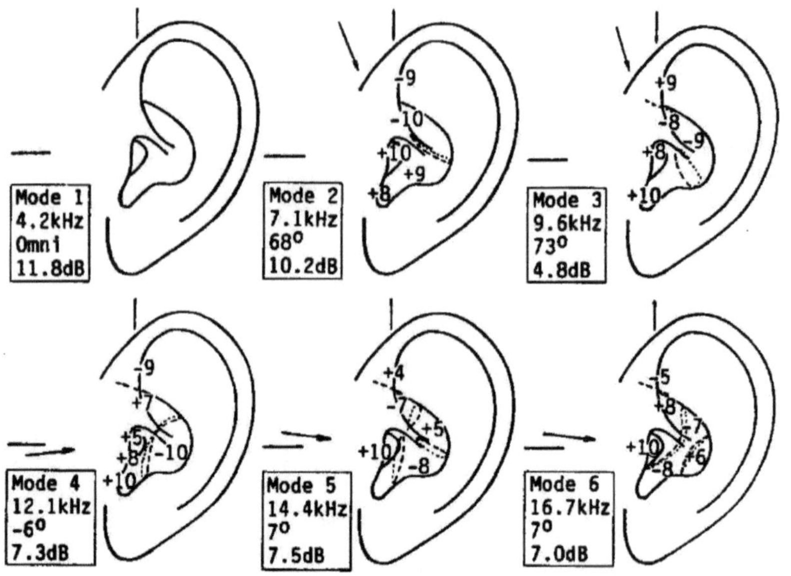

Fig. 4.11 Average characteristics of six modes under blocked meatus conditions based on data for ten subjects. Numerals indicate relative values of sound pressure, on a linear scale, measured at base of the concha. Signs (+/−) indicate relative phase (0°/180°). *Broken and dotted lines* show positions of nodal surfaces. *Arrow* indicates most favorable source direction. Data at *left* show mode number, mode frequency, most favorable source angle, and mode response [31]

The mode patterns, resonance frequencies, directions of maximum response, and response levels are average values based on data for ten subjects. The first mode (4.2 kHz) shown in Fig. 4.11 is a simple quarter-wavelength depth resonance with uniform sound pressure across the base of the concha. It is strongly excited from all directions. The other modes are essentially transverse and fall into two groups: a "vertical" pair (modes 2 and 3) and a "horizontal" triplet (modes 4, 5, and 6). Geronazzo et al. [13] have used this knowledge about the concha to model pinna-related transfer functions (PRTFs) for 3D sound rendering. The resulting model could be suitable for integration into a structural HRTF model, and for parametrization over anthropometrical measurements of a wide range of subjects. The effect of the pinna flange, for example, is to increase eardrum pressure by a few dB at high frequencies for sounds arriving from the front and to reduce substantially the eardrum pressure for high-frequency sounds arriving from the rear due to interference between the direct wave and a scattered wave from the edge of the pinna flange [30]. This plays a big role for sound localization in the horizontal plan. Many studies have been done within this and as an example the plots in Fig. 4.12 show average transformation of sound pressure level from free-field to the ear canal across different frequencies as a function of azimuths presented from frontal, lateral, and posterior sectors. Common to all the curves in Fig. 4.12 is the peak at 2.6 kHz, the primary resonance frequency of the external ear, where pressure gain attains its largest values. To summarize Shaw [30] estimated the effects of various structures on the pressure gain. He suggested that the total pressure, "T,"

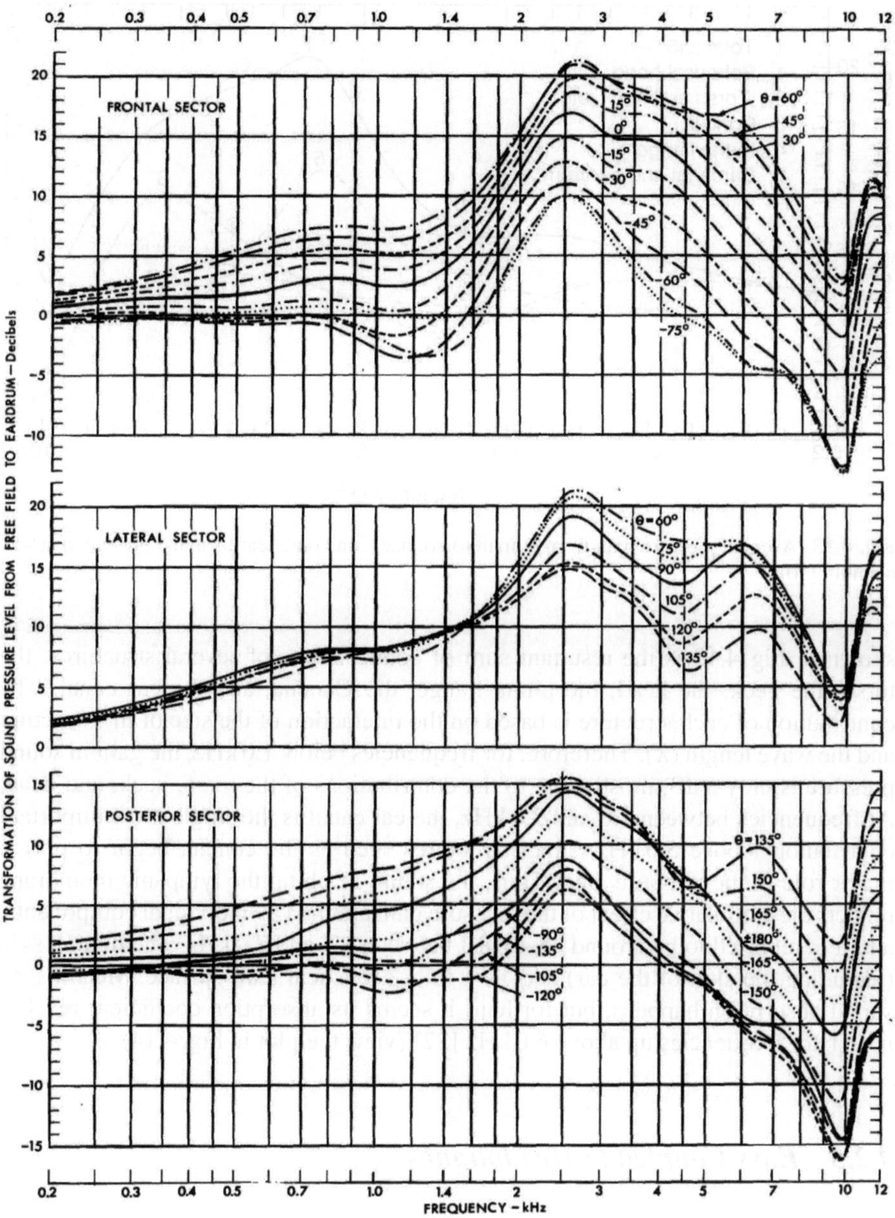

Fig. 4.12 Average transformation of sound pressure level from free-field to the ear canal across different frequencies as a function of azimuths presented from frontal, lateral, and posterior sectors [30]

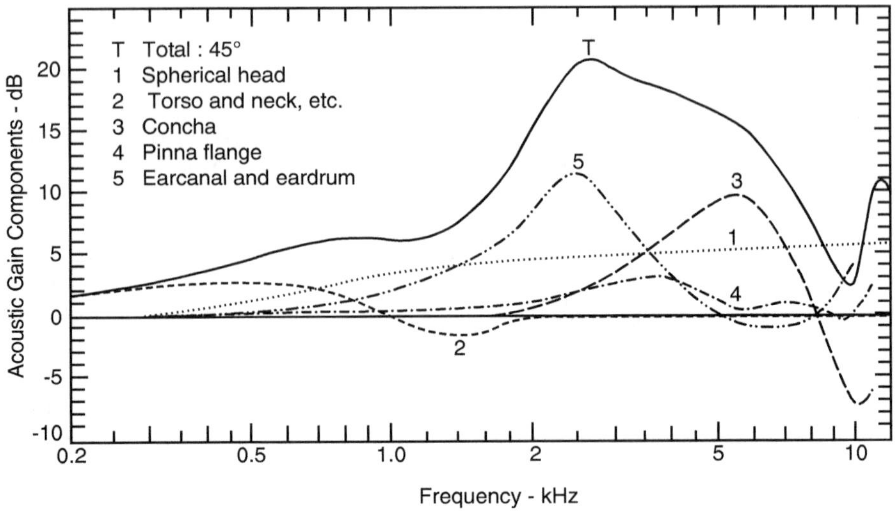

Fig. 4.13 Average pressure gain from human torso, neck, and outer ear to sound presented at 45° azimuth [30]

shown in Fig. 4.13 is the resultant sum of contributions of several structures: the torso, the neck, the head, the pinna flange, the Concha, and the ear canal. The contribution of each structure is based on the interaction of the size of the structure and the wave length (λ). Therefore, for frequencies below 1.0 kHz, the gain in sound pressure is only 5 dB, mostly due to the contributions of the torso, neck, and head. At frequencies between 1.0 and 3.0 kHz, the ear canal is the single most important contributor. Above 3.0 kHz, small structures such as the concha begin to play a major role in the pressure gain. Thus, the sound reaching the tympanic membrane reflects the cumulative effect of the pressure gain attained by individual components, and it is reported to be around 15–20 dB between 1.5 and 7.0 kHz. The human skin (including the skin of the ear) acts as a fully acoustical hard surface. Meaning no sound absorption happens, but for hair, it seems its absorption coefficient reaches near 1 for frequencies up above 6.0 kHz [22] (view the plot in Fig. 4.14).

4.2.5 Ears Coupled to Headphones

When the external ear is coupled to a headphone, the free-field characteristics of the ear are replaced by very different characteristics that are dependent on the interaction between the individual earphone and the individual ear. The nature of this interaction varies greatly with the class of headphone, as can be seen in Fig. 4.15. In general, there is a low-frequency regime where the system can be described in terms of lumped acoustical elements and a high-frequency regime where wave effects are predominant [30]. As the frequency increases, the transition between the two regimes can be said to commence when $\lambda/10$ becomes comparable with

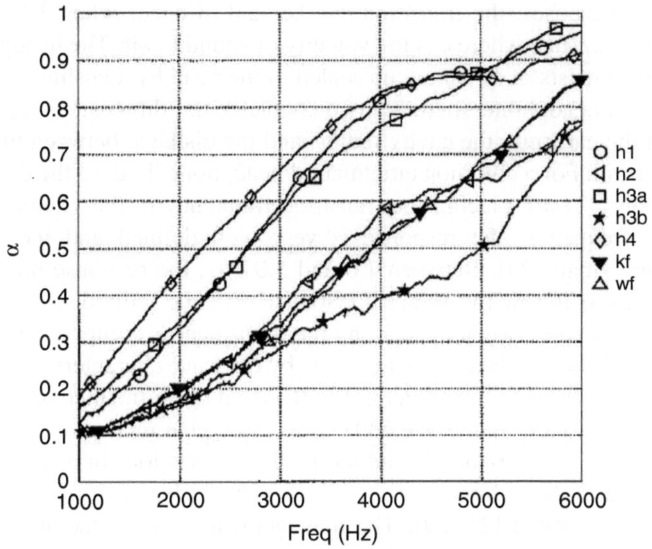

Fig. 4.14 Absorption coefficient measurements for hair (h1, h2, h3a, h3b, h4) and samples of two foams (kf, wf). Taken from [22]

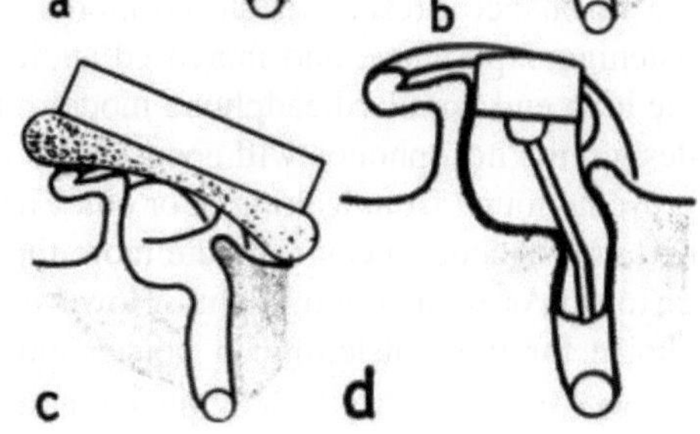

Fig. 4.15 External ear coupled to four classes of earphone: (**a**) Circumaural. (**b**) Supra-aural with firm airtight cushion. (**c**) Supra-aural with soft porous cushion. (**d**) Insert

a characteristic dimension, l, and can be considered complete when $\lambda/4 \approx l$, where λ is the wavelength of sound. In other words, for sound waves to interact with objects (to have wave effects such as reflections, standing waves, diffractions, scatterings, and shadowings), the one tenth of the wavelength of the sound must be comparable with the objects' dimension, or in terms of resolution (if having a microphone to record a signal's frequency response), one can say that the higher the frequency of the sound becomes, the greater the resolution it will have of picking up slightly changes for an object's dimension, which is revealed in its spectrum.

For practical purposes, the transition can be said to occur when $\lambda/6 \approx l$ or at a frequency $f \approx c/(6l)$, where c is the velocity of sound in air. The headphone shown in Fig. 4.15a consists of an outer cup sealed to the head by a cushion. For this class of circumaural headphone, there are two characteristic dimensions, in terms of the response at the eardrum, the cavity radius, and the distance between the transducer and the eardrum. For a common circumaural headphone, both of the dimensions are approximately 40 mm, placing the transition frequency in the vicinity of 1.5 kHz. Below this frequency, the response is very well defined and the inter-subject variations are small [29]. Between 1.0 and 5.0 kHz, the response measured at the eardrum tends to follow the primary resonances of the individual ear because the acoustical coupling between the ear and the transducer is comparatively weak. The response may, however, be greatly modified by the wave characteristics of the cavity that encloses the ear. The mode patterns in the cavity become increasingly complex, as the frequency rises, and above 5 kHz, the transverse modes of the Concha also come into play. The inter-subject and intra-subject variations in response between 5 and 10 kHz can be very large, unless the interactions between these wave systems are properly controlled [21, 29]. The management of these factors has generally proved to presents formidable challenges in headphone design.

Other characteristic dimensions account for the supra-aural and insert headphones. They are in general more tightly coupled to the ear than the circumaural headphone. Since the headphones chosen for the acoustic ear recognition experiments is of circumaural class, the description and discussion of the other headphone classes will be left out for now.

Circumaural and supra-aural headphones can both also be further differentiated by the type of design of their earcups. If it as an "open" design (sometimes not even covering the entire ear), more sound will leak out of the headphone simultaneously letting more ambient sounds into the headphone. This extra air flow is said to reduce the resonance present in closed-model headphones, resulting in a more natural listening experience, and increased audio fidelity. Therefore it seems that most of the high end popular headphone models are open. If the ear cups are of a "closed" design, the headphones will acoustically isolate your ears from the surrounding. It prevents sound from leaking in or out. The closed nature of such models means that certain frequencies can resonate more than others, which might result in unwanted distortion or poor sound fidelity. However, the isolating qualities make them a good choice for those listening in noisier environments, or looking for at completely immersive experience free of outside noise or distraction.

Many studies have treated the measuring of the characteristics of a system consisting of the ears of test subjects coupled to headphones, in terms of headphone transfer functions [20, 21, 23]. In [21] through a model of sound transmission, the open headphone types described above is reported as FEC headphones, abbreviation for headphones with free-air equivalent coupling to the ear. In [23] Kulkarni et al. specifically investigate the variability in the characterization of the headphone transfer function. Their purpose was to report the variability in the response of supra-aural headphones arising from the positioning of the headphone cushion during normal usage. The headphone responses were obtained on a KEMAR acoustical mannequin. They reported the spectral features in headphone transfer functions (the introduced coloration on the acoustical stimuli by the varying placement of the headphone cushion) were in fact very similar to the directional features reported in HRTFs. Choosing to compute inverse filters to equalize for the headphone response, by measuring the headphone transfer function for each individual listener was argued to be insufficient to solve this ambiguity, and rather may amplify it. They also suggested a simple model which qualitatively explained the phenomenon, but an actual solution was not proposed.

4.3 Measuring Device

The measuring equipment within our work comprised of a pair of open circumaural headphones (with frequency response from 17 to 26,000 Hz), AKG K530, an electret condenser microphone (with frequency response from 20 to 20,000 Hz), Sennheiser KE 4-211, and a battery driven amplifier for the electret microphone. The AKG headphones have the dimensions ⌀100 mm for the ear-cup including the cushion width and ⌀60 mm excluding the cushion width. The thickness of the cushion is about 20 mm. When wearing the headphones, they do more or less enclose the whole external ear, though at some spots, the edges of the pinna flange touch the inner surfaces of the cup. The headband of the headphones do not rest very tightly on top of one's head, making the sealing of the soft cushions around one's ear not totally airtight (Fig. 4.16).

The choice of the microphone for the measurements was a compromise between having a satisfying sensitivity and acceptable dimensions. The Sennheiser KE 4-211 is an omnidirectional microphone which has a sensitivity of approximately 10 mV/Pa and are cylindrical in shape, having diameter of ⌀4.75 mm and height of 4.00 mm without the feet. Coupled with a gain as suggested in the data sheet, the sensitivity increases to approximately 35 mV/Pa. The battery driven amplifier amplifies the signal with around 40 dB.

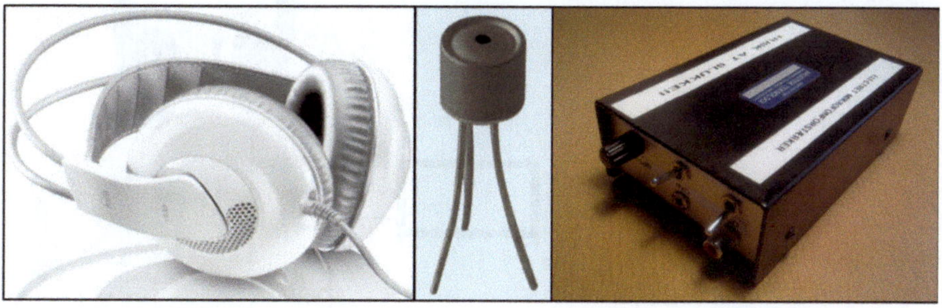

Fig. 4.16 Images of the AKG K530 headphones, Sennheiser KE4-211 microphone, and battery driven amplifier for the microphone

At first the microphone was attached to the headphone just by taping it to the inner side of the cup. It was placed in line with the opening of the ear canal and it was made sure that it did not touch the ear while wearing the headphones. But after some trials of letting test subjects wearing them, it became obvious that this could not last as a solution for a bigger experiment, since the tape would loosen after some time and the microphone could thus vary in position. Therefore a small hole was drilled in the shield of the headphone-cup, so the microphone could fit into there and be more stable. The image in Fig. 4.17 shows how the microphone now is integrated into the headphone.

The measuring devices described above were hooked on an ordinary laptop computer through mini-jack ports as illustrated in Fig. 4.18. The computer both collected and generated the signals, through a software SIGVIEW. SIGVIEW is a real-time signal analysis software package with wide range of FFT spectral analysis tools, statistics functions, and a comprehensive visualization system where all functions are parameters are custom set. An example of the results, shown in the plot windows in Fig. 4.19, gives the signal in its raw time domain (the blue colored

Fig. 4.17 Image of the microphone integrated into the headphone by a small hole drilled on the outer shield of the headphone-cup

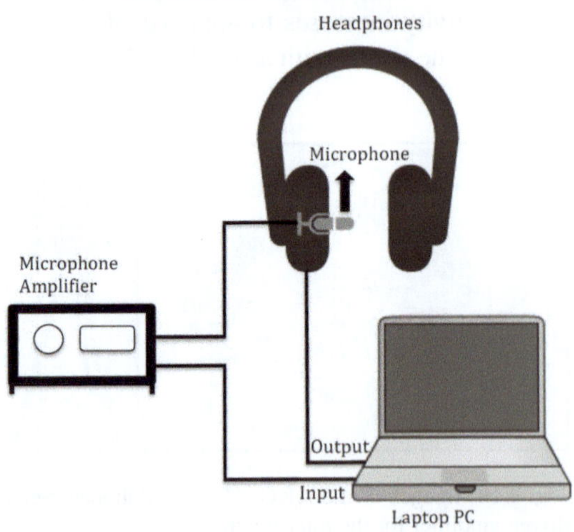

Fig. 4.18 The physical setup of the measuring device

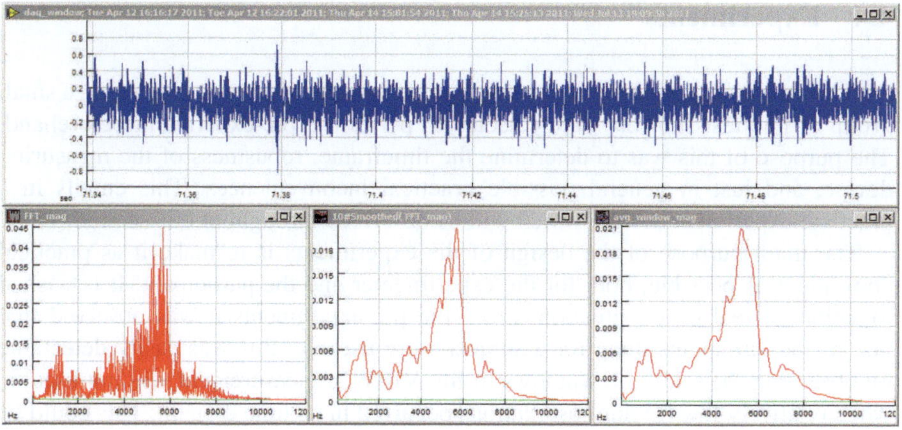

Fig. 4.19 Screenshots of the plot windows. The *blue colored* shows the raw signal in time domain, the *red colored* shows the signal in the processing states in the frequency domain

plot), and in its processing states in the frequency domain (the red colored plots). The first of these shows the signal right after FFT, which is a very rough signal. The second spectrum window shows the signal right after applying the smoother, and the third is the final processed spectrum window showing the averaged signal, which becomes very stable within a couple of seconds. The final data result is the averaged signal spectrum with 2048 points/samples (in amplitude $|a + ib|$), representing the frequency components from 6 to 12,000 Hz. This frequency range was chosen based on the results in [1] and the findings in [30]. In the acoustic ear recognition study it was proved that authentication performed well for a large frequency range with high frequencies, even up to 22.0 kHz, but in more precise measurement studies of the headphone characteristics, it was reported that intra-subject variance would become relatively large towards higher frequencies (above 5.0 kHz). Therefore 12,000 Hz seems to be an appropriate threshold for the compensation between the two issues. This signal spectrum composing of 2048 points/samples is four times more than the samples of the signal spectrum used in [1]. No studies have shown the effect of using a larger frequency resolution, so it is only encouraging to investigate this matter, besides if lower frequency resolution is wanted, it is always easier (and sometimes even only possible) to down-bringing a resolution than up-bringing a resolution. No calibrations have been made for the measuring devices, simply because it is not really necessary for the purpose of authentication. As long as the setup remains the same for all tests and experiments, and the environment, mostly in terms of low background noise, do not change radically, there is no need to calibrate the measuring devices.

4.4 Experiments

Besides the validation of the measuring device described in previous section, a small group of five test subjects were selected to perform a pre-experiment beforehand. The purpose of this was to determine the timeframe, robustness of the measuring device, and just in general possible practical inconveniences. This entails in a practical experiment design, which will be described throughout this section.

The main purpose of the design of the experiments is to make it as practical convenient as possible, both for the experimenter and the participant. It is wished to imitate a real usage situation. From the pre-experiments it was observed that not all measurements done for a subject were useable; this was easily detectable by observing the real-time windows in SIGVIEW, a measurement was not useable when relatively low (to non) oscillations occurred in its time domain. This could be due to that the microphone was somehow blocked by the pinna flange or earrings1 during the measurement. Therefore extra measurements besides the predetermined could be needed, because it has been decided in the data analysis to only include useable measurements. This meant errors in terms of failed measurements, or in biometric terminologies "failure to capture (FC)," would not be taken into consideration.

4.4.1 Design Considerations

The time it took to make one measurement, meaning recording the frequency response of an acoustic signal followed by saving it into an appropriately named file, would take around 30 s. Fifteen seconds to do the recording and 15 s to export, save, and naming the file appropriately. The goal is to make 50 measurements per subject (the more measurements, the better the statistical estimates will become). Fifty measurements would take 25 min, not including extra measurements due to unusable measurements and giving instructions to the participant. Further on it was wished to have as many participating test subjects as possible, with a goal of 50 test subjects. This was practically too overwhelming to achieve with the current manual method of making a measurement. Therefore automatization was needed.

SIGVIEW supports various command-line parameters enabling one to use some of SIGVIEW functions from external applications or batch files. Based on this, a GUI based application was developed in Microsoft Visual Studio (in C#), calling SIGVIEW and its functionalities for signal management and file manipulation. Eventually the application was able to automate a whole measurement process with a single push of a button. This reduced the time it took to make one measurement to 7 s, meaning 50 measurements could be completed in under 6 min. Including instruction explanation and extra measurements, the time of an experiment per subject is estimated to be about 10–15 min, which was practically completable.

The participants handled the wearing and off-wearing of the headphones themselves, and it was stressed to them, that the headphones had to be fully removed between each measurement. No further instructions were told on how they should wear the headphones (since that is more or less intuitive). The following was instructed to the test subjects about the measurements:

- They were asked to not chewing on chewing gum during the measurements. This was due to that jaw movements could cause volume changes in the ear canal, thus could affect the measurements.
- They were asked to remove their hair if it shielded their ear.
- They were asked to not making any sudden head movements, since that could affect the placement/position of the headphones.
- They were asked to not yawn during the measurements, again this jaw movement could affect the measurements.
- They were asked to just in general sit still during the measurements.

4.4.2 Data Collection

The measurements took place in a small conference room at the Norwegian Information Security Laboratory in the Norwegian University of Science and Technology (NTNNU) and was accomplished in a period of about 2 weeks. The measuring device setup was as illustrated in Fig. 4.18 and SIGVIEW was used as signal acquisition and processing software.

Fifty test subjects participated, where 50–60 measurements were carried out for each subject, where eventually 50 would be chosen as applicable measurements. They were 39 male and 11 females, from the age of 15–43, with an average age of 25.16 and 7 of the 50 subjects were wearing glasses and earrings, respectively, during the measurements. Most of them were international students (from all over Europe) at NTNU. All subjects were controlled for no physical abnormalities on their outer ear that could affect the measurements. During the measurements it was observed that different subjects worn the headphones in different ways. Some had the headband of the headphones leaned more towards the forehead, while some had it leaned more towards the back of one's head. Perhaps more importantly, some of the subjects seemed to now and then fall into a behavioral rhythm when taking the headphones on and off. This could be seen in the real-time frequency responses displayed in SIGVIEW having high similarity between some consecutive measurements.

4.5 Data Analysis

In previous sections it has been shown how the measuring device collected biometric samples from the subjects, and how these raw samples have been processed by SIGVIEW. It is now time to present the machine interpretation of the biometric samples and the appertaining analysis. For this, terminologies, methods, and in general areas within pattern recognition will be utilized.

4.5.1 Preprocessing

The frequency responses measured on test subjects contain three characteristics, or if regarding the measured responses as signals, three filters can be said to have been applied to the source signal (white noise). The first filter is caused by the loudspeaker of the headphone, and it is common that this electroacoustic transducer rarely has an even frequency response. The second filter is caused by the outer ear, where the acoustic wave effects of the pinna, concha, and ear canal characterize its spectrum. This is what supposed to be unique for each individual. However, one cannot neglect the inherent alterations caused by the enclosing of the headphone on the ear. It consists (for each side of the head) of a cup-formed shield sealed to the head by a cushion. Even though some parts are made of sound absorbent material, the several perforated holes and smaller hard surfaces (which can be seeing in Fig. 4.17) could cause uncertain wave reflections. Together with the cavity between the ear and the sealed headphone, there could occur standing waves at higher frequencies (as the findings in Sect. 4.2.5), thus the second filter is also characterized by the interaction between the headphone and the individual ear. The third filter is caused by the electret microphone itself, but it is known that this transducer has a flat frequency response. Figure 4.20 shows the described frequency characteristics in terms of a cascade of transfer functions. Having the frequency response characterized only by the outer ear (and its interaction with the headphone) is achieved by eliminating the characteristics of the loudspeaker and microphone (though the microphone response has a flat spectrum, thus does not alter anything as a filter). This is done by deriving a transfer function, where the numerator is the response characterized by a measurement on a test subject (Fig. 4.20), and the denominator is the response characterized by a measurement in free air. Figure 4.21 shows the response characteristics, when the measurement is done in free air (just have the loudspeaker of the headphone facing upwards with no obstacles nearby to cause reflections). It includes the characteristics of the headphone and microphone,

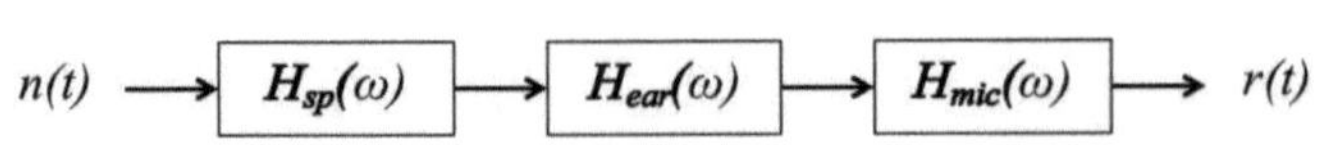

Fig. 4.20 Characteristics of frequency responses measured on test subjects

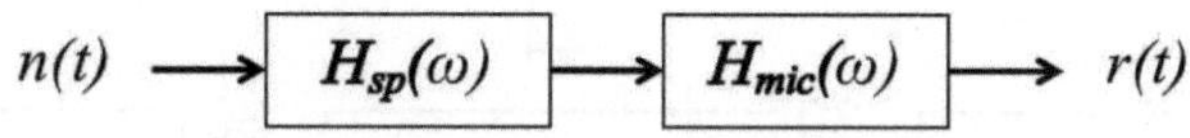

Fig. 4.21 Characteristics of frequency responses measured in free air

thus the derived transfer function will solely include the characteristics of the outer ear and its interaction with the headphone. This is preferable, because by having the machine interpretation to only include the characteristics of the ear (and some coloration in terms of the interaction between the ear and the headphone), one is able to detect and perhaps extract rich information packed features based on the acoustics of the ear. The fewer the characteristics of other objects in the machine interpreted data, the easier it should be to analyze and make use of the data for recognition tasks. The following equation has been applied to the frequency responses in order to obtain the transfer functions expressed in decibels

$$H_i(\omega) = 20 \log_{10} \frac{r_i(\omega)}{n_i(\omega)} \tag{4.3}$$

$H_i(\omega)$ are the desired transfer function, $r_i(\omega)$ are the frequency responses measured on test subjects, and $n_i(\omega)$ are the frequency responses measured in free air. This has been applied for all 50 measurements done on each test subject.

4.5.2 *Initial Data Analysis*

A first glimpse of the plotted data reveals that there exists differences between the data across different subjects. The frequency location of the peaks and valleys of the transfer functions, and their number of occurrence, seems to vary between subjects, at least for frequencies greater than 5 kHz. The amplitude range for most of the plots varies from −10 to 30 dB, though some few individual measurements reach beyond this threshold. This upper amplitude limit is comparable to the measurement summary of Shaw [30], which stated that the combined resonance of the pinna and ear canal provides a gain of 20–30 dB between 1.5 and 7.0 kHz. However, this frequency range does not match the data, since most of the high amplitude peaks (peaks over 20 dB) seem to occur at around 10.0 kHz. This shift of the gains towards higher frequencies must be due to the differences in the measurement setup. Shaw measured the transformation of sound pressure level from free-field to the ear canal, where the sound source was placed at a distance away from the ear, and the microphone was placed in the ear canal. In the measurement setup of the collected data, the sound source is the loudspeakers of the headphones, placed a few centimeters from the ears, and the microphone was placed on the inner wall of the headphone-cup sealing the ear. The influence of the headphone caused wave effect interactions in the cavity between the external ear and the headphone, which

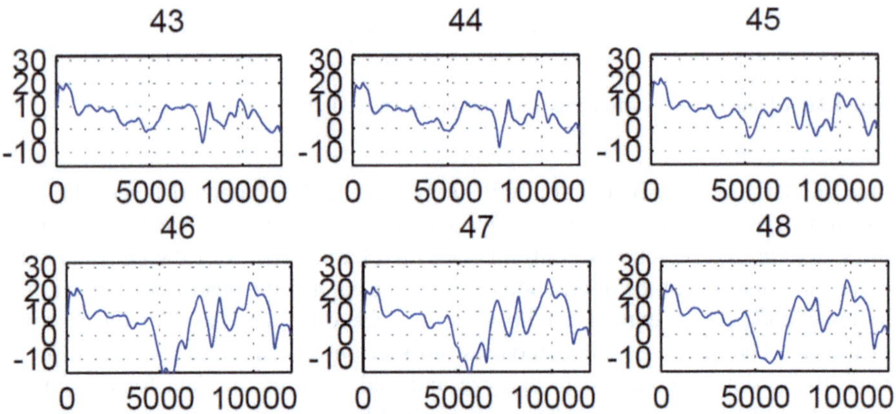

Fig. 4.22 Measurements of a subject running from number 43 to 48

was not present in Shaw's measurements, thus it could somehow only be assumed that there would be differences in the collected data and Shaw's data. The nature of Shaw's measurements was to determine head related transfer functions, whereas in the collected data it is to determine the outer ear/headphone related transfer functions.

Looking at the plots of the 50 measurements within a subject, one sees there are somehow an amount of intra-subject variance. Some subjects have better comparable intra measurements than others such as the measurements for subject 3, where all 50 measurements are very consistent. Then there are the measurements of subject 26, where all 50 measurements are very inconsistent. But in general there are more consistency than inconsistency in the intra measurements, which seems promising at this stage of data analysis. Something additional to notice is that at several places there arise some patterns in the individually measured data which seem to consecutively repeat. A good example is the measurements of subject 36 running from number 43 to 48. The plots in Fig. 4.22 show how the first three somehow look similar, followed by the next three which differentiate from the first three, but look similar within one another. This could indicate the test subjects to fall into some mechanical rhythm during the measurements where they had to wear the headphones on and off. Behavioral characteristics in terms of putting a pair of headphones on one's head might need to be taken into considerations. This pattern in the data will be addressed further in later sections. Next it is wished to look upon some statistical attributes of the data.

4.5.3 Statistical Attributes

To begin with a sample of five test subjects (A, B, C, D, E) are looked upon. From those five subjects two (subject B and C) were chosen because of their notable consistency and inconsistency, respectively, and the rest are randomly chosen.

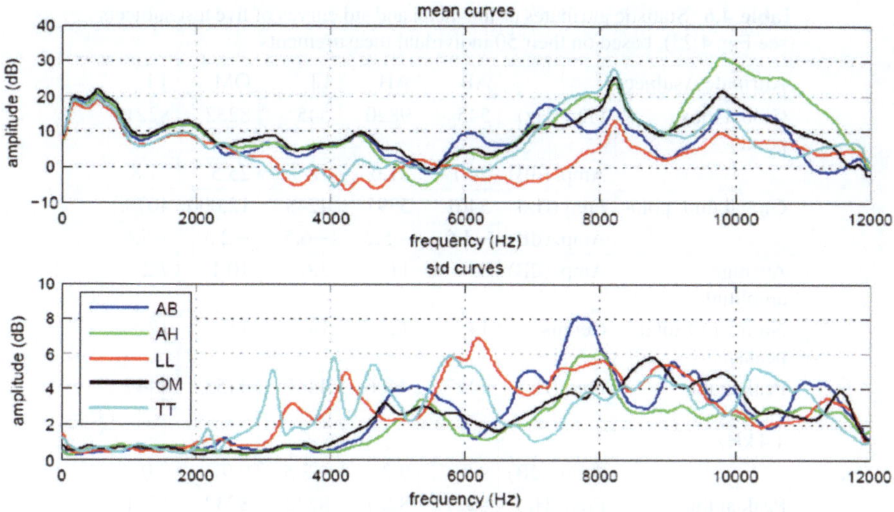

Fig. 4.23 Plots of the mean and standard deviation of five test subjects, based on their 50 individual measurements

The plots[1] in Fig. 4.23 shows the mean and standard deviation for each subjects' 50 measurements. Viewing the mean curves for all five subjects, one sees they are much alike from 0 to around 2 kHz. This range can be described by a steep upward going curve from about 10 dB into two peaks, then going down to a relatively broad valley followed by a shallow peak. From 2 kHz and upwards the curves become more diverse; however, there seems to be peaks at common frequencies, though with different amplitudes. These frequencies are in the vicinity of 4.40, 8.25, and 9.75 kHz.

Viewing the std curves for all five subjects one sees they are steady and under 2 dB in amplitude for frequencies under 2.0 kHz. This is somehow coincidental consistent with the comparability of the mean curves described above. Above 2.0 kHz the curves overall increase in amplitude; with the highest peak being 8 dB for subject A around 7.7 kHz. The peaks and valleys do not seem to be comparable at either frequency nor amplitude for the five subjects. Table 4.6 shows the sum of the std values across frequencies for each subject, in this case it also corresponds as the areas below each std curve. Even though the std curves show that subject E commerce with higher peaks than subject C, Table 4.6 shows that subject B overall has the highest std values, which is in agreement with the presumed inconsistency. It also holds true for subject 3 having the lowest std sum, thus being the most consistent, at least amongst this sample of five subjects. The other attributes in

[1]Usually the frequency axis for such transfer functions is logarithmic; for mirroring the way human perceive sound, but since the purpose of this analysis deals with data classification, letting the frequency axis being linear represents the data better in this case.

Table 4.6 Statistic attributes of the mean and std curves of five test subjects (see Fig. 4.23), based on their 50 individual measurements

Attributes\subjects		AB	AH	LL	OM	TT
Global max. point	Freq (Hz)	545	9820	545	8232	8221
	Amp. (dB)	20.7	21.0	18.9	25.5	27.8
Global min. point	Freq (Hz)	5461	5391	4248	12,000	4078
	Amp. (dB)	−1.9	−5.2	−6.5	−2.3	−5.6
Average amplitude	Amp. (dB)	8.4	11.7	3.6	10.1	7.2
Number of total peaks	Counts	14	12	14	11	12
Peak at the vicinity of 4.4 kHz	Freq (Hz)	4447	4406	4447	4453	4447
	Amp. (dB)	9.5	9.2	−3.5	9.9	2.0
Peak at the vicinity of 8.2 kHz	Freq (Hz)	8227	8227	8232	8232	8221
	Amp. (dB)	16.8	23.9	13.3	25.5	27.8
Peak at the vicinity of 9.8 kHz	Freq (Hz)	9826	9820	9797	9814	9807
	Amp. (dB)	16.2	31.0	9.8	21.4	16.7
Total sum of std values across freq.	Amp. (dB)	5396.2	3965.4	6058.0	4766.0	5717.3

Table 4.6 such as the global maximum and minimum point, average amplitude level, and number of total peaks do not quite reveal any further useful information.

Next it is wished to break down the data for all 50 test subjects with their 50 measurements each, as a totality. The most obvious is to somehow compare the intra-subject data to inter-subject data. Perhaps, in other words, compare the variance of subjects own 50 measurements with the variance of each subjects mean amplitude curve. With the inspiration of [34] a functional box-plot of the std's of the subjects based on their individual 50 measurements (corresponds to the std curves in Fig. 4.23, just done for all 50 subjects instead of only 5) and a std curve derived from the mean curves of subjects (corresponds to deriving the std curve from the mean curves in Fig. 4.23; again for all 50 subjects instead of only 5) are made in Fig. 4.24. Basically the functional box-plot[2] corresponds to a point box-plot by having the edges of the box, as the blue colored curves and the median as

[2]Based on the center outwards ordering induced by band depth for functional data, the five descriptive statistics of a functional box-plot are: the upper and lower boundaries of the central 50 % region, the median, the minimum and the maximum boundaries. In addition, outliers can be

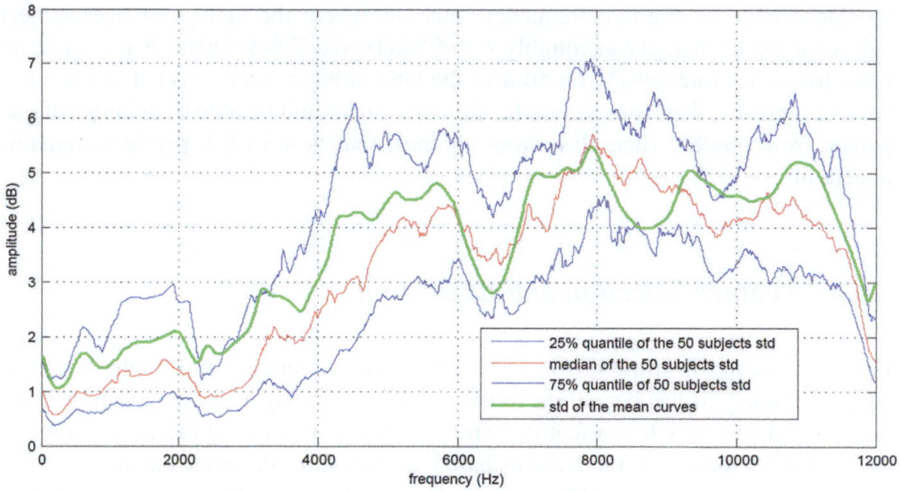

Fig. 4.24 Comparison between the intra- and inter-subject standard deviation values

the red colored curve. However, the whiskers which extend to the most extreme data points and outliers have been omitted for the sake of visuality. The functional box-plot represents the variance of intra-subject data whereas the green std curve represents the variance of the inter-subject data. Following the behavioral of the inter-subject std curve across frequency, it can be said that it is steady[3] (around 1 dB) at the low frequencies, until around 3.0 kHz, where an upward going slope commerces. It tops around 5.6 kHz, where an immediate valley emerges, with local minimum at around 6.5 kHz, followed by another valley with local minimum at around 8.6 kHz, until it descends to around 3 dB at the right end of frequencies. Now following the behavioral of the median of the intra-subject std curves across frequency, one sees it more or less follows the envelope of the inter-subject std curve, perhaps not having the same degree of valleys occurring at the frequencies given above. It can be inferred that at lower frequencies, approximately below 3.0 kHz, the performance of intra-class verification will be better than at higher frequencies (above 3 kHz). However, the performance of between-class verification will be poorer at lower frequencies than at higher frequencies. This is also reversely valid for the performance of between-class verification at higher frequencies. This issue seems to be in agreement with colorizations caused by the use of headphones, occurring at the higher frequencies.

Comparing the median of the intra-subject std curves with the inter-subject std curve, it can be said that the latter in general lies approximately 0.5 dB above the

detected in a functional box-plot by a proposed empirical rule analog to the 1.5 times the inter-quartile range rule for classical box-plots [34].

[3]Though by viewing the quantiles the std's vary in a magnitude of 3 dB within this 50 %.

median except for the two frequency intervals where the valleys of inter-subject std curve occur, namely at roughly 6.0–6.8 kHz and 7.8–9.2 kHz. It means at all other frequency intervals,[4] over 50 % of the intra-subject variance is below the inter-subject variance, therefore a classification system should be able to distinguish one person from another (one class from another), but how well it performs remains uncertain.

4.5.4 Feature Selection and Extraction

Feature selection is straightforward based on all frequency components; however, additional analysis has been done to this set of features by applying principle component analysis, which is a feature extraction tool for reducing the dimensionality of the feature space. For the features based on octave bands and acoustic properties of the ear, the principle of adding bands of noise plus curve-fitting and finding their peaks are utilized.

4.5.4.1 All Frequency Components

This extraction of all frequency components are trivial since the features in this case are just the calculated frequency components of the transfer function given by Eq. (4.3). The number of features is 2048, which compared to other biometric systems is tremendous. This triggered a second round of analysis with the request of reducing the number of features, wherefore PCA has been applied. Figure 4.25 shows plots of the PCA-applied data for the different composed matrix designs. One is not able to recognize them as being transfer functions any longer. They are all quite similar, only the two on the top separate slightly from the other three, in terms of local peak and valley values for dimension numbers above 10.

4.5.4.2 Octave Bands

The selection of frequency components of the transfer function for octave bands creation is based on the lower and upper cut-off frequencies of the octave bands' center frequencies. They include the bandwidth informations about one-third octave bands. The center frequencies are the usually preferred frequencies for octave bands and one-third octave bands. The lower and upper cut-off frequencies, f_1 and f_2, respectively, have the following relations:

$$f_2 = 2f_1, \qquad \text{bandwidth} = f_2 - f_1 \tag{4.4}$$

[4] Maybe with the exception of frequencies around 10.7 kHz.

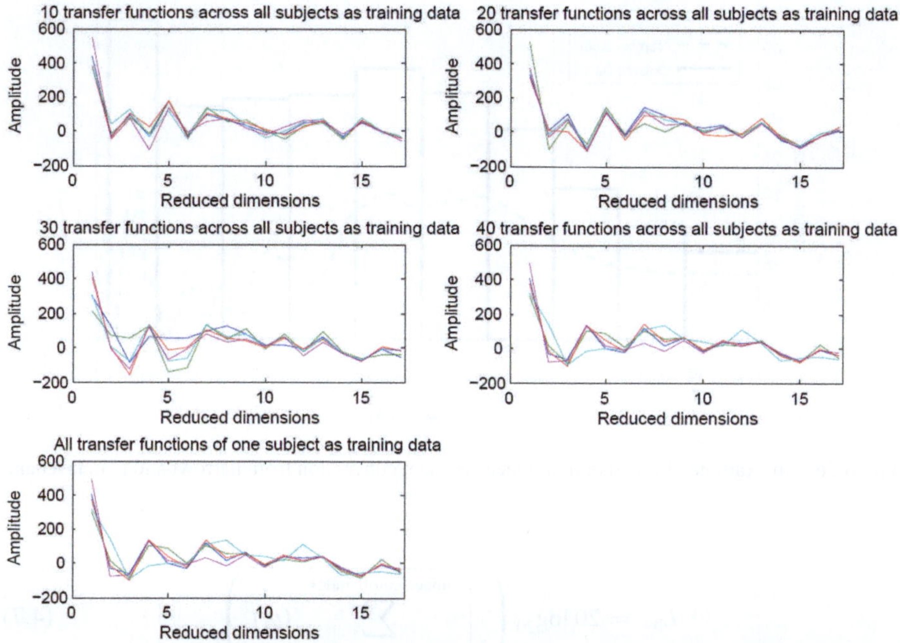

Fig. 4.25 Five example plots showing the PCA-applied data for different composed matrix designs. The first five test data of subject A is used

and they are calculated based on its center frequency, f_0, by the following formula:

$$f_1 = \frac{f_0}{\sqrt{2}}, \quad f_2 = \sqrt{2} f_0 \tag{4.5}$$

However, the frequency pass bands are a compromise; rather than follow a strictly octave sequence which would not repeat, they are adjusted slightly so that they repeat on a logarithmic scale. The merging of the frequency components of the transfer function into octave bands is based on adding bands[5] of noise with random phases (which is also the current case). The following accounts for adding incoherent sounds on a linear energy (pressure squared) basis:

$$p_{\text{total}}^2 = \sum_{i=1}^{n} p_i^2 \tag{4.6}$$

Once utilizing this for converting the narrow bands of the transfer function into octave bands, by finding the correct amount and indices of the narrow bands to match octave band passes, the unified amplitude level (expressed in decibels) of each octave bandpass is calculated by

[5]The frequency components of the transfer function can be regarded as a narrow bands.

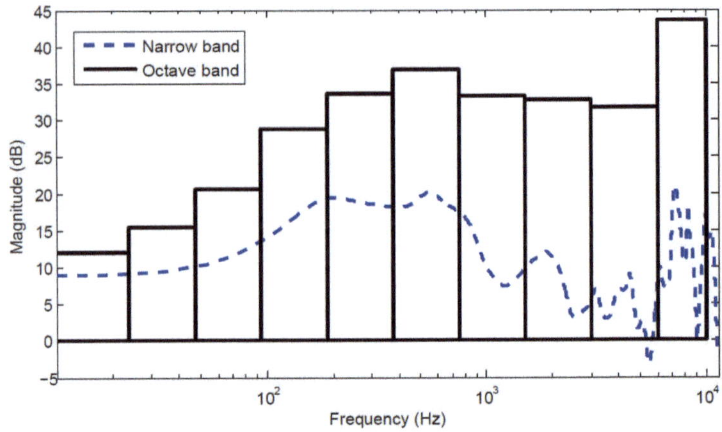

Fig. 4.26 An example of one individual measurement conversion from narrowband to octaveband

$$L_{\mathrm{bw}} = 20 \log_{20} \left(\sqrt{\sum_{i=\text{lower cutoff index}}^{\text{upper cutoff index}} (p_i)^2} \right) \tag{4.7}$$

where p_i are the pressure frequency components of the transfer function within the corresponding octave bandwidth, the start and end number of the sum then corresponds to the lower and upper cut-off frequency index, respectively, and L_{bw} is the unified amplitude level in dB, for each octave band. Figure 4.26 shows a semi logarithmic plot of the octave bands, on top of the former transfer function of a subject. It is seen how tall the last octave band stands (even though the global peak value of the transfer function may not lie in that octave band), which is due to the natural behavior of the octave bandwidths; they increase exponentially in size towards higher frequencies. In fact, if displaying unfiltered white noise in octave bands, the level of the bands would steadily increase by 3 dB per octave. A way to hinder this is by using pink noise, which has a downward going slope in its spectrum towards higher frequencies, thus causing its display in octave bands to stay even. The plot in Fig. 4.27 compares all 50 transfer functions converted to octave bands for one individual within their appurtenant center frequency. The largest variations between them seem to happen at the octave band with 8.0 kHz as center frequency. The plot in Fig. 4.28 compares the average of all octave bands-converted transfer functions across subjects. It seems to agree with the previous plot, namely that the largest variance is found in the 8.0 kHz octave band. This is no surprise since the original frequency components of the transfer functions behaved similarly. Comparing those two figures with each other seems to reveal that the between-class variance is greater than the within-class variance. Breaking down the octave bands-converted for all 50 test subjects their 50 measurements each, it is wished to investigate the between-class variance versus the within-class variance. A

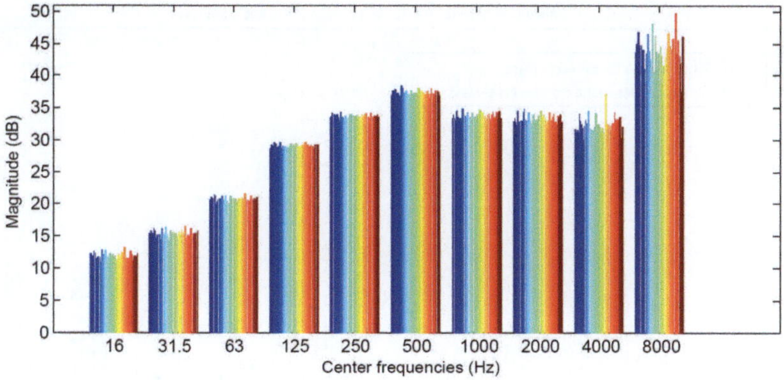

Fig. 4.27 All 50 measurements converted to octave bands for one individual

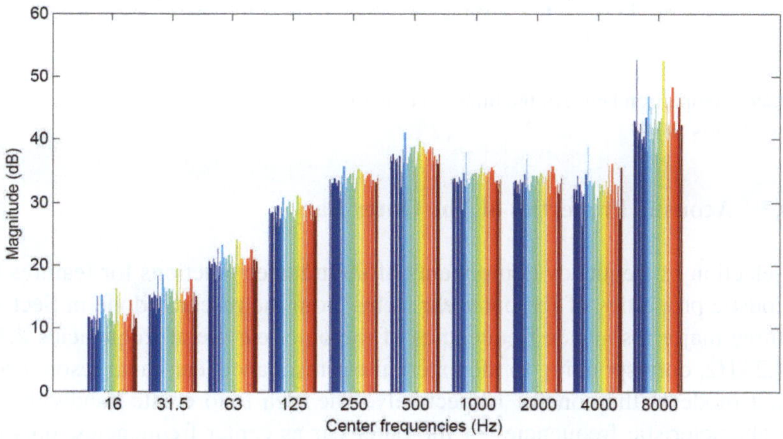

Fig. 4.28 The average of all 50 measurements for each subject, plotted in each frequency band. The octave-bands are plotted for all 50 subject, each *bar color* represents a subject

box-plot of the std's of the subjects based on their individual 50 measurements and std values derived from the average values of the subjects is made in Fig. 4.29. This corresponds to the plots done in Fig. 4.24, where instead of a functional box-plot, the box-plot here is done the usual point-wise way, and the whiskers and outlier points are included. The between-class variance is higher than the within-class median variance for all octave bandpasses, approximately 0.5 dB above. For the first three bandpasses, the between-class variance is even greater than the third (75 %) quantile of the within-class variance. But again, nothing more definite than that the system will be able to differentiate the subjects from one another can be concluded on the performance of the octave bands. The degree of authentication performance can only be answered by the results of the evaluation of the system.

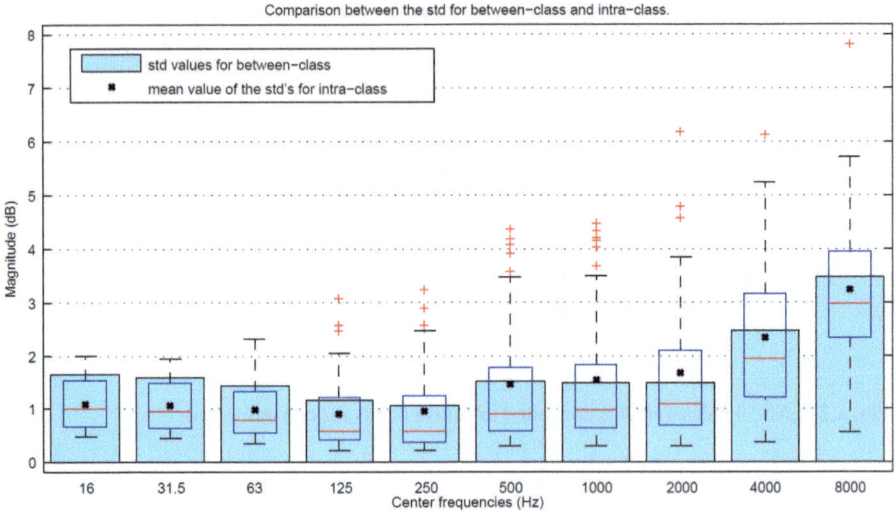

Fig. 4.29 Comparison between the intra- and inter-subject standard deviation values for features based on octave bands

4.5.4.3 Acoustic Properties of The Outer Ear

The selection of frequency components of the transfer functions for features based on acoustic properties of the outer ear stems from the research done in Sect. 4.2.4. The three major resonance frequencies of the outer ear are at frequencies 2.5, 3.4, and 4.2 kHz, corresponding to the external ear resonance, ear-canal resonance, and the first mode of the concha, respectively. The idea is to create bandwidths with these characteristic frequencies of the outer ear as center frequencies, plus detect the peak frequency at the vicinity of these frequencies. The unified energy level of the bands together with the peak frequencies shall be the features for a subject. Figure 4.30 shows the chosen band-limits, which lie right next to one another. The calculations of the unified energy level of the bands are identical to the calculations presented in previous section for octave bands, namely Eq. (4.7), just substituting the new band-limits instead of the octave band-limits. The detection of the peak frequencies within the vicinity of the center frequencies is done on the basis of curve-fittings. A polynomial curve-fitting function of degree 5 has been used in MATLAB on the piece of transfer functions within the bandpasses in Fig. 4.30. The found coefficients have then been substituted into the polynomial of degree 5 and the plots in Fig. 4.31 shows their comparability with the measured transfer functions. As one can see, the polynomials fit the measured transfer functions quite well. The benefits of finding the peaks by curve-fitting are that you avoid the ambiguity in terms of the exact peak frequency, in case of the measured transfer functions being non-smooth, but irregular at the local peak. In addition to curve-fitting, the peak frequencies for detection will have better resolution than the processed frequencies

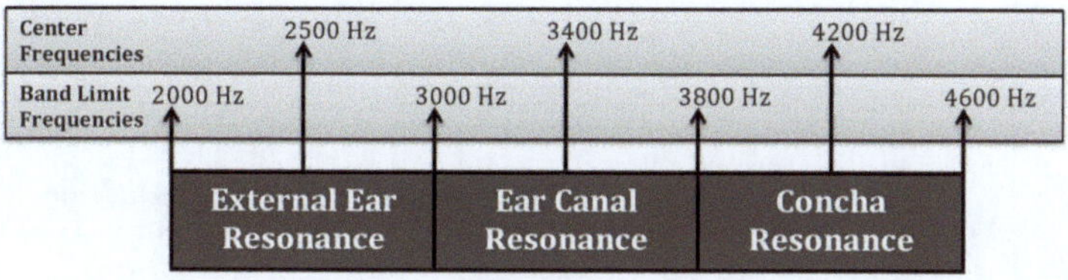

Fig. 4.30 Bandwidths and their limit and center frequencies based on acoustic properties of the outer ear

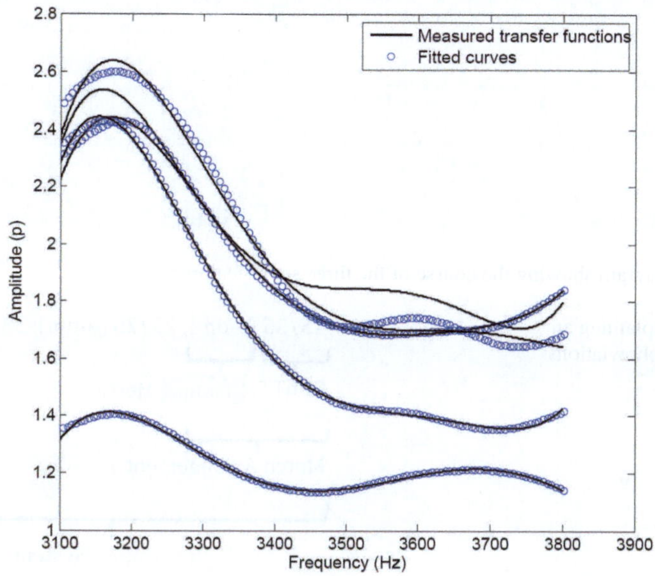

Fig. 4.31 Plots of five fitted curves based on their measured transfer functions within the ear-canal resonance bandwidth

of the measured transfer functions, given by the program SIGVIEW (also the curve-fitting gives dependencies on the slopes of the peak, so they are in general just better to use than the peaks of the direct measured data).

The diagram in Fig. 4.32 has been made for clarifying the analysis and processing of the data with respect to concepts of the stages in the classification system. It illustrates the extracted features and which routes they are following through in order to be evaluated as a system. As one can see, up to six systems can be evaluated specified by the three set of extracted features combined with match engines and classifiers, in which four of them being verification systems and two being identification systems.

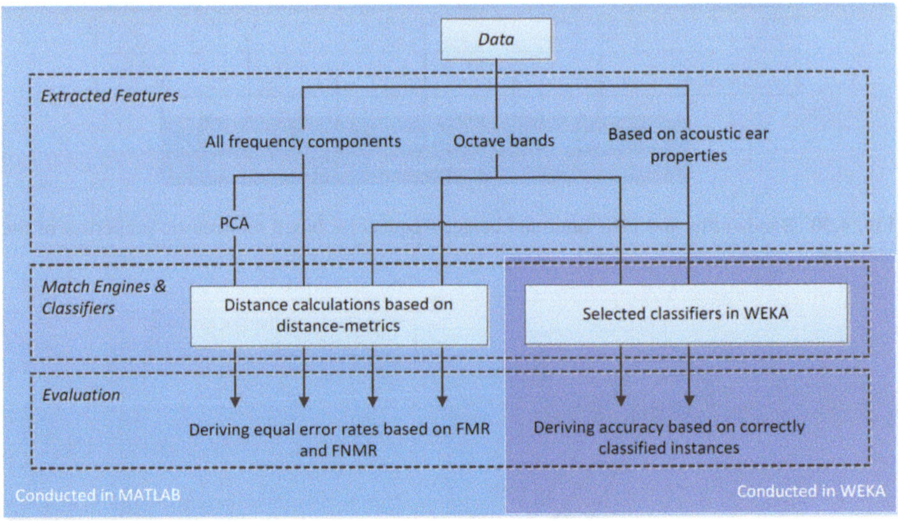

Fig. 4.32 Diagram showing the course of the three set of features

Fig. 4.33 Explaining an example of abbreviations

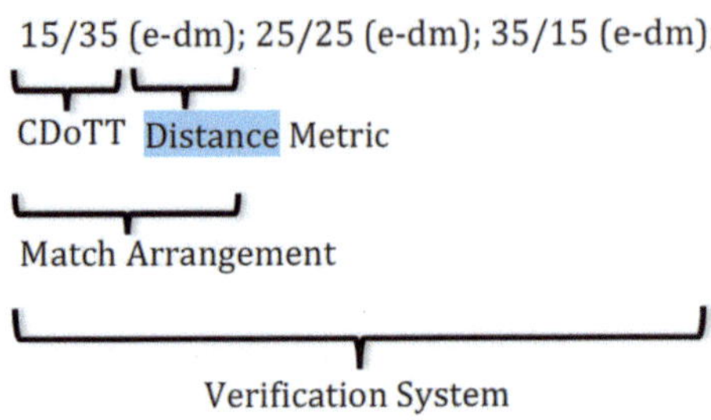

4.6 Results

This section presents the results of the data analysis and processing in the previous section.

The following explains what is meant by the abbreviations used and what they constitute in biometric terms throughout this section. Figure 4.33 describes the concepts. The set of numbers represents the composed design of template and test data (CDoTT), where in cases of the PCA, training data will also be included listed in the beginning in the following manner <measurements for training>/<measurements for template>/<measurements for test>. The following abbreviations in parenthesis are the employed distance metrics. Pairing a CDoTT with a distance metric forms a match arrangement, which calculate distance scores. Pairing different CDoTTs with a distance metric constitute a system, or in other words, a system is specified by the distance metric, with exception of the systems for the PCAs, since the CDoTTs include two different splits of the training data.

Further on the all frequency component, octave band, and ear characteristic bands and peaks features will be referred to as AFC, OB, and ECBP features, respectively.

The plots and figures basically reflect the findings of the system evaluation methods listed in Table 4.7. Based on the extracted features, the equal error rates and/or percentage correct classified instances will be given for the following: Comparing the EERs with each other corresponds to comparing their system performances in terms of verification, if it involved distance calculations of a match engine, or in terms of identification, if it involved employment of a classifier. One is then able to evaluate the systems within each extracted feature and also across the features themselves.

4.6.1 Performance of All Frequency Components

Figure 4.34 shows the box-plots of 30 EERs calculated from 24 match engines (constituting eight systems), corresponding to those listed in the beginning of this section. Further on the mean and standard deviation value for each box-plot is listed in Table 4.8. The lower the EERs, the better the performance of the system. First thing to notice is how the mean and median of the EERs drop in value within a system when using more data for the template in the CDoTT. Unfortunately the boxes also become wider, which mean the variance of the 30 EERs within each match arrangement increases with more data used for the template. The worst performing system is undoubtedly the one using the (1std-dm) distance metric, giving the highest mean EER of 24.34 %. The best performing system is the one using the (2std-sm-dm), giving the lowest mean EER of 14.9 %. Investigating all these 24 match arrangements has revealed a significant gap between the best and worst performance by an EER difference of 9.44 %, which stress the importance of considering various match arrangements in terms of CDoTTs and distance metrics.

4.6.2 Performance of PCA

Figure 4.35 shows the box-plots of 30 EERs calculated from 14 match arrangements (constituting 4 systems), corresponding to those listed in the beginning of this section. Further on the mean and standard deviation value for each box-plot is listed in Table 4.9. Besides data for the template and test, these CDoTTs (maybe in this case called CDoTTTs) also contain training data. The more data used for the training, the less data will be available for the template and test. In this case, the remaining template and test data are distributed equally for every CDoTTs. It is seen how the 30 EERs for each match arrangement raises when more data is used for training within the CDoTTs, where the split has been done vertically through the summary matrix, meaning splitting the number of measurements across all subjects

Table 4.7 Overview of the most important aspects in data analysis

			Extracted features			
			All frequency components			
Data processing				PCA	Octave bands	Ear characteristic bands and peaks
Feature dimensional space			2048	17	10	10
Verification system	Match arrangement	Composed designs of (training), template and test data	15/35, 25/25, 35/15	10/20/20, 20/15/15, 30/10/10, 40/5/5, 1s/15/35, 1s/25/25, 1s/35/15	15/35, 25/25, 35/15	15/35, 25/25, 35/15
		Distance metrics employed	e-dm, m-dm, sm-dm, 1std-dm, 2-std-dm, 2std-m-dm, 2std-sm-dm, 4std-dm	2std-sm-dm	2std-sm-dm	2std-sm-dm
	System evaluation method		EER of the FMR and FNMR	EER of the FMR and FNMR	EER of the FMR and FNMR	EER of the FMR and FNMR
Identification system	Applied test option		–	–	Cross-validation, ten folds	Cross-validation, ten folds
	Classifiers employed		–	–	BayersNaive, LibSVM, RandomForest, DecisionTable	BayersNaive, LibSVM, RandomForest, DecisionTable
	System evaluation method		–	–	Percentage correct classified instances	Percentage correct classified instances

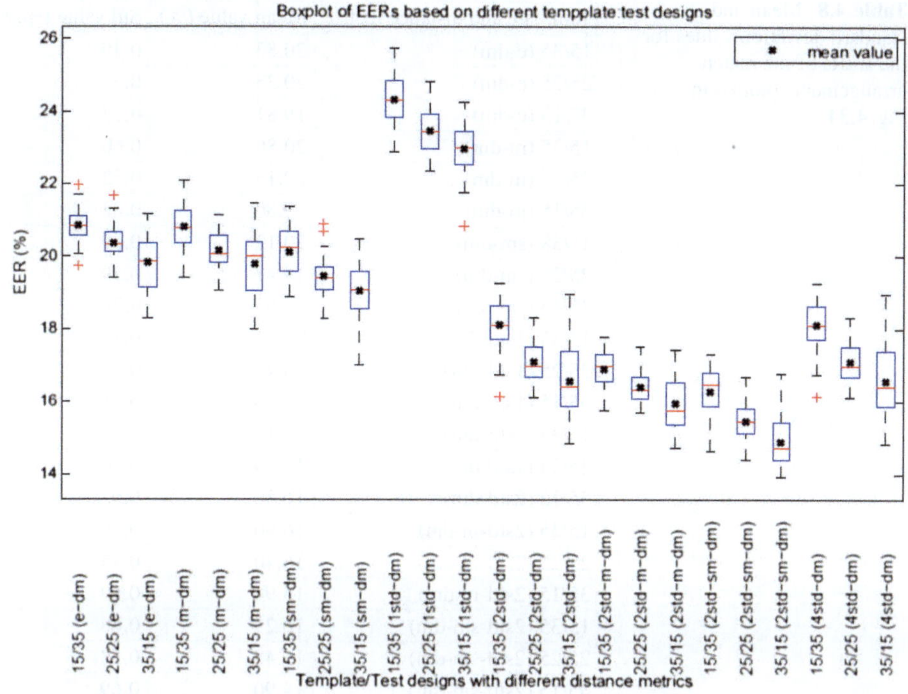

Fig. 4.34 Box-plot of the 30 EERs per match arrangement for AFC features

(the first eight match arrangements). However, it might be more correct to say, that the 30 EERs for each match arrangement drop in values when quantitatively using more data for the template, in which case is also consistent with the previous finding. This indicates the amount of data used for training do not impact the system performance, whether it is 10 or 40 measurements. This is also confirmed by the EERs for match arrangements where the training data is split horizontally through the summary matrix, meaning in this case only using one subject's 50 measurements (the last six match arrangements). The EERs of these match arrangements lay in proximity of all the other EERs, even though only data representing one subjects has been used as training for the system. They behave similarly, the mean and median drops as more data is used for the template. Surprisingly the lambda weighted distance metrics (lw) have not proven to produce better performance, on the contrary they are slightly worse. This only stress the important statement of any biometric systems, namely that the true values of the various error rates cannot be computed or theoretically established; it is only possible to obtain statistical estimates of the errors using test databases of biometric samples. The best performing match arrangement is 10/20/20 (2std-sm-dm), with a mean EER value of 17.1 %, and the worst performing is 40/5/5 (2std-sm-dm), with a mean EER value of 23.72 %, unfortunately these two math arrangements both belong to

Table 4.8 Mean and standard deviation values for the EERs of the match arrangements plotted in Fig. 4.34

Template and distance metric	Mean value (%)	Std value (%)
15/35 (e-dm)	20.83	0.49
25/25 (e-dm)	20.35	0.52
35/15 (e-dm)	19.81	0.77
15/35 (m-dm)	20.80	0.60
25/25 (m-dm)	20.15	0.52
35/15 (m-dm)	19.80	0.82
15/38 (sm-dm)	20.12	0.69
25/25 (sm-dm)	19.47	0.59
35/15 (sm-dm)	19.05	0.74
15/35 (1std-dm)	24.34	0.78
25/25 (1std-dm)	23.47	0.62
35/15 (1std-dm)	22.96	0.81
15/35 (2std-dm)	18.13	0.69
25/25 (2std-dm)	17.09	0.60
35/15 (2std-dm)	16.56	0.93
15/35 (2std-m-dm)	16.90	0.56
25/25 (2std-m-dm)	16.40	0.45
35/15 (2std-m-dm)	15.94	0.69
15/35 (2std-sm-dm)	16.29	0.68
25/25 (2std-sm-dm)	15.48	0.58
35/15 (2std-sm-dm)	14.90	0.69
15/35 (4std-dm)	18.13	0.69
25/25 (4std-dm)	17.09	0.60
35/15 (4std-dm)	16.56	0.93

the same system. Therefore overall this system may not be the best performing one. When taking everything into consideration, the third system4 appears to be the best performing one.

4.6.3 *Performance of Octave Bands*

Figure 4.36 shows the box-plots of 30 EERs calculated from four match arrangements (constituting only one system), corresponding to those listed in the beginning of this section. Further on the mean and standard deviation value for each box-plot is listed in Table 4.10. These box-plots behave quite odd comparing to the box-plots of EERs based on other features. The mean and median value of the EERs lay relatively high, with the worst performance being the 35/15 (2std-sm-dm) match arrangement, with a mean EER of 35.47 %, and the best performance being the 15/35 (2std-sm-dm) match arrangement, with a mean EER of 23.95 %. Some of the box-plots have

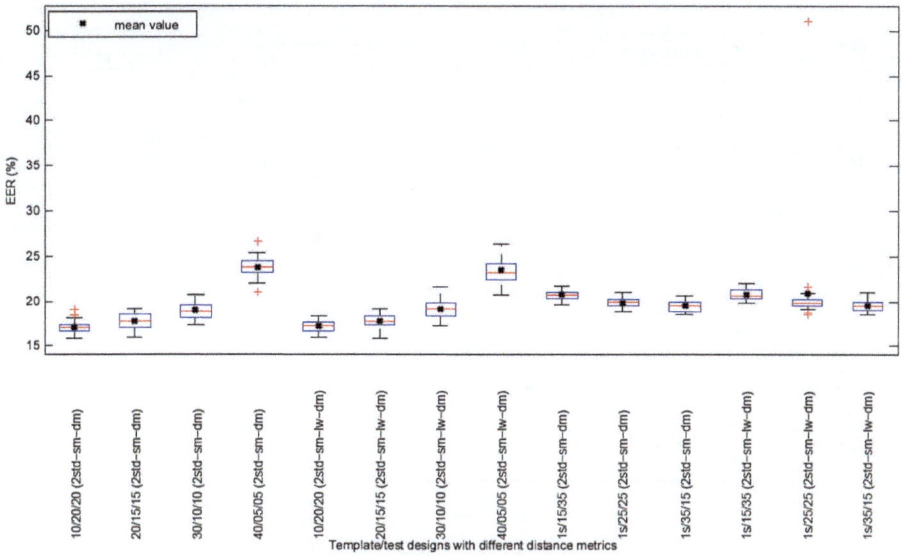

Fig. 4.35 Box-plot of the 30 EERs per match arrangement for AFC processed through PCA as features

Table 4.9 Mean and standard deviation values for the EERs of the match arrangements plotted in Fig. 4.35

Template and distance metric	Mean value (%)	Std value (%)
10/20/20 (2std-sm-dm)	17.10	0.71
20/15/15 (2std-sm-dm)	17.77	0.92
30/10/10 (2std-sm-dm)	18.95	0.90
40/5/5 (2std-sm-dm)	23.72	1.14
10/20/20 (2std-sm-lw-dm)	17.17	0.61
20/15/15 (2std-sm-lw-dm)	17.75	0.86
30/10/10 (2std-sm-lw-dm)	19.07	1.05
40/5/5 (2std-sm-lw-dm)	23.45	1.40
1s/15/35 (2std-sm-dm)	20.78	0.56
1s/25/25 (2std-sm-dm)	19.87	0.47
1s/35/15 (2std-sm-dm)	19.51	0.55
1s/15/35 (2std-sm-lw-dm)	20.79	0.60
1s/25/25 (2std-sm-lw-dm)	20.89	5.72
1s/35/15 (2std-sm-lw-dm)	19.56	0.64

outliers, which lie over 10 % in amounts to its extreme values, and the lower and upper box edges of 25/25 (2std-sm-dm) seem to span over EER values of 22.2 % and 36 % (13.8 % amount in difference). This might be due to calculation errors in the match engines, or just odd behavior of the biometric system using these features. Surprisingly though the performance of 5/45 (2std-sm-dm) seems to be better than the 25/25 (2std-sm-dm) and 35/15 (2std-sm-dm).

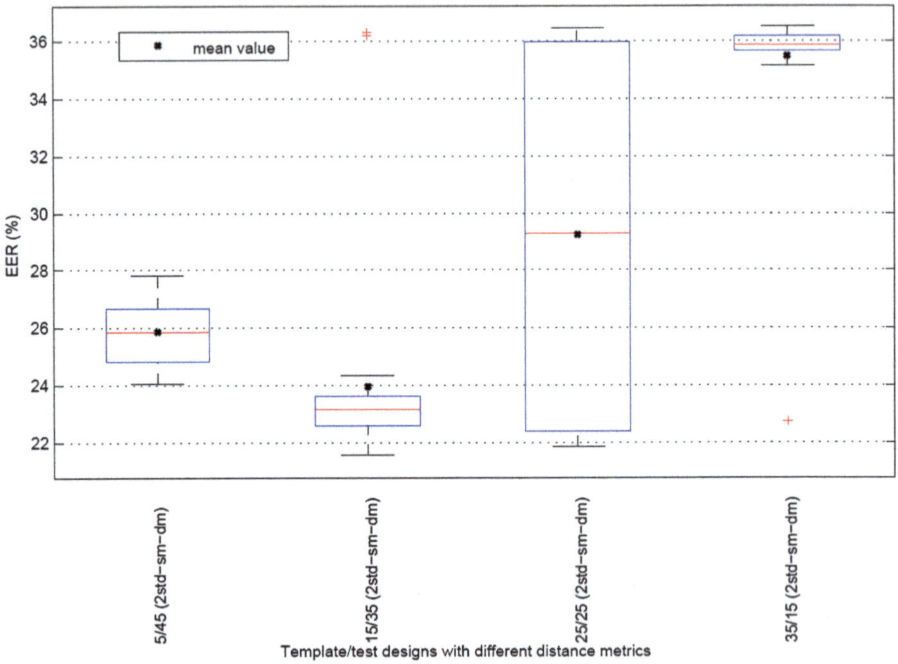

Fig. 4.36 Box-plot of the 30 EERs per match arrangement for OB features

Table 4.10 Mean and standard deviation values for the EERs of the match arrangements plotted in Fig. 4.36

Template and distance metric	Mean value (%)	Std value (%)
05/45 (2std-sm-dm)	25.86	1.06
15/35 (2std-sm-dm)	23.94	3.41
25/25 (2std-sm-dm)	29.24	6.85
35/15 (2std-sm-dm)	35.47	2.43

Table 4.11 Cross-validated correct classified instances of the four selected classifiers, using OB features

Classifiers	Correct classified instances (%)
BayersNaive	42.96
LibSVM	67.88
RandomForest	63.40
DecisionTable	36.96

Table 4.11 shows the cross-validated (ten folds) correct classified instances (in percent) for the four selected classifiers. The support vector machine classifier proves to have the best performance with 67.88 % correct classified instances.

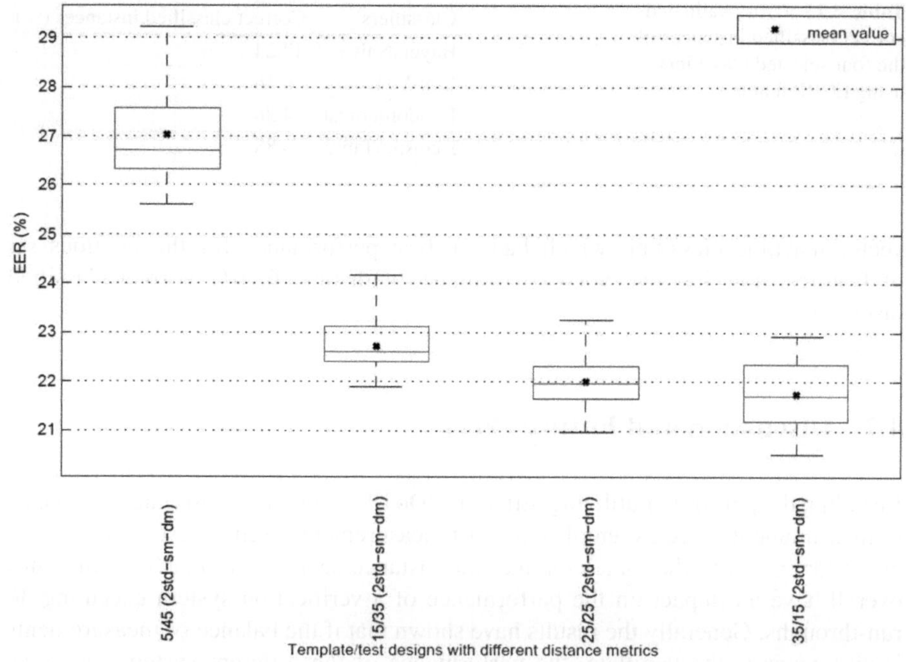

Fig. 4.37 Box-plot of the 30 EERs per match arrangement for ECBP features

Table 4.12 Mean and standard deviation values for the EERs of the match arrangements plotted in Fig. 4.37

Template and distance metric	Mean value (%)	Std value (%)
05/45 (2std-sm-dm)	27.04	0.90
15/35 (2std-sm-dm)	22.73	0.55
25/25 (2std-sm-dm)	22.01	0.51
35/15 (2std-sm-dm)	21.77	0.70

4.6.4 Performance of Ear Characteristic Bands and Peaks

Figure 4.37 shows the box-plots of 30 EERs calculated from four match arrangements (constituting only one system), corresponding to those listed in the beginning of this section. Further on the mean and standard deviation value for each box-plot is listed in Table 4.12. The behavior of these box-plots are in accordance to the box-plots of EERs based on AFC features. The match arrangement having the best performance seems to be 35/15 (2std-sm-dm), with a mean EER of 21.77 %, and the match arrangement with the worst performance seems to be 05/45 (2std-sm-dm), with a mean EER of 27.04 %.

Table 4.13 shows the cross-validated (ten folds) correct classified instances (in percent) for the four selected classifiers. The random forest classifier proves to have the best performance with 74.36 % correct classified instances. However support

Table 4.13 Cross-validated correct classified instances of the four selected classifiers, using ECBP features

Classifiers	Correct classified instances (%)
BayersNaive	47.24
LibSVM	6.16
RandomForest	74.36
DecisionTable	44.88

vector machine classifier, which had the best performance for the previous set of features, now has the worst performance with only 6.16 % correct classified instances.

4.7 Discussion and Future Work

Recalling the purpose of utilizing different CDoTTs, it was to investigate the balance of measurements used as template against measurements used as test, which would affect (as inputs to the match engine) the distance scores of a match engine, thus overall have an impact on the performance of a verification system executing 30 run-throughs. Generally the results have shown that if the balance of measurements incline towards the template, the performance of the different systems increase. However, the variance of the 30 EERs also increase relatively (compared to the other systems where the balance of measurements incline towards the test), which could suggest that the 30 EERs of the 30 run-throughs are not sufficient enough for generalizing the statistical estimates. The 30 times of run-throughs were mainly decided upon the available timeframe, but it actually turned out to be much more time consuming than estimated, which has caused omitting some of the planned data processings.

From a practical biometric point of view with regard to the enrollment process, having to make 35 measurements in order to create the template, is not very convenient. In an ideal situation for recognizing physiological characteristics, one measurement per subject for enrollment is hypothetically sufficient enough for a biometric system to authenticate the subjects. But in case of recognizing behavioral characteristics, one measurement per subject for enrollment will never be as sufficient as for systems recognizing physiological characteristics. As for the acoustic method used for ear recognition in this thesis, it has proved that the system recognizes on the basis of both behavioral and physiological characteristics, thus is only in agreement with the system to perform better by using more measurements (to make the template) during the enrollment phase.

As for the future work there are bulletpoints to be performed which did not get included in this chapter, such as to employ different distance metrics for the OB and ECBP features, to extract and test out new set of features based on specific selections of frequency components in terms of good inter-subject and intra-subject conditions and to investigate concrete biometric verification methods. First mentioned would be important to investigate to confirm that the properties of a set of features are not necessarily similar to another set of features (even if it is extracted from the first set of features), in terms of verification. In other words, the match arrangement with the best performance of the first set of features do not necessarily have the best performance for a second set of features, even if extracted from the first set. But then it could be more complicated to compare and explain their performance evaluations, since they might no longer be based on the same match arrangement. The third mentioned came as an idea when since there exists two frequency regions where the inter-subject variance is above the 75 % quantile of the intra-subject variance distribution. These areas would be an interesting place to extract frequency components for new features, since there exists good condition for verification performance. The last mentioned was also originally planned in the beginning of the data analysis stage. The idea here is to investigate concrete biometric verification methods, such as for how many trails should a subject provide to the system before getting authenticated? All the current performance evaluations are based on giving one trail to the system for authentication, would it maybe help on the performance if two or three trials were delivered to the system before it makes a decision. This would correspond to (in biometric terms) biometric fusion. The use of fused biometric measurement often gives a substantially improved technical performance.

Finally a very obvious future work would be to go back and redesign the measuring device, making it more customized for the purpose of acoustic ear recognition. The position of the microphone could be placed with more precision. For instance, it could be based on simulations of the sound-field within the cavity of the headphone, to identify where the optimal sound image would be. However it would be faced with the same issues within creating authentic sound reproductions in headphones, namely the variability in the characterization of the headphone transfer function due to positioning of the headphone cushion during normal usage, in which still no solutions exist.

4.8 Conclusion

This chapter has treated the holistic development of a biometric system based on acoustic ear recognition, from the assembly and setup of the measuring device to the authentication performance of a database, consisting of biometric samples in terms of transfer functions collected from 50 test subjects.

Commencing with obtaining an outline of the relevant research and studies done within forensics and biometrics, it has indeed been proved that the human ear is very unique and applicable as a biometric modality, even with benefits over other more traditional biometric modalities. Investigating the acoustical properties of the human outer ear have emphasized its variations among people, with respect to different lengths and sizes of the ear canal, and its many characteristic cavity resonances. This has become to appear logical, since the nature of the physiology design of the outer ear is to provide rich cues for spatial sound detection. When examining the ears coupled to headphones, the free-field characteristics of the outer ear are suddenly replaced by different characteristics that are dependent on the interaction between the individual headphones and the individual ear. These different characteristics have proved to be not easy determinable, due to the many complicated factors in play, especially for the higher frequencies where the wave effects/phenomena are predominant. The common issue, which is found both in acoustics for sound reproduction and within the work of this thesis, is the variability in the frequency response (for the higher frequencies) of headphones arising from the positioning of the headphone cushion during normal usage. This resulted in undeniable intra-subject variance, which in connection with observations made during the measuring experiments, is considered relating to as a behavioral characteristic of the subjects.

At first a larger than necessary number of feature candidates were generated directly from the raw biometric samples (which was the smoothed and averaged FFT of a recorded white noise stimulation), known as the AFC features with a dimension of 2048. Then it was attempted to reduce this feature dimensionality while still keeping its rich information in terms of inter-subject distinctions and intra-subject similarities, which resulted in applying PCA and LDA. In addition two other set of features were extracted on the basis of acoustical principals, known as the OB features and the ECBP features, both with a dimension of 10. The analysis of AFC features using 35/15 (2std-sm-dm) match arrangement has yield the overall best verification performance with an EER of 14.9 %. However, because of AFC's relatively tremendous dimensional feature space, an alternative suggestion would be to consider the ECBP or PCA extracted features as better applicable features.

Acoustic ear recognition is still a newcomer to the field of biometrics. This chapter has given a broader perspective on acoustic ear recognition, both in terms of its biometrics and acoustics. It has mapped some fundamental potentials and deficiencies, thus making acoustic ear recognition ever slightly more familiar in the jargon of biometrics.

References

1. A.H.M. Akkermans, D.W.E. Schobben, T.A.M. Kevenaar, Acoustic Ear Recognition for Person Identification, in *IEEE Workshop on Automatic Identification Advanced Technologies,* vol. 00 (2005), pp. 219–223
2. A.J. Hoogstrate, H. van den Heuvel, and E. Huyben, Ear identification based on surveillance camera images, Science and Justice **41**, 167–172 (2001)
3. M. Ali, M.Y. Javed, A. Basit, Ear recognition using wavelets, in *Proceedings of Image and Vision Computing*, New Zealand (2007), pp. 83–86
4. B.B. Ballachanda, *The Human Ear Canal* (Singular Publishing Group, Inc., San Diego, 1995)
5. N.A. Bel Moreno, J.F. Vlez, Use outer ear images for personal identification in security applications, in *Proceedings of the 33rd Annual 1999 International Carnahan Conference*, Madrid (1999), pp. 469–476
6. A. Bertillon, *La photographie judiciaire, avec un appendice sur la classification et l'identification anthropometriques* (Gauthier-Villars, Paris, 1890)

7. J. Bohn, V. Coroamă, M. Langheinrich, F. Mattern, M. Rohs, Social, economic, and ethical implications of ambient intelligence and ubiquitous computing, in *Ambient Intelligence* (Springer, Berlin/Heidelberg, 2005), pp. 5–29
8. R.M. Bolle, J.H. Connell, S. Pankanti, N.K. Ratha, A.W. Senior, *Guide to Biometrics* (Springer, New York, 2004)
9. C. Boyer, U. Merzbach, *History of Mathematics* (Wiley, New York, 1991)
10. M. Burge, W. Burger, Ear biometrics in computer vision, in *Proceedings of the 15th International Conference of Pattern Recognition* (2000), pp. 822–826
11. K. Chang, K. Bowyer, S. Sarkar, B. Victor, Comparison and combination of ear and face images in appearance-based biometrics. IEEE Trans. Pattern Anal. Mach. Intell. **25**, 1160–1165 (2003)
12. H. Chen, B. Bhanu, Human ear recognition in 3d. IEEE Trans. Pattern Anal. Mach. Intell. **29**(4), 718–737 (2007)
13. M. Geronazzo, S. Spagnol, F. Avanzini, Estimation and modeling of pinna-related transfer functions, in *Proceedings of the 13th International Conference on Digital Audio Effects (DAFx-10)* (2010)
14. J.A. Heathcote. Why do old men have big ears. **311**(7021), BMJ (British Medical Journal), p. 1668 (1995)
15. A.J. Hoogstrate, H. van den Heuvel, E. Huyben, Ear identification based on surveillance camera's images. Sci. Justice **41**, 167–172 (2001)
16. D.J. Hurley, M.S. Nixon, J.N. Carter, Force field feature extraction for ear biometrics. Comput. Vis. Image Underst. **98**, 491–512 (2005)
17. A. Iannareli, *Ear Identification*. Forensic Identification Series (Paramount Publishing Company, Fremont, 1989)
18. A. Jain, R. Bolle, S. Pankanti, *Biometrics: Personal Identification in Networked Society*. The Springer International Series in Engineering and Computer Science (Springer, New York, 1999). https://books.google.no/books?id=XPC9ucFbddsC
19. A.K. Jain, A. Ross, S. Prabhakar, An introduction to biometric recognition. IEEE Trans. Circuits Syst. Video Technol. **14**(1), 4–20 (2004). doi:10.1109/TCSVT.2003.818349. http://dx.doi.org/10.1109/TCSVT.2003.818349
20. C.B. Jensen, M.F. Srensen, H. Møller, D. Hammershøi, Transfer characteristics of headphones: measurements on 40 human subjects. J. Audio Eng. Soc. **40** (1992)
21. C.B. Jensen, M.F. Srensen, H. Møller, D. Hammershøi, Transfer characteristics of headphones measured on human ears. J. Acoust. Soc. Am. **43** (1995)
22. B.F.G. Katz, Acoustic absorption measurement of human hair and skin within the audible frequency range. J. Acoust. Soc. Am. **108** (2000)
23. A. Kulkarni, H.S. Colburn, Variability in the characterization of the headphone transfer-function. J. Acoust. Soc. Am. **107** (2000)
24. J.-P. Linnartz, P. Tuyls, New shielding functions to enhance privacy and prevent misuse of biometric templates, in *Proceedings of 4th International Conference on Audio- and Video-Based Biometric Person Authentication* (2003), pp. 393–402
25. Z. Mu, L. Yuan, Z. Xu, Shape and structural feature based ear recognition, in *Advances in Biometric Person Authentication*, Guangzhou (2004), pp. 663–670
26. US Patent. 5787187. (1998)
27. C.J. Plack, *The Sense of Hearing*, 1st edn. (Psychology Press, Hove, 2005)
28. M. Rahman, R. Islam, N.I. Bhuiyan, B. Ahmed, A. Islam, *Person Identification Using Ear Biometrics* (Computer Science and Engineering Discipline, Khulna University, Khulna, 2007)
29. E.A.G. Shaw, Ear canal pressure generated by circumaural and supra-aural earphones. J. Acoust. Soc. Am. **39** (1966)
30. E.A.G. Shaw, Transformation of sound pressure level from the free field to the eardrum in the horizontal plane. J. Acoust. Soc. Am. **56**, 1848–1861 (1974)
31. E.A.G. Shaw, The external ear: new knowledge, in *Earmolds and Associated Problems: Proceeding of the Seventh Danavox Symposium,* ed. by S.C. Dalsgaard. Scandinavian Audiology, Suppl. 5 (1975), pp. 24–50

32. E.A.G. Shaw, The acoustics of the external ear, in *Acoustical Factors Affecting Hearing Aid Performance*, ed. by G.A. Studebaker, I. Hochberg (University Park Press, Baltimore, 1980)
33. M.R. Stinson, B.W. Lawton, Specification of the geometry of the human ear canal for the prediction of sound-pressure level distribution. J. Acoust. Soc. Am. **85** (1989)
34. Y. Sun, M.G. Genton, J. Comput. Graph. Stat. **20**, 313–334 (2011)
35. R. Teranishi, E.A.G. Shaw, External-ear acoustic models with simple geometry. J. Acoust. Soc. Am. **44**, 257–263 (1968)
36. P. Tuyls, J. Goseling, Capacity and examples of template protection in biometric authentication systems, in *Biometric Authentication Workshop* (2004), pp. 158–170
37. P. Tuyls, E. Verbitskiy, T. Ignatenko, D. Schobben, T. Akkermans, Privacy protected biometric templates: acoustic ear identification. Proc. SPIE **5404**, 176–182 (2004)
38. P. Tuyls, A. Akkermans, T. Kevenaar, G. Schrijen, A. Bazen, R. Veldhuis, Practical biometric template protection system based on reliable components, in *Proceedings of the 5th International Conference on Audio and Video-Based Biometric Person Authentication* (2005) pp. 436–446
39. F.E. Uzoka, T. Ndzinge, An investigation of factors affecting biometric technology adoption in a developing country context. Int. J. Biom. **1**(3), 307–328 (2009). doi:10.1504/IJBM.2009.024276. http://dx.doi.org/10.1504/IJBM.2009.024276
40. B. Victor, K.W. Bowyer, S. Sarkar, An evaluation of face and ear biometrics, in *Proc. ICPR* (2002), pp. 429–432
41. P. Yan, K.W. Bowyer, Biometric recognition using three-dimensional ear shape. IEEE Trans. Pattern Anal. Mach. Intell. (2006)
42. L. Yan, Z.C. Mu, Ear recognition based on 2D images, in *Biometrics: Theory, Applications, and Systems* (2007)

Chapter 5
Eye Blinking EOG Signals as Biometrics

Sherif N. Abbas and M. Abo-Zahhad

5.1 Introduction

In the last decade, biometric authentication has gained an increased attention in privacy and security related issues. This is due to the uniqueness of these traits and also due to the development of applications that needs high security like e-banking, remote access control, tele-medicine, etc. However, due to the advanced hacking and spoofing techniques, conventional biometric traits like finger-print and facial characteristics have been shown that they can be easily forged. In [1], Matsumoto et al. showed that a commercial finger-print system can be easily breached using fake finger-prints, where a true acceptance rate ranging from 65 % up to 100 % was achieved using gummy fingers tested over 11 different commercial finger-print systems. Moreover, in [2], it was shown that face recognition technique can be easily spoofed using printed face models. This leads the scientific community to investigate the feasibility of other physiological or behavioral traits for the purpose of biometric authentication like Electro-Encephalo-Gram (EEG) (brainwaves) signals [3], Electro-Cardio-Gram (ECG), and Phono-Cardio-Gram (PCG) signals [4]. One of the bio-electrical signals that has not been much investigated in previous works is the Electro-Oculo-Gram (EOG) signals.

EOG signal is the electrical recording of the eyeball and eyelid movements by means of electrode placed near the eye. EOG signals have some advantages over conventional biometric traits such as using finger-print and facial characteristics. These signals cannot be easily forged nor be captured at a distance like finger-print and face. Moreover, they are one dimensional, low frequency signals that can be easily processed. Regarding uniqueness, Jesper Rønager (an expert neurologist)

S.N. Abbas (✉) • M. Abo-Zahhad
Department of Electrical and Electronics Engineering, Faculty of Engineering, Assiut University, Assiut, Egypt
e-mail: sherif.siha13@eng.au.edu.eg; zahhad@yahoo.com

R. Jiang et al. (eds.), *Biometric Security and Privacy*, Signal Processing for Security Technologies, DOI 10.1007/978-3-319-47301-7_5

stated that eye blinking patterns are distinct because when the eye blinks, many electrical signals flows through the neurons of the brain that can be collected easily by electrodes placed near the eyes [5]. Since every person has a unique complicated neural network that consists of billions of neurons, eye blinking signals are unique. All these reasons motivate the authors to investigate the distinctiveness of the eye blinking signals and the feasibility of using these signals as biometric traits. The next paragraph presents a brief review of using EOG signals as biometric traits.

EOG signals have been employed previously in biometric authentication systems, however, only eye movement EOG signals have been investigated for human identification task. EOG signals were recorded from users while following a moving target with their eyes that is displayed on a screen in front of the user. This produces rapid vertical or horizontal eye movements known as saccades [6]. In [7], saccades from 30 subjects were collected and features based on amplitude, accuracy, latency, and maximum velocity were computed. Using different classifiers and different verification protocols, correct verification results in the range 96–100 % were achieved. In [8], a similar approach was adopted for biometric verification based on eye movements. The system was build using a database of 40 subjects (19 healthy subjects and 21 otoneurological patients) recorded with electro-oculography. Achieved verification results were in range 86–97 %. Although eye movement EOG signals showed high performance as biometrics, however, recording eye movement EOG signals requires a lot of effort to be done by users following the moving target which makes this technique impractical for biometric authentication. In this chapter, a new biometric authentication technique is investigated using EOG signals that are recorded while performing eye blinking tasks.

The remainder of this chapter is organized as follows. Section 5.2 provides a detailed description about the dipole model of the eye and the origin of the eye blinking waveform. Section 5.3 describes the adopted algorithms for the main components of the proposed system; pre-processing, feature extraction, feature selection, and classification. The achieved results are presented in Sect. 5.4 for identification and verification tasks. Finally, Sect. 5.5 summarizes the main conclusions and future directions.

5.2 Origin of Eye Blinking EOG Signals

As mentioned earlier, electro-oculography is the electrical recording of the potential generated due to eyeball or eyelid movements. EOG can be recorded by skin electrodes placed around the eye. The amplitude of the EOG signals ranges between 10 and 200 μV with a frequency falling in the range 0.5–15 Hz. The eyes are electrically charged; positive at the cornea and negative at the retina [9]. When the eyes or eyelids move, the potential field arising from the charge changes generates a large amplitude electric signal, which is detectable by any electrodes near the eye [10].

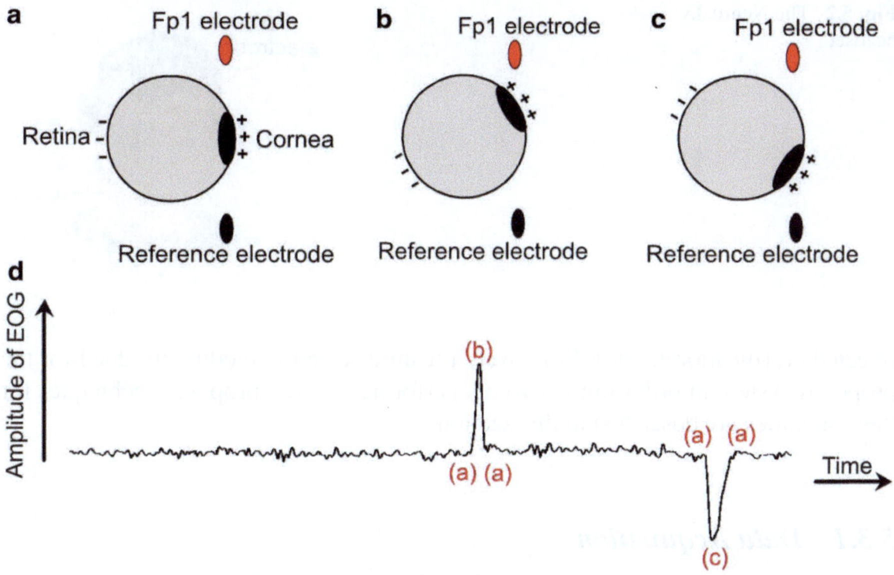

Fig. 5.1 The dipole model of eyeball with waveform of EOG signal when the cornea is in the center (**a**), rotating upwards (**b**), and downwards (**c**), EOG signal (**d**)

As shown in Fig. 5.1, when the eyeball rotates upwards, the positive pole (cornea) becomes closer to Fp1 electrode and produces a positive deflection. Similarly, when the eyeball rotates downwards, the positive pole (cornea) becomes far away from Fp1 (closer to the reference electrode) producing a negative deflection. This is similar to what happens when the eye blinks. When the eyelid closes, the cornea becomes closer to Fp1 and a positive pulse is produced. But, when the eyelid opens, the cornea rotates away from Fp1 and a negative pulse is produced. The eye blinking duration usually ranges from 300 to 400 ms. In the next section, the proposed system for human authentication using eye blinking EOG signals is described in details.

5.3 Proposed Approach for Eye Blinking EOG Biometric System

Any biometric authentication system consists of four basic modules [11]: (1) data acquisition device: which is the sensor used to acquire the biometric data, (2) pre-processing module: where the acquired data is processed and made ready for feature extraction, (3) feature extraction module: where the features discriminating between the individuals are extracted from the pre-processed data, and classification module: where features extracted are compared against the stored template, then the user's identity is established (identification mode) or the claimed identity is accepted or

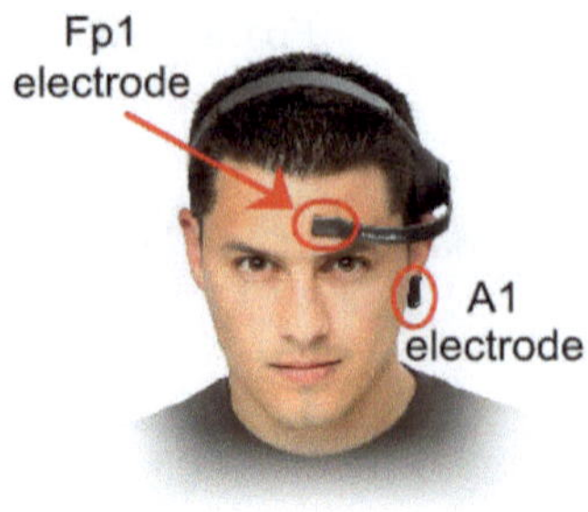

Fig. 5.2 The Neurosky headset

rejected (verification mode). Moreover, a feature selection module is added for the proposed system in order to improve its performance. The proposed techniques for these modules are described in this section.

5.3.1 Data Acquisition

All EOG signals used for the proposed system were recorded using Neurosky Mindwave headset as shown in Fig. 5.2. The headset consists of an ear clip and a sensor arm. This headset is actually used for recording EEG signals; however, it can be used to measure EOG signals as the arm sensor is resting on the forehead above the left eye (Fp1 position). The reference electrode is on the ear clip (A1 position). The sensor of Neurosky headset is made of dry electrode which does not require any skin preparation or conductive pastes. Also, the headset is wireless which makes it suitable for practical implementation of biometric authentication systems. The sampling rate of the device is 512 Hz.

The raw signal was collected from 40 users. Only 11 users out of 40 performed two session recordings with different time separation between the two sessions, however, the 29 remaining users performed only one session (1 day) recordings. More information about the recorded database is provided in Table 5.1. In one session, 6–10 trials were recorded with duration of 20 s each. Each user was asked to make 8–12 natural eye blinks in each trial. The users were asked not to do any eye movements as possible. Figure 5.3 shows the recorded eye blinking signal using Neurosky headset. MATLAB software was used for recording the raw signal from Neurosky headset and further processing. The ThinkGear Communication Driver (TGCD) [12] was used to connect Neurosky headset with Matlab through Bluetooth connection. The relation between the recorded raw values, V_{raw}, and the actual voltage, V_{actual}, is given by the following equation:

$$V_{\text{actual}} = \frac{1.8 V_{\text{raw}}}{4096 \times 2000} \tag{5.1}$$

where 1.8 is the reference voltage, 4096 is the maximum digital range, and 2000 is the amplifier gain of the Neurosky headset.

Table 5.1 Description of the recorded database (S1: first session and S2: second session)

User ID	Age	Number of sessions	Number of trials per sessions	Duration between sessions
1	24	1	7	–
2	34	1	9	–
3	74	1	9	–
4	40	1	8	–
5	23	1	9	–
6	26	2	S1: 7	16 months
			S2: 6	
7	30	1	7	–
8	23	1	10	–
9	23	1	8	–
10	27	1	8	–
11	26	1	7	–
12	28	1	10	–
13	28	1	9	–
14	31	1	7	–
15	27	1	8	–
16	29	1	10	–
17	24	1	10	–
18	27	2	S1: 7	45 days
			S2: 7	
19	28	2	S1: 8	–
			S2: 7	–
20	27	1	7	–
21	26	2	S1: 8	16 months
			S2: 10	
22	26	1	9	–
23	30	2	S1: 7	12 days
			S2: 7	
24	30	1	8	–
25	28	2	S1: 9	9 months
			S2:7	
26	24	1	10	–
27	32	2	S1: 10	13 months
			S2: 8	
28	22	2	S1: 6	18 days
			S2: 7	
29	23	1	8	–
30	24	2	S1: 7	2 months
			S2: 7	

(continued)

Table 5.1 (continued)

User ID	Age	Number of sessions	Number of trials per sessions	Duration between sessions
31	23	1	7	–
32	22	1	8	–
33	22	1	6	–
34	23	1	7	–
35	22	1	9	–
36	23	2	S1: 6	18 days
			S2: 7	
37	22	1	7	–
38	22	1	6	–
39	25	1	7	–
40	26	2	S1: 6	16 days
			S2: 7	

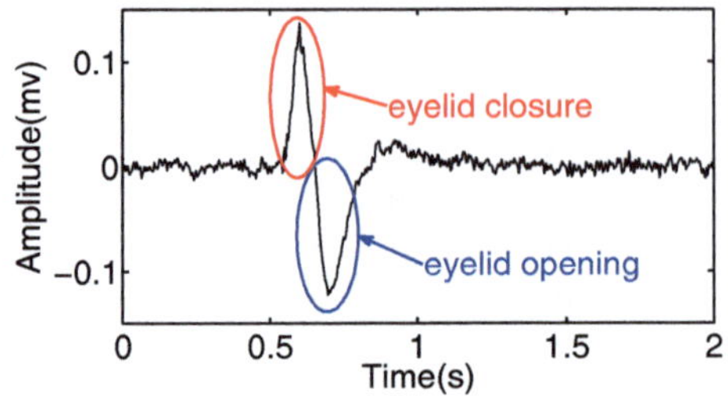

Fig. 5.3 The recorded eye blinking signal

5.3.2 *Pre-processing*

As mentioned in Sect. 5.3.1, the Neurosky headset is used to measure EEG signals. Therefore, the main purpose of the pre-processing stage is to isolate EOG signals from brainwaves. Also, pre-processing stage involves eye blinking waveforms extraction.

For EOG isolation, the recorded data is decomposed up to the third level using discrete wavelet decomposition. The mother wavelet used for decomposition is the Daubechies wavelet of the second order (db2) because it resembles the eye blinking waveform as shown in Fig. 5.4. Figure 5.5 shows the decomposed signals up to the third level where the approximation coefficients at the third (a_3) represent the isolated EOG signal.

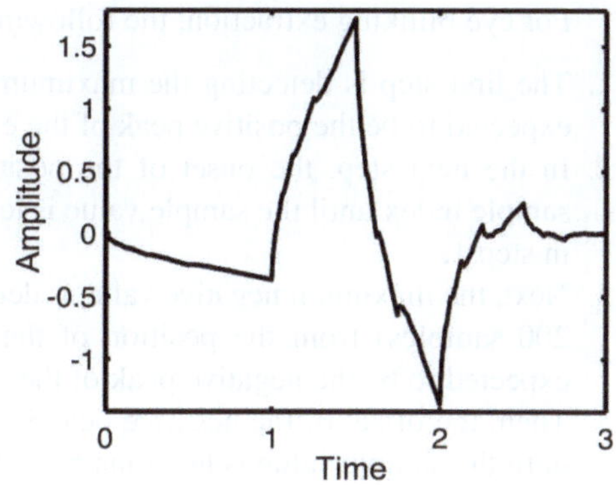

Fig. 5.4 The second order Daubechies (db2) wavelet function

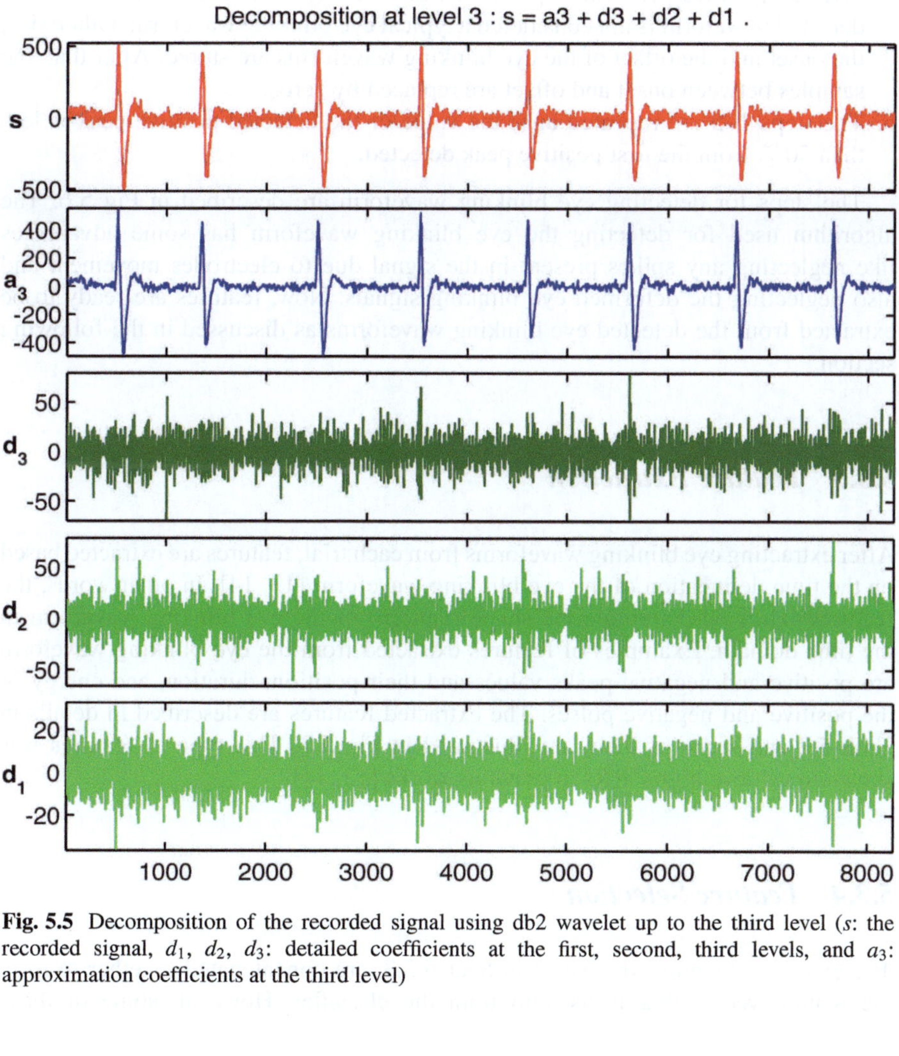

Fig. 5.5 Decomposition of the recorded signal using db2 wavelet up to the third level (s: the recorded signal, d_1, d_2, d_3: detailed coefficients at the first, second, third levels, and a_3: approximation coefficients at the third level)

For eye blinking extraction, the following procedure is followed:

1. The first step is detecting the maximum positive value in the EOG signal. It is expected to be the positive peak of the eye blinking waveform.
2. In the next step, the onset of the positive peak is detected by decreasing the sample index until the sample value is less than 5 % of the detected positive peak in step 1.
3. Next, the maximum negative value is detected within a duration of 400 ms (about 200 samples) from the position of the positive peak detected in step 1. It is expected to be the negative peak of the eye blinking waveform.
4. Then, the offset of the negative peak is detected by increasing the sample index until the sample value is less than 5 % of the detected negative peak in step 3.
5. If the value of the detected negative peak (in step 3) is less than 35 % of the detected positive peak (in step 1), then this waveform is discarded because the detected waveform is not considered a typical eye blinking waveform. Otherwise, the onset and the offset of the eye blinking waveforms are stored. After that, the samples between onset and offset are replaced by zeros.
6. The steps 1–5 are repeated until the value of the detected positive peak is less than 50 % from the first positive peak detected.

The steps for detecting eye blinking waveform are described in Fig. 5.6. The algorithm used for detecting the eye blinking waveform has some advantages like neglecting any spikes present in the signal due to electrodes movement and also neglecting the deformed eye blinking signals. Now, features are ready to be extracted from the detected eye blinking waveforms as discussed in the following section.

5.3.3 Feature Extraction

After extracting eye blinking waveforms from each trial, features are extracted based on the time delineation of the eye blinking waveform [13, 14]. In other words, the features extracted determine the shape (pattern) of the eye blinking waveform in the time domain. Examples of features extracted from the eye blinking waveform are positive and negative peaks values and their position, duration, and energy of the positive and negative pulses. The extracted features are described in details in Table 5.2 and some of them are illustrated in Fig. 5.7. This type of features was shown to achieve high CRRs in previous works [13, 14].

5.3.4 Feature Selection

The extracted features, described in Sect. 5.3.3, are concatenated together to form the feature vector that is used to train the classifier. However, some of these

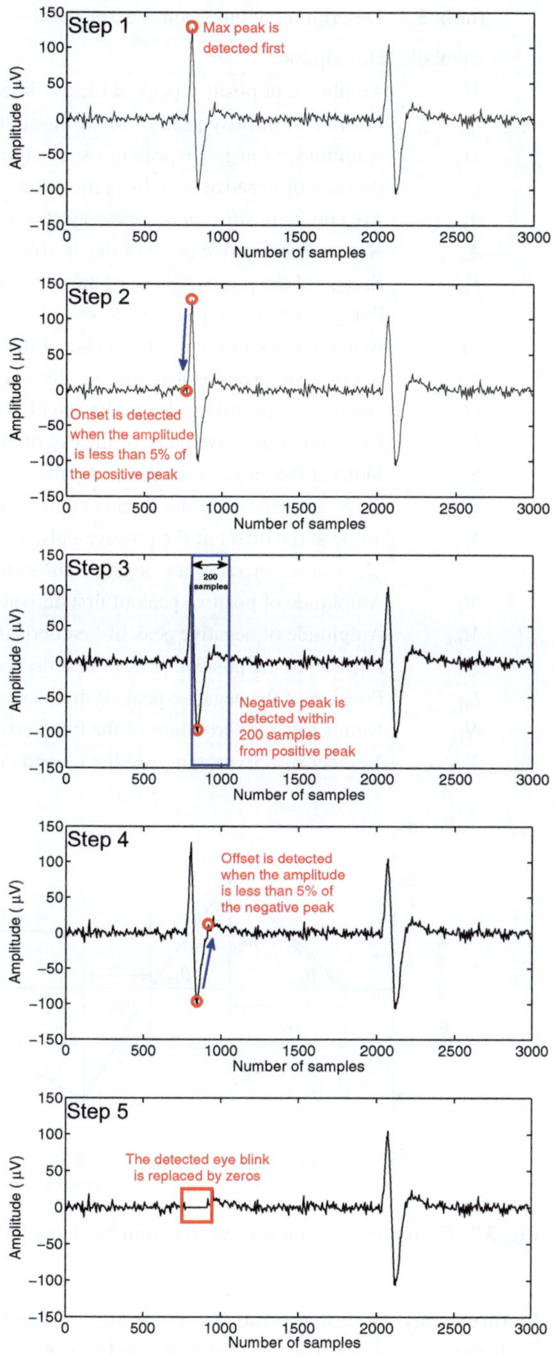

Fig. 5.6 Illustration of the steps for eye blinking waveform detection

Table 5.2 Description of the features extracted from the eye blinking waveform

Symbol	Description
M_p	Amplitude of positive peak of the eye blink
I_p	Position of positive peak from the onset of positive pulse
M_n	Amplitude of negative peak of the eye blink
I_n	Position of negative peak from the onset of negative pulse
A_p	Area under positive pulse of the eye blink
A_n	Area under negative pulse of the eye blink
E_p	Energy of the positive pulse of the eye blink
E_n	Energy of negative pulse of the eye blink
Av_p	Average value of the positive pulse of the eye blink
Av_n	Average value of negative pulse of the eye blink
D_p	Duration of positive pulse of the eye blink
D_n	Duration of negative pulse of the eye blink
S_{op}	Slope at the onset of the positive pulse ($\tan(\theta_{\mathrm{op}})$)
S_{on}	Slope at the onset of the negative pulse ($\tan(\theta_{\mathrm{on}})$)
S_{fp}	Slope at the offset of the positive pulse ($\tan(\theta_{\mathrm{fp}})$)
S_{fn}	Slope at the offset of the negative pulse ($\tan(\theta_{\mathrm{fn}})$)
M_{p1}	Amplitude of positive peak of first derivative of the eye blink signal
M_{n1}	Amplitude of negative peak of first derivative of the eye blink signal
I_{p1}	Position of the positive peak of first derivative of the eye blink signal
I_{n1}	Position of the negative peak of first derivative of the eye blink signal
N_{z1}	Number of zero crossings of the first derivative of the eye blink signal
N_{z2}	Number of zero crossings of the second derivative of the eye blink signal

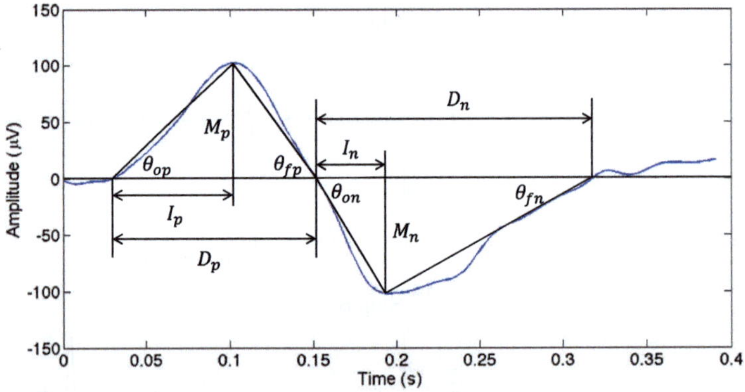

Fig. 5.7 Examples of extracted features from eye blinking waveform

features may be not unique for each subject and so it will degrade the classifier performance. Therefore, a Feature Selection (FS) technique is added to select a subset of the available features to minimize the classifier error (or maximize the classifier accuracy) and remove redundant or irrelevant features. Assuming an original feature vector of length, N_f, which is equal to 22 in our case, the objective of feature selection is to identify the most informative subset of M_f features ($M_f < N_f$).

The most important factors that should be considered for a feature selection technique are the accuracy and the search strategy. In [15], a new feature selection technique is proposed that was based on Differential Evolution (DE) with a new statistical measure to aid in the selection of the most relevant features. DE is a population-based method for minimizing a cost function. The first step in DE is generating a population of members (random feature vectors) of dimension to be optimized. The next step is to generate mutant population based on a weighted difference of the original population. In order to overcome real number problem and to apply DE for feature selection problem, a roulette wheel weighting scheme is utilized where a cost weighting is implemented in which the probabilities of each feature are calculated from the distribution factors associated with it. More details about Differential Evolution Feature Selection (DEFS) can be found in [15–17]. DEFS technique showed better performance in accuracy than other familiar FS techniques as stated in [16]. Therefore, the DEFS MATLAB program, available in [18], is employed for the proposed biometric authentication system.

5.3.5 Classification

The classifier adopted for the proposed system is the Discriminant Analysis (DA) classifier. DA assumes that the features extracted from every user, s, in the database have a multivariate Gaussian distribution as follows [19]:

$$P_s(x_f) = \frac{1}{(2\pi)^{\frac{N_f}{2}}\,|\Sigma_s|} e^{-\frac{1}{2}(x_f-\mu_s)^T \Sigma_s^{-1}(x_f-\mu_s)} \tag{5.2}$$

where N_f is the dimension of the testing feature vector, x_f. μ_s and Σ_s are the mean and the covariance of the feature vectors of the user s. The mean and covariance for each user are calculated using the following criteria.

The training feature vectors, $t_s^i, i = 1, 2, \ldots, n_{ts}$ (n_{ts} is the total number of feature vectors extracted for user s), extracted from every eye blinking pattern are concatenated together to form the training feature matrix, T_s, where T_s is the training feature matrix for the user s and $T_s \in R^{N_f \times n_{ts}}$. The mean is then calculated using the following equation:

$$\mu_s = \frac{1}{n_{ts}} \sum_{i=1}^{n_{ts}} t_s^i, \quad \mu_s \in R^{N_f \times 1} \tag{5.3}$$

Then, the covariance matrix is estimated according to the following equation:

$$\Sigma_s = \frac{1}{n_{ts}} (T_s - \mu_s)(T_s - \mu_s)^T, \quad \Sigma_s \in R^{N_f \times N_f} \tag{5.4}$$

The classifier decision can be carried using two decision rules as discussed in the following sections.

5.3.5.1 Linear Decision Rule

Linear Discriminant Analysis (LDA) assumes that all the users (classes) have the same covariance Σ. The classifier decision is performed using the optimum Bayes rule which maximizes the posterior probability equation or its logarithm which is given by

$$D_{s(\mathrm{LDA})}(x_f) = -\frac{1}{2}(x_f - \mu_s)^T \Sigma^{-1}(x_f - \mu_s) + \log(f_s), \quad s = 1, 2, 3, \ldots, S \tag{5.5}$$

where the covariance matrix, Σ, is calculated using Eq. (5.4) after concatenating all the training feature matrices from all users. f_s is the prior probability for the user s which is assumed to be uniform and S is the total number of users registered in the system. The unknown feature vector is assigned to the user s if it has the highest posterior probability (i.e., max $D_{s(\mathrm{LDA})}, s = 1, 2, 3, \ldots, S$).

5.3.5.2 Mahalanobis Decision Rule

The decision of the classifier in this case is estimated by calculating the Mahalanobis distance between the unknown feature vector x_f and the mean and the variance for each user, s, as follows:

$$D_{s(\mathrm{Mahal})}(x_f) = \sqrt{(x_f - \mu_s)^T \Sigma_s^{-1}(x_f - \mu_s)}, \quad s = 1, 2, 3, \ldots, S \tag{5.6}$$

The unknown feature vector is assigned to the user s if it has the minimum Mahalanobis distance (i.e., min $D_{s(\mathrm{Mahal})}, s = 1, 2, 3, \ldots, S$). As a summary, Fig. 5.8 shows a brief description of the proposed system.

5.4 Experimental Setup and Results

In general, the performance of the proposed system is evaluated in identification and verification modes using K-fold cross validation. For the proposed system, only six trials, with the largest number of eye blinking waveforms, were selected for each user from the whole set of trials. Then, the proposed system is tested under two protocols: one session test (S1 protocol) and two session test (S1–S2 protocol).

In S1 protocol, five trials are selected for the training from S1 session and the remaining trial from the same session is used for testing. This is repeated six times, where every time a different trial is chosen for testing and the remaining five trials are used for training. In S1–S2 protocol, the six trials in S1 session are used for training and one trial is selected from S2 session for testing. Again, this is repeated six times, where every time a different trial is chosen from S2 session

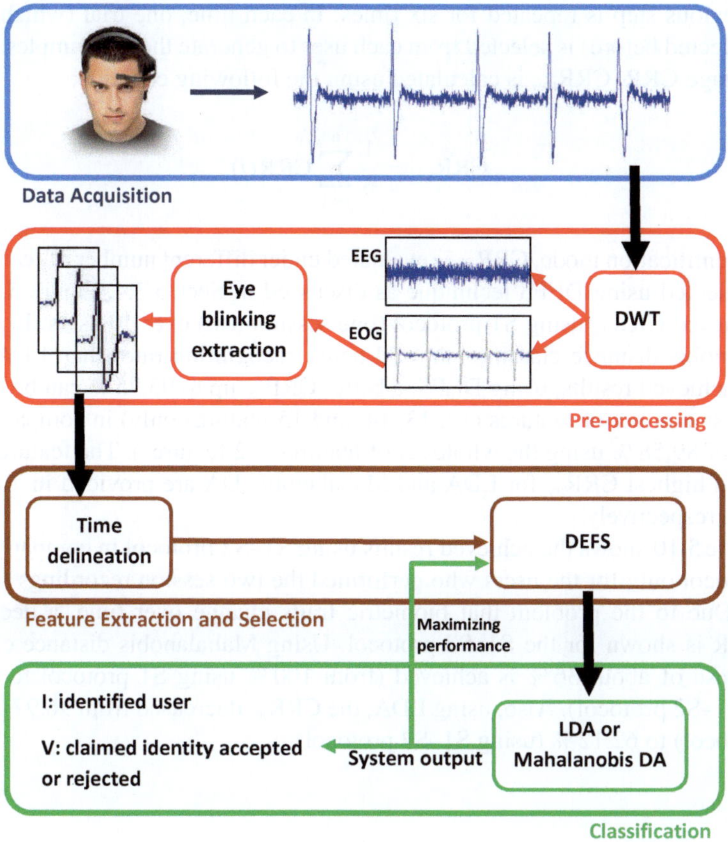

Fig. 5.8 Block diagram of the proposed eye blinking EOG-based biometric authentication system

for testing. Finally, the average CRR (in identification mode) or the average error rates (in verification mode) are calculated. This will be discussed in details in the following two sections.

5.4.1 Identification Mode

For evaluating the proposed system in identification mode, one trial is selected from each user to generate the test samples. Then, the proposed system generates the identity of each testing sample for each user according to Eq. (5.5) or Eq. (5.6). The CRR for this experiment is defined as follows:

$$\text{CRR}(j) = \frac{\text{Number of correctly identified users in step } j}{\text{Total number of users}} \tag{5.7}$$

The previous step is repeated for six times. In each time, one trial (which has not been selected before) is selected from each user to generate the test samples. Finally, the average CRR, CRR_{av}, is calculated using the following equation:

$$\text{CRR}_{\text{av}} = \frac{1}{6}\sum_{j=1}^{6}\text{CRR}(j) \tag{5.8}$$

In identification mode, CRR_{av} is evaluated under different number of features that were selected using DEFS technique as discussed in Sect. 5.3.4. Figure 5.9 shows the achieved CRR_{av} using S1 protocol (one session test) over 40 users. In this test, Mahalanobis distance classifier showed better recognition rates than LDA. Based on the achieved results, using DEFS, a better CRR_{av} up to 93.75 % can be obtained using less number of features (12, 13, 14, and 15 features only) in comparison to a CRR_{av} of 89.58 % using the whole set of features (22 features). The feature subsets achieved highest CRR_{av} for LDA and Mahalanobis DA are provided in Tables 5.3 and 5.4, respectively.

Figure 5.10 shows the achieved results using S1–S2 protocol in comparison with S1 protocol only for the users who performed the two session recordings (11 users only). Due to the problem that biometric traits change over time, a decrease in the CRR is shown for the S1–S2 protocol. Using Mahalanobis distance classifier, a decrease of about 36 % is achieved (from 100 % using S1 protocol to 63.64 % using S1–S2 protocol). Also, using LDA, the CRR_{av} decreased from 96.97 % (using S1 protocol) to 62.12 % (using S1–S2 protocol).

5.4.2 *Verification Mode*

For evaluating the proposed system in verification mode, each user in the database tries to access the system with his true identity (genuine) and with a false identity (the remaining identities in the database). For the claimed identity, I, provided by each user, s, the database is divided into two classes; the I-related database and the I-non-related database. Then, the claimed identity is accepted or rejected according to the following relation:

$$V_s = \begin{cases} 1, & \text{if } \frac{D_I(x_f)}{D_{\text{non}-\text{I}}(x_f)} \geq T_s \\ 0, & \text{if } \frac{D_I(x_f)}{D_{\text{non}-\text{I}}(x_f)} < T_s \end{cases} \tag{5.9}$$

where V_s is the output of the system in verification mode, "1" means that the system accepts the claimed identity, and "0" means that the system rejects the claimed identity. Also, $D_I(x_f)$ represents the probability that the claimed identity belongs to I-related class [estimated using Eq. (5.5)] or the Mahalanobis distance between the testing sample and the I-related class [estimated using Eq. (5.6)]. Similarly,

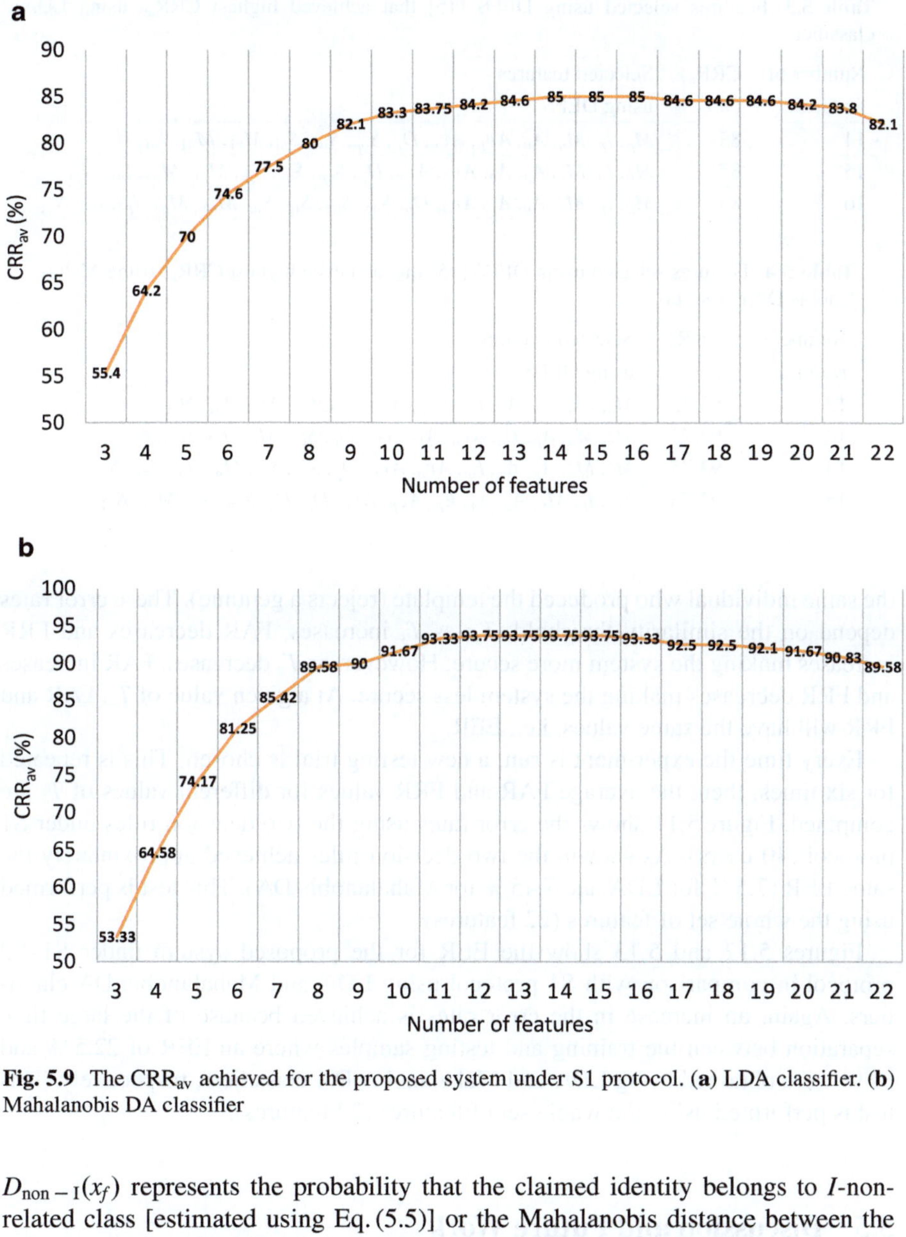

Fig. 5.9 The CRR_{av} achieved for the proposed system under S1 protocol. (**a**) LDA classifier. (**b**) Mahalanobis DA classifier

$D_{\text{non}-\text{I}}(x_f)$ represents the probability that the claimed identity belongs to I-non-related class [estimated using Eq. (5.5)] or the Mahalanobis distance between the testing sample and the I-non-related class [estimated using Eq. (5.6)].

The similarity threshold, T_s, defines the degree of security of the system. In this mode, two error rates are defined; False Acceptance Rate (FAR) and False Rejection Rate (FRR). FAR defines the number of times the system incorrectly matches a sample from one person to a template from another (accepts an imposter). FRR defines the number of times the system does not recognize a sample as coming from

Table 5.3 Features selected using DEFS [15] that achieved highest CRR_{av} using LDA classifier

Number of features	CRR_{av} (%)	Selected features using DEFS
14	85	$M_p, I_p, M_n, A_n, Av_p, Av_n, D_p, S_{op}, S_{on}, S_{fn}, M_{p1}, M_{n1}, I_{n1}, N_{z1}$
15	85	$M_p, I_p, M_n, A_p, A_n, Av_p, Av_n, D_p, S_{op}, S_{on}, S_{fn}, M_{p1}, M_{n1}, I_{n1}, N_{z1}$
16	85	$M_p, I_p, M_n, A_p, A_n, Av_n, D_p, S_{op}, S_{on}, S_{fp}, S_{fn}, M_{p1}, M_{n1}, I_{p1}, I_{n1}, N_{z1}$

Table 5.4 Features selected using DEFS [15] that achieved highest CRR_{av} using Mahalanobis DA classifier

Number of features	CRR_{av} (%)	Selected features using DEFS
12	93.75	$M_p, M_n, A_p, A_n, E_p, Av_p, Av_n, S_{on}, S_{fp}, M_{p1}, I_{p1}, N_{z1}$
13	93.75	$M_p, M_n, A_p, E_p, Av_p, Av_n, D_p, D_n, S_{on}, M_{p1}, I_{n1}, N_{z1}, N_{z2}$
14	93.75	$M_p, M_n, A_p, A_n, E_p, Av_p, Av_n, D_p, S_{on}, S_{fp}, M_{p1}, I_{p1}, I_{n1}, N_{z1}$
15	93.75	$M_p, I_p, M_n, A_p, A_n, E_p, Av_p, Av_n, D_p, D_n, S_{on}, S_{fn}, M_{p1}, N_{z1}, N_{z2}$

the same individual who produced the template (rejects a genuine). These error rates depend on the similarity threshold, T_s; as T_s increases, FAR decreases and FRR increases making the system more secure. However, as T_s decreases, FAR increases and FRR decreases making the system less secure. At a given value of T_s, FAR and FRR will have the same values, i.e., EER.

Every time the experiment is run, a new testing trial is chosen. This is repeated for six times, then, the average FAR and FRR values for different values of T_s are computed. Figure 5.11 shows the error rates using the two decision rules under S1 protocol (40 users). As shown, the two decision rules achieved approximately the same EER (7.5 % for LDA and 7.45 % for Mahalanobis DA). This test is performed using the whole set of features (22 features).

Figures 5.12 and 5.13 show the EER for the proposed system under S1–S2 protocol in comparison with S1 protocol using LDA and Mahalanobis DA classifiers. Again, an increase in the error rates is achieved because of the large time separation between the training and testing samples where an EER of 22.5 % and 32 % were obtained using LDA and Mahalanobis DA classifiers, respectively. This test is performed using the whole set of features (22 features).

5.5 Discussion and Future Work

In this chapter, the authors presented a new technique for human authentication using eye blinking waveforms extracted from EOG signals. The proposed system used time delineation to extract important features from the eye blinking waveform. Over a population of 40 users, the proposed system successfully detected the identity

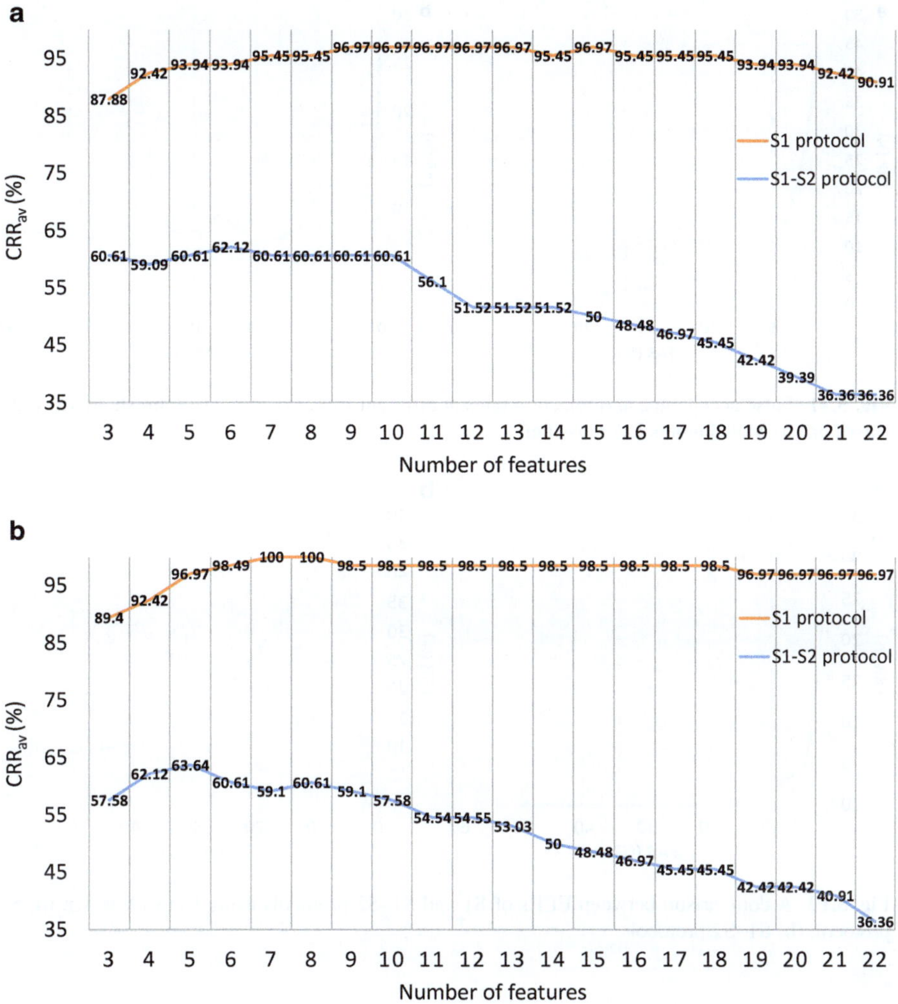

Fig. 5.10 A comparison between CRR_{av} achieved for the proposed system under S1 and S1–S2 protocols. (**a**) LDA classifier. (**b**) Mahalanobis DA classifier

of the users with accuracy up to 93.72 % in identification mode (S1 protocol). In verification mode, the proposed system rejects the true identity and accepts false identity with an EER of 7.45 % (S1 protocol). Also, the issue of permanence was tested for the proposed system using S1–S2 protocol. Although the achieved results for the proposed system look promising, there are some issues that need to be addressed in the future as stated in the following points:

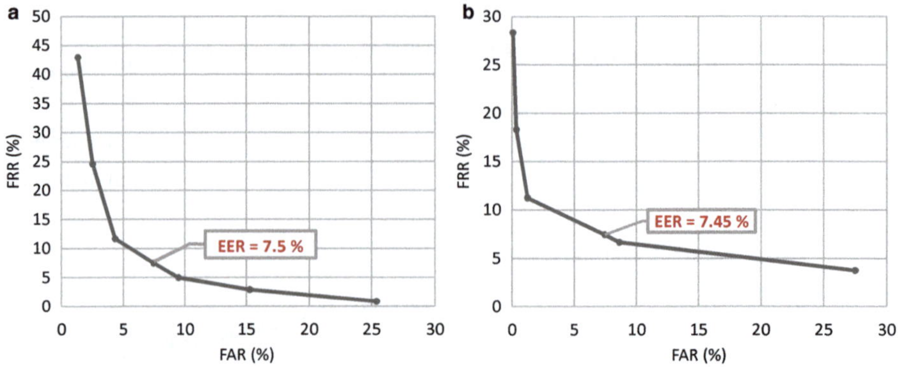

Fig. 5.11 False acceptance and rejection rates at different values of similarity threshold. (**a**) LDA classifier. (**b**) Mahalanobis DA classifier

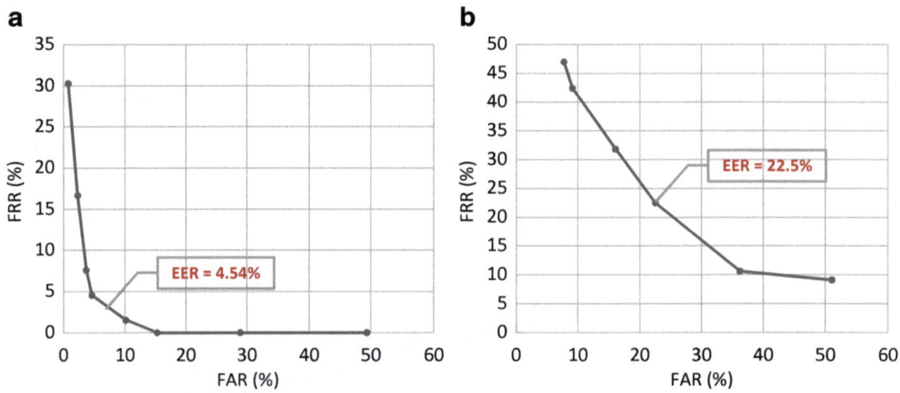

Fig. 5.12 A comparison between EERs of S1 and S1–S2 protocols using LDA classifier. (**a**) S1 protocol. (**b**) S1–S2 protocol

- Population (number of users): For building practical biometric system, the discrimination ability of the eye blinking waveform should be tested over large population, may be thousands of users.
- Permanence (stability of the biometric trait over time): As discussed in Sect. 5.4, using testing samples that were recorded after a large time separation from the training samples degrades the system's performance. This issue needs to addressed in the future. A possible solution for this issue is to use training samples recorded at different time intervals for the same user. This may enhance the system's performance.
- Factors affecting eye blinking: The degree of concentration or fatigue may affect the strength of the eye blinking waveform, hence, affecting the performance of the system. Therefore, these factors should be addressed in future work.

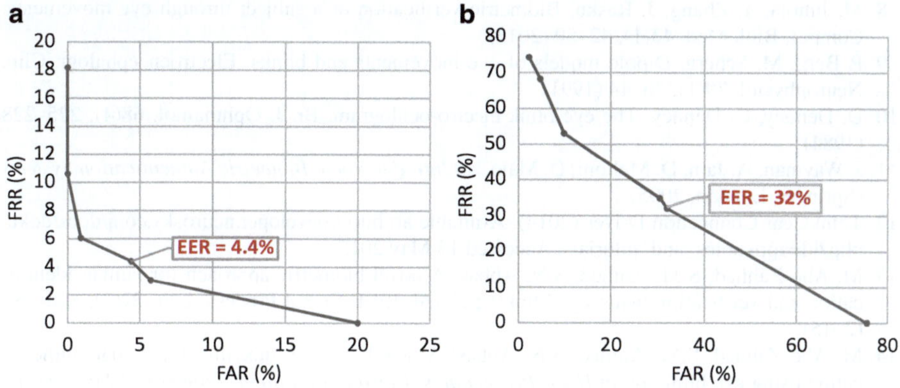

Fig. 5.13 A comparison between EERs of S1 and S1–S2 protocols using Mahalanobis DA classifier. (**a**) S1 protocol. (**b**) S1–S2 protocol

- Feature extraction technique: Different techniques for feature extraction may be tested and compared with the proposed one.
- Fusion with other traits: Eye blinking EOG biometric trait can be fused with other traits like EEG signals to build a multi-modal system to improve the performance of the EEG-based biometric authentication systems [20].

Although the achieved simulation results are promising, eye blinking EOG signals still need a lot of investigation to discuss the previous issues before considering these signals as biometric traits for identifying humans.

References

1. T. Matsumoto, H. Matsumoto, K. Yamada, S. Hoshino, Impact of artificial gummy fingers on fingerprint systems, in *SPIE Proceedings*, vol. 4677. Optical Security and Counterfeit Deterrence Techniques IV (2002), pp. 275–289
2. N. Erdogmus, S. Marcel, Spoofing face recognition with 3D masks. IEEE Trans. Inf. Forensics Secur. **9**(7), 1084–1097 (2014)
3. M. Abo-Zahhad, S.M. Ahmed, S.N. Abbas, State-of-the-art methods and future perspectives for personal recognition based on electroencephalogram signals. IET Biometrics **4**(3), 179–190 (2015)
4. M. Abo-Zahhad, S.M. Ahmed, S.N. Abbas, Biometric authentication based on PCG and ECG signals: present status and future directions. Signal Image Video Process. **8**(4), 739–751 (2014)
5. J. Klonovs, C.K. Petersen, Development of a mobile EEG-based feature extraction and classification system for biometric authentication. M.Sc. thesis, Aalborg University Copenhagen, 2012
6. U. Saeed, A survey of automatic person recognition using eye movements. Int. J. Pattern Recognit. Artif. Intell. **28**(8), 1456015 (2015)
7. Y. Zhang, J. Rasku, M. Juhola, Biometric verification of subjects using saccade eye movements. Int. J. Biometrics **4**(4), 317–337 (2012)

8. M. Juhola, Y. Zhang, J. Rasku, Biometric verification of a subject through eye movements. Comput. Biol. Med. **43**(1), 42–50 (2013)
9. P. Berg, M. Scherg, Dipole models of eye movements and blinks. Electroencephalogr. Clin. Neurophysiol. **79**(1), 36–44 (1991)
10. D. Denney, C. Denney, The eye blink electro-oculogram. Br. J. Ophthamol. **68**(4), 225–228 (1984)
11. J. Wayman, A. Jain, D. Maltoni, D. Maio, *An Introduction to Biometric Authentication Systems* (Springer, London, 2005)
12. ThinkGear Connection Driver (2014). Available at: http://developer.neurosky.com/docs/doku.php?id=app_notes_and_tutorials. Accessed 13 May 2015
13. M. Abo-Zahhad, S.M. Ahmed, S.N. Abbas, A novel biometric approach for human identification and verification using eye blinking signal. IEEE Signal Process. Lett. **22**(7), 876–880 (2015)
14. M. Abo-Zahhad, S.M. Ahmed, S.N. Abbas, A new biometric modality for human authentication using eye blinking, in *IEEE Proceedings. Cairo International Biomedical Engineering Conference (CIBEC)* (2014), pp. 174–177
15. R.N. Khushaba, A. Al-Ani, A. Al-Jumaily, Differential evolution based feature subset selection, in *IEEE Proceedings. International Conference Pattern Recognition (ICPR)* (2008), pp. 1–4
16. R.N. Khushaba, A. Al-Ani, A. Al-Jumaily, Feature subset selection using differential evolution and a statistical repair mechanism. Expert Syst. Appl. **38**(9), 11515–11526 (2011)
17. A. Al-Ani, A. Alsukker, R.N. Khushaba, Feature subset selection using differential evolution and a wheel based search strategy. Swarm Evol. Comput. **9**, 15–26 (2013)
18. MATLAB Program for Differential Evolution Based Feature Selection (DEFS) (2015). Available at http://www.mathworks.com/matlabcentral/fileexchange/30877. Accessed 13 May 2015
19. S. Theodoridis, K. Koutroumbas, *Pattern Recognition*, 4th edn. (Academic, New York, 2008)
20. M. Abo-Zahhad, S.M. Ahmed, S.N. Abbas, A new multi-level approach to EEG based human authentication using eye blinking. Pattern Recogn. Lett. (2015). doi:10.1016/j.patrec.2015.07.034

Chapter 6
Improved Model-Free Gait Recognition Based on Human Body Part

Imad Rida, Noor Al Maadeed, Gian Luca Marcialis, Ahmed Bouridane, Romain Herault, and Gilles Gasso

6.1 Introduction

In the past years with frequent terrorist attacks, a considerable number of surveillance cameras have been installed in public places, train stations, and airports and many research efforts have been devoted to build intelligent systems able to analyze the visual data in order to extract information about the behavior of humans in scenes. Ideal intelligent monitoring system should be able to automatically analyze the collected video data, detect the suspicious or endangering human behavior, and give out an early warning before the adverse event happens. A system which detects abnormal behavior should also be able to identify all the suspicious persons in the scene, and track them across the zones. Monitoring system requires not only to estimate the location and behavior, but also to obtain the identity information.

I. Rida (✉) • R. Herault • G. Gasso
LITIS EA 4108 - INSA de Rouen, Avenue de l'Université, Saint Etienne du Rouvray 76801, France
e-mail: imad.rida@insa-rouen.fr; romain.herault@insa-rouen.fr; gilles.gasso@insa-rouen.fr

N. Al Maadeed
Department of Computer Science and Engineering, Qatar University, Doha, Qatar
e-mail: n.alali@qu.edu.qa

G.L. Marcialis
Department of Electrical and Electronic Engineering (DIEE), University of Cagliari, 09123 Cagliari, Italy
e-mail: marcialis@diee.unica.it

A. Bouridane
Department of Computer Science and Digital Technologies, Northumbria University, Newcastle upon Tyne, UK
e-mail: ahmed.bouridane@northumbria.ac.uk

R. Jiang et al. (eds.), *Biometric Security and Privacy*, Signal Processing for Security Technologies, DOI 10.1007/978-3-319-47301-7_6

Gait is the most suitable biometric modality in the case of intelligent video surveillance. In monitoring scenes, people are usually distant from cameras, which makes most of biometric features not suitable even the use of face for identification. The drawbacks are obvious, for example, view angle variations and occlusions cause the impossibility to capture the full faces and distance brings low-resolution face images. Therefore, face cannot always achieve good performances in practice. In contrast, gait is a behavioral biometric, including not only individual appearance, such as limb, leg length, and width, but also the dynamic information of individual walking. Gait as a biometric trait can be seen as advantageous over other forms of biometric identification techniques for the following reasons:

- The gait of a person walking can be extracted and analyzed from distance without any contact with the sensor.
- The images used in gait recognition can be easily provided by low-resolution, video-surveillance cameras.
- The gait of an individual is difficult to disguise, by trying to do so the individual will probably appear more suspicious.

Gait recognition consists on discriminating among people by the way or manner they walk. Techniques can be classified into two main categories: model based and model-free approach. Model based approach [5, 27] models the person body structure that estimates static body parameters over time (i.e., trajectory, limb lengths, etc.). This process is computationally intensive since it needs to model and track the subject body. The model-free approach does not recover a structural model of human motion. It uses the features extracted from the motion or shape and hence requires much less computation (see Table 6.1). Furthermore, dynamic information results in better recognition performance than its static counterpart [34]. These motivate researchers to develop new feature representations in model-free approach context.

There exists a considerable amount of work in the context of model-free approach. Benabdelkader et al. [4] introduced a self similarity representation to measure the similarity between pairs of silhouettes. Collins et al. [7] proposed a template based silhouette matching in some key frames. Hayfron-Acquah et al. [11] suggested a contour based representation by analyzing the symmetry of human motion using the Generalized Symmetry Operator. Lee et al. [23] introduced a novel spatio-temporal representation called Shape Variation Based Frieze Pattern which aims to capture the motion information over time. Kobayashi and Otsu [20] used Cubic Higher-order Local Auto-Correlation to extract gait features. Lu and Zhang [25] used multiple gait feature representations based on Independent Component

Table 6.1 Comparison between model based and model-free approach gait recognition

	Model-free	Model based
Complexity	✓	✗
Covariates	✗	✓
Calculation	✓	✗

Analysis and Genetic Fuzzy Support Vector Machine. Huang et al. [16] presented a manifold based approach for cross-speed recognition. Hu et al. [13] proposed an incremental framework based on optical flow, including dynamics learning, pattern retrieval, and recognition. Liu et al. [24] integrated gait recognition in person re-identification. Hu et al. [14] suggested a View-invariant Discriminative Projection method by a unitary linear projection to improve the discriminative ability of multiview gait features. Hu et al. [12] introduced a gait modeling method for gender classification.

Recent trends seem to favor Gait Energy Image (GEI) representation suggested by Han and Bhanu [10]. It is a spatio-temporal representation of the gait obtained by averaging the silhouettes over a gait cycle. For the recognition step, Component Discriminant Analysis (CDA) was applied, which applies Principal Component Analysis (PCA) followed by Multiple Discriminant Analysis (MDA). A considerable amount of works use GEI representation. Yu et al. [36] applied a template matching on GEI without any dimensionality reduction and feature selection. Tao et al. [32] used Gabor filters to extract information from GEI and a General Tensor Discriminant Analysis for recognition. Xu et al. [35] presented an extension of Marginal Fisher analysis to address the problem of gait recognition.

The main challenge of model-free gait recognition is coping with various intra-class variations caused by the presence of shadows, clothing variations, and carrying conditions. Segmentation and view angle are further causes of recognition error [10, 26, 36]. To overcome the limitations of GEI presentation, several approaches have been proposed. Bashir et al. [3] introduced a feature selection method named Gait Entropy Image (GEnI). It computes entropy for each pixel to distinguish static and dynamic pixels of GEI. The GEnI represents a measure of feature significance. In the same context Bashir et al. [2] suggested a gait representation by a weighted sum of the optical flow corresponding to each direction of human motion. An unsupervised method is used to select GEI pixels based on their intensity value [1]. Dupuis et al. [8] introduced a feature selection method based on Random Forest feature rank algorithm. Rida et al. [29] estimated a mask based on pixel variations. In the same context, Rida et al. [28] proposed a method which selects the human dynamic body part. Jeevan et al. [17] introduced a gait representation called Gait Pal and Pal Entropy Image. Kusakunniran [21, 22] proposed a framework to construct gait feature directly from a raw video. Rokanujjaman et al. [31] introduced a novel frequency-domain gait entropy representation. Choudhury and Tjahjadi [6] proposed a View-Invariant Multiscale Gait Recognition (VI-MGR) method. Zeng and Wang [38] introduced a novel method to cope with the problem of walking speed. Recently, Rida et al. [30] used the Modified Phase Only Correlation which is an improved version of the Phase Only Correlation algorithm using a band-pass-type spectral weighting function in order to achieve superior performances.

This chapter proposes a new framework to mitigate the effect of the intra-class variation of GEI representation. Contributions are summarized as follows:

- A horizontal motion vector is proposed that is more reliable and better characterizes the gait than the pixel-wise motion.

- A human body-part selection method is proposed based on group Lasso to cluster the individual dynamic lines into homogeneous parts of human body.
- Feature selection set is separated from the training set to enhance the generalization of body-part selection.
- For view angle variations, a pose estimation method is proposed which is capable to compare a query gait sample without prior knowledge of its view angle with the corresponding gait sequences with the same view angle in the training dataset.

6.2 Proposed Method

Among the available feature representations we choose GEI that is an effective representation, a good compromise between the computational cost and the recognition performance [3]. Figure 6.1 shows our framework of part selection, training and testing, divided into two modules. The first one estimates the human body parts based on motion and group Lasso and selects the discriminative part that is also robust to the intra-class variation. The estimated body parts should not be overspecialized for a particular training set [8]. Therefore, we perform it on a separated feature selection set. The second module applies CDA to the part of GEI features of the training data selected in the first module. Gait recognition performance is measured by Correct Classification Rate (CCR) on the testing dataset.

It has been found that the gait of an individual is characterized much more by the horizontal than the vertical motion [9]. Therefore, instead to estimate the motion of each pixel [3], we propose to estimate the horizontal motion by taking the Shannon entropy of each row from the GEI. The resulting column vector is named as motion based vector.

To generalize the contiguous human body parts from the motion based vector, we further propose to apply group Lasso learning algorithm to segment the motion based vector into shared blocks with similar motion value. The body part with the highest average motion value over the selection dataset is selected, which is discriminative and robust to the intra-class variation.

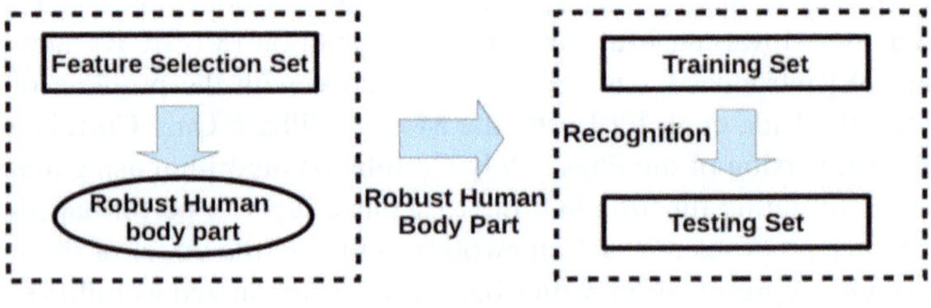

Fig. 6.1 Scheme of our part selection, training and testing

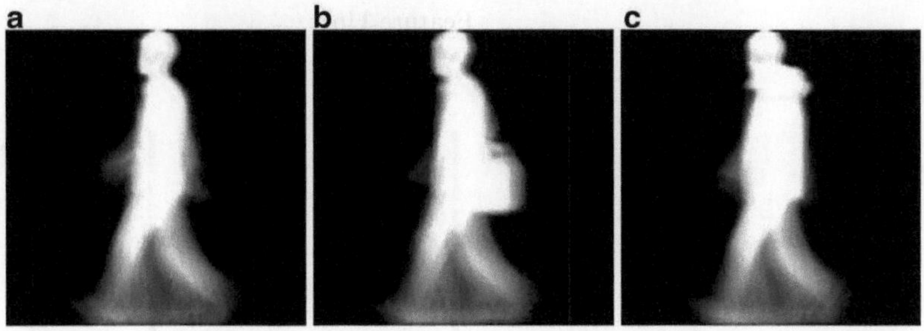

Fig. 6.2 Gait energy image of an individual under different conditions. (**a**) Normal walk. (**b**) Carrying bag. (**c**) Wearing coat

6.2.1 The Proposed Motion Based Vector

GEI is a spatio-temporal representation of gait pattern. It is a single grayscale image (see Fig. 6.2) obtained by averaging the silhouettes extracted over a complete gait cycle [10] as

$$\mathbf{G} = \frac{255}{T} \sum_{t=1}^{T} \mathbf{B}(t) \tag{6.1}$$

where $\mathbf{G} = \{g_{i,j}\}$ is GEI, $1 \leq i \leq N$ and $1 \leq j \leq M$ are the spatial coordinates, T is the number of the frames of a complete gait cycle, and $\mathbf{B}(t)$ is the silhouette image of frame t.

For each GEI, a motion based vector $\mathbf{e} \in \mathbb{R}^N$ shown in Fig. 6.3 is generated by computing the Shannon entropy of each row of GEI. The element of the motion based vector $\mathbf{e}$ is given by:

$$e_i = -\sum_{k=0}^{255} p_k^i \log_2 p_k^i \tag{6.2}$$

where p_k^i is the probability that the pixel value k occurs in the ith row of image $\mathbf{G}$, which is estimated by:

$$p_k^i = \frac{\#(g_{i,j} = k)}{M}; \ \forall j \in [1, M] \tag{6.3}$$

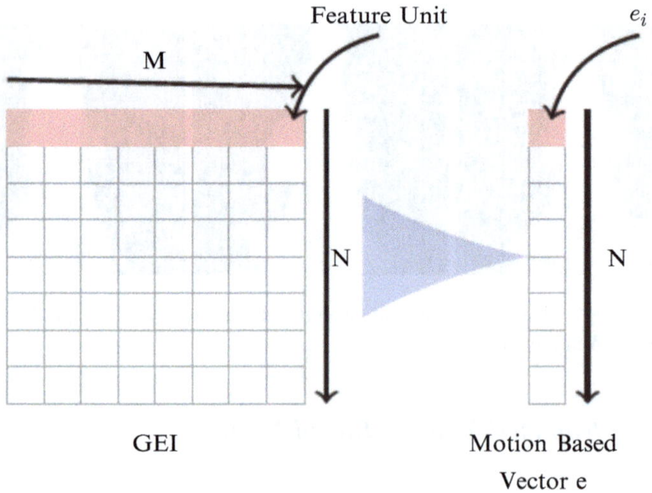

Fig. 6.3 Illustration of the motion based vector

6.2.2 *Group Lasso for Multiple Change-Point Detection*

Let P motion based vectors $\{\mathbf{e}_k\}_{k=1}^{P}$ of P GEIs stored in $N \times P$ matrix $\mathbf{E}$. The aim is to detect the shared change-point locations across all motion based vectors $\{\mathbf{e}_k\}_{k=1}^{P}$ by approximating matrix $\mathbf{E} \in \mathbb{R}^{N \times P}$ by a matrix $\mathbf{V} \in \mathbb{R}^{N \times P}$ of piecewise-constant vectors that share change-points. This can be achieved by resolving the following convex optimization problem:

$$\min_{\mathbf{V} \in \mathbb{R}^{N \times P}} \|\mathbf{E} - \mathbf{V}\|_F^2 + \lambda \sum_{i=1}^{N-1} \|\mathbf{v}_{i+1} - \mathbf{v}_i\|_1 \tag{6.4}$$

where $\mathbf{v}_i$ is the ith row of $\mathbf{V}$ and $\lambda > 0$. Intuitively, when increasing λ enforces many increments $\mathbf{v}_{i+1} - \mathbf{v}_i$ to converge to zero. This implies that the position of nonzeros increments will be same for all vectors. Therefore, the solution of (6.4) provides an approximation of $\mathbf{E}$ by a matrix $\mathbf{V}$ of piecewise-constant vectors which share change-points. The problem (6.4) is reformulated as a group Lasso regression problem as follows:

$$\min_{\boldsymbol{\beta} \in \mathbb{R}^{(N-1) \times P}} \left\| \bar{\mathbf{E}} - \bar{\mathbf{X}}\boldsymbol{\beta} \right\|_F^2 + \lambda \sum_{i=1}^{N-1} \|\boldsymbol{\beta}_{\mathbf{i}}\|_1 \tag{6.5}$$

where $\bar{\mathbf{X}}$ and $\bar{\mathbf{E}}$ are obtained by centering each column from $\mathbf{X}$ and $\mathbf{E}$ knowing that:

$$\begin{cases} \mathbf{X} \in \mathbb{R}^{N\times(N-1)};\ x_{i,j} = \begin{cases} 1 \text{ for } i > j \\ 0 \text{ otherwise} \end{cases} \\ \\ \beta_i = \mathbf{v}_{i+1} - \mathbf{v}_i \end{cases} \tag{6.6}$$

The problem (6.5) can be solved based on the group LARS described in [37] which approximates the solution path with a piecewise-affine set of solutions and iteratively finds change-points independently of λ value. The full derivation of the method can be found in [33].

6.2.3 Canonical Discriminant Analysis

On the training dataset, Canonical Discriminant Analysis (CDA) is applied to the GEI features of the robust human body part determined by the group Lasso on the feature selection dataset. The CDA applies PCA followed by a MDA. PCA removes unreliable dimensions that adversely affect the robustness of the classification [18, 19] and hence improves the classification accuracy. MDA maximizes the distance between classes and preserves the distance inside the classes. As suggestion in [10] we retain $2C$ eigenvectors after applying PCA, where C corresponds to the number of classes (the full explanation is found in [15]). The performance of our method is measured by the CCR that is the ratio of the number of correctly classified samples over the total number of samples.

Let n d-dimensional training GEI templates $\{\mathbf{g}_1, \ldots, \mathbf{g}_n\}$, where each template is a column vector obtained by concatenating the rows of the corresponding GEI. The discriminative human body part with highest motion is selected using the group Lasso to obtain n d'-dimensional GEI templates $\{\mathbf{x}_1, \ldots, \mathbf{x}_n\}$ where $d' < d$. PCA aims to minimize the following objective function:

$$J_{d''} = \sum_{k=1}^{n} \left\| \left(\mathbf{m} + \sum_{l=1}^{d''} a_{kl}\mathbf{u}_l \right) - \mathbf{x}_k \right\|^2 \tag{6.7}$$

where $d'' < d' < d$, $\mathbf{m} = \frac{1}{n}\sum_{k=1}^{n} \mathbf{x}_k$, $\{\mathbf{u}_1, \ldots, \mathbf{u}_{d''}\}$ set of orthogonal unit vectors representing new coordinate system of the subspace and a_{kl} is the projection of the kth data over $\mathbf{u}_l$.

$J_{d''}$ is minimized when $\mathbf{u}_1, \ldots, \mathbf{u}_{d''}$ are eigenvectors of the largest eigenvalues of the covariance matrix Σ given by:

$$\Sigma = \sum_{k=1}^{n} (\mathbf{x}_k - \mathbf{m})(\mathbf{x}_k - \mathbf{m})^T \tag{6.8}$$

The d''-dimensional feature vector $\mathbf{y}_k$ obtained from $\mathbf{x}_k$ is given by:

$$\mathbf{y}_k = \mathbf{M}_{\mathrm{PCA}}\mathbf{x}_k = [a_1, \ldots, a_{d''}]^T = [\mathbf{u}_1, \ldots, \mathbf{u}_{d''}]^T \mathbf{x}_k, \quad k = 1, \ldots, n \tag{6.9}$$

Suppose that the n d'-dimensional principal vectors $\{\mathbf{x}_1, \ldots, \mathbf{x}_n\}$ belong C classes, as suggestion in [2] we retain $d'' = 2C$ eigenvectors after applying PCA. MDA is a supervised learning method which seeks a transformation matrix $\mathbf{W}$ that maximizes the ratio of the between-class scatter matrix $\mathbf{S}_B$ to the within-class scatter matrix $\mathbf{S}_W$ given by:

$$J(\mathbf{W}) = \frac{|\mathbf{W}^T\mathbf{S}_B\mathbf{W}|}{|\mathbf{W}^T\mathbf{S}_W\mathbf{W}|} \tag{6.10}$$

The within-class scatter matrix in the PCA subspace $\mathbf{S}_W$ is defined as $\mathbf{S}_W = \sum_{c=1}^{C} \mathbf{S}_c$ where:

$$\begin{cases} \mathbf{S}_c = \sum_{\mathbf{y} \in \mathscr{D}_c} (\mathbf{y} - \mathbf{m}_c)(\mathbf{y} - \mathbf{m}_c)^T \\ \mathbf{m}_c = \frac{1}{n_c} \sum_{\mathbf{y} \in \mathscr{D}_c} \mathbf{y} \\ \{\mathscr{D}_c\}_{c=1}^{C} \text{ training data of class } c \text{ of size } n_c \end{cases} \tag{6.11}$$

The between-class scatter in the PCA subspace $\mathbf{S}_B$ is given by:

$$\mathbf{S}_B = \sum_{c=1}^{C} n_c(\mathbf{m}_c - \mathbf{m})(\mathbf{m}_c - \mathbf{m})^T \tag{6.12}$$

where $\mathbf{m} = \frac{1}{n}\sum_{\mathbf{y} \in \mathscr{D}} \mathbf{y}$. $J(\mathbf{W})$ is maximized when the columns of $\mathbf{W}$ are the generalized eigenvectors that correspond to $C - 1$ nonzero eigenvalues in:

$$\mathbf{S}_B\mathbf{w}_r = \lambda_r \mathbf{S}_W \mathbf{w}_r \tag{6.13}$$

where $\mathbf{w}_r$ is the rth column of the matrix $\mathbf{W}$. The corresponding generalized eigenvectors are denoted by $\mathbf{v}_1, \ldots, \mathbf{v}_{C-1}$. The $(C-1)$-dimensional feature vector $\mathbf{z}_k$ in the MDA subspace is obtained from the d''-dimensional principal component vector $\mathbf{y}_k$:

$$\mathbf{z}_k = \mathbf{M}_{\mathrm{MDA}}\mathbf{y}_k = [\mathbf{v}_1, \ldots, \mathbf{v}_{C-1}]^T \mathbf{y}_k, \quad k = 1, \ldots, n \tag{6.14}$$

For each training gait template, its gait feature vector is obtained as follows:

$$\mathbf{z}_k = \mathbf{M}_{\mathrm{MDA}}\mathbf{M}_{\mathrm{PCA}}\mathbf{x}_k \quad k = 1, \ldots, n \tag{6.15}$$

6.3 Experiments and Results

In this section, we will introduce the dataset, the different experiments performed on it as well as the obtained results.

6.3.1 Dataset

The proposed method is tested on CASIA dataset B [36] to evaluate its ability to handle the carrying, clothing, and view angle variations. CASIA dataset B is a large multiview gait database created in January 2005 containing 124 subjects captured from 11 different view angles using 11 USB cameras around the left-hand side of the walking subject starting from 0° to 180° (see Fig. 6.4).

Each subject has six normal walking sequences (SetA), two carrying conditions sequences (SetB), and two clothing variations sequences (SetC). The first four sequences of setA noted as SetA1 are used for training. The two remaining sequences of SetA noted as SetA2 as well as SetB and SetC are used for testing normal, carrying, and clothing conditions, respectively. For each sequence, GEI of size 64*64 is computed (see Figs. 6.5 and 6.6).

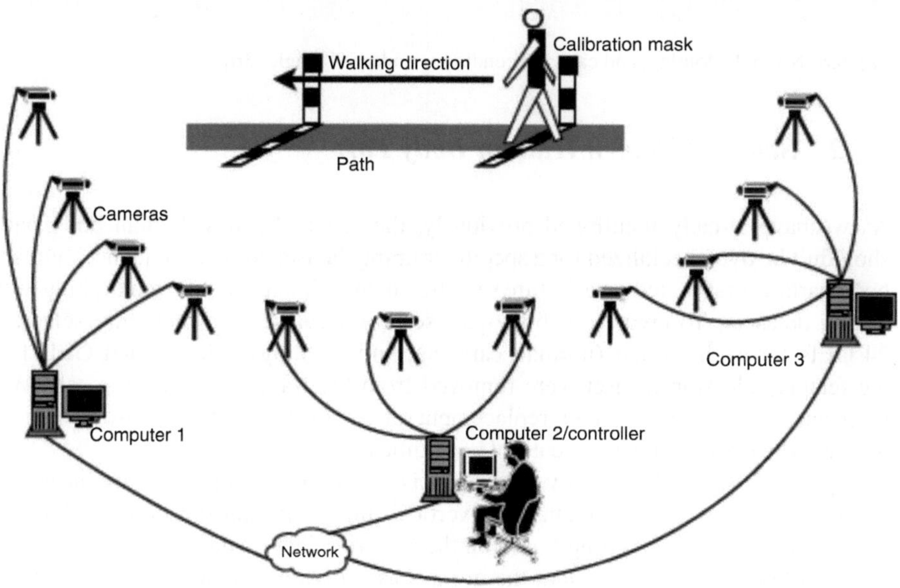

Fig. 6.4 Set-up for gait data collection in CASIA [36]

Fig. 6.5 Normal walking conditions under different view angles [36]

Fig. 6.6 Normal, clothing, and carrying conditions under 90° angle [36]

6.3.2 Selected Robust Human Body Part

As we have already mentioned previously, the selected robust human body part shouldn't be overspecialized for a specific training dataset, as a consequence, human body parts are estimated on a feature selection dataset independent from training and testing datasets. To create our body-part selection dataset, we randomly selected 24 GEIs for each variant (normal, carrying, and clothing). All selected GEIs for the feature selection dataset were removed from the training and testing sets. We performed a bagging without replacement of 45 GEIs on the feature selection dataset. The operation was repeated $L = 5$ times.

Figure 6.7 shows the entropy value (y-axis) of all GEIs against feature index (x-axis) for the $L = 5$ experiments The vertical lines represent the limits of human body parts learnt by the group Lasso on the feature selection dataset.

From Fig. 6.7 we can see that the group Lasso divides the horizontal motion of human body into four parts (the corresponding parts of GEI are shown in Fig. 6.8). It can be seen also in Fig. 6.7 that the part formed by feature units (rows of GEI) from 46 to 64 has the highest mean motion value. It corresponds to the most dynamic part from the human body which contains discriminative information to

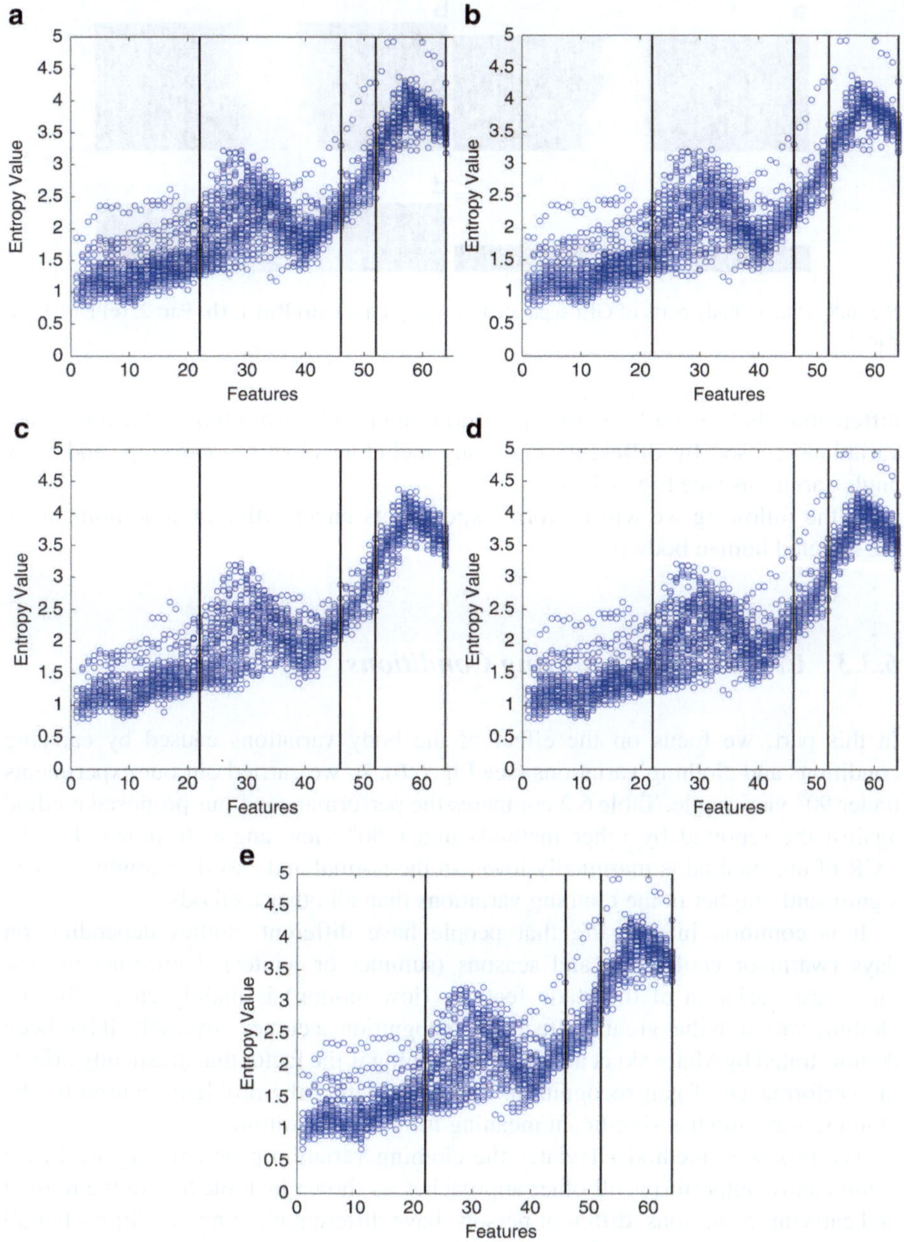

Fig. 6.7 Values of motion based vectors in selection datasets and parts of shared motion value separated by group Lasso. (**a**) Experiment 1. (**b**) Experiment 2. (**c**) Experiment 3. (**d**) Experiment 4. (**e**) Experiment 5

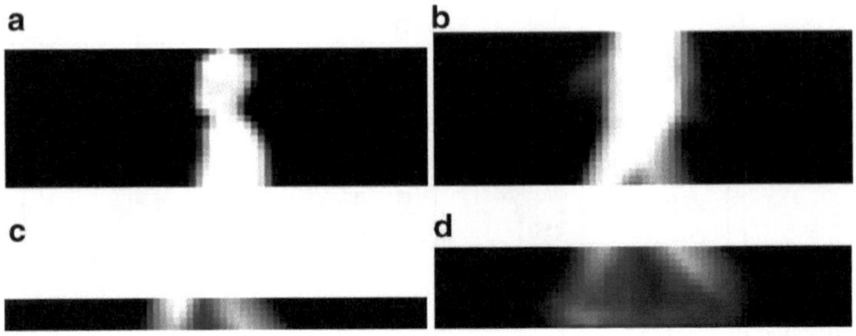

Fig. 6.8 Human body parts of GEI separated by group Lasso. (**a**) Part 1. (**b**) Part 2. (**c**) Part 3. (**d**) Part 4

differentiate between different people, furthermore it is also robust to the intra-class variations caused by different conditions including clothing, carrying, and view angle variations (see Fig. 6.8c, d).

In the following we will perform experiments under different conditions using the selected human body part.

6.3.3 Clothing and Carrying Conditions

In this part, we focus on the effect of the body variations caused by carrying conditions and clothing variations (see Fig. 6.6), so we carried out our experiments under 90° view angle. Table 6.2 compares the performance of our proposed method against the reported by other methods under 90° view angle. It shows that the CCR of our method is marginally lower in the normal and carrying conditions and significantly higher in the clothing variations than all other methods.

It is common in real life that people have different clothes depending on days (warm or cool days) and seasons (summer or winter). Unfortunately, the intra-class variation of the static features (low motion) is mainly caused by the clothing variation that greatly affects the recognition accuracy adversely. It has been demonstrated by Matovski et al. [26] that clothing is the factor that drastically affects the performance of gait recognition. Thus, alleviating the problems caused by the clothing variation has significant meaning for gait recognition.

The proposed method alleviates the clothing variation problem very well as it significantly outperforms all other approaches as shown in Table 6.2. In the normal and carrying conditions, different persons have different clothing conditions but all samples of a same person always have the same clothing condition in the dataset. Thus, the cloths in the normal and carrying conditions in fact undesirably contribute to differentiate persons. Therefore, these recognition rates could be misleading as they do not well reflect the real gait recognition performance. In the next sections, we will further see the problems of testing the gait recognition performance using the training and test data in the same cloth for the same persons. Nevertheless, the

Table 6.2 Comparison of CCRs (in %) from several different algorithms on CASIA database using 90° view

Method	Normal conditions	Carrying conditions	Clothing conditions	Overall	Std
Yu et al. [36]	97.60	32.70	52.00	60.77	33.33
Han and Bhanu [10]	99.60	57.20	23.80	60.20	37.99
Bashir et al. [3]	**100.00**	78.30	44.00	74.10	28.24
Bashir et al. [2]	97.50	**83.87**	48.80	76.63	25.09
Bashir et al.[1]	99.40	79.90	31.30	70.20	35.07
Dupuis et al. [8]	98.80	73.80	63.70	78.77	18.07
Rida et al. [30]	93.60	81.70	68.80	81.37	12.40
Rida et al. [29]	95.97	63.39	72.77	77.38	16.77
Hu et al. [13]	94.00	45.20	42.90	60.70	28.86
Kusakunniran [22]	95.40	60.90	52.00	69.43	22.92
Rakanujjaman et al. [31]	97.61	83.87	51.61	77.70	23.61
Kusakunniran [21]	94.50	60.90	58.50	71.30	20.13
Jeevan et al. [17]	93.36	56.12	22.44	57.31	35.47
Our proposed method	98.39	75.89	**91.96**	**88.75**	**11.59**

The bold values correspond to the best results

proposed method performs the best among all approaches on the whole test dataset that contains one-third samples with cloth variation and two-third samples without the cloth variation.

6.3.4 Cross-View Gait Recognition

In real life, subjects are often captured under different view angles; to simulate these conditions we perform experiments in the so-called cross-view gait recognition. In this case, different view angle combinations between training and testing data are used to estimate the recognition performances. Tables 6.3, 6.4, and 6.5 show the results of the body-part cross-view under normal, carrying conditions, and clothing variations, respectively, when Tables 6.6, 6.7, and 6.8 show the same results of whole-body under the same conditions.

The results demonstrate that our body-part method significantly outperforms the whole-body one under cloth variations; however, it has marginally lower performances in normal conditions due to the undesirable contribution of clothing in recognition which was already pointed out previously. From the same results it can be seen that both the whole-body and body-part give good performances when the training view angle is similar to the testing one; however, the performances significantly decrease when the difference between the training view angle and the

Table 6.3 Cross-view body-part recognition under normal conditions (%)

	Probe angle (°)										
Gallery angle (°)	0	18	36	54	72	90	108	126	144	162	180
0	**98.37**	5.24	1.61	1.21	0.40	0.81	0.81	1.61	0.81	0.81	9.27
18	6.10	**98.79**	17.74	1.61	0.81	0.81	1.21	1.61	4.44	2.42	2.82
36	3.66	23.79	**95.97**	32.66	5.65	0.81	1.21	0.81	0.40	3.63	2.42
54	2.03	5.24	33.87	**96.77**	11.69	4.84	1.61	1.21	0.40	1.61	2.02
72	1.22	2.02	3.23	10.08	**98.39**	82.26	20.16	1.21	0.81	1.61	2.02
90	1.22	1.21	2.82	7.66	67.74	**98.39**	48.79	4.84	3.23	1.61	1.21
108	2.03	2.82	4.44	4.44	23.79	67.34	**97.18**	30.24	4.84	3.63	1.61
126	0.81	2.42	2.42	4.03	5.65	7.26	29.03	**95.56**	38.31	3.63	1.61
144	0.81	2.02	1.21	2.42	5.24	4.44	6.05	47.18	**97.18**	2.02	0.81
162	3.66	3.23	0.81	0.81	0.81	0.81	0.81	0.81	1.21	**97.98**	6.85
180	10.57	2.42	1.61	0.40	0	0.40	0.81	1.61	2.42	3.63	**97.58**

Bold value correspond to CCR when gallery angle is similar to probe angle

Table 6.4 Cross-view body-part recognition under carrying conditions (%)

	Probe angle (°)										
Gallery angle (°)	0	18	36	54	72	90	108	126	144	162	180
0	**72.36**	2.02	0.81	0.81	0.40	0	0.40	2.02	1.62	2.04	8.50
18	5.28	**73.79**	9.68	2.03	2.02	1.79	1.61	2.02	1.62	3.67	2.02
36	4.07	16.94	**77.02**	27.64	4.44	1.34	2.02	0.81	0	5.31	1.62
54	1.63	6.45	25.40	**75.61**	10.48	3.57	1.21	1.21	0.81	2.04	2.02
72	1.63	1.61	1.61	10.16	**75.00**	56.70	15.32	2.02	0.81	2.04	2.83
90	0.81	1.61	2.42	5.69	45.16	**75.89**	25.00	4.86	2.43	0.82	1.21
108	0.81	0.81	4.03	3.66	14.92	53.57	**75.00**	22.27	6.88	3.27	2.43
126	1.22	1.21	2.42	2.44	6.85	6.25	29.84	**76.52**	28.34	2.04	1.21
144	1.22	0.81	1.61	2.03	4.84	4.46	5.24	33.60	**77.33**	0	0.81
162	2.85	1.21	1.21	1.22	1.21	1.34	0.81	0.81	0.40	**74.69**	3.24
180	9.76	2.42	0.81	0.81	0.40	0.89	0.81	2.02	1.62	4.08	**75.71**

Bold value correspond to CCR when gallery angle is similar to probe angle

testing one increases. This makes us conclude that there is an invert relationship between the view angle difference between training and testing data and the performance.

Based on the obtained results, we can clearly understand that conventional methods fail to give good recognition performances in case of the large view angle variations between the training and testing data. Unfortunately, the latter is frequently encountered in real life gait recognition applications. This clearly shows the mandatory to introduce new methods capable to address the very challenging problem of view angle variations.

Table 6.5 Cross-view body-part recognition under clothing variations (%)

	Probe angle (°)										
Gallery angle (°)	0	18	36	54	72	90	108	126	144	162	180
0	**80.89**	4.03	2.42	1.62	0.81	0.89	0.81	2.43	2.02	0.82	9.27
18	5.28	**83.06**	12.90	2.02	0.81	0.89	0.81	1.62	2.83	2.04	3.23
36	2.44	19.35	**85.08**	29.55	6.85	2.68	1.61	1.62	0.40	2.45	1.21
54	1.63	5.65	30.24	**87.04**	10.08	4.02	1.21	0.81	0	0.82	0.81
72	1.22	1.61	2.42	12.96	**91.13**	62.95	18.55	0.40	0	0.82	0.81
90	0.41	1.61	3.23	6.07	60.48	**91.96**	40.32	4.05	2.43	1.63	1.61
108	1.63	3.23	1.61	3.64	18.95	56.25	**88.71**	31.58	4.45	3.67	1.61
126	1.22	1.61	1.61	4.05	4.44	4.91	22.18	**87.04**	40.08	3.67	1.61
144	2.03	1.21	1.61	2.02	5.65	1.79	4.03	27.13	**90.28**	2.86	1.61
162	3.25	2.82	2.02	1.62	1.21	1.34	1.21	1.62	1.21	**86.94**	6.85
180	9.35	2.02	2.02	0.81	0.81	0.89	0.81	1.62	0.81	2.86	**84.27**

Bold value correspond to CCR when gallery angle is similar to probe angle

Table 6.6 Cross-view whole-body recognition under normal conditions (%)

	Probe angle (°)										
Gallery angle (°)	0	18	36	54	72	90	108	126	144	162	180
0	**100**	70.16	14.92	5.24	2.42	2.02	0.81	0.81	4.44	15.32	40.32
18	82.11	**100**	92.74	16.13	3.63	1.21	2.42	4.84	15.32	21.77	31.85
36	38.21	94.76	**99.19**	85.89	30.24	15.73	12.50	22.58	20.97	21.77	9.27
54	9.76	27.82	92.34	**99.19**	70.97	35.48	21.77	27.42	23.79	6.05	6.45
72	6.10	4.03	16.13	63.31	**99.19**	98.79	74.19	14.92	4.84	5.24	4.44
90	2.03	2.02	6.45	17.34	98.79	**100**	97.18.79	22.98	6.05	2.82	2.42
108	2.44	0.81	8.06	33.06	79.84	97.98	**99.60**	91.53	22.58	3.63	2.42
126	6.50	4.84	12.10	31.45	47.58	50.81	90.73	**98.39**	94.76	15.32	6.45
144	13.01	15.73	27.02	19.35	8.87	6.45	31.45	95.16	**99.19**	34.68	11.29
162	20.73	25.00	15.32	6.05	0.81	0.81	1.21	2.42	6.05	**99.60**	70.56
180	52.44	18.55	12.10	4.84	3.23	1.61	0.81	2.42	9.27	77.42	**100**

Bold value correspond to CCR when gallery angle is similar to probe angle

Starting from the observation that the view angle similarity between the training and testing data helps to give good recognition performances, we introduce in the following section a novel method named "gait recognition without prior knowledge of the view angle" capable to estimate the view angle of the testing samples to compare them to training samples with similar view angle and as a consequence improve the recognition performances.

Table 6.7 Cross-view whole-body recognition under carrying conditions (%)

Gallery angle (°)	Probe angle (°) 0	18	36	54	72	90	108	126	144	162	180
0	**83.74**	45.56	14.92	6.50	4.44	2.23	1.61	2.02	2.83	6.53	21.46
18	54.07	**79.44**	54.03	11.79	4.44	0.45	1.21	4.45	5.67	10.20	10.53
36	27.64	55.24	**74.60**	46.34	16.13	6.70	3.63	7.69	6.48	8.98	5.26
54	4.88	14.52	48.79	**69.11**	37.90	23.21	10.08	11.74	9.31	8.98	5.67
72	5.69	4.44	7.66	24.80	**59.68**	47.77	23.79	8.91	4.86	3.67	5.26
90	2.03	2.42	3.63	11.79	47.98	**55.80**	39.92	9.72	4.05	2.86	2.43
108	2.44	0.81	4.44	15.45	40.73	50.89	**59.27**	35.22	12.55	4.08	2.83
126	4.07	3.23	9.68	20.73	27.02	28.57	38.31	**62.35**	43.32	8.57	4.45
144	5.69	8.87	15.32	11.38	5.24	5.36	8.47	48.58	**70.45**	17.96	8.10
162	10.98	13.71	5.24	2.44	1.61	1.79	1.61	2.43	4.05	**67.35**	31.17
180	29.27	13.71	6.05	3.66	2.42	0.45	2.02	2.02	6.48	34.29	**76.11**

Bold value correspond to CCR when gallery angle is similar to probe angle

Table 6.8 Cross-view body-part recognition under clothing variations (%)

Gallery angle (°)	Probe angle (°) 0	18	36	54	72	90	108	126	144	162	180
0	**28.05**	14.52	5.65	2.02	1.21	0.45	1.21	1.62	3.64	6.94	7.66
18	11.38	**25.81**	21.37	6.48	4.03	3.57	2.82	4.05	6.07	6.94	5.65
36	8.94	18.95	**31.05**	23.48	8.87	6.70	4.44	6.88	5.26	7.76	2.42
54	1.22	7.66	20.97	**28.34**	16.53	7.59	6.85	6.88	4.45	2.45	0.40
72	0.81	1.61	2.42	9.31	**29.44**	22.32	12.50	4.86	1.62	1.63	2.02
90	2.85	1.61	2.02	7.29	16.53	**25.45**	14.92	5.67	1.62	2.04	0
108	0.81	1.61	3.23	5.26	13.71	17.86	**24.60**	12.96	5.26	1.63	0.40
126	1.22	2.02	3.23	5.26	10.48	11.61	23.39	**31.58**	19.43	1.22	1.21
144	5.28	5.65	7.26	8.50	6.45	3.13	6.05	25.91	**37.25**	4.08	3.23
162	5.28	6.45	7.26	5.67	1.21	1.34	0.81	2.02	4.45	**31.02**	12.10
180	10.16	7.66	5.24	1.21	1.61	1.79	2.02	2.83	4.45	12.24	**30.65**

Bold value correspond to CCR when gallery angle is similar to probe angle

6.3.5 *Gait Recognition Without Prior Knowledge of the View Angle*

The framework in Fig. 6.9 is designed to recognize individuals without a prior knowledge of the viewpoint. Towards this end, the first step consists on estimating the pose of the query subject using the selected human body part, i.e., row 46–64 (it has been explained above how the body part is selected using the group Lasso of motion). For this aim, a simple knn with $k = 1$ is used to find the group of training samples that have the similar pose to that of the query subject. The results of pose

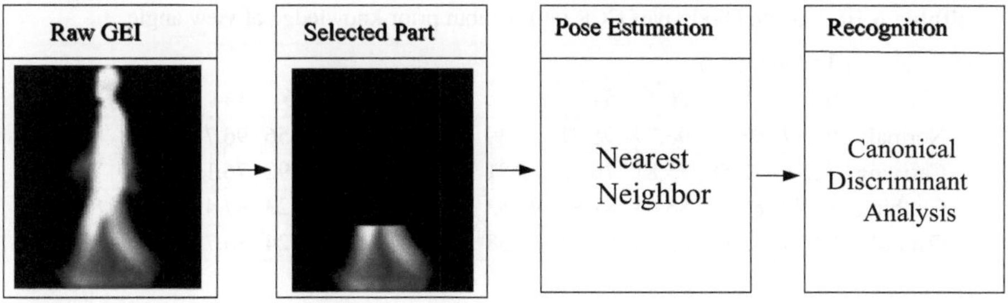

Fig. 6.9 Framework of view angle variation without prior knowledge of the view angle

Table 6.9 Pose estimation–confusion matrix (%)

Real angle (°)	Predicted angle (°)										
	0	18	36	54	72	90	108	126	144	162	180
0	**98.78**	0.27	0	0	0	0	0	0	0.40	0	0.54
18	0.40	**97.58**	1.34	0	0	0.13	0	0.13	0.26	0	0.13
36	0.26	1.20	**97.31**	0.80	0	0	0	0	0.40	0	0
54	0.13	0.13	0.8	**98.65**	0	0	0.13	0	0.13	0	0
72	0	0.26	0.13	0	**98.92**	0.13	0.40	0.13	0	0	0
90	0	0.14	0	0.43	0.43	**98.41**	0.57	0	0	0	0
108	0	0	0	0.13	0	1.34	**97.71**	0.53	0	0.26	0
126	0	0	0	0.13	0	0	0.40	**98.92**	0	0.26	0.26
144	0	0.13	0.13	0	0	0	0.13	0.26	**97.57**	1.48	0.26
162	0	0.27	0.13	0.13	0	0	0	0	1.62	**97.83**	0
180	1.07	0.26	0	0	0	0	0	0.13	0	0	**98.51**

Bold values correspond to performance of well-predicted angles

estimation are shown in Table 6.9, it can be seen that the human body part selected by the group Lasso is very discriminative and we are able to estimate the pose of the query subjects of all the dataset with an error less than 3 % for all view angles from 0° to 180°.

The next step consists in identifying the query subject among the group of training samples with the same pose using CDA, which corresponds to PCA followed by MDA (it has been well introduced above in Sect. 6.2.3). Results are shown in Tables 6.10, 6.11, and 6.12, which, respectively, record the CCR of our proposed body-part selection approach, the approach that uses the whole-body, and the VI-MGR methodwhich has been introduced by Choudhury and Tjahjadi in [6] especially to deal with the problem of view angle variations.

Results in these tables clearly show that our proposed body-part selection method significantly outperforms VI-MGR and the approach without the part selection (whole GEI template) for all 11 view angle variations in the case of the clothing variation (see Fig. 6.10). On the whole test dataset that contains one-third samples

Table 6.10 Selected body-part CCR (%) without prior knowledge of view angle

	Test angle (°)										
	0	18	36	54	72	90	108	126	144	162	180
Normal	97.97	98.79	96.37	96.77	98.39	97.98	97.18	95.56	96.77	97.98	97.58
Carrying	72.76	72.58	75.81	76.42	75.81	73.66	74.60	76.92	76.11	75.10	76.11
Clothing	80.49	83.47	85.08	87.85	91.53	91.07	87.90	86.23	87.45	84.90	83.06
Overall	83.74	84.95	85.75	87.02	88.58	87.57	86.56	86.24	86.78	85.99	85.59

Table 6.11 Whole-body CCR (%) without prior knowledge of view angle

	Test angle (°)										
	0	18	36	54	72	90	108	126	144	162	180
Normal	100	100	99.19	99.19	99.19	100	99.60	97.98	99.60	99.19	100
Carrying	82.11	77.42	75.81	68.70	59.68	52.68	53.63	60.73	68.02	66.53	72.47
Clothing	26.83	25.81	28.63	27.53	28.63	22.77	23.79	31.17	34.01	29.39	29.84
Overall	69.65	67.74	67.88	65.14	62.50	58.48	59.01	63.30	67.21	65.04	67.44

Table 6.12 VI-MGR CCR (%) without prior knowledge of view angle

	Test angle (°)										
	0	18	36	54	72	90	108	126	144	162	180
Normal	100	99	100	99	100	100	99	99	100	100	99
Carrying	93	89	89	90	77	80	82	84	92	93	89
Clothing	67	56	70	80	71	75	77	75	65	64	66
Overall	86.66	81.33	86.33	89.66	82.66	85	86	86	85.66	85.66	84.66

with cloth variation and two-third samples without the cloth variation, the proposed approach outperforms the no-part selection approach for all view angle variations and outperforms VI-MGR in 8 of the 11 view angle variations (see Fig. 6.11).

The problems of the CCR for normal and carrying conditions are shown in Tables 6.11 and 6.12. It is well known that the maximum gait information is captured for the view angle near 90° and the minimum gait information is captured for the view angle near 0° or 180°. However, while perfect or near perfect CCR is achieved by almost all view angles in normal condition, in carrying condition, visibly higher CCR is achieved for view angles near 0° or near 180° than that for view angles near 90°. This shows that the cloths in the normal and carrying conditions in fact undesirably contribute to differentiate persons. Therefore, these recognition rates could be misleading as they do not well reflect the real gait recognition performance. Figure 6.10 shows CCR of the three approaches on all test data with cloth and view angle variations, which clearly shows the significant performance gain achieved by the proposed approach.

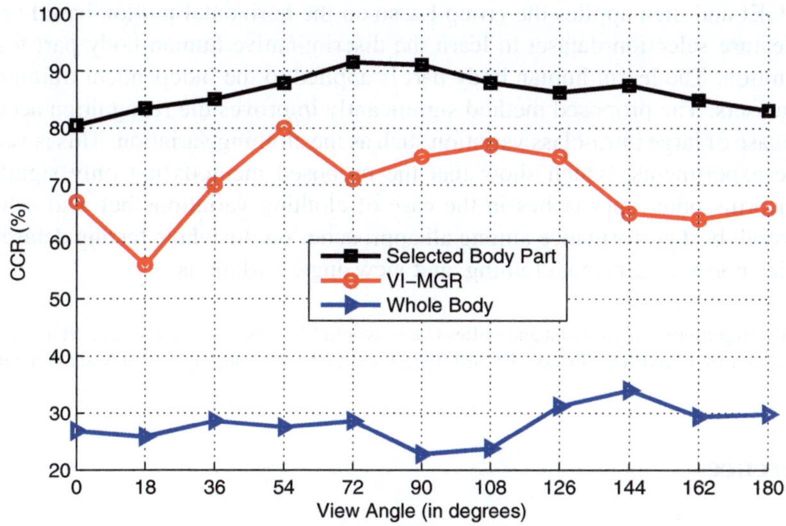

Fig. 6.10 CCR of different approaches on test data with cloth and view angle variations

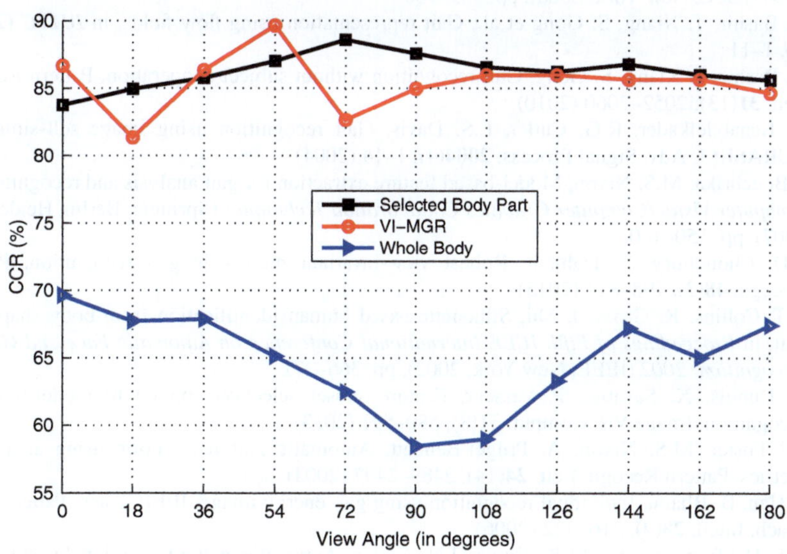

Fig. 6.11 The mean CCR of different approaches on test data under different conditions

6.4 Conclusion

In this chapter we proposed a method that finds the discriminative human body part that is also robust to the intra-class variations for improving the human gait recognition. The proposed method first generates a horizontal motion based vector from GEI and then applies the group Lasso on the horizontal motion based vectors of a feature selection dataset to learn the discriminative human body part for gait recognition. The learnt human body part is applied to the independent training and test datasets. The proposed method significantly improves the recognition accuracy in the case of large intra-class variation such as the clothing variation. This is verified by the experiments, which show that the proposed methods not only significant outperforms other approaches in the case of clothing variations but also achieves the overall best performance among all approaches on the whole testing dataset that contains normal, carrying, clothing, and view angle variations.

Acknowledgements Imad Rida and Gilles Gasso would like to acknowledge support and funding from French FUI AAP 15 - Project RCSM, Risk Credit Chain & Supply Chain Management.

References

1. K. Bashir, T. Xiang, S. Gong, Feature selection on gait energy image for human identification, in *IEEE International Conference on Acoustics, Speech and Signal Processing, 2008. ICASSP 2008* (IEEE, New York, 2008), pp. 985–988
2. K. Bashir, T. Xiang, S. Gong et al., Gait representation using flow fields, in *BMVC* (2009), pp. 1–11
3. K. Bashir, T. Xiang, S. Gong, Gait recognition without subject cooperation. Pattern Recogn. Lett. **31**(13), 2052–2060 (2010)
4. C. Benabdelkader, R.G. Cutler, L.S. Davis, Gait recognition using image self-similarity. EURASIP J. Adv. Signal Process. **2004**(4), 1–14 (2004)
5. I. Bouchrika, M.S. Nixon, Model-based feature extraction for gait analysis and recognition, in *Computer Vision/Computer Graphics Collaboration Techniques* (Springer, Berlin, Heidelberg, 2007), pp. 150–160
6. S.D. Choudhury, T. Tjahjadi, Robust view-invariant multiscale gait recognition. Pattern Recogn. **48**(3), 798–811 (2015)
7. R.T. Collins, R. Gross, J. Shi, Silhouette-based human identification from body shape and gait, in *Proceedings of Fifth IEEE International Conference on Automatic Face and Gesture Recognition, 2002* (IEEE, New York, 2002), pp. 366–371
8. Y. Dupuis, X. Savatier, P. Vasseur, Feature subset selection applied to model-free gait recognition. Image Vis. Comput. **31**(8), 580–591 (2013)
9. J.P. Foster, M.S. Nixon, A. Prügel-Bennett, Automatic gait recognition using area-based metrics. Pattern Recogn. Lett. **24**(14), 2489–2497 (2003)
10. J. Han, B. Bhanu, Individual recognition using gait energy image. IEEE Trans. Pattern Anal. Mach. Intell. **28**(2), 316–322 (2006)
11. J.B. Hayfron-Acquah, M.S. Nixon, J.N. Carter, Automatic gait recognition by symmetry analysis. Pattern Recogn. Lett. **24**(13), 2175–2183 (2003)
12. M. Hu, Y. Wang, Z. Zhang et al., Gait-based gender classification using mixed conditional random field. IEEE Trans. Syst. Man Cybern. B Cybern. **41**(5), 1429–1439 (2011)
13. M. Hu, Y. Wang, Z. Zhang et al., Incremental learning for video-based gait recognition with LBP flow. IEEE Trans. Cybernet. **43**(1), 77–89 (2013)
14. M. Hu, Y. Wang, Z. Zhang et al., View-invariant discriminative projection for multi-view gait-based human identification. IEEE Trans. Inf. Forensics Secur. **8**(12), 2034–2045 (2013)
15. P.S. Huang, C.J. Harris, M.S. Nixon, Recognising humans by gait via parametric canonical space. Artif. Intell. Eng. **13**(4), 359–366 (1999)
16. S. Huang, A. Elgammal, J. Lu et al., Cross-speed gait recognition using speed-invariant gait templates and globality? Locality preserving projections. IEEE Trans. Inf. Forensics Secur. **10**(10), 2071–2083 (2015)

17. M. Jeevan, N. Jain, M. Hanmandlu et al., Gait recognition based on gait pal and pal entropy image, in *2013 20th IEEE International Conference on IEEE Image Processing (ICIP)* (2013), pp. 4195–4199
18. X. Jiang, Asymmetric principal component and discriminant analyses for pattern classification. IEEE Trans. Pattern Anal. Mach. Intell. **31**(5), 931–937 (2009)
19. X. Jiang, Linear subspace learning-based dimensionality reduction. IEEE Signal Process. Mag. **28**(2), 16–26 (2011)
20. T. Kobayashi, N. Otsu, Action and simultaneous multiple-person identification using cubic higher-order local auto-correlation, in *Proceedings of the 17th International Conference on Pattern Recognition, 2004. ICPR 2004* (IEEE, New York, 2004), pp. 741–744
21. W. Kusakunniran, Attribute-based learning for gait recognition using spatio-temporal interest points. Image Vis. Comput. **32**(12), 1117–1126 (2014)
22. W. Kusakunniran, Recognizing gaits on spatio-temporal feature domain. IEEE Trans. Inf. Forensics Secur. **9**(9), 1416–1423 (2014)
23. S. Lee, Y. Liu, R. Collins, Shape variation-based frieze pattern for robust gait recognition, in *IEEE Conference on Computer Vision and Pattern Recognition, 2007. CVPR'07* (IEEE, New York, 2007), pp. 1–8
24. Z. Liu, Z. Zhang, Q. Wu et al., Enhancing person re-identification by integrating gait biometric. Neurocomputing **168**, 1144–1156 (2015)
25. J. Lu, E. Zhang, Gait recognition for human identification based on ICA and fuzzy SVM through multiple views fusion. Pattern Recogn. Lett. **28**(16), 2401–2411 (2007)
26. D.S. Matovski, M.S. Nixon, S. Mahmoodi et al., The effect of time on gait recognition performance. IEEE Trans. Inf. Forensics Secur. **7**(2), 543–552 (2012)
27. S.A. Niyogi, E.H. Adelson, Analyzing and recognizing walking figures in XYT, in *1994 IEEE Computer Society Conference on Computer Vision and Pattern Recognition, 1994. Proceedings CVPR'94* (IEEE, New York, 1994), pp. 469–474
28. I. Rida, S. Almaadeed, A. Bouridane, Improved gait recognition based on gait energy images, in *2014 26th International Conference on Microelectronics (ICM)* (IEEE, New York, 2014), pp. 40–43
29. I. Rida, A. Bouridane, G.L. Marcialis et al., Improved human gait recognition, in *Image Analysis and Processing-ICIAP 2015* (Springer International Publishing, Cham, 2015), pp. 119–129
30. I. Rida, S. Almaadeed, A. Bouridane, Gait recognition based on modified phase-only correlation. Signal Image Video Process. **10**(3), 463–470 (2016)
31. M. Rokanujjaman, M.S. Islam, M.A. Hossain et al., Effective part-based gait identification using frequency-domain gait entropy features. Multimedia Tools Appl. **74**(9), 3099–3120 (2015)
32. D. Tao, X. Li, X. Wu et al., General tensor discriminant analysis and gabor features for gait recognition. IEEE Trans. Pattern Anal. Mach. Intell. **29**(10), 1700–1715 (2007)
33. J.-P. Vert, K. Bleakley, Fast detection of multiple change-points shared by many signals using group LARS, in *Advances in Neural Information Processing Systems* (2010), pp. 2343–2351
34. L. Wang, H. Ning, T. Tan, W. Hu, Fusion of static and dynamic body biometrics for gait recognition. IEEE Trans. Circuits Syst. Video Technol. **14**(2), 149–158 (2004)
35. D. Xu, S. Yan, T. Dacheng et al., Marginal fisher analysis and its variants for human gait recognition and content-based image retrieval. IEEE Trans. Image Process. **16**(11), 2811–2821 (2007)
36. S. Yu, D. Tan, T. Tan, A framework for evaluating the effect of view angle, clothing and carrying condition on gait recognition, in *IEEE 18th International Conference on Pattern Recognition, ICPR 2006* (2006), pp. 441–444
37. M. Yuan, Y. Lin, Model selection and estimation in regression with grouped variables. J. R. Stat. Soc. Ser. B (Stat Methodol.) **68**(1), 49–67 (2006)
38. W. Zeng, C. Wang, Gait recognition across different walking speeds via deterministic learning. Neurocomputing **152**, 139–150 (2015)

Chapter 7
Smartphone User Authentication Using Touch Dynamics in the Big Data Era: Challenges and Opportunities

Lijun Jiang and Weizhi Meng

7.1 Introduction

According to the latest preliminary release from the International Data Corporation, vendors shipped a total of 334.9 million smartphones worldwide in the first quarter of 2016 (1Q16), up slightly from the 334.3 million units in 1Q15 [1]. With the worldwide popularity of smartphones, users start storing more sensitive and private information on their phones. Therefore, user authentication has become very crucial on these portable devices to defend against various threats [2]. Existing authentication mechanisms can be generally classified based on *what you know* (e.g., password), *what you have* (e.g., keys), and *what you are* (e.g., fingerprint).

The first methodology utilizes the knowledge in users' memory and should be the most widely used techniques such as passwords and personal identification numbers. However, long-term memory limitations are one of the weaknesses [3], which may result in simple passwords in practice [4]. In addition, passwords can be easily captured by direct observation or various malware techniques. As an alternative, user authentication can be conducted through what users have such as keys and tokens. But such methodology has not been widely used in smartphone authentication, as it is required to deploy additional hardware on phones' side, which would be expensive for smartphone users [5].

In recent years, research has been focused more on biometric-based authentication on mobile devices, since biometric features are mostly unique and not duplicable or transferable [6]. This methodology can use either *physiological* or

L. Jiang
Department of Computer Science, City University of Hong Kong, Hong Kong, China

W. Meng (✉)
Department of Applied Mathematics and Computer Science, Technical University of Denmark, Kongens Lyngby, Denmark
e-mail: yuxin.meng@my.cityu.edu.hk

R. Jiang et al. (eds.), *Biometric Security and Privacy*, Signal Processing for Security Technologies, DOI 10.1007/978-3-319-47301-7_7

behavioral features for authentication. Physiological features use measurements from the human body such as fingerprint, face, teeth, iris, and retina. By contrast, behavioral features use measurements from human actions such as signature, gait, and keystroke. With the rapid adoption of smartphones, touch dynamics has become a notable topic for smartphone authentication. For example, touchscreens have already become the leading input method on the mobile platform, with more than 78 % of all phones using a touchscreen [6]. Touch dynamics can be described as the characteristics of the inputs received from a touchscreen when a user is interacting with a device (e.g., a touchscreen mobile phone). Intuitively, touch dynamics is different from keystroke dynamics in that touch dynamics has more input types such as multi-touch and touch movement. On the other hand, the inputs of press button up and press button down in keystroke dynamics are similar to the actions of touch press up and touch press down (e.g., single-touch) in touch dynamics. Due to its characteristics, touch dynamics received more attention from the literature.

Although more research has been done on touch dynamics-based authentication on smartphones, few studies pay attention to its application and challenges in the context of big data. Compared to a traditional mobile platform, the concept of *big data* makes the situation more complicated and involves more characteristics. For instance, big data could be represented by a mathematical relationship among three independent variables: volume, velocity, and variety [7]. Thus, big data can be defined as the following situation: when the volume, variety, and velocity of the data are increased, causing the current techniques and technologies to be unable to handle storage and processing of the data. When applying to mobile platforms, it means that mobile traffic/information under big data will show high volume of data with large varieties of information types arriving in high velocity. Subsequently, traditional techniques that have been used in smartphone user authentication may encounter new challenges since authentication is time sensitive on mobile devices.

In this chapter, we focus on two aspects: touch dynamics and big data, and the impact of big data on touch dynamics-based user authentication on smartphones. Our purpose is to provide insights about touch dynamics and its future trends in the era of big data. The contributions of the chapter can be summarized as below:

- We begin by providing a brief introduction related to the background of both Big Data (e.g., characteristics) and touch dynamics-based user authentication on smartphones.
- Then, we review the recent studies regarding the design of touch dynamics-based user authentication on mobile devices, especially smartphones, and identify major challenges and future trends of designing proper authentication schemes in the big data era.

The remaining parts of this chapter are organized as follows. Section 7.2 introduces the background of big data and touch dynamics-based user authentication. Section 7.3 gives a review of user authentication using touch dynamics on mobile phones. Then Sect. 7.4 identifies several major challenges and future trends of designing such authentication schemes in the era of big data. Finally, Sect. 7.5 makes a conclusion.

7.2 Background

In this section, we introduce the concept of big data and the background of touch dynamics-based user authentication on smartphones.

7.2.1 Big Data

Nowadays, many organizations are facing more data issues. For example, companies may not know how to get value of the stored information, so that it is hard to decide whether to maintain such data or not. Generally, *big data* means all the data presented ever, whether categorized or not. It is agreed that big data cannot be manipulated using normal spreadsheets and regular tools of database management [7]. In other words, it is a collection of large and complex datasets that is difficult to use traditional data processing applications.

Big Data can be represented in terms of three independent characteristics: *volume*, *velocity*, and *variety* [7]. Figure 7.1 describes these characteristics from a view of IBM. These new characteristics create the need for a new way to provide better control over existing knowledge domain and the ability to act on them.

- *Volume*. This term means the quantity of generated and stored data. The size of the data determines the value and potential insight and whether it can actually be considered big data or not. Taking Twitter as an example, it may generate 7 terabytes (TB) of data every day [7]. As shown in Fig. 7.1, it presents that data volumes have changed from terabytes to petabytes with an inevitable shift to zettabytes. In such case, some data cannot be analyzed in time or even will not be analyzed. For this gap between the increased data volumes and the relatively decreased processing capability, it is hard to predict the potential consequences.

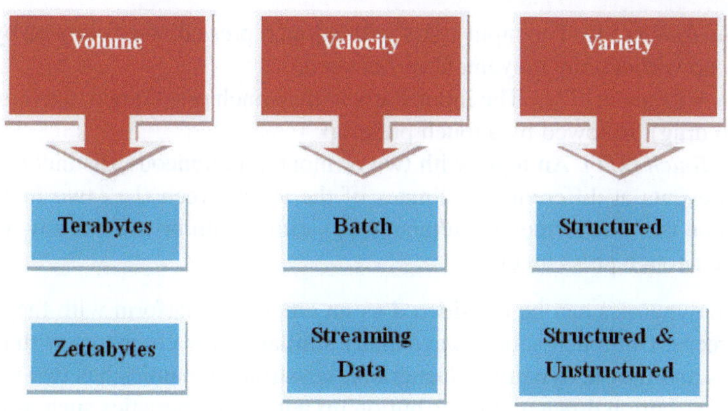

Fig. 7.1 The view from IBM on big data: its volume, velocity, and variety

- *Variety.* The term indicates the type and nature of the data. With existing collaboration technologies, data in an organization has already become complex as it contains not only traditional relational data, but also raw, semi-structured, and unstructured data from other sources such as web pages, email messages, sensor data, and search indexes. Figure 7.1 shows a fundamental shift in analysis requirements from traditional structured data to unstructured data. Subsequently, traditional analysis tools cannot handle this new characteristic of data volume.
- *Velocity.* This term means the speed at which the data is generated and processed to meet the demands and challenges that lie in the path of growth and development. Figure 7.1 shows a shift from merely batch insight to streaming insight. For instance, current organizations usually deal with petabytes of data instead of terabytes and the increasing information streams have made a constant flow of data at a pace, in which a traditional system is hard to handle [7]. As a result, handling the data requires conducting analysis against the volume and variety of data while it is still in the motion.

There also existing other characteristics such as variability and veracity. The former means inconsistency of the data can hamper processes to handle and manage while the latter means that the quality of captured data can vary greatly, affecting accurate analysis. Hence, big data brings new challenges in the aspects of security and privacy [8]. As data processing is critical for smartphone users authentication, big data would bring many new challenges and issues as well.

7.2.2 Touch Dynamics-Based User Authentication

Touch dynamics can be defined as the characteristics of the inputs received from a touchscreen when a user is interacting with a touchscreen device. There are many types of features regarding touchscreen inputs, which can be classified into one of the following categories [9]:

- *Single-Touch (ST).* The input starts with a touch press down, followed by a touch press up without any movement in-between.
- *Touch Movement (TM).* The input starts with a touch press down, movement (also called drag), followed by a touch press up.
- *Multi-Touch (MT).* An input with two or more simultaneous, distinct touch press down events at different coordinates of the touchscreen (i.e., two fingers press down on the touchscreen simultaneously), either with or without any movement before a touch press up event.

Touch dynamics can be considered as an emerging platform with the advent of touchscreens. But touch actions have many similarities as compared with keystroke dynamics and mouse dynamics. Generally, keystroke dynamics has two main input types: press button down and press button up with characteristics such as key hold-time and digraph time. Mouse dynamics has a bit more features like mouse-move, drag-drop, and move-click with characteristics such as movement speed, movement direction, and movement distance. The differences and similarities among touch dynamics, keystroke dynamics, and mouse dynamics can be summarized as below [9].

- *Differences.* Intuitively, touch dynamics is different from keystroke dynamics due to more input types available such as multi-touch and touch movement, and is different from mouse dynamics due to multi-touch inputs. In particular, keystroke dynamics only has buttons as input methods, so there is no movement feature. Regarding mouse dynamics, the trace in mouse dynamics is continuous (i.e., mouse inputs can start from the last point where the last mouse input was terminated) while the trace in touch dynamics can be non-continuous (i.e., a touch input can start at a different point than the point where the last touch input ended). Overall, touch dynamics offers movement features and can provide more behavioral characteristics.
- *Similarities.* Each dynamics has their own features, while touch dynamics has some similarities with the others in certain aspects. For example, keystroke dynamics enables press button up and press button down, which are similar to the actions of touch press up and touch press down (e.g., single-touch) in touch dynamics. For mouse dynamics and touch dynamics, both of them offer a movement input (i.e., mouse movement versus touch movement). More specifically, a single-touch input can be considered similar to a click action in mouse dynamics. In the context of inputs, touch dynamics can be considered as a combination of keystroke dynamics and mouse.

Figure 7.2 shows a typical architecture of touch dynamics-based authentication systems on smartphones, including data collection, profile construction, and profile comparison.

In particular, the *data collection component* is used for collecting raw data from the touchscreen (i.e., recording all touch gestures) and converting the collected data into meaningful information (i.e., identifying sessions for users and filtering out noise data [9]). The *profile construction component* is responsible for analyzing collected raw data, extracting features, and building a user's normal profile. The *profile comparison component* aims to compare the current user's actions with the stored profile and decide whether the current user is a legitimate user or not.

7.3 A Review of Touch Dynamics-Based User Authentication on Smartphones

Started nearly from 2012, touchscreens have rapidly become a leading input method among various mobile devices like smartphones and ipads. Hence, research has been moved to touch dynamics and how to design proper touch dynamics-based user authentication schemes. In this section, we provide a review of touch dynamics-based user authentication on smartphones.

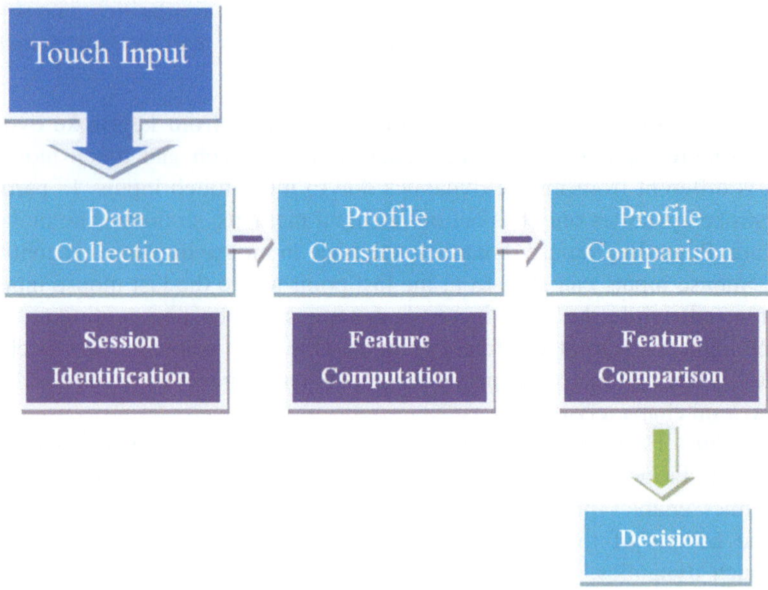

Fig. 7.2 Typical architecture of touch dynamics-based user authentication systems on smartphones

7.3.1 *Year: 2012*

Touch dynamics (or touch gestures) related authentication has received more attention from 2012. Feng et al. [10] noticed that securing the sensitive data stored in mobile devices is a problem of paramount importance, so they proposed *FAST*, a touch-based user authentication mechanism that was able to support both passive and continuous authentication for mobile users based on their touch gestures. One advantage of *FAST* is its transparency to the user: namely, the touch data is captured by sensors without disrupting normal user–device interactions. In the login stage, the traditional explicit authentication process is triggered only when FAST detects an anomaly (i.e., the current user is likely different from the smartphone owner).

For implementation, they designed a digital sensor glove with IMU digital combo boards ITG3200/ADXL345, which can provide six degrees of freedom and allow collecting fine-grained biometric information of finger movements. More specifically, *FAST* collects defined gestures and features including gesture type, *X* and *Y* coordinates, directions of the finger motion, finger motion speed, pressure at each sampled touch point, and the distance between multi-touch points. They then constructed 53 features for each touch gesture. They mainly considered six frequent and useful gestures: down to up swipe, up to down swipe, left to right swipe, right to left swipe, zoom-in, and zoom-out. In the user study with a total of 40 participants, they achieved a False Accept Rate (FAR) of 4.66 % and False Reject Rate of 0.13 % in the login phase.

In the same year, Meng et al. [9] presented the first work that defined the concept of touch dynamics and proposed a touch dynamics-based user authentication scheme for smartphones. They particularly extracted 21 different features to construct an authentication signature for user authentication. The features are the following: average touch movement speed per direction (eight directions), fraction of touch movements per direction (eight directions), average single-touch time, average multi-touch time, number of touch movements per session, number of single-touch events per session, and number of multi-touch events per session.

For data collection, they used a Google/HTC Nexus One Android phone with a capacitive touchscreen (resolution 480×800 px) and gave to users. The advantage of this particular phone is that the stock Android operating system can be replaced with a modified custom version of the Android OS. The modification consists of changes to the application framework layer to record raw input data from the touchscreen, such as the timing of touch inputs, the coordinates x and y, and the type of the inputs (e.g., single-touch, multi-touch, and movement). They also installed a separate application, which allows extracting the recorded data from the phone. In the study, they have 20 participants and designed a hybrid classifier called PSO-RBFN. With this classifier, they could achieve an average error rate of 2.92 % (FAR of 2.5 % and FRR of 3.34 %).

7.3.2 Year: 2013

From 2013, touch dynamics received more attention. Frank et al. [11] presented an investigation on whether a classifier can continuously authenticate users based on the way they interact with the touchscreen of a smartphone. That is, they tried to check if it is possible to authenticate users while performing basic navigation steps on a touchscreen device and without any dedicated and explicit security action that requires attention from the user. To achieve this goal, they designed a proof-of-concept classification framework to collect a set of 30 behavioral features that can be extracted from the touchscreen input of commodity mobile devices. The framework can train user profiles based on vertical and horizontal strokes using a nearest neighbor classifier and a Gaussian RBF kernel support vector machine (SVM). These classifiers could achieve robust authentication results, with equal error rates (EERs) between 0 and 4 %, depending on the application scenario, which demonstrate the feasibility of their proposed authentication scheme based on natural navigation gestures (i.e., a median EER of 0 % for intra-session authentication, 2–3 % for intersession authentication, and below 4 % when the authentication test was carried out 1 week after the enrollment phase).

Then, Li et al. [12] presented a re-authentication system for smartphones based on users' finger movements. Their system could keep running in the back end, learn the owner's finger movement patterns, continuously monitor the current user's finger movement, and compare the current user's movement patterns against the owner's patterns. In particular, they used 13 metrics to measure a sliding gesture such as first touch position, first touch pressure, first touch area, and first moving direction. They implemented the monitoring program, which needs the root privilege, into the lower layer of an Android system. They then conducted a study with 75 users and carried out the process of data collection and all the experiments on two Motorola Droid phones, with 550 MHz A8 processor, 256 MB memory, 16 GB sdcard, and Android 2.2.2 OS. With SVM classifier, they could achieve the best accuracy of 95.78 % for sliding up gestures.

7.3.3 Year: 2014

The research on touch dynamics maintained its trend, but focused more on multi-touch in 2014. Sae-Bae et al. [13] targeted on multi-touch gestures for user authentication on touchscreen-based devices. They constructed up to 22 multi-touch gestures that can be extracted from hand and finger movement. They considered one important advantage of gesture-based user authentication schemes is revocability. For example, when a gesture is compromised or no longer effective, it can be replaced by another gesture. They also implemented a position and rotation invariant multi-touch gesture classifier to distinguish different users. Two studies were conducted to validate the performance and usability. The first study aims to test biometric performance and in terms of usability, the experimental results show strong correlation between positive user experience, in terms of ease of use, pleasure, and excitement, and biometric performance of the gestures. In terms of memorability, they found that it is feasible for a user to recall user-defined gestural passwords and it is observed that the recall rate increases over time.

Sun et al. [14] argued that current mobile authentication techniques were often ineffective to use and also vulnerable to several attacks such as shoulder-surfing and smudge attacks. For better authentication, they presented and designed *TouchIn*, a two-factor authentication system for multi-touch mobile devices. In the enrollment phase, a user performs one or multiple fingers to draw arbitrary geometric curves of his/her own choice (called a curve password) on the touchscreen. Then, an authentication template is created based on the features extracted from this input, including x-coordinate, y-coordinate, direction, curvature, x-velocity, y-velocity, x-acceleration, y-acceleration, finger pressure, and hand geometry. The first four features can together accurately define the geometric characteristics of the drawn curve. Under *TouchIn*, it is nearly impossible for attackers to infer and forge even if they may manage to know the correct curve password. Their user study with 30 participants showed that *TouchIn* has very low false positives and negatives, denying unauthorized users for 97.7 % (97.8 %) of the time and admitting authorized users for 97.5 % (99.3 %) of the time for a password involving a single curve (multiple curves).

Meng et al. [15, 16] then identified that the selection of classifiers may impact the authentication performance, so they designed an adaptive mechanism that can periodically select a better classifier to maintain the authentication accuracy during user authentication. In their work, they proposed a lightweight touch-dynamics-based user authentication scheme on a touchscreen mobile phone, which consists of only eight touch gesture related features. In addition, they implemented a cost-based metric that enables this mechanism to choose a less costly classifier. In the evaluation with 50 participants, the experimental results demonstrated that the adaptive authentication scheme can achieve an average error rate of 2.46 %. Then, Xu et al. [17] aimed to study how to effectively model multiple types of touch data and perform continuous authentication. They employed SVM classifier and recruited 32 users in their study. They found that their designed biometric system could generally achieve EER values lower than 10 % for all operation types, where the slide operation performs the best by achieving an EER lower than 1 %.

7.3.4 Year: 2015

The research work on touch dynamics keeps increasing in 2015. Jain and Kanhangad [18] presented a user authentication scheme for smartphones based on behavioral biometrics. Similar to other work, they also deployed an android application that we run on the smartphones and acquire a set of behavioral data when the user interacts with the device, including seven gestures: left to right swipe (L2R), right to left swipe (R2L), scroll up (SU), scroll down (SD), zoom in (ZI), zoom out (ZO), and single tap (ST). They further employed 104 users and conducted feature matching using the modified Hausdorff distance (MHD). The results showed that the proposed MHD-based matching could achieve better performance (in terms of both error rate and computational time) than the approach based on DTW. In the best scenario, their approach could achieve the lowest EER of 0.31 % when all the gestures performed by the users are combined at the score level for user authentication.

Yang et al. [19] presented *TapLock*, a smartphone password system that exploits the finger tap events on capacitive touchscreens for increasing the password's resilience to shoulder-surfing attacks. They found that the size and major/minor axis length of the finger touch area on touchscreens are useful to create a multi-dimensional password that is shoulder-surfing resistant.

7.3.5 Year: 2016

Research on touch dynamics is still being popular in 2016. For instance, Meng et al. [20] argued that some additional mechanisms should be added to improve the security of unlock patterns. For this sake, they conducted two studies. The first one showed that users would perform a touch movement differently when interacting with the touchscreen and that users would perform somewhat stably for the same pattern after several trials. In the second study, they developed a touch movement-based security mechanism, called *TMGuard*, to enhance the authentication security of Android unlock patterns by verifying users' touch movement during pattern input. In the evaluation with 75 participants, they found that *TMGuard* can positively improve the security of Android unlock patterns without compromising its usability. Taking 9-dot patterns as an example, the successful rate is decreased from 97.8 % to 91.1 % for males and from 97.8 % to 88.9 % for females, respectively.

7.3.6 Discussion

Overall, touch dynamics has promising potentials to strengthen the security of mobile devices or on-line services accessible via mobile devices without additional hardware. With the rapid development of touchscreen techniques, the availability of various sensors in recent mobile devices provides added opportunities for such potentials to be explored [21]. Owing to the potential of touch dynamics biometrics, there have been increasing research efforts in this topic area.

Thanks to this, touch dynamics will become much more important in designing future biometric authentication mechanisms on mobile phones. For instance, the use of multi-touch gestures can provide more choices to enlarge password entropy and increase the difficulty of observation and smudge attacks [6]. We summarize the related work mentioned above in Table 7.1. More details about recent research studies on this topic can be referred to relevant surveys [6, 21].

7.4 Challenges and Future Trends in the Era of Big Data

Existing research on touch dynamics has not considered big data context; however, with the rapid increase of data volumes, the performance of touch dynamics-based authentication will be significantly affected, especially on mobile phones (i.e., with limited resources). In this section, we describe several potential challenges and point out future directions in this field.

7.4.1 Challenges Under Big Data

7.4.1.1 Volume Issue: Large Data Volume

User authentication is a performance-sensitive task, which should balance between both security and usability. For this sake, it is extremely important to establish an appropriate model for users' profile. Traditionally, users' actions can be represented with several features or vectors, which mean that relevant datasets are manageable.

Table 7.1 A summary of touch dynamics-based user authentication schemes on smartphones

Work	Proposed mechanism	Results
Feng et al. [10] in 2012	FAST with 53 features	• FAR: 4.66 % • FRR: 0.13 %
Meng et al. [9] in 2012	A touch dynamics-based authentication scheme with 21 features	• FAR: 2.5 % • FRR: 3.34 %
Frank et al. [11] in 2013	Touchalytics with 30 features	• ERR: nearly 4 %
Li et al. [12] in 2013	Sliding gesture with 13 features	• Best Acc.: 95.78 % for sliding up
Sae-Bae et al. [13] in 2014	Multi-touch based authentication with 22 features	• Best EER: 8.86 %
Sun et al. [14] in 2014	TouchIn	• FAR: 2.5 % • FRR: 2.3 %
Meng et al. [15] in 2014	Adaptive authentication scheme with eight features	• FAR: 2.55 % • FRR: 2.37 %
Xu et al. [17] in 2014	User authentication with touch operations	• EER: 10 %
Jain and Kanhangad [18] in 2015	User authentication scheme with seven features	• Best EER: 0.31 %
Meng et al. [20] in 2016	TMGuard	• EER: 1–3 %

However, in the era of big data, datasets would become very large, making profile modeling a difficult task to complete. Since most existing touch dynamics-based authentication schemes utilize machine learning techniques in profile classification, their performance can be greatly decreased under large datasets (i.e., resulting in large delays).

Such issues have already appeared in intrusion detection when dealing with large datasets [22, 23]. Taking Snort as an example, it usually spends about 30 % of its total processing time in signature comparison, but the time consumption may reach over 80 % when in a heavy traffic scenario [24]. It is the same for biometric authentication; large data volumes can greatly exhaust authentication applications and cause huge delays. As a result, existing touch dynamics-based schemes would not be effective on smartphones in the era of big data.

7.4.1.2 Velocity Issue: Real-Time Requirement

User authentication is a time-sensitive task as well, where any delay in login process may cause usability issues. On the other hand, the early detection of an imposter can prevent data leakage and reduce the loss. Thus, real-time requirement is extremely crucial for touch dynamics-based authentication on smartphones in the aspects of usability and security. However, under big data, large data can continuously pass through an authentication application, which can exhaust its processing capability. Thus, it is a big challenge for touch dynamics-based schemes to meet the real-time requirement, especially on smartphone, in the era of big data.

This situation is similar in intrusion detection. For the signature-based detection, the process of signature matching is a key limiting factor to hinder the performance of detection, in which the matching burden is at least linear to the size of an input string [25]. In the era of Big Data, more packets can arrive per unit time so that the process of signature matching will become a main bottleneck.

7.4.1.3 Variety Issue: Profile Establishment

After reviewing existing touch dynamics-based authentication schemes, machine learning techniques should be the most commonly used means for building users' profile. However, big data may cause many issues for the use of machine learning techniques in traditional user authentication due to data variety:

- Traditionally, a learning classifier should be trained on a labeled dataset, and it may not be suitable for another dataset. This means that a traditional machine learning algorithm cannot handle *variety* since it cannot be robust and scalable over different datasets (i.e., needing adjustment).
- To train a machine learning classifier, the first work is to decide class types and obtain a certain number of labeled items. But under big data, the number of class types can be dynamically growing, which makes modeling a hard task and results in inaccurate classification.

7.4.1.4 Discussion

Overall, potential challenges can be caused by the distinct features of big data such as volume, velocity, and variety. Under big data, it seems to be unlikely for a biometric authentication scheme to perform analytics and complex queries on large, unstructured datasets with incomplete and noisy features. The management of large data has usually been expensive, and their deployment may require strong business support. More specifically, big data may cause many security and usability issues in designing a proper touch dynamics-based authentication.

- *Feature selection and extraction.* As mentioned, most authentication schemes including touch dynamics-based authentication take advantage of machine learning classifiers for user verification. However, big data makes traditional machine learning techniques difficult to model users' profile. Thus, it becomes a big challenge to decide features with big data volumes.
- *Benchmark.* The literature reveals that the availability of public datasets is still very limited in this field. Without properly documented and realistic datasets, cross-comparison between different methodologies may not be conclusive [21].

Under big data, this issue will become more severe. One reason is that collecting such dataset is time-consuming and requires a large storage space. Another reason is that it is hard to control the environment to collect such a dataset (i.e., how to set a real environment to collect large data).

To sum up, big data may increase the error rates in user authentication, which will harm both security and usability. For security, big data may delay authentication process and reduce the classification accuracy. For usability, users are likely to quit installing any user authentication scheme due to inconvenience such as authentication delay and inaccuracy. Therefore, developing new authentication applications and schemes to handle big data is very challenging.

7.4.2 *Future Trends in the Era of Big Data*

7.4.2.1 Big Data Tuning

In the era of big data, an important thing is to clearly define particular targets for touch dynamics-based authentication on smartphones. Big data can provide many new characteristics; thus, it is very important to tune it in the specific context. For instance, there is a need to define authentication goals in terms of three variables such as volume, velocity, and variety.

7.4.2.2 Machine Learning Improvement

Machine learning techniques may suffer from many issues under big data, but they are a common and powerful tool for existing user authentication schemes. In this case, improving existing machine leaning classifiers is very essential [26]. For example, machine learning algorithms with appropriate evaluation metrics should be designed, which are able to handle the large data volumes in an effective way.

In addition, traditional machine learning classifiers can only deal with pre-defined features. But in the era of big data, the feature of *variety* causes data even complex. Thus, it is necessary to design new features for particular user authentication schemes based on touch gestures and operations.

7.4.2.3 Advanced Techniques Development

With the large data volumes (big data), it is difficult for a touch dynamics-based user authentication scheme to perform real-time detection on mobile devices, resulting in high false acceptance rate and false rejection rate. Hence, more efficient and advanced analysis techniques should be developed in the era of big data [27]. For example, new tools such as the Hadoop ecosystem, stream mining, complex-event processing, and NoSQL databases can help enable the analysis of large-scale, heterogeneous datasets at unprecedented scales and speeds. Tuning those existing big data related tools is one of the promising solutions.

In addition to common tools like machine learning, statistical analysis could be used in user authentication to assist in performance improvement. As an example, Meng et al. [20] have developed a statistic-based scheme in *TMGuard*, called *proportional matching*, aiming to make a better balance between security and usability. This method specifically uses a confidence threshold in the authentication. That is, users are only allowed to perform a touch movement within a defined deviation. For instance, if we set the confidence threshold to 0.98; thus, it is allowed a deviation less than 0.02 as compared to the stored normal profile. Such statistical method can help speed up the authentication and provide more flexibility in settings.

7.4.2.4 Discussion

The term of big data is still a hot topic in many fields, whereas most existing research on touch dynamics-based authentication has not paid much attention to it. Similar to the area of intrusion detection, the followings are the major goals of designing future user authentication based on touch dynamics in the era of big data [28].

- Handling large touch data volumes or data inputs on smartphones.
- Developing more efficient user authentication schemes using touch dynamics to meet real-time requirement.
- Designing standards of selecting and extracting appropriate data features from touch operation and actions.
- Developing more robust machine learning algorithms for modeling users' profile, which fit mobile platforms.

Several technologies have been proposed to handle big data including massively parallel-processing databases, distributed file systems and databases, and cloud-based infrastructure. In future, such technologies can be tuned in touch dynamics-based user authentication to meet the above goals.

7.5 Conclusion

In this chapter, we mainly focus on the impact of big data on touch dynamics-based user authentication. In particular, we present a review of recent development relating to smartphone touch dynamics in the literature, and provide insights about its future trends in the big data era. We identified that big data may increase the error rates in user authentication, which will harm both security and usability. Then, we point out future trends regarding touch dynamics-based authentication in the era of big data. To the best of our knowledge, this chapter provides an early view on the impact of big data on user authentication using touch dynamics. It is worth noting that the identified challenges can be applicable to other user authentication domains. Our effort attempts to simulate more research studies to these challenges and issues in this area.

References

1. Worldwide Smartphone Growth Goes Flat in the First Quarter (2016) Preliminary Report from the International Data Corporation (IDC). http://www.idc.com/getdoc.jsp?containerId=prUS41216716
2. N.K. Ratha, J.H. Connell, R.M. Bolle, Enhancing security and privacy in biometrics-based authentication systems. IBM Syst. J. **40**(3), 614–634 (2001)
3. R.N. Shepard, Recognition memory for words, sentences, and pictures. J. Verbal Learn. Verbal Behav. **6**(1), 156–163 (1967)
4. A.S. Brown, E. Bracken, S. Zoccoli, K. Douglas, Generating and remembering passwords. Appl. Cogn. Psychol. **18**(6), 641–651 (2004)
5. N.L. Clarke, S.M. Furnell, Authentication of users on mobile telephones—a survey of attitudes and practices. Comput. Secur. **24**(7), 519–527 (2005)
6. W. Meng, D.S. Wong, S. Furnell, J. Zhou, Surveying the development of biometric user authentication on mobile phones. IEEE Commun. Surv. Tutorials **17**(3), 1268–1293 (2015)
7. P.C. Zikopoulos, C. Eaton, D. deRoos, T. Deutsch, G. Lapis, *Understanding Big Data—Analytics for Enterprise Class Hadoop and Streaming Data* (McGraw-Hill, New York, 2012)
8. R. Lu, H. Zhu, X. Liu, J.K. Liu, J. Shao, Toward efficient and privacy-preserving computing in big data era. IEEE Netw. **28**(4), 46–50 (2014)
9. Y. Meng, D.S. Wong, R. Schlegel, L.-F. Kwok, Touch gestures based biometric authentication scheme for touchscreen mobile phones, in *Proceedings of the 8th China International Conference on Information Security and Cryptology (INSCRYPT)*. Lecture Notes in Computer Science, vol. 7763 (Springer, 2012), pp. 331–350, Beijing
10. T. Feng, Z. Liu, K.-A. Kwon, W. Shi, B. Carbunar, Y. Jiang, N. Nguyen, Continuous mobile authentication using touchscreen gestures, in *Proceedings of the 2012 IEEE Conference on Technologies for Homeland Security (HST)*, Waltham (2012), pp. 451–456
11. M. Frank, R. Biedert, E. Ma, I. Martinovic, D. Song, Touchalytics: on the applicability of touchscreen input as a behavioral biometric for continuous authentication. IEEE Trans. Inf. Forensics Secur. **8**(1), 136–148 (2013)
12. L. Li, X. Zhao, G. Xue, Unobservable re-authentication for smartphones, in *Proceedings of the 20th Annual Network and Distributed System Security Symposium (NDSS)*, San Diego (2013), pp. 1–16
13. N. Sae-Bae, N. Memon, K. Isbister, K. Ahmed, Multitouch gesture-based authentication. IEEE Trans. Inf. Forensics Secur. **9**(4), 568–582 (2014)
14. J. Sun, R. Zhang, J. Zhang, Y. Zhang, TouchIn: sightless two-factor authentication on multitouch mobile devices, in *Proceedings of the 2014 IEEE Conference on Communications and Network Security (CNS)*, San Francisco (2014), pp. 436–444
15. Y. Meng, D.S. Wong, L.F. Kwok, Design of touch dynamics based user authentication with an adaptive mechanism on mobile phones, in *Proceedings of the 29th Annual ACM Symposium on Applied Computing (ACM SAC)*, Gyeongju (2014), pp. 1680–1687
16. W. Meng, D.S. Wong, L.F. Kwok, The effect of adaptive mechanism on behavioural biometric based mobile phone authentication. Inf. Manage. Comput. Secur. **22**(2), 155–166 (2014)
17. H. Xu, Y. Zhou, M.R. Lyu, Towards continuous and passive authentication via touch biometrics: an experimental study on smartphones, in *Proceedings of the Tenth Symposium on Usable Privacy and Security (SOUPS)*, Menlo Park (2014), pp. 187–198

18. A. Jain, V. Kanhangad, Exploring orientation and accelerometer sensor data for personal authentication in smartphones using touchscreen gestures. Pattern Recogn. Lett. **68**(2), 351–360 (2015)
19. H. Yang, L. Chen, K. Bian, Y. Tian, F. Ye, W. Yan, T. Zhao, X. Li, TapLock: exploit finger tap events for enhancing attack resilience of smartphone passwords, in *Proceedings of the IEEE International Conference on Communications (ICC)*, London (2015), pp. 7139–7144
20. W. Meng, W. Li, D.S. Wong, J. Zhou, TMGuard: a touch movement-based security mechanism for screen unlock patterns on smartphones, in *Proceedings of the 14th International Conference on Applied Cryptography and Network Security (ACNS 2016)*, Guildford (2016), pp. 629–647
21. P.S. Teh, N. Zhang, A.B.J. Teoh, K. Chen, A survey on touch dynamics authentication in mobile devices. Comput. Secur. **59**, 210–235 (2016)
22. Y. Meng, L.F. Kwok, W. Li, Towards designing packet filter with a trust-based approach using Bayesian inference in network intrusion detection, in *Proceedings of the 8th International ICST Conference on Security and Privacy in Communication Networks (SecureComm)*, Padua (2012), pp. 203–221
23. A. Papadogiannakis, M. Polychronakis, E.P. Markatos, Improving the accuracy of network intrusion detection systems under load using selective packet discarding, in *Proceedings of the Third European Workshop on System Security (EUROSEC)*, Paris (2010), pp. 1–7
24. Y. Meng, L.F. Kwok, Adaptive context-aware packet filter scheme using statisticbased blacklist generation in network intrusion detection, in *Proceedings of the 7th International Conference on Information Assurance and Security (IAS)*, Melacca (2011), pp. 74–79
25. H. Dreger, A. Feldmann, V. Paxson, R. Sommer, Operational experiences with high-volume network intrusion detection, in *Proceedings of the 11th ACM Conference on Computer and Communications Security (ACM CCS)*, Washington (2004), pp. 2–11
26. X. Wu, X. Zhu, G.-Q. Wu, W. Ding, Data mining with big data. IEEE Trans. Knowl. Data Eng. **26**(1), 97–107 (2014)
27. T.-F. Yen et al., Beehive: large-scale log analysis for detecting suspicious activity in enterprise networks, in *Proceedings of the Annual Computer Security Applications Conference (ACSAC)*, New Orleans (2013), pp. 199–208
28. W. Meng, W. Li, A review of network intrusion detection in the big data era: challenges and future trends, in ed. by S. Yu, X. Lin, J. Misic, Xuemin (Sherman) Shen. *Networking for Big Data (Book)* (CRC Press, Taylor & Francis, Boca Raton, 2015), pp. 195–213

Chapter 8
Enhanced Biometric Security and Privacy Using ECG on the Zynq SoC

Amine Ait Si Ali, Xiaojun Zhai, Abbes Amira, Faycal Bensaali, and Naeem Ramzan

8.1 Introduction

Human identification systems are crucial nowadays. Such systems can be used to monitor patients in hospitals, to check identities in airports, or to access restricted areas in work places. Those recognition systems which are deployed in sensitive zones require a high level of security and privacy as well as a fast response time. Human recognition systems can be based not only on simple passwords or photo identity cards, but also on more secure biometrics [1]. Fingerprint, iris, voice, and face are among the most common used biometrics for identification. Population of old people is the most dominated in the highest developed countries and regions, which requires governments to deal with the aging problems in the health-care sector. One of the most promising technologies is found in the field of wireless monitoring of vital signs, in which it embraces the measuring and digitization of vital signs such as the blood pressure or electrocardiogram (ECG), transmitting packets over a wireless network, and delivering this medical information to health-care professionals. It can not only be used in clinics and hospitals to monitor patients, but also outside of medical facilities, giving elderly or sick people the opportunity to be part of their social communities by being simultaneously monitored and/or even in contact with their health-care provider. However, since the health-care data contains highly sensitive and personal data, strict security measures

A. Ait Si Ali (✉) • A. Amira • F. Bensaali
KINDI Center for Computing Research, Qatar University, P. O. Box: 2713, Doha, Qatar
e-mail: amine.aitsiali@qu.edu.qa

X. Zhai
College of Engineering and Technology, University of Derby, Derby DE22 1GB, UK

N. Ramzan
School of Computing and Engineering, University of the West of Scotland,
Paisley PA1 2BE, UK

R. Jiang et al. (eds.), *Biometric Security and Privacy*, Signal Processing for Security Technologies, DOI 10.1007/978-3-319-47301-7_8

must be applied for protection and encryption. Therefore in order to apply the strict security measure to protect the sensitive data, there is a high interest in using the health-care monitored biometrics signals to identify patients.

One of the new emerging techniques for human identification consists of the use of ECG signals as biometrics as they are less prone to frauds [2]. It has been widely used in the field of health monitoring where it can provide an automatic living identification system. The drawback of using ECG signals for human identification is that there are a large number of signals coming from the sensor that need to be processed in real time due to the high sensitivity of the application, but also the fact that data in connected health is exchanged between various platforms, people, and organizations which lead to a privacy and security issue. Therefore, new solutions are necessary to increase both the data stream processing efficiency and privacy level in this area.

Heterogeneous platforms such as the Xilinx ZC702 Zynq system-on-chip (SoC) [3] are increasingly being used by researchers for hardware acceleration of computationally intensive algorithms to meet real-time requirements. The Zynq SoC is a programmable SoC tightly integrated dual-core ARM Cortex-A9 processors with Xilinx 7-series field programmable gate array (FPGA) logic, which has been widely used by researchers to accelerate digital signal processing (DSP) algorithms to meet real-time requirements. Researchers could take the advantages of parallelism, pipelining, and the hardware/software co-design offered by the Zynq SoC and create a low-cost heterogeneous platform to provide the users with high performance and flexibility.

A security solution is presented in this book chapter for the identification of individuals using ECG biometric as well as for the encryption and decryption of ECG signals using the advanced encryption standard (AES) algorithm. Three different ECG databases representing 2372 samples are used. The identification is performed using principal component analysis (PCA) along with Euclidian distance, k-nearest neighbors (KNN), and extended nearest neighbor (ENN) [4], a comparison between all those classifiers is performed. The Xilinx ZC702 Zynq based prototyping board is used for the hardware implementation of the proposed system. The hardware implementation has shown promising results as it outperformed current FPGA based systems in terms of power consumption, processing time, and hardware resources usage. Using the most optimized hardware implementation, one ECG sample can be processed in 10.71 ms with a power consumption of 107 mW while using only 30 % of all available on-chip resources. Moreover, the classification accuracy is between 94 % and 100 % depending on the classifier and on the dataset used.

The remaining parts of this book chapter are organized as follows. The simulation of the proposed approach for AES encryption and ECG identification technique is presented in Sect. II. Section III is concerned with the system architecture and its implementation on the Zynq SoC. Experimental results and various analyses are then given in Sect. IV. Finally, Sect. V concludes the book chapter.

8.2 Related Work

Connected health system is the one of the emerging application areas within the internet of things (IoT) domain, which has been increasingly attracting many researcher. A heart electrical impulses are introduced to monitor the heart of cardiac patients based on the ECG measurements, where the collected ECG signals are sent to a personal digital assistant (PDA) for diagnosis purpose [5]. The visualized medical data is displayed on the PDA, and medical staff can be informed in the case of an emergency or critical situation. The connected health system presented in [6] consists of a glove and a set of configured sensors to measure hand finger flexion for rehabilitation purposes. The sensors are placed in specific places and connected to an acquisition unit, which aims to give feedback to the rehabilitation system based on the positions of the fingers. However, the major issues with both applications are the security and privacy which are not well considered, due to the fact that the processing power of the IoT devices are limited.

AES which is considered as the most secure encryption algorithms is being extensively used not only in various applications in general but also in healthcare related applications [7]. Various hardware implementations of the AES algorithm have been performed. A low-area implementation with a low-throughput was presented in [8]. This implementation is not suitable for sensitive applications such as connected health where the response time is very important. Alternatively, a high-throughput solution was discussed in [9]; however, a high amount of hardware resources were used and much design efforts were needed. Another FPGA implementation of AES was presented in [10], the claimed throughputs are 58.13 and 36.5 Mbps for the implementations on a Spartan 6 and a Spartan 3, respectively.

Various techniques have been applied with different implementations to identify human using ECG signals. In [11], the fiducial based approach is presented, which consists in the search for specific points of interest to extrapolate timing and magnitude measurements. The extracted features are linked to peaks and timing duration of the P, QRS, and T waves which are the representation of the heart activity in terms of depolarization and repolarization of the atria and ventricles. The statistical and analytical features from the morphology of the signal waveforms are considered during the identification process. In [12], a fiducial approach based on the multi resolution Daubechies D4 and D6 wavelet transforms is proposed to detect the QRS complex and the onsets and offsets of the P and T waves on the MIT-BIH database [13]. A positive recognition rate of 98 % was achieved. Another approach used to extract features from ECG signals for human identification is called the fiducial independent approach. The identification algorithm presented in [14] achieves a positive recognition rate of 99.6 %.

Various ECG based identification systems were implemented on FPGA. An ECG-QRS recognition system has been implemented on the Xilinx Virtex-II pro FPGA in [15], and a recognition rate of 99.681 % was realized when using the MIT-BIH database [13]. Another SoC implementation was presented in [16], it detects the arrhythmia patterns using ECG signals.

In this chapter, an efficient architecture and implementation of the AES-128 encryption and ECG based identification system for protecting personal medical data privacy are presented. Various ECG based identification algorithms (i.e., PCA, KNN, and ENN) have been implemented and evaluated, the achieved results show that the speed, resources, and power consumption of the proposed implementation are sufficient for the use of the proposed architectures in the connected health system.

8.3 Proposed Solution

The aim of the proposed system is to collect ECG signals in a safe and secure environment such as a hospital, home, or ambulance. An overview of the proposed system is shown in Fig. 8.1. The ECG signals are first collected using two types of wireless ECG sensors which are VS100 [17] and Shimmar3 [18]. The collected data can then be sent wirelessly to the local processing unit based on the Zynq SoC where all information will be processed in real-time. Identification of the patient is performed based on the ECG signals collected locally, other ECG based examinations can be performed in the hospital or in the ambulance. In the case where the data need to be sent to a different location for further examination or for exchange of information, the data will be encrypted first using the AES-128 algorithm and then decrypted when received in the final destination.

The original ECG signals are divided into two sub-sets, one set is used for training and the other set is stored in text files and used to test the cipher encryption, decipher decryption, and identification. In the case of the identification using PCA and Euclidian distance the ECG training set is used in training the phase to generate the other data sets used in the testing phase including the Eigen ECG vectors, projected training vectors, and mean vector. In order to calculate the features of

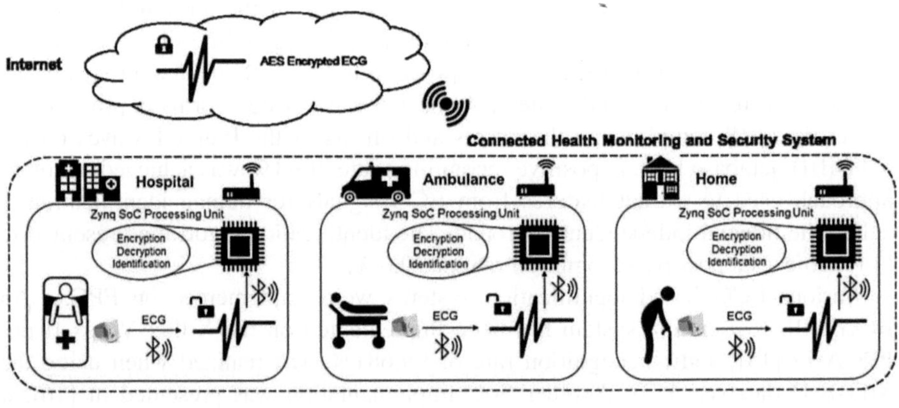

Fig. 8.1 System overview

the ECG signals used for testing, the mean of the training database must first be subtracted from the test signal. Subsequently, the projected vector is then calculated via a multiplication of the Eigen ECG matrix by the mean subtracted signal. This process has been summarized as follows:

1. Data Acquisition
   A training set of ECG signals are stored in a training matrix T with size $(N \times P)$, where each row of the matrix is a different ECG signal. All signals must have the same length and resolution. If necessary the signals have to be resampled before putting them all together into the training matrix. It is also very important that all samples are centered at the same point. For example, the R-wave is a significant point for centering the signal. For a good recognition rate it is also necessary that the ECG samples represent a normal stress level of the person, which can be easily recovered.
2. Mean Subtraction
   First the mean over all training signals is calculated, before subtracting it from each signal separate:

$$m = \frac{1}{p} * \sum_{j=1}^{p} \left(\frac{1}{n} * \sum_{i=1}^{n} T_j(n) \right) \tag{8.1}$$

   where n is number of samples of one ECG signal and p is number of training signals. The centered signals are stored in the matrix A with size $(N \times P)$.

$$A_i = T_i - m \tag{8.2}$$

3. Calculation of the Covariance Matrix
   This is basically done like in the example of the basic mathematics, where all covariances are calculated with respected to one another.

$$C = A * A^T \tag{8.3}$$

   This calculation is quite computation-intensive, but can be optimized by applying a mathematical rule. For a $P \times Q$ matrix, the maximum number of non-zero eigenvalues that the matrix can have is the smaller value of $P - 1$ and $Q - 1$. Since the number of training signals (P) is usually less than the number of signal values N, the most non-zero eigenvalues that can be found are equal to $P - 1$. So we can calculate eigenvalues of $A^T * A$ with size of $P \times P$ instead of $A * A^T$ with size of $(P \times N)$, which has a much smaller dimension. The matrix for calculating the eigenvalues and eigenvectors is a surrogate of the covariance matrix.

$$L = A^T * A \tag{8.4}$$

4. Calculation of the Eigenvectors and Eigenvalues of the Surrogate Covariance Matrix
The eigenvectors and eigenvalues are calculated from the L matrix. The diagonal elements are the eigenvalues for L. At this stage a computer based algorithm is used for computing the values. The outcomes are D (a matrix that contains the eigenvalues) and V (a matrix that contains the eigenvectors).
5. Choosing Components and Forming a Feature Vector
All eigenvalues of matrix L are sorted and those who are less than a defined threshold are eliminated. Once the eigenvectors are calculated the next step is to order them by eigenvalue, highest to lowest. The result is the components in order of significance. The components with lower significance (below threshold) can be ignored, and only a small amount of information is lost. If the low significance data is left out, the final data set will have fewer dimensions than the original. The final step is to form a feature matrix, which contains the selected eigenvectors.

$$Feature\_Matrix\_L = (v_1, v_2, \ldots, v_n) \tag{8.5}$$

However the aim was not to get the features of the L matrix, the aim was to calculate the features of the C matrix. But fortunately this matrix is easy to get from the *Feature_Matrix_L*. A simple multiplication with the matrix A, that contains the centered ECGs, and the *Feature_Matrix_L* gives the eigenvectors of the covariance matrix C. This feature matrix is also called the *EigenECG* matrix in this project because it contains the eigenvectors of the training ECGs.

$$Feature\_Matrix\_L = A * Feature\_Matrix\_L \tag{8.6}$$

$$EigenECG = Feature\_Matrix\_C \tag{8.7}$$

6. Project a Signal into the "ECGspace"
The *EigenECG* matrix contains all of the principal components of the training matrix. This data is useful for data reduction but not for an identification process, where each signal needs a unique representation. To represent a single signal it is projected into the so-called ECGspace. In the "ECGspace" a signal is represented by its main eigenvalues. To get this data the first step is to subtract the mean from signal before multiplying it with the transposed *EigenECG* matrix. This data gives a single vector for each signal that can be used for comparing with other signals, the basic of the identification process.

$$p = E^T \times I \tag{8.8}$$

where E is the Eigen ECG matrix, I is the mean subtracted signal, and p is projected test vector. In the case of ENN and KNN, the training data is stored and used in the testing phase directly without modification. It is worth mentioning that 2372 ECG signal samples are used in total for the evaluation and validation

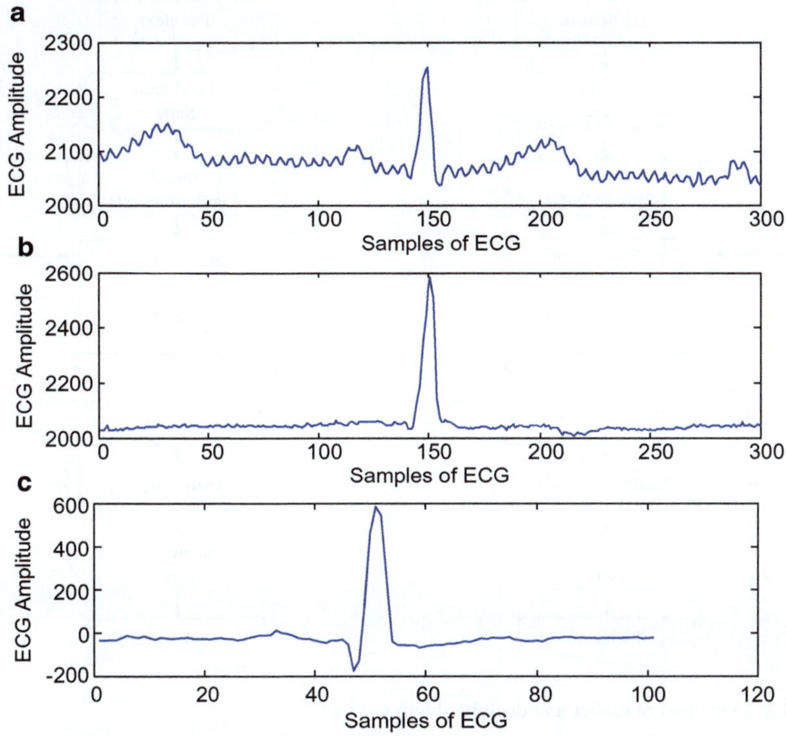

Fig. 8.2 Samples from used ECG database. (**a**) Sample data collected from VS100 ECG sensor; (**b**) Sample collected from Shimmer ECG sensor; (**c**) Sample collected from MIT-BIH database

of the system, the ECG signals are obtained from two different sensors, the VS100 [17] and the Shimmer3 [18]. ECG signals from the MIT-BIH database [13] were also used. The collected ECG signals represent 30 individuals from different age groups and genders. The ECG signals are represented by 300 discrete samples in the case of the VS100 and Shimmer and 100 discrete samples in the case of the MIT-BIH database. Figure 8.2 shows three samples from the used ECG databases.

8.3.1 AES Encryption and Decryption

The text files that contain the second sub-set of the original ECG signals are firstly used in the cipher block. The encrypted text files are then used in decipher block. Finally, the decrypted text files are used in the identification block. The algorithms

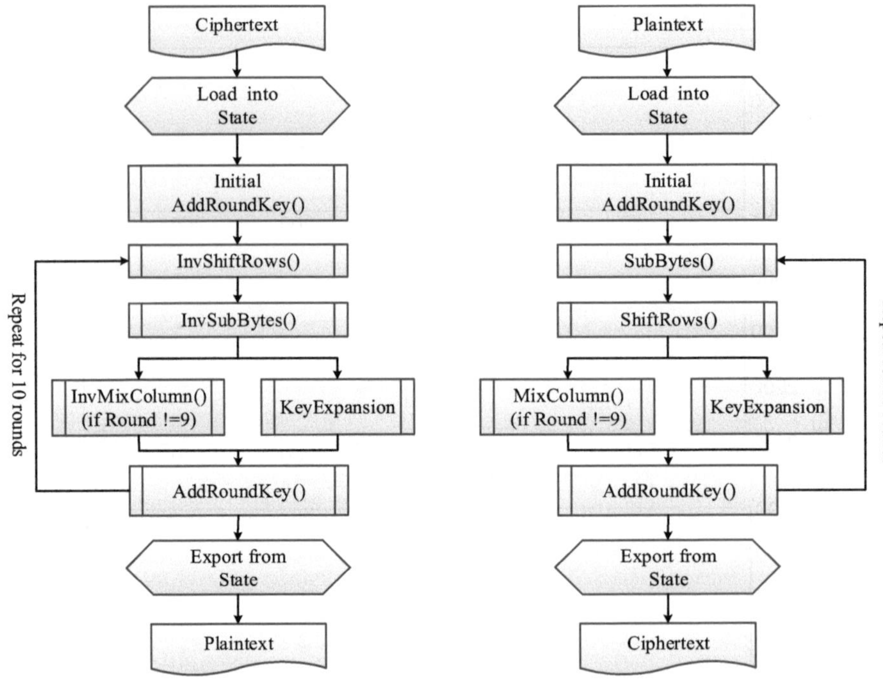

Fig. 8.3 Flowchart of cipher and decipher blocks

of the cipher and decipher blocks, presented in [19], have been implemented on the Zynq SoC using C++. The encryption and decryption algorithm supports AES-128 standard, and the structures of the algorithms are shown in Fig. 8.3. The cipher consists of four basic steps, generally referred to as AddRoundKey(), SubBytes(), ShiftRows(), and MixColumns(). AddRoundKey() takes the round key and XORs every byte of the key with the corresponding byte of the state. SubBytes() substitutes the byte in the state using S-Box. ShiftRows() shifts the rows of the state by different offsets. Row one is not shifted, row two is shifted by one, row three by two, and row four by three. MixColumns() mixes the columns of the matrix, where the state is therefore multiplied with a fixed matrix to gain a new state S'. Decipher is an inverted version of cipher.

8.3.1.1 The State and Key Expansion/Key Scheduling

When loading the plaintext into the cipher, the plaintext vector is transformed to a matrix called State, consisting of four rows and four columns (see Fig. 8.4). AES is a key-iterated cipher. The algorithm takes therefore the AES key to generate round keys with the same vector size as the input plaintext. Therefore it can be transformed into a matrix equal to the size of the state. Each round key applies then to one particular execution round where it is used to manipulate the state. For a further description the following conventions apply:

State	$S = s_{0,0} \ldots s_{3,3}$
Round key	$K = k_{0,0} \ldots k_{3,3}$
Transformed state	$S' = s'_{0,0} \ldots s'_{3,3}$

To generate the round keys, the AES algorithm performs a key expansion routine (also known as key scheduling). The algorithm generates then an expanded key with the overall length of (128 (Nr + 1)) bits, where Nr is the number of rounds. Therefore, the key expansion routine provides for every round plus for the initial ARK a proper key with the fixed length of 128 bits, independently of the AES symmetric key length itself.

The key expansion is a separate task which is not included in the cipher itself and therefore the place where it is finally implemented is open to the designer.

8.3.1.2 AddRoundKey() Transformation

This transformation takes the round key and XORs every byte of the key with the corresponding byte of the state (see Fig. 8.4).

8.3.1.3 SubBytes() Transformation

The next transformation is the substitution of the bytes in the state. Every byte in the state is substituted by a corresponding byte. It is the only non-linear transformation. The transformation itself is shown in Eq. (8.9):

$$S(x) = A * x^{-1} + b \tag{8.9}$$

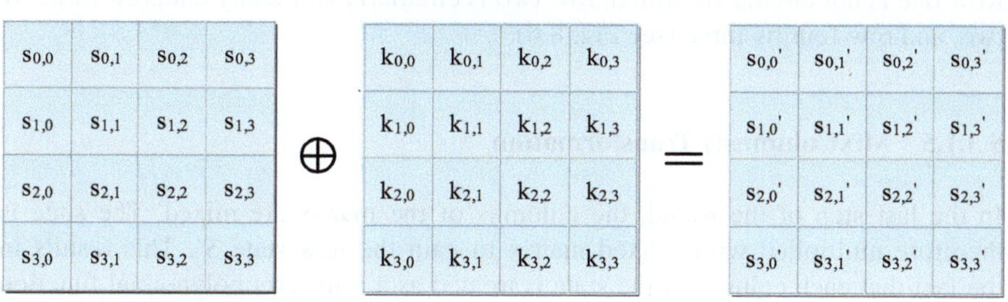

Fig. 8.4 AddRoundKey() transformation

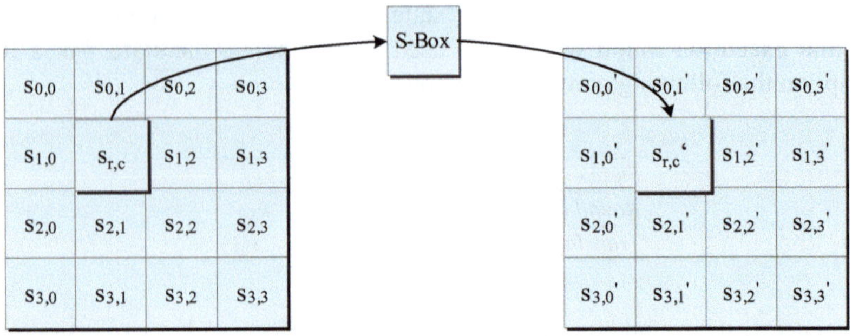

Fig. 8.5 SubBytes() transformation

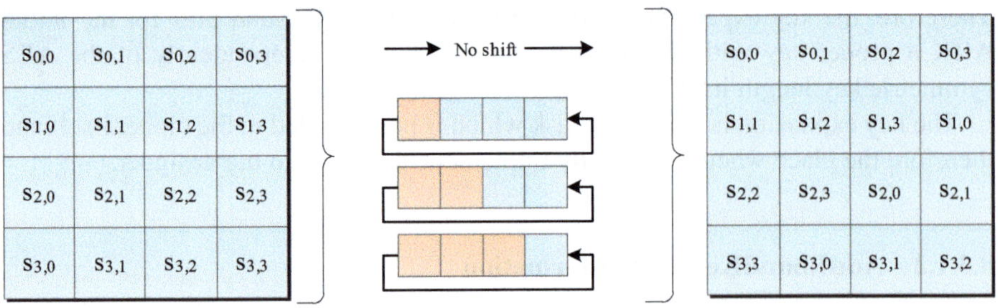

Fig. 8.6 ShiftRows() transformation

The transformation function *S* is called S-Box and computationally hard to derive. To save resources and increase the performance, the S-Box is normally implemented as a look-up table (LUT) that holds for every possible input value of the state a corresponding S-Box output value (see Fig. 8.5)

8.3.1.4 ShiftRows() Transformation

In this transposition, the rows of the state are circularly shift by different offsets. Row one is not circularly shifted, row two is circularly shifted by one, row three by two, and row four by three (see Fig. 8.6).

8.3.1.5 MixColumns() Transformation

In the last step of the round, the columns of the matrix are mixed. The state is therefore multiplied with a fixed matrix to gain the new state *S'*. This results in the fact that each column of the state is treated as a four-term polynomial function with fixed coefficients. The four polynomials can then be treated for simplicity as a matrix multiplication (see Fig. 8.7).

$$\begin{bmatrix} 02 & 03 & 01 & 01 \\ 01 & 02 & 03 & 01 \\ 01 & 01 & 02 & 03 \\ 03 & 01 & 01 & 02 \end{bmatrix} \times \begin{bmatrix} s_{0,0} & s_{0,1} & s_{0,2} & s_{0,3} \\ s_{1,0} & s_{1,1} & s_{1,2} & s_{1,3} \\ s_{2,0} & s_{2,1} & s_{2,2} & s_{2,3} \\ s_{3,0} & s_{3,1} & s_{3,2} & s_{3,3} \end{bmatrix} = \begin{bmatrix} s_{0,0}' & s_{0,1}' & s_{0,2}' & s_{0,3}' \\ s_{1,0}' & s_{1,1}' & s_{1,2}' & s_{1,3}' \\ s_{2,0}' & s_{2,1}' & s_{2,2}' & s_{2,3}' \\ s_{3,0}' & s_{3,1}' & s_{3,2}' & s_{3,3}' \end{bmatrix}$$

Fig. 8.7 MixColumns() transformation

8.3.2 *ECG Identification*

8.3.2.1 PCA Based ECG Identification

In the ECG identification block, which is based on PCA and Euclidian distance, the same process of generating the database is repeated, where the testing ECG signal is projected to the ECG space using (8.1). The next step is to calculate the Euclidian distance between each projected training vector and the projected test vector using Eq. (8.10) where d_i is the Euclidian distance between the ith training vector and the projected test vector. p and p' are the projected test vector and the ith projected training vector, respectively. The index of the minimum error between the feature vectors represents the ID of the identified ECG.

$$d_i = \sqrt{\sum_{i=1}^{n} (p - p'_i)^2} \tag{8.10}$$

The process of identification could be called just once for a single recognition using one test sample or for multiple recognitions using all test samples for evaluating the algorithm. The single identification process is justified as it allows greater scrutiny of a test signal that was falsely recognized, because a plot shows three signals: the test ECG; the expected training ECG; and the identified ECG for further analysis, such as the difference between the signals. Furthermore, the single identification process can be used for security identification, as the displayed ID can be used for verification.

8.3.2.2 KNN Based ECG Identification

KNN is a classification algorithm used to classify objects by a majority vote of the k-nearest reference objects. There are two main stages in KNN classification: distance computing and distance ranking. KNN uses a training data set with known class label to identify query vector to a class label. Training data consists of an

$N \times M$ matrix with N is the training samples and M is the number of features with an additional column representing the class label of each training sample. The first step is to calculate the distances between the query and all of the samples using a particular distance metric such as Euclidian or the Manhattan distance given by (8.11) or (8.12):

$$d(x, y) = \sqrt{\sum_{i=1}^{M} (x_{Ni} - y_i)^2} \tag{8.11}$$

$$d(x, y) = \sum_{i=1}^{M} (|x_{Ni} - y_i|) \tag{8.12}$$

where x_{Ni} and y_i are the training samples and the query, respectively.

After distance computation, the distance between the query and all the training samples are sorted according to the number of neighborhoods (k), where the proper selection of k is important to the performance of the KNN classifier. The classifier performs a voting on those KNNs to assign the query to the KNN that has the closest distance.

8.3.2.3 ENN Based ECG Identification

Unlike the KNN method, the ENN classifier predicts input patterns based on the maximum gain of intra-class coherence, where the ENN method makes a prediction in a two-way communication approach, which means it considers not only the nearest neighbors of the test sample, but it also considers the test sample as their nearest neighbor. The rule of the ENN is shown in Eq. (8.13), where the test sample Z is classified based on the decision results in the greatest intra-class coherence among all possible classes.

$$f_{\mathrm{ENN}} = \underset{j \in 1,2,\ldots,N}{\arg\max} \sum_{i=1}^{N} T_i^j \tag{8.13}$$

where j, $j = 1,2,\ldots,N$, is each possible class, and T_i^j is the generalized class-wise statistic for each class i, $i = 1,2,\ldots,N$.

In order to avoid the recalculation of the generalized class-wise statistics in the testing stage, a simplified version of ENN is proposed in [11]. Given an unknown sample Z to be classified, a series of possible class j, $j = 1, 2, \ldots, N$, are defined. The class membership is predicted based on:

$$f_{\mathrm{ENN}} = \underset{j \in 1,2,\ldots,N}{\arg\max} \left\{ \left(\frac{\Delta n_i^j + k_i - kT_i}{(n_i + 1)\,k} \right)_{i=j} - \sum_{i \neq j}^{N} \frac{\Delta n_i^j}{n_i k} \right\} \tag{8.14}$$

Table 8.1 Comparison of identification results from PCA, KNN, and ENN classifiers

Databases	VS			Shimmer			MIT		
Classifiers	PCA	KNN	ENN	PCA	KNN	ENN	PCA	KNN	ENN
Classification accuracy (%)	99.5	95.71	94.28	95	100	100	100	97.5	100

where k is the user-defined parameter of the number of the nearest neighbors, n_i is the number of training data for class i, k_i is the number of the nearest neighbors of the test sample Z from class i, Δn_i^j represents the change of the KNN for class i when the test sample Z is assumed to be class j, and T_i represents the generalized class-wise statistic of original class i.

8.3.2.4 Classification Accuracy

The proposed encryption and identification algorithms are simulated in both MATLAB and C/C++ environment. The ECG signals dataset is generated in MATLAB and used to test the AES encryption cipher block, AES decryption decipher block, and ECG identification block. All three identification methods have been validated using three different databases (i.e., VS, Shimmer, and MIT), the performance of each identification method for each database is compared in Table 8.1. It is worth mentioning that in the case of KNN and ENN, only the best results which are in the case of $k = 1$ are presented in the table because the identification rate has been decreasing when k is increased. This is due to the fact that the number of training and testing samples is relative small, a greater number of k is not suitable in this type of application.

As it can be seen from Table 8.1, the PCA identification method is the best in the case of VS with a classification accuracy of 99.5 %. However, the PCA identification method is the worst in the case of the Shimmer with an accuracy of 95 %. It can also be noted that both KNN and ENN reach a classification accuracy of 100 % in the case of the Shimmer while in the MIT case the 100 % accuracy is reached with PCA and ENN. In summary, the PCA based approach has better performance on more complicated environment than the other two approaches, this is due to the factor of extra feature extraction process. However, PCA is not good at handling some simple cases. Therefore, for a database that has a large numbers of individuals, the PCA based approach should be considered.

8.4 Hardware Implementation

Since the implementations of each algorithm for individual database are similar, and the only difference would be the training data, and for the purpose of proof of concept, the proposed system has been implemented based on the VS dataset.

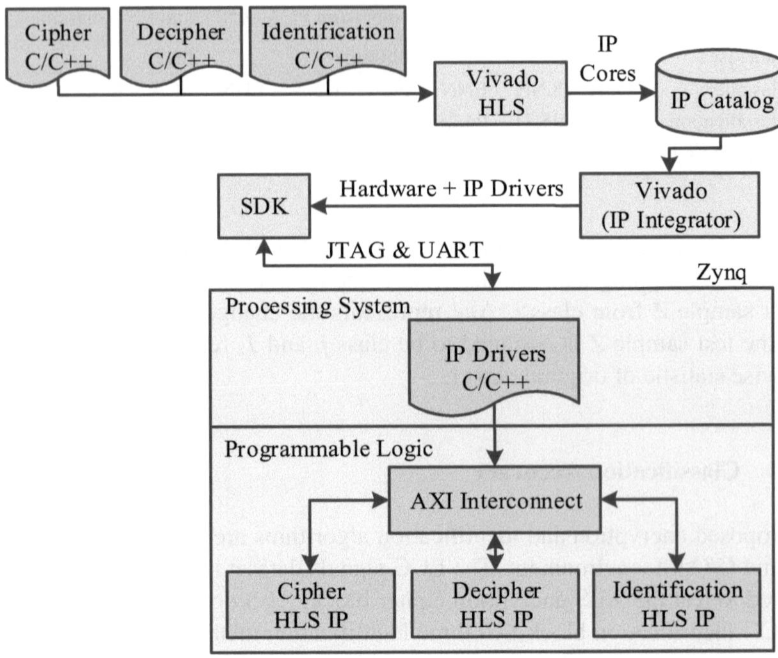

Fig. 8.8 Block diagram for the implementation of the system on the Zynq

Since ENN is an extended version of KNN that requires more computation without a clear improvement in the classification accuracy, it has been decided to only implement PCA and KNN for the identification and compare them. The hardware implementations of the cipher, decipher, and identification blocks are implemented on the heterogeneous Zynq SoC platform [3] using high level synthesis (HLS) [20]. A summary of the design flow is presented in Fig. 8.8. The starting point is Vivado HLS where the three blocks are converted from C/C++ implementation to a register transfer level (RTL) based IP-cores. This allows a considerable gain in development time since the blocks were already implemented and tested in C/C++ without scarifying on high parallelism characteristics because Vivado HLS provides a large number of powerful optimization directives [20]. The generated IP-cores are then exported and stored in the Xilinx IP Catalog before being used in Vivado IP Integrator to create the hardware block design with all needed components and interconnections. The following step is to export the generated hardware along with IP drivers to the software development kit (SDK) tool. The SDK tool is used to program the Xilinx ZC702 prototyping board via joint test action group (JTAG) interface and the terminal in SDK is used to communicate with the board via universal asynchronous receiver/transmitter (UART) interface. All IPs are implemented on the programmable logic (PL) of the Zynq SoC and communicate with the processing system part via the Xilinx AXI-Interconnect IP. A software is written in C/C++ and executed on the PS to manage the IPs present in the PL.

In the cipher and decipher as well as ECG identification blocks, the accumulation/multiplication operations will occupy the most of the computation time. So we will focus on accelerating loops in the above blocks. The objective of computational optimization is to enable efficient loop unrolling or pipelining while fully utilization of all computational resources provided by the PL.

Loop Unrolling The loop unrolling strategy can be used to increase the utilization of massive computation resources in FPGA devices. Depending on the way to unroll along different loop dimensions, the implementation variants will be generated. The complexity of the generated hardware will be affected by the data dependency of the unrolled execution instances as well as the affection from the unrolled hardware copies and the hardware operation frequency. An independent data sharing relation generates direct connections between buffers and computation elements. However, a dependent data sharing relation generates interconnects with complex hardware topology. Loop dimension *col* is selected to be unrolled to avoid complex connection topologies for all arrays.

Loop Pipelining Loop pipelining is another key optimization techniques in HLS to improve the throughput of the system, where the execution of operations from different loop iterations are overlapped in an organized way. Similar to the loop unrolling techniques, the throughput achieved is limited by resource constraints and data dependency in the loop.

8.4.1 AES Cipher Encryption Block

There are two data interfaces in the cipher block, which include an input and output data array. Both input and output data arrays are designed as 16×8-bit arrays, thus the block size for each calculation is 128 bits. The algorithm of the cipher block has been implemented using C++. Directives have been used in Vivado HLS to optimize the codes for hardware implementation. The used optimization directives are *s_axilite* to create AXI-Lite slave interface, *Array_Reshape* to reshape a given array to a specific dimension, and *unroll* to unroll a given loop by creating dedicated resources for each loop iteration.

The pseudo code of AES Cipher Encryption Block can be written as that in Code 1.

Code 1: Pseudo code of AES Cipher Encryption Block

```
1. for (cpr_iteration = 0; cpr_iteration < 10; cpr_iteration ++){
2.    Sub_Bytes;
3.    Shift_Row;
4.    Mix_Cloumn;
5.    Key_Expansion;
6.    Add_Round_Key
7. }
```

Code structure after optimization for loop unrolling and loop pipelining is shown in Code 2.

Code 2: Pseudo code of AES Cipher Encryption Block (optimized structure)

```
for (cpr_iteration = 0; cpr_iteration < 10; cpr_iteration ++){
#pragma HLS UNROLL
    Sub_Bytes;
    Shift_Row;
    Mix_Cloumn;
    Key_Expansion;
    Add_Round_Key;
}
```

8.4.2 AES Decipher Decryption Block

Similar to the cipher block, the decipher block is an inverse calculation of the cipher block. For this reason, its input and output arrays are the same as in the cipher block, and the directives used in Vivado HLS are also the same.

8.4.3 ECG Identification Block

8.4.3.1 PCA Based ECG Identification

The proposed ECG identification architecture consists of enrolment and hardware acceleration parts. The enrolment part is completed on software. However, the hardware acceleration part is performed on the PL, which mainly consists of PCA values and Euclidian distance calculators. The PCA values calculator is mainly used to perform the PCA projection [i.e., (8.1)]. The Euclidean Distance calculator is used to calculate the Euclidean distance [i.e., (8.2)]. In the PCA projection block, the test input signal is first subtracted from the mean signal of the complete training matrix T. Subsequently, the projected vector is then calculated via a multiplication of the Eigen ECG matrix by the mean subtracted signal. In the Euclidian distance calculator block, Euclidian distance of the projected training vector and the projected test vector is calculated as stated in (8.2). Since the calculation of the square root would not affect the search of the minimum value of the Euclidian distance, this calculation has been eliminated to reduce the hardware usage. Once the calculated Euclidean distance vector is available, the next step is to search for the minimal Euclidian distance that represents the ID of the identified ECG. The identification block is also implemented in C/C++ in Vivado HLS where multiple optimization directives have been used. In addition to the same directives used in the cipher and decipher block, *pipelined* was also used to increase the throughput of PCA and Euclidean distance calculator.

The pseudo code of calculating the deviation of each ECG from mean ECG Block after optimization for loop unrolling can be written as that in Code 3.

Code 3: Pseudo code of Calculating the deviation of each ECG from mean ECG

```
for (diffindex = 0; diffindex < TEST_SIZE; diffindex++){
//TEST_SIZE is the number of samples in one ECG signal
#pragma HLS UNROLL factor = 30
    Diff[diffindex] = ECG_Input[diffindex] - MEAN[diffindex];
}
```

The “factor = 30” would optimize the loop body to be repeated 30 times, and the iteration information will be adjusted accordingly. The reason of this setting is to balance the utilization of the hardware resources and the throughput of the loop body.

The pseudo code of Projection of centered ECGs into ECG space block after optimization for loop unrolling can be written as that in Code 4.

Code 4: Pseudo code of Projection of centered ECGs into ECG space block

```
while (y < PCA_FEATURES)
// PCA_FEATURES is the number of PCA features in database
{
            h = 0;
            while (h < TEST_SIZE)
//TEST_SIZE is the number of samples in one ECG signal
            {
            #pragma HLS UNROLL factor = 30
            PCA_TEST [y] = EigenMatrix[y][h]*Diff
[h] + PCA_TEST [y];
                h++;
            }
            PCA_TEST[y] = PCA_TEST[y]>>9;
            y++;
}
```

Similar to the code 3, the directives is to balance the utilization of the hardware resources and the throughput of the loop body.

The pseudo code of Calculating Euclidean distances block after optimization for loop unrolling can be written as that in Code 5.

Similar to the code 4, the directive “pragma HLS UNROLL” in code 5 is used to unroll the inner loop, and the directive “pragma HLS PIPELINE” is used to pipeline the outer loop. Since the number of PCA features is small number, the inner loop can be fully unrolled in parallel.

Code 5: Pseudo code of Calculating Euclidean distances block

```
do
{
#pragma HLS PIPELINE
    y = 0;
    do
    {
    #pragma HLS UNROLL
Euc_dist[z] += (PCA_TEST[y]-
Proj_ECG[z][y])*(PCA_TEST[y]-Proj_ECG[z][y]);
    y++;
    }while (y < PCA_FEATURES);
// PCA_FEATURES is the number of PCA features in database
    z++;
}while (z < TEST_NUMBER);
// TEST_NUMBER is the number of PCA features in database
```

8.4.3.2 KNN Based ECG Identification

In the case of the identification using KNN, the main optimization directive is the “Pipeline” one. It is worth mentioning that in this scenario the entire training data has to be stored on chip to be used for classification.

The pseudo code of identification using KNN after optimization for loop pipelining can be written as that in Code 6.

Code 6: Pseudo code of identification using KNN

```
for (i = 0, i < No Training Data, i++)
{
#pragma HLS PIPELINE
            Compute Euc_Dist [i]
}
for (i = 0, i < No Training Data, i++)
{
#pragma HLS PIPELINE
            // KNN, k = 1
             Find the nearest neighbour
}
Assign the ID of the nearest neighbour to the test sample
}
```

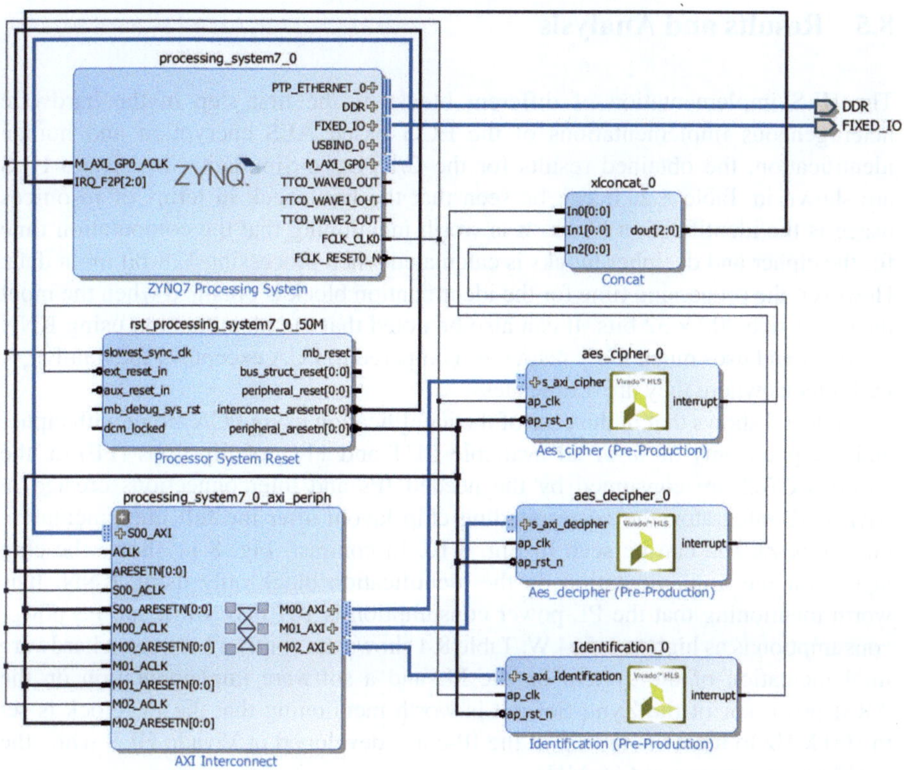

Fig. 8.9 Hardware block design

8.4.4 Implementation on the Zynq SoC

The hardware block design required for the implementation of the cipher, decipher, and identification blocks is shown in Fig. 8.9. The hardware block design is created in Vivado IP Integrator. It contains seven blocks in total, the three HLS IPs, the ZYNQ7 Processing System IP which is used for configuration purpose and is not implemented on the PL, the Processing System Reset IP, The AXI Interconnect IP, and finally the Concat IP.

8.5 Results and Analysis

The HLS implementation of different blocks is the first step in the hardware heterogeneous implementations of the ECG based AES encryption and human identification, the obtained results for the C/RTL Co-Simulation in Vivado HLS are shown in Table 8.2. It can be seen that the bottleneck in terms of resources usage is the identification block. It is worth mentioning that the computation time for the cipher and decipher blocks is calculated when processing 128-bit input data. However, the processing time for the identification block is obtained when the input data is of size 300×32 bits. It can also be noted that the identification using KNN is slower and uses much more resources compared to PCA except for the number of DSP blocks where they are 4 times less.

Table 8.3 shows that in the case of the identification using PCA along with cipher and decipher, only 30 % of all available LUT and 11 % of flip-flops (FF) on the Xilinx ZC702 are consumed by the needed IPs and interconnections created in Vivado IP integrator, the corresponding chip layout after the full implementation on Xilinx ZC702 can be seen in Fig. 8.10. In contrast, Fig. 8.11 shows the chip layout for the implementation on the identification block only using KNN. It is worth mentioning that the PL power consumption is 107 mW while the PS power consumption is as high as 1.564 W. Table 8.4 shows the contrast between a hardware implementation of the system on the PL and a software implementation on the ARM processor of the Zynq SoC. It is worth mentioning that the PL clock is set to 100 MHz to meet the timing of the IP-cores developed in Vivado HLS while the ARM processor runs at 666 MHz.

It can be seen that the software implementation is 44 times slower than its hardware counterpart in the case of PCA and ten times slower in the case of KNN. In addition, the proposed heterogeneous implementation on the Zynq outperformed

Table 8.2 Processing time and hardware utilization of each block in Vivado HLS

	Processing time (ms)	LUT (%)	FF (%)	DSP48E (%)	BRAM_18K (%)
Cipher	0.022	5	0.8	0	3
Decipher	0.050	6	0.8	0	3
Identification (PCA)	0.397	29	9	69	29
Identification (KNN)	0.550	78	34	14	46

Table 8.3 Overall hardware resources usage for cipher, decipher, and PCA identification

	Usage	Total available resources	Utilization (%)
LUT	16,133	53,200	30.33
FF	11,797	106,400	11.09
BRAM_18K	17	140	12.14
DSP48E	12	220	5.45

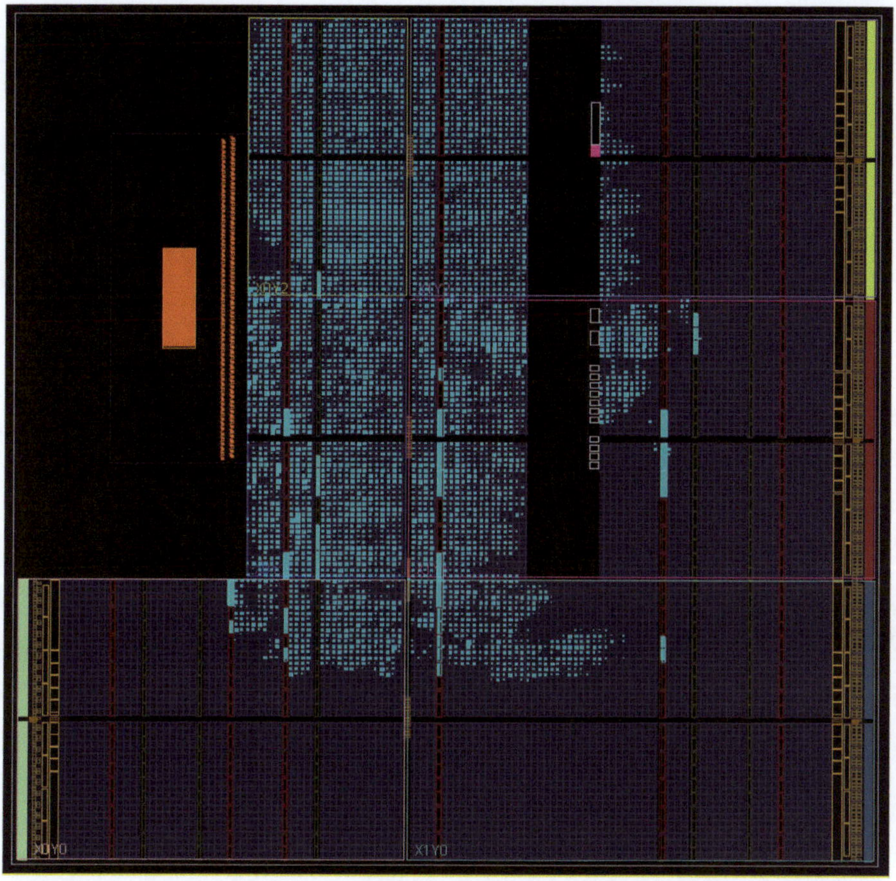

Fig. 8.10 Chip layout for the implementation of cipher, decipher, and PCA identification

the hybrid solution presented in [21] by a factor of 305 and the throughputs of the proposed cipher and decipher implementations which are 5.1 Gbit/s and 2.2 Gbit/s, respectively, have outperformed the existing work in [19] and [22] by a factor of 1.6 and 3.7, respectively.

8.6 Conclusion

Combining ECG biometrics with other security technologies allows the development of a promising connected health solution. This book chapter presented an encryption, decryption, and identification system using ECG signals. Three different ECG databases representing 2372 samples are used. The encryption and decryption is based on the secure AES algorithm while the identification uses PCA

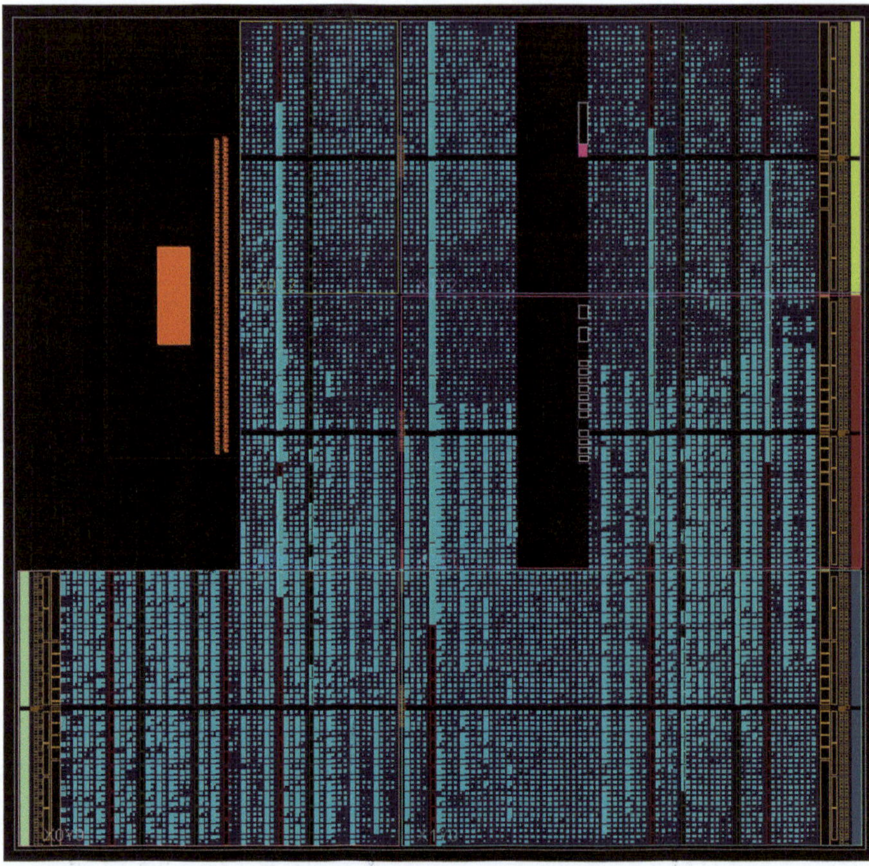

Fig. 8.11 Chip layout for the implementation of the KNN identification

Table 8.4 Software vs hardware processing time on the Zynq

	Cipher	Decipher	Identification (PCA)	Identification (KNN)	Total (PCA/KNN)
Hardware (ms)	3.21	7.41	0.09	0.55	10.71/11.17
Software (ms)	108.31	361.09	0.32	5.16	469.72/474.56

and Euclidian distance after a comparison with KNN and ENN. A heterogeneous hardware implementation on the Zynq SoC is performed to meet the real-time requirements. It has been noted that KNN can outperform PCA in the classification accuracy in some cases but at the cost of more resources and slower execution time on hardware. The best proposed implementation on the Xilinx ZC702 consumes only 107 mW and uses 30 % of all available resources on the chip to process a single ECG signal in 10.71 ms. The hardware implementation outclasses the existing FPGA based systems in multiple key performance metrics. Precious development time is saved by using Vivado HLS for the implementations of different blocks without compromising on achieving a high level of parallelism, this is possible with the use of HLS optimization directives. In addition, the classification accuracy is between 94 % and 100 % depending on the classifier and on the dataset used.

Acknowledgment This book chapter was made possible by National Priorities Research Program (NPRP) grant No. 5 – 080 – 2 – 028 from the Qatar National Research Fund (a member of Qatar Foundation). The statements made herein are solely the responsibility of the authors.

References

1. A.K. Jain, A. Ross, S. Prabhakar, An introduction to biometric recognition. IEEE Trans. Circuits Syst. Video Technol. **14**, 4–20 (2004)
2. F.M. Bui, F. Agrafioti, D. Hatzinakos, Electrocardiogram (ECG) biometric for robust. Biom. Theory Methods Appl. **9**, 383 (2009)
3. Xilinx Inc. *Zynq-7000 All Programmable SoC.* Available: http://www.xilinx.com/products/silicon-devices/soc/zynq-7000.html. Accessed Jan 2016
4. B. Tang, H. He, ENN: extended nearest neighbor method for pattern recognition [research frontier]. IEEE Comput. Intell. Mag. **10**, 52–60 (2015)
5. C. De Capua, A. Meduri, R. Morello, A smart ECG measurement system based on web-service-oriented architecture for telemedicine applications. IEEE Trans. Instrum. Meas. **59**, 2530–2538 (2010)
6. M. Borghetti, E. Sardini, M. Serpelloni, Sensorized glove for measuring hand finger flexion for rehabilitation purposes. IEEE Trans. Instrum. Meas. **62**, 3308–3314 (2013)
7. H. Kim, Y. Kim, H.-J. Yoo, A low energy bio sensor node processor for continuous healthcare monitoring system, in *IEEE Asian Solid-State Circuits Conference* (2008), pp. 317–320
8. L. Krukowski, J. Sugier, Organization of AES cryptographic unit for low cost FPGA implementation, in *Third International Conference on Dependability of Computer Systems* (2008), pp. 347–354
9. T. Good, M. Benaissa, Pipelined AES on FPGA with support for feedback modes (in a multi-channel environment). IET Inf. Secur. **1**, 1–10 (2007)
10. J. Chu, M. Benaissa, Low area memory-free FPGA implementation of the AES algorithm, in *22nd International Conference on Field Programmable Logic and Applications (FPL)* (2012), pp. 623–626
11. S. Karpagachelvi, M. Arthanari, M. Sivakumar, ECG feature extraction techniques-a survey approach. Int. J. Comput. Sci. Inf. Secur. IJCSIS, **8**(1), 76–80 (2010). ISSN:1947 5500
12. S. Mahmoodabadi, A. Ahmadian, M. Abolhasani, M. Eslami, J. Bidgoli, ECG feature extraction based on multiresolution wavelet transform, in *2005 IEEE Engineering in Medicine and Biology 27th Annual Conference* (2006), pp. 3902–3905
13. G.B. Moody, R.G. Mark, The impact of the MIT-BIH arrhythmia database. IEEE Eng. Med. Biol. Mag. **20**, 45–50 (2001)
14. C. Ye, M.T. Coimbra, B.V. Kumar, Investigation of human identification using two-lead electrocardiogram (ECG) signals, in *2010 Fourth IEEE International Conference on Biometrics: Theory Applications and Systems (BTAS)* (2010), pp. 1–8
15. Y. Li, H. Yu, L. Jiang, L. Ma, Z. Ji, Adaptive lifting scheme for ECG QRS complexes detection and its FPGA implementation, in *2010 3rd International Conference on Biomedical Engineering and Informatics* (2010), pp. 721–724

16. Y. Jewajinda, P. Chongstitvatana, FPGA-based online-learning using parallel genetic algorithm and neural network for ECG signal classification, in *2010 International Conference on Electrical Engineering/Electronics Computer Telecommunications and Information Technology (ECTI-CON)* (2010), pp. 1050–1054
17. Intelesens Ltd. *VS100 patch-based vital signs monitor*. Available: http://www.intelesens.com/pdf/vitalsensoem.pdf. Accessed Jan 2016
18. Shimmer. *Shimmer 3 ECG sensor*. Available: http://www.shimmersensing.com/shop/shimmer3-ecg-unit. Accessed Jan 2016
19. A. Amira, M.A. Saghir, N. Ramzan, C. Grecos, F. Scherb, A reconfigurable wireless environment for ECG monitoring and encryption. Int. J. Embed. Real Time Commun. Syst. **4**, 72–87 (2013)
20. Xlinx Inc. *Vivado HLS User Guide* (2015). Available: www.xilinx.com. Accessed Jan 2016
21. M. Cvikl, A. Zemva, FPGA-oriented HW/SW implementation of ECG beat detection and classification algorithm. Digit. Signal Process. **20**, 238–248 (2010)
22. H. Tana, A.N. Sazish, A. Ahmad, M.S. Sharif, A. Amira, Efficient FPGA implementation of a wireless communication system using Bluetooth connectivity, in *Proceedings of 2010 IEEE International Symposium on Circuits and Systems* (2010), pp. 1767–1770

Chapter 9
Offline Biometric Signature Verification Using Geometric and Colour Features

Abdelaali Hassaine, Somaya Al Maadeed, and Ahmed Bouridane

9.1 Introduction

With the increasing need of reliable authentication and verification systems, biometric modalities continue to be preferred due to their high accuracy and their relatively low vulnerability to breaches. Biometric signature is used in many domains because it is fast and easy to obtain. Moreover, people are usually willing to share their signatures in contrast to other biometric modalities such as iris or fingerprint.

Biometric signatures are usually acquired using two different modalities known as offline and online. In the offline modality, the image of the signature is acquired using a scanner, while in the online modality, the trajectory of the pen throughout the signature process is acquired using an adequate tablet and a stylus. Online signatures provide more information as they contain temporal information of the trajectory and the pressure at each point, this generally leads to a higher verification performance than that of offline signatures. However, offline signatures remain more popular because of their ease of acquisition and their vast usage in many domains including checks, letters and almost all kind of forms.

Offline signature verification is a challenging pattern recognition problem that performs the following steps common to all other pattern recognition tasks: data acquisition, pre-processing, feature extraction and classification/verification. The data-acquisition stage captures the scanned images using a digital scanner. The

A. Hassaine (✉) • S. Al Maadeed (✉)
Computer Science and Engineering Department, College of Engineering,
Qatar University, Doha, Qatar
e-mail: hassaine@qu.edu.qa; s_alali@qu.edu.qa

A. Bouridane
Department of Computer Science and Digital Technologies, Northumbria University,
Newcastle upon Tyne, UK
e-mail: ahmed.bouridane@northumbria.ac.uk

R. Jiang et al. (eds.), *Biometric Security and Privacy*, Signal Processing
for Security Technologies, DOI 10.1007/978-3-319-47301-7_9

pre-processing stage involves noise reduction, resizing, binarisation and thinning, among others. Then, an appropriate feature extraction method would extract the ideal features that maximises the interpersonal distance among the signatures of various persons and simultaneously minimises the intrapersonal distances of the same person. Global features, local features and geometric features are used for this purpose. In the classification step, there are template-matching approaches, statistical approaches such as the hidden Markov model and the Bayesian model, structural approaches such as graph- and tree-matching techniques, spectrum-analysis approaches and neural network approaches, among others. Various systems have been developed with specific algorithms to go through each stage.

In this paper, we introduce a set of geometric and colour features which can characterise signatures and subsequently combine those features using several classifiers. This paper is organised as follows: Sect. 9.2 describes the existing work related to offline signature verification, Sect. 9.3 describes our method with both its feature extraction and classification steps, Sect. 9.4 provides detailed evaluation results on different datasets, a comparison with other methods as well as the result analysis, and Sect. 9.5 concludes this work and provides some perspectives.

9.2 Previous Work

Offline signature verification is a highly active research field. In this section, we present an overview of the existing approaches in this field.

Local histogram representation of signature is used in the offline signature verification system proposed in [1]. Zone based division of signature is performed using rectangular and polar grids. It is followed by the calculation of HOG and LBP features. Features from both grid zones get concatenated and a combined feature vector is generated. Verification is done using two separate SVM, global and also user-dependent. A fusion of classifiers is also tested to verify the performance. The feature selection in [2] provides a cost-effective method for writer independent offline Signature verification systems. Its multiple feature extraction stage allows alleviating the issues of interpersonal variability. Dichotomy transformation in the feature space helps to overcome the challenges faced with a smaller set of reference signatures from numerous users. Finally the boosting feature selection collects only the significant features during training. The characterisation of signature images in [3] was based on the descriptor, Pyramid histogram of oriented gradients (PHOGs). Local image shape here was represented by calculating the histogram of edge orientations for the individual subregions, followed by quantisation into bins. The system recognised signatures with classification accuracies of 99 % and 96 %, respectively, for GPDS and DAVAB dataset.

Zouari et al. [4] proposed a signature identification as well as verification system for handwritten signatures. The fractal approach based feature extraction algorithm identified signatures with 95 % accuracy and verification was at a rate of 83 %. The SURF algorithm [5] intends to identify the significant areas from which the features are to be extracted prior to signature verification. The experiments were run on publicly available database of competition 4NSigComp2010 containing forged, genuine and also disguised signatures. And the system achieved an equal error rate (EER) of 15 %.

An automatic offline signature verification system [6] with Gradient local binary pattern (GLBP) extracted the texture information whereby the gradient description is obtained. Verification step relies on writer dependent framework with the help of an SVM classifier. Serdouk et al. [7] implemented an Artificial Immune Recognition System (AIRS) for developing an automatic offline verification system for handwritten signatures. The AIRS provided memory cells that can be used with an SVM classifier. For signature characterisation, a novel gradient was also proposed. Experiments conducted on CEDAR, GPDS-300 corpuses showed that the algorithm overcame the traditional AIRS-kNN by more than 9 % for the average error rate metric. The feature extraction phase of the offline signature verification system in [8] involves the partitioning of signature image into 8 equal vertical grids and derives the morphological pattern spectrum for all the grids. It is represented as 10-bin histogram and then normalised. The 8D feature vector is then obtained by combining all 8 histograms. Verification stage was experimented with SVM and MLP classifiers on the signature datasets CEDAR, MUKOS and, GPDS-160 a local language, Kannada dataset.

The One Class-SVM method is usually effective when the quantity of samples available is large. But the misclassification occurring with a reduced dataset is avoided by modifying the decision function used for the OC-SVM, which is implemented in [9]. Only a single class will be modelled here. Experiments conducted on CEDAR as well as GPDS handwritten signature datasets showed good results. A novel triangulation geometric feature set consisting sides, angles, perimeter of triangle which derived from the centre of gravity of a signature image is proposed in [10] for the signature verification. Euclidean classifier together with Voting based classifier verifies the forgery signature. This classification was experimented using triangular feature and also special global features. In Grupo de Senales 960 (GPDS-960) signature database, the proposed triangular feature method yielded a small average error rates (AER) with a percentage of 34 % in contrast to 43 % of the identified global feature. An approach for offline signature verification in [11] uses polar feature descriptors such as Radon Transform as well as Zernike Moments. The verification is performed with a Multiclass SVM. Once a signature is identified as a member of a specific class, Partial Least Mean Square Regression is applied to the sample to compare it with the entire samples in the dataset of the verified user and hence to obtain the regression score. The log likelihood of the sample against all samples of the user is computed using the Hidden Markov Model. Then, the classification authenticity is justified by checking whether the distance deviation between the regression score and the log likelihood is lower than 5 %. The system verified the signature with an accuracy of 98 % and a false-acceptance rate of .8 %. It detected skilled forgery and random forgery with accuracies of 71 % and 76 %, respectively.

An offline signature verification system in [12], which uses a feature set of concentric squares based on geometric features (slope and slope angle) with an SVM classifier, claims to achieve a false-acceptance rate of 1.25 % and a false-rejection rate of 1.66 %. Directional transitional features are trained using an SVM for offline signature verification in [13]. Vargas et al. used the co-occurrence matrix and the local binary pattern as features in [14], which worked at the global-image level and measured the grey-level variations in the image using statistical texture features. In addition, a background-removal method is proposed here. Genuine samples and random forgeries were used to train an SVM model, whereas random and skilled forgeries were used for testing. The method was improved in [15] by introducing the wavelet transform, which allows estimating and alleviating the global influence of the ink type. Further improvement was proposed in [16] to make it robust to grey-level distortion and segmentation errors because of the complex background. In [17], four directional chain code histograms of each grid on the contour of the signature image were extracted as the features and subsequently enhanced using a Laplacian of the Gaussian filters. With an SVM as the verification tool, a false-acceptance rate of 9.64 % was obtained.

An optical-flow approach was used to estimate the local stability among the signatures in [18]. Optical-flow vectors between the test signature and the reference specimens were considered to verify the genuineness of the test signature using an Alternating Decision Tree.

Paper [19] presented an offline signature recognition and verification method that relied on Fourier Descriptor and Chain Code features. The recognition process was based on the Principal Component Analysis. For verification, a multilayer feed-forward artificial neural network was designed. The verification process achieved a false-acceptance rate of 2.6 % and a false-rejection rate of 1.6 %. Another offline signature verification system based on the fusion of grid and global features that used neural networks was presented in [20].

Sigari et al. used the Gabor wavelet features in [21] to develop signature verification with minimal dependency on the nationality of the signer. Here, the nearest neighbour was used as the classifier, and the algorithm was tested on four signature databases with different nationalities. An alternative approach was proposed in [22] to generate features from the contourlet transform of the signature image. Each signature segment was coded with its dominant direction. Thus, the final coded image was characterised by the directions in the form of localisation and occurrences of the signatures of the direction using a co-occurrence matrix. Hence, the contourlet transform and the co-occurrence matrix features were conjointly used for verification using an SVM.

Table 9.1 summarises the existing methods, the databases on which they have been validated as well as their corresponding performance measures.

Table 9.1 Summary of the existing methods

Method	Database	Performance (%)	
[1]	GPDS-160 signature dataset	6.97 and 7.98 % EERs (12 and 5 genuine references)	
[3]	GPDS, DAVAB dataset	Verification for skilled forgery:- GPDS: Accuracy of 96.38 % Signature recognition:- GPDS: 99 % accuracy, DAVAB: 96 % accuracy	
[4]	FUM-PHSDB	Identification rate: 95 % Verification rate: 83 %	
[5]	Dataset of forensic signature verification competition, 4NSigComp2010	EER: 15 %	
[11]	Dataset 1: 600 signatures from 20 Persian writers Stellenbosch dataset (English)	FAR: 17, FRR: 4 2 (CF) FRR: 19 22 (SF)	
[12]	1000 signatures Training: 800 Testing: 200	FAR: 1.25 FRR: 1.66	
[13]	GPDS Database	EER: 4	
[14]	CORPUS1 (69 writer sets) CORPUS2 (20 writer sets) GPDS300 signature database	EER: 7 (SF), 5.48 (RF) EER: 3 (SF), 2 (RF) EER: 3.8 (SF), 5.7 (RF)	
[15]	MCYT-75 GDPS-100	EER: 12.82 (SF), 3.81 (RF) EER: 13.38 (SF), 4.59 (RF)	
[16]	CEDAR GDPS-100 MUKOS	FAR: 7.84, FRR:9.36 FAR: 9.64, FRR:13.16 FAR: 8.62, FRR:5.84	
[19]	600 signatures from 5 writers (60 genuine, 40 SF, 20 RF per writer)	FAR: 4 (RF), 18 (SF) FRR:3.3	
[20]	Persian Database English Database (Dolfing's)	FAR:14.5(SF), FRR:12.5(SF) FAR: 22.72 (SF), 2.27 (CF) FRR: 23.18(SF), 23.18(CF)	
[21]	MCYT-75 GDPS-100 GDPS-750	FAR: 0.15(RF), 2.43 (SF) 0.09(RF), 5.87 (SF) 0.07(RF), 7.75 (SF)	FRR: 19.41(RF), 30.11(SF) 19.72(RF), 24.77(SF) 21.14(RF), 24.72(SF)
[22]	Stellenbosch data set Dolfing's data set	EER: 18 (SF), 4.5 (CF) EER: 12.2 (SF)	

FAR false-acceptance rate, *FRR* false-rejection rate, *AER* average error rate, *EER* equal error rate, *RF* random forgery, *CF* casual forgery, *SF* skilled forgery

9.3 Method Description

Like all signature verification systems, our method comprises a feature extraction step followed by classification. Feature extraction step involves the extraction of characterising features from both the questioned signatures and the reference signatures. In the classification step, the difference between these two set of features is used to decide whether or not the questioned signature is genuine. Figure 9.1 summarises these two steps.

9.3.1 Feature Extraction

Several features are generally known for characterising signatures or handwritings well. These features are used by forensic document investigators to analyse the authenticity of handwritings and signatures [23]. Some of the features are extracted from the binarised signature image that is obtained using the Otsu thresholding algorithm [24]. Some other features are extracted from the signature's foreground colour. Note that some of these features are defined by a probability distribution function (PDF), which is retrieved from the handwriting images to characterise the writer's individuality [25]. The PDF indicates the relative likelihood for a typical feature to have a certain given value. Apart from the directions, curvature, tortuosity, chain code and edge based directional features used in our previous work [26], the following features are also extracted.

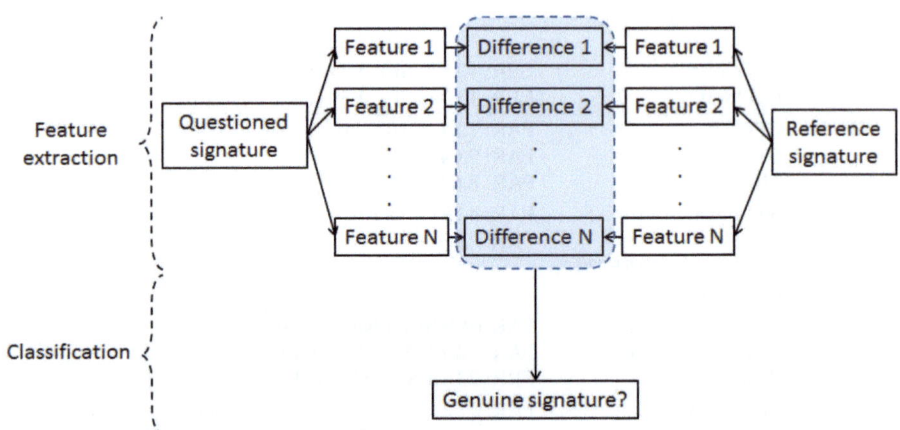

Fig. 9.1 Global scheme of our signature verification system

9.3.1.1 Global Moments (f27–f63)

Moments are often used as features for offline signature verification [27–31]. They provide a quantitative measure of the signature. Several moments have been computed, including:

- Spatial moments:

$$M_{x\_order,y\_order} = \sum_{x,y} \left(I(x,y) \,.x^{x\_order} .y^{y\_order} \right) \tag{9.1}$$

where $I(x,y)$ is the value of the signature image at pixel (x,y). Several values of *x_order* and *y_order* are implemented such that each order pair corresponds to a different feature: f27(0,0), f28(1,0), f29(2,0), f30(3,0), f31(0,1), f32(1,1), f33(2,1), f34(0,2), f35(1,2) and f36(0,3).

- Central moments:

$$\mu_{x\_order,y\_order} = \sum_{x,y} \left(I(x,y) \,.(x - x_c)^{x\_order} .(y - y_c)^{y\_order} \right) \tag{9.2}$$

where x_c and y_c are the coordinates of the gravity centre: $x_c = \frac{M_{1,0}}{M_{0,0}}$ and $y_c = \frac{M_{0,1}}{M_{0,0}}$. Several values of *x_order* and *y_order* are implemented such that each order pair corresponds to a different feature: f37(0,0), f38(1,0), f39(2,0), f40(3,0), f41(0,1), f42(1,1), f43(2,1), f44(0,2), f45(1,2) and f46(0,3).

- Normalised central moments:

$$\eta_{x\_order,y\_order} = \frac{\mu_{x\_order,y\_order}}{M_{0,0}^{(x\_order+y\_order/2)+1}} \tag{9.3}$$

Several values of *x_order* and *y_order* are implemented such that each order pair corresponds to a different feature: f47(0,0), f48(1,0), f49(2,0), f50(3,0), f51(0,1), f52(1,1), f53(2,1), f54(0,2), f55(1,2) and f56(0,3).

- Hu moments:

Seven Hu moments, which correspond to features f57–f63, are computed for the signature images [32].

Fig. 9.2 Computing local moment using a moving window for a given signature

9.3.1.2 Local Moments (f64–f396)

The above 37 global moments are computed locally for a moving window that is placed on the edges of the signature. The size of the moving window is empirically set to 19 (see Fig. 9.2). For each moment category, 9 statistical measures are computed: the average, the median, the variance, the kurtosis, the skewness, the minimum, the maximum, the mode and the entropy, resulting in a total of 333 local moment features (f64–f369).

9.3.1.3 Colour Features (f397–f399)

Colour can be a notably informative and discriminative feature because it characterises the pressure at each point for each signature. As per our knowledge, the usage of colour features is a novel idea for signature verification.

In this study, we extracted the hue, the saturation and the luminosity histograms of the foreground pixels of each signature. Those histograms are computed on vectors of 256 elements (f397–f399). Figure 9.3 illustrates those features.

9.3.2 *Classification*

Once the features are extracted from a specific questioned signature, the absolute difference between them and the set of reference signature features is computed:

$$|f_i\,(\text{questioned}) - f_i\,(\text{reference})|\,. \tag{9.4}$$

For n reference signatures, the average absolute difference between the features is considered instead:

$$\frac{\sum_{k=1}^{n} |f_i\,(\text{questioned}) - f_i\,(\text{reference}_k)|}{n} \tag{9.5}$$

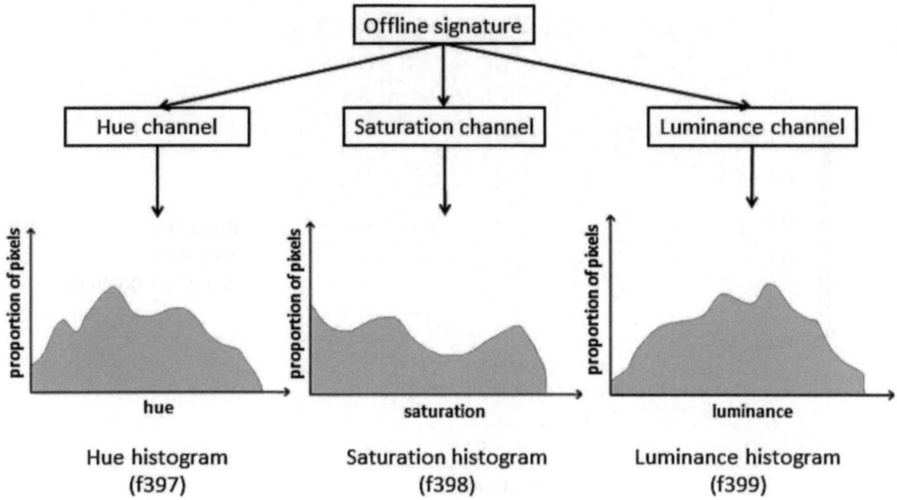

Fig. 9.3 Extraction of colour features

For a feature with length L, the following formula is applied:

$$\frac{\sum_{k=1}^{n} \frac{\sum_{l=1}^{L} |f_{i,l}(\text{questioned}) - f_{i,l}(\text{reference}_k)|}{L}}{n} \tag{9.6}$$

where $f_{i,l}$ is the lth element of feature i.

In this study, the following classifiers were examined:

9.3.2.1 Logistic Regression (LR)

In logistic regression classifier, the probabilities that describe the authenticity of the signature are modelled as a function of the differences between the features [33]. Figure 9.4 illustrates fitting a logistic regression classifier for a given feature.

9.3.2.2 Random Forests

Random forests are an ensemble learning method for the classification task. Their basic principle is the construction of several decision trees during the training and finally to output the class which is the mode of class each of the trees has produced. The construction of each decision tree is based on the following strategy:

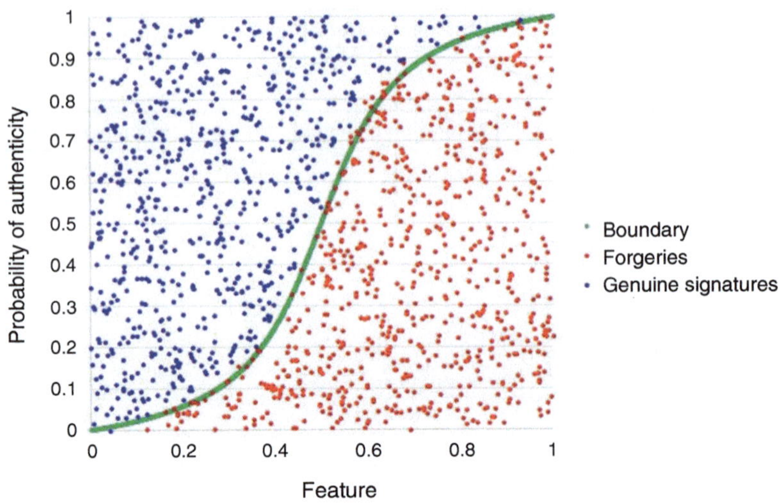

Fig. 9.4 Illustrating logistic regression for a given feature

1. When there are N cases in the training set, sample n cases such that $n < N$ at random but there should be replacement from the actual data. The resultant sample will be the taken as the training set to grow the tree.
2. When there are M input variables, a number $m \ll M$ is specified and at each node, m variables are selected at random. The nodes are split in such a way that it is the best split on these m variables. During the forest-growing process, m is kept fixed.
3. Each tree is grown to the maximum possible level. No trimming will be done.

In our study, we have used the R random forest library [34].

Figure 9.5 illustrates the use of random forest classifier in our case.

9.3.2.3 Generalised Linear Models

The generalised linear model (GLM) is basically a flexible generalisation of the normal linear regression, where response variables with error distribution models other than a normal distribution are allowed. The GLM will generalise the linear regression whereby the linear model is related to the response variable with the help of a link function. The magnitude of the variance of each measurement will be function of its predicted value.

In this study, we used a binomial distribution within the R GLMNET implementation of the GLM [35]. This package produces the regularisation path over a grid of values for the tuning parameter lambda. The optimal value of lambda is obtained using 10 folds cross validation.

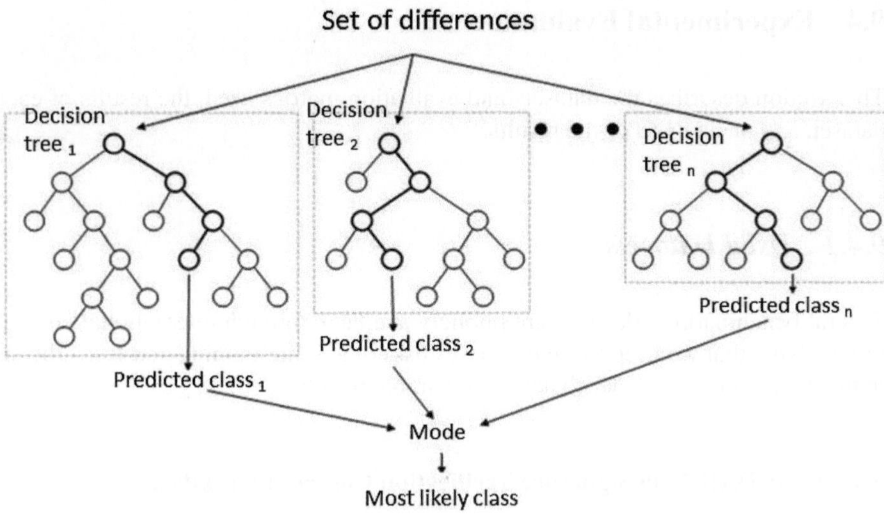

Fig. 9.5 Illustrating the random forests classifier

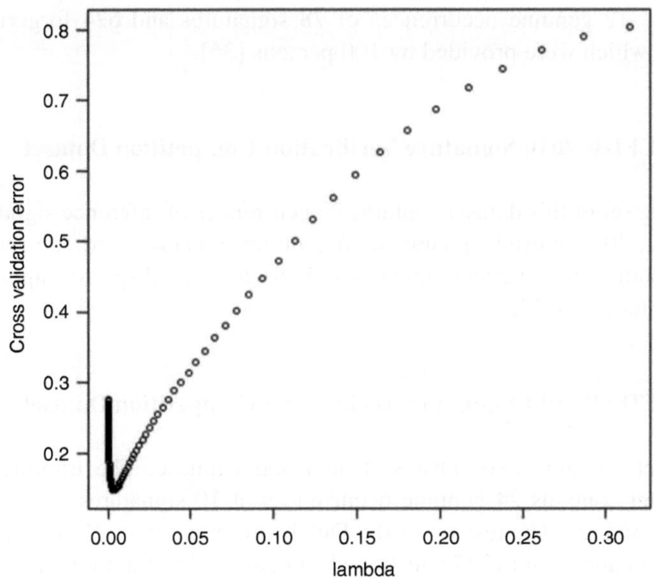

Fig. 9.6 Regularisation parameter lambda for generalised linear model obtained through cross validation

Figure 9.6 illustrates an example of determining the lambda parameter using cross validation.

9.4 Experimental Evaluation

This section describes the datasets and evaluation metrics used, the results of each dataset, and an analysis of the results.

9.4.1 *Used Datasets*

Several benchmarking datasets are publicly available through international competitions. Note that we used those databases using the same training and test splits as in the competitions in order to have comparable results.

9.4.1.1 ICDAR 2009 Signature Verification Competition Dataset

The training set of this dataset contains 5 genuine occurrences of 12 signatures and 1860 forgeries of those signatures, which were provided by 31 persons. The test set contains 12 genuine occurrences of 78 signatures and 624 forgeries of those signatures, which were provided by 100 persons [36].

9.4.1.2 ICFHR 2010 Signature Verification Competition Dataset

The training set of this dataset contains 9 occurrences of reference signatures of the same writer, 20 disguised signatures, 76 genuine signatures and 103 forgeries. The test set contains 25 references signatures, 90 forgeries, 7 disguised signatures and 3 genuine signatures [37].

9.4.1.3 ICDAR 2011 Signature Verification Competition Dataset

This dataset contains two subsets: Dutch and Chinese. The training set of the Dutch subset contains 24 genuine occurrences of 10 signatures and 123 forgeries of those signatures. The test set of the Dutch subset contains 12 occurrences of 54 reference signatures and 1287 questioned signatures. The training set of the Chinese subset contains 24 genuine occurrences of 10 signatures and 340 forgeries of those signatures. The test set of the Chinese subset contains 12 occurrences of 10 reference signatures and 487 questioned signatures [38].

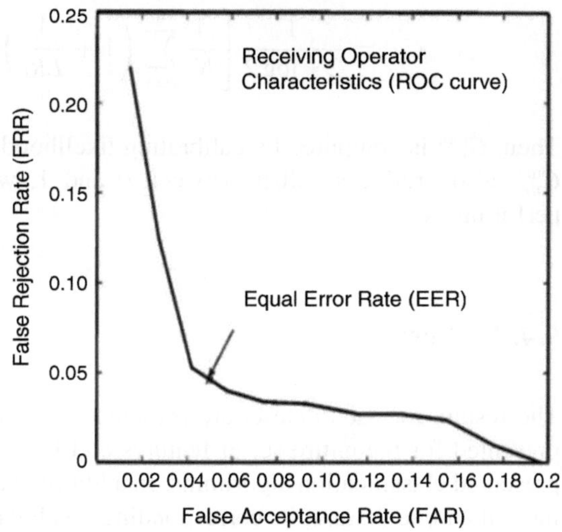

Fig. 9.7 Obtaining equal error rate

9.4.1.4 ICFHR 2012 Signature Verification Competition Dataset

The training set of this dataset is the combination of the training set and the testing set of the ICFHR 2010 signature verification datasets. The test set contains the signatures of 3 people in a total of 501 images [39].

9.4.2 Evaluation Metrics

EER and cost of log likelihood ratio evaluation metrics were used in those competitions dataset.

EER is obtained by thresholding the probabilities returned by the system at various thresholds and computing the false-positive rate and the false-negative rate for each threshold value. The EER is the value where those two errors are equal. Figure 9.7 shows how we obtain the EER on an illustrative example. ERR varies between 0 and 1. A perfect prediction algorithm would score 0, and in real systems, smaller values indicate a better system.

Cost of log likelihood ratio ($\hat{C}_{llr}^{min}$) This metric is described in detail in [40]. It is implemented using the FoCal MATLAB toolkit and aims at giving a high penalty for being both confident and wrong. C_{llr} is an application-independent evaluation metric proposed by Brümmer [40]. It is calculated using the following equation where N_0 is the number of trials corresponding to hypothesis H_0 (the questioned signature is an authentic signature of the reference writer) and N_1 is the number of trials corresponding to hypothesis H_1 (the questioned signature is written by a writer other than the reference writer).

$$C_{llr} = \frac{1}{2 * \log 2}\left[\frac{1}{N_0}\sum_{i=1}^{N_0}\left(1 + \frac{1}{LR_i}\right) + \frac{1}{N_1}\sum_{i=1}^{N_1}(1 + LR_i)\right] \tag{9.7}$$

Then, $\hat{C}_{llr}^{\min}$ is computed by calibrating likelihood ratios using the logistic function. $\hat{C}_{llr}^{\min}$ also produces values between 0 and 1, where smaller values mean better performance.

9.4.3 Results

The results for each dataset are presented in this section. Note that the results are presented for the entire set of features and for each category of features using the presented classification algorithms. In addition, because the GLM classifier requires more than one feature, its corresponding result cannot be determined for individual features. For each competition dataset, the results of the other participating methods are also given here for comparison purposes.

9.4.4 Discussion and Analysis

Table 9.2 shows that colour features produce the optimal results for the ICDAR 2009 dataset. This is true because the same pen was used for producing all the data. Consequently, colour histograms are indicative of the writing pressures applied on the pen by each person. Moreover, the performance of colour feature outperforms by a large extent the performance of next optimal feature, which means that the colour information needs to be exploited whenever available, specifically in cases where all data has been produced using similar pens.

Table 9.2 EER results for the ICDAR2009 competition dataset

Our method	LR classifier	RF classifier	GLM classifier	Participating methods	
All features	19.85 %	10.75 %	10.5 %	CMM	9.15 %
Global moments	24.22 %	28.81 %	24.12 %	Anonymous	15.50 %
Local moments	45.07 %	19.86 %	50 %	DFKI	15.78 %
Colour	8.91 %	11.10 %	9.36 %	DFKI	16.10 %
Tortuosity	26.05 %	39.34 %	NA	BRG	18.27 %
Direction	23.38 %	37.04 %	NA	CEDAR	23.00 %
Curvature	31.23 %	42.37 %	NA	CVC	41.12 %
Chain codes	23.94 %	29.85 %	23.71 %	Anonymous	43.02 %
EBDF	19.49 %	23.92 %	18.85 %		

Table 9.3 EER results for the ICFHR2010 competition dataset

Our method	LR classifier	RF classifier	GLM classifier	Participating methods	
All features	50 %	32.65 %	2.58 %	AVTS	80 %
Global moments	43.13 %	28.01 %	6.70 %	LISIC	58 %
Local moments	50 %	32.65 %	3.09 %	NifiSoft (full)	70 %
Colour	33.16 %	33.16 %	33.16 %	Parascript	70 %
Tortuosity	25.43 %	22.34 %	NA	Sabanci	55 %
Direction	26.46 %	21.82 %	NA	Anonymous	60 %
Curvature	25.95 %	25.77 %	NA	NifiSoft (partial)	70 %
Chain codes	27.49 %	28.01 %	28.01 %		
EBDF	33.33 %	18.73 %	4.12 %		

Table 9.4 $\hat{C}_{llr}^{\min}$ results for the ICDAR2011 Dutch signature competition dataset

Our method	LR classifier	RF classifier	GLM classifier	Participating methods	
All features	7.52 %	33.00 %	47.83 %	Sabanci	57.32 %
Global moments	79.25 %	85.97 %	78.86 %	Anonymous-1	67.40 %
Local moments	100.00 %	98.14 %	99.01 %	HDU	38.61 %
Colour	19.63 %	18.89 %	19.19 %	DFKI	72.20 %
Tortuosity	75.25 %	82.85 %	NA	Anonymous-2	79.40 %
Direction	74.49 %	85.31 %	NA		
Curvature	74.19 %	81.16 %	NA		
Chain codes	75.30 %	82.50 %	75.20 %		
EBDF	67.16 %	66.39 %	66.69 %		

Table 9.3 shows instead that colour information is not very discriminative in the ICFHR2010 competition dataset. This is particularly true for databases in which each person used a different pen. However, due to the cursive nature of the data, the edge based directional features as well as the moment features outperform other features. This table shows also that a different classifier may lead to a much higher performance as it is the case for many features.

Table 9.4 confirms the findings regarding the colour information for the ICDAR 2011 Dutch competition database. Clearly, the data has been produced by a similar pen in this database as well. Moreover, we can notice that the combination of all features does not lead necessarily to an improved performance over individual features. This is because many features are not discriminative and most classifiers do not handle them appropriately.

Table 9.5 shows that most results on the ICDAR 2011 Chinese competition database are outperformed by those of other competing methods. This is because the signature images contain many delimiters originally present in the sheet containing the signatures. Those delimiters (mainly lines) were intended for helping contributors signing within a specific region; however, they are not easily separable from signatures themselves. The features are therefore extracted from images containing both signatures and delimiters. This causes classifiers to give misleading results.

Table 9.6 shows that tortuosity produces the optimal performance for the ICFHR2012 competition dataset because of the tortuous nature of signatures in this database. However, this dataset contains signatures of three persons only; therefore results are not well representative of strength of the features.

More generally, by analysing the results, we note the following (Fig. 9.8):

- In most cases, the proposed method outperforms all the participating methods, which proves the efficiency of our method in determining forged and disguised signatures.
- In the datasets where our method is outperformed by the other methods, the images contain some delimiters and background text that cannot be easily isolated from the signature.
- We plotted the average EER value for each feature for all datasets. The results are shown in Fig. 9.2. The edge based directional features outperform all other features, which confirms that this feature is notably discriminative for both writer identification and signature verification.
- We computed the average of the best EERs that were obtained using each classifier. The results are shown in Fig. 9.3. The generalised linear model generally makes the difference on our features (Fig. 9.9).

9.5 Conclusion

We presented in this paper a method which relies on several geometric as well as colour features in the verification of a given signature. The method can detect both forgeries and disguised signatures and is applicable for any script. The method has been tested on several international competition datasets and in many cases outperforms the other participating methods. The results were reported using the EERs and the log likelihood ratios.

Experiments show that edge based directional features generally outperform other features and colour features are also notably discriminative. The results also suggest that the generalised linear model is generally preferred over random forests and logistic regression, but these two classifiers may perform better for some datasets. Future work includes exploring other methods to combine the features proposed here with other classifiers such as kernel discriminant analysis, neural networks and deep learning methods.

Table 9.5 $\hat{C}_{llr}^{min}$ results for the ICDAR2011 Chinese signature competition dataset

Our method	LR classifier	RF classifier	GLM classifier	Participating methods	
All features	97.60 %	97.66 %	96.22 %	Sabanci	69.33 %
Global moments	93.14 %	94.98 %	93.05 %	Anonymous-1	76.50 %
Local moments	100.00 %	100.00 %	100.00 %	HDU	78.99 %
Colour	100.00 %	100.00 %	100.00 %	DFKI	92.66 %
Tortuosity	91.93 %	97.88 %	NA	Anonymous-2	91.85 %
Direction	94.09 %	100.00 %	NA		
Curvature	94.25 %	98.82 %	NA		
Chain codes	95.71 %	98.39 %	95.65 %		
EBDF	97.48 %	96.07 %	97.51 %		

Table 9.6 $\hat{C}_{llr}^{min}$ results for the ICFHR2012 signature competition dataset

Our method	LR classifier	RF classifier	GLM classifier	Participating methods	
All features	100.00 %	76.27 %	84.42 %	Griffith University	43.00 %
Global moments	100.00 %	96.49 %	80.34 %	Sabanci University	58.00 %
Local moments	83.71 %	88.41 %	78.96 %	Anonymous-1	83.00 %
Colour	86.94 %	86.38 %	85.36 %	Anonymous-2	74.00 %
Tortuosity	66.31 %	67.86 %	NA		
Direction	78.04 %	82.23 %	NA		
Curvature	98.25 %	100.00 %	NA		
Chain codes	83.71 %	91.48 %	83.81 %		
EBDF	85.12 %	63.83 %	71.31 %		

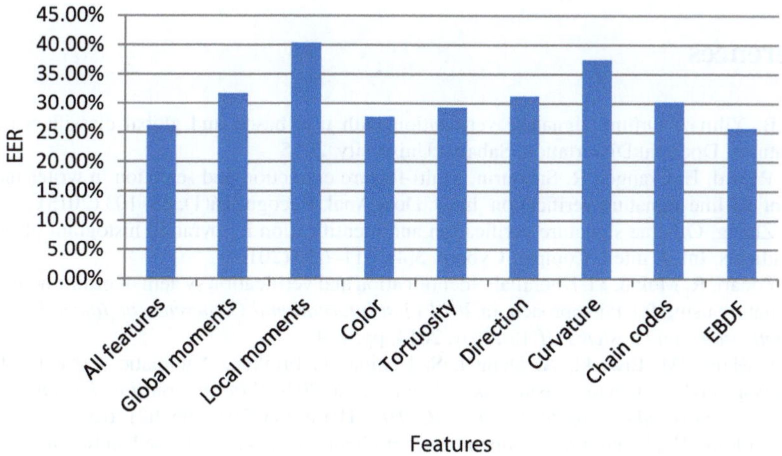

Fig. 9.8 Average EER value for each category of features

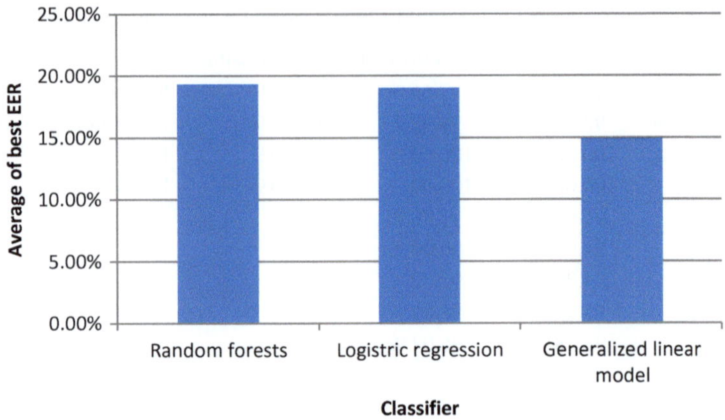

Fig. 9.9 Average best EER for each classifier

There are also on-going studies on the combination of the signature verification with other biometric modalities such as fingerprint or iris recognition in order to improve the verification performance. There is also an on-going research on the extraction of the trajectory of offline signatures in order to generate online signatures which are proven to produce more accurate results.

Acknowledgments This publication was made possible by a grant from the Qatar National Research Fund through National Priority Research Program (NPRP) No. 09-864-1-128. Its contents are solely the responsibility of the authors and do not necessarily represent the official views of the Qatar National Research Fund or Qatar University.

References

1. M.B. Yilmaz, Offline signature verification with user-based and global classifiers of local features, Doctoral Dissertation, Sabancı University, 2015
2. D. Rivard, E. Granger, R. Sabourin, Multi-feature extraction and selection in writer independent off-line signature verification. Int. J. Doc. Anal. Recogn. **16**(1), 83–103 (2013)
3. B. Zhang, Off-line signature verification and identification by pyramid histogram of oriented gradients. Int. J. Intell. Comput. Cybern. **3**(4), 611–630 (2010)
4. R. Zouari, R. Mokni, M. Kherallah, Identification and verification system of offline handwritten signature using fractal approach, in *2014 First International Conference on Image Processing, Applications and Systems (IPAS)*, Nov. 2014, pp. 1–4
5. M.I. Malik, M. Liwicki, A. Dengel, S. Uchida, V. Frinken, Automatic signature stability analysis and verification using local features, in *2014 14th International Conference on Frontiers in Handwriting Recognition (ICFHR)*, Heraklion, 2014, pp. 621–626
6. Y. Serdouk, H. Nemmour, Y. Chibani, New gradient features for off-line handwritten signature verification, in *2015 International Symposium on Innovations in Intelligent SysTems and Applications (INISTA)*, Madrid, 2015, pp. 1–4

7. Y. Serdouk, H. Nemmour, Y. Chibani, An improved Artificial Immune Recognition System for off-line handwritten signature verification, in *2015 13th International Conference on Document Analysis and Recognition (ICDAR)*, Tunis, 2015, pp. 196–200
8. B.H. Shekar, R.K. Bharathi, J. Kittler, Y.V. Vizilter, L. Mestestskiy, Grid structured morphological pattern spectrum for off-line signature verification, in *2015 International Conference on Biometrics (ICB)*, Phuket, 2015, pp. 430–435
9. Y. Guerbai, Y. Chibani, B. Hadjadji, The effective use of the one-class SVM classifier for handwritten signature verification based on writer-independent parameters. Pattern Recogn. **48**(1), 103–113 (2015)
10. Z. Zulkarnain et al., Triangular geometric feature for offline signature verification. World Academy of Science, Engineering and Technology. Int. J. Comput. Elect. Autom. Control Inf. Eng. **10**(3), 485–488 (2016)
11. K.N. Pushpalatha, A.K. Gautham, R. Dhashikumar, K.B. ShivaKumar, R. Das, Offline signature verification with random and skilled forgery detection using polar domain features and multi stage classification-regression model. Int. J. Adv. Sci. Technol. **59**, 27–40 (2013)
12. M.K. Randhawa, A.K. Sharma, R.K. Sharma, Off-line signature verification with concentric squares and slope based features using support vector machines, in *2013 IEEE 3rd International Advance Computing Conference (IACC)*, Ghaziabad, Feb. 2013, pp. 600–604
13. K. Tselios, G. Economou, E.N. Zois, A. Nassiopoulos, Fusion of directional transitional features for off-line signature verification, in *Proceedings of International Joint Conference on Biometrics*, Washington, DC, Oct. 2011, pp. 1–6
14. J.F. Vargas, M.A. Ferrer, C.M. Travieso, J.B. Alonso, Off-line signature verification based on grey level information using texture features. Pattern Recogn. **44**, 375–385 (2011)
15. J.F. Vargas, C.M. Travieso, J.B. Alonso, M.A. Ferrer, Off-line signature verification based on gray level information using wavelet transform and texture features, in *Proceedings of International Conference on Frontiers in Handwriting Recognition*, Kolkata, India, Nov. 2010, pp. 587–592
16. M. Ferrer-Aguilar, F. Vargas, A. Morales, A. Ordoñez, Robustness of off-line signature verification based on gray level features. IEEE Trans. Inf. Forensics Secur. **7**(3), 966–977 (2012)
17. R.K. Bharathi, B.H. Shekar, Off-line signature verification based on chain code histogram and support vector machine, in *2013 International Conference on Advances in Computing, Communications and Informatics (ICACCI)*, Mysore, India, Aug. 2013, pp. 2063–2068
18. G. Pirlo, D. Impedovo, Verification of static signatures by optical flow analysis. IEEE Trans. Hum. Mach. Syst. **43**(5), 499–505 (2013)
19. I.A. Ismail, M.A. Ramadan, T.S. El-Danaf, A.H. Samak, An efficient off-line signature identification method based on Fourier descriptor and chain codes. IJCSNS Int. J. Comput. Sci. Netw. Secur. **10**(5), 29–35 (2010)
20. D.R. Shashi Kumar, K.B. Raja, R.K. Chhotary, S. Pattanaik, Off-line signature verification based on fusion of grid and global features using neural networks. Int. J. Eng. Sci. Technol. **2**(12), 7035–7044 (2010)
21. M.H. Sigari, M.R. Pourshahabi, H.R. Pourreza, Offline handwritten signature identification and verification using multi-resolution gabor wavelet. Int. J. Biom. Bioinf. **5**(4), 234–248 (2011)
22. A. Hamadene, Y. Chibani, H. Nemmour, Off-line handwritten signature verification using contourlet transform and co-occurrence matrix, in *2012 International Conference on Frontiers in Handwriting Recognition*, Bari, Italy, Sep. 2012, pp. 343–347
23. K. Koppenhaver, *Forensic Document Examination: Principles and Practice* (Humana Press, 2007)
24. N. Otsu, A threshold selection method from gray-level histograms. IEEE Trans. Syst. Man Cybern. **9**(1), 62–66 (1979)
25. A. Hassaine, S. Al-Maadeed, J. Alja'am, A. Jaoua, A. Bouridane, The ICDAR2011 Arabic writer identification contest, in *Proceedings of Eleventh International Conference on Document Analysis and Recognition*, Beijing, China, Sep. 2011, pp. 1470–1474

26. A. Hassaïne, S. Al-Maadeed, A. Bouridane, A set of geometrical features for writer identification, in *Neural Information Processing*, 2012, pp. 584–591
27. M. Bulacu, L. Schomaker, Text-independent writer identification and verification using textural and allographic features. IEEE Trans. Pattern Anal. Mach. Intell. **29**(4), 701–717 (2007)
28. A. Al-Maadeed, Text-dependent writer identification for Arabic handwriting. J. Electr. Comput. Eng. (2012). doi:10.1155/2012/794106
29. I. Siddiqi, N. Vincent, Text independent writer recognition using redundant writing patterns with contour-based orientation and curvature features. Pattern Recogn. **43**(11), 3853–3865 (2010)
30. F. Leclerc, R. Plamondon, Automatic signature verification: the state of the art—1989–1993. Int. J. Pattern Recogn. Artif. Intell. **8**(3), 643–660 (1994)
31. R. Plamondon, G. Lorette, Automatic signature verification and writer identification—the state of the art. Pattern Recogn. **22**(2), 107–131 (1989)
32. M.K. Hu, Visual pattern recognition by moment invariants. IRE Trans. Inf. Theory **8**(2), 179–187 (1962)
33. C.M. Bishop, N.M. Nasrabadi, *Pattern Recognition and Machine Learning*, vol. 1 (Springer, 2006)
34. A. Liaw, M. Wiener, Classification and regression by random forest. R News **2**(3), 18–22 (2001). http://CRAN.R-project.org/doc/Rnews
35. J. Friedman, T. Hastie, R. Tibshirani, Regularization paths for generalized linear models via coordinate descent. J. Stat. Softw. **33**(1), 1–22 (2010)
36. V.L. Blankers, E.V.D. Heuvel, K.Y. Franke, L.G. Vuurpijl, ICDAR 2009 signature verification competition. *2009 10th International Conference on Document Analysis and Recognition*, 2009, pp. 1403–1407
37. M. Liwicki, E.V.D. Heuvel, B. Found, M.I. Malik, Forensic signature verification competition 4nsigcomp2010-detection of simulated and disguised signatures, in *2010 International Conference on Frontiers in Handwriting Recognition (ICFHR)*, Kolkata, India, Nov. 2010, pp. 715–720
38. M. Liwicki, M.I. Malik, E.V.D. Heuvel et al. Signature verification competition for online and offline skilled forgeries (sigcomp2011), in *2011 International Conference on Document Analysis and Recognition (ICDAR)*, Beijing, China, Sep. 2011, pp. 1480–1484
39. M. Liwicki, M.I. Malik, L. Alewijnse, E.V.D. Heuvel, B. Found, ICFHR 2012 competition on automatic forensic signature verification (4NsigComp 2012), in *2012 International Conference on Frontiers in Handwriting Recognition (ICFHR)*, Bari, Italy, Sep. 2012, pp. 823–828
40. N. Brümmer, J. du Preez, Application-independent evaluation of speaker detection. Comput. Speech Lang. **20**(2), 230–275 (2006)

Chapter 10
Non-cooperative and Occluded Person Identification Using Periocular Region with Visible, Infra-Red, and Hyperspectral Imaging

Muhammad Uzair, Arif Mahmood, and Somaya Ali Al-Maadeed

10.1 Introduction

Existing face recognition techniques perform well in controlled environments when the subjects are cooperative. However, the performance of face recognition algorithms degrades if the subjects are non-cooperative resulting in less controlled over the imaging conditions and illumination. The performance degradation is exacerbated by facial occlusions due to head gear, headphone, head scarf, or hair style. The complete face of offenders is often not visible in surveillance videos. Helmets, that additionally cover faces, are commonly used by motorcyclists and sportsmen. Helmets and skiing masks have been exploited by offenders to cover their faces during crimes. Finally, in many countries, women partially cover their faces due to religious and cultural reasons. In most of these cases, the region around the eyes, that is the *periocular region*, is the only visible biometric (see Fig. 10.1).

In addition to the robustness to occlusion, periocular region based person identification also offers some additional advantages over the full face biometrics. This face region is less affected by expression variations, aging effects, and the changes due to growth of male facial hair [18, 25]. Moreover, full face recognition performance degrades significantly in the presence of pose variations whereas the periocular region based identification is likely to perform better in the case of

M. Uzair
Department of Electrical Engineering, COMSATS Institute of Information Technology, Wah Cantonment, Pakistan
e-mail: uzair@ciitwah.edu.pk

A. Mahmood (✉) • S. Ali Al-Maadeed
Department of Computer Science and Engineering, Qatar University, Doha, Qatar
e-mail: arif.mahmood@qu.edu.qa; rfmahmood@gmail.com; s_alia@qu.edu.qa

R. Jiang et al. (eds.), *Biometric Security and Privacy*, Signal Processing for Security Technologies, DOI 10.1007/978-3-319-47301-7_10

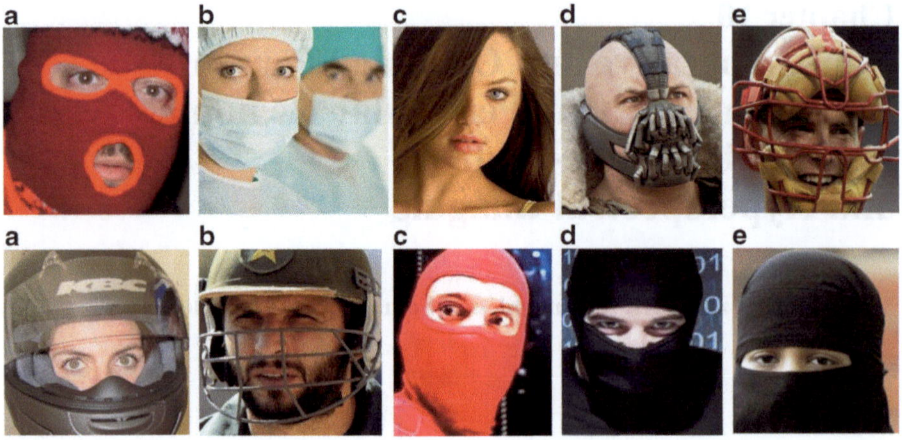

Fig. 10.1 Some real world cases where periocular biometrics are more effective than the full face biometrics. From the *left of top row* (**a**) skiing mask (**b**) doctors in the operation theater (**c**) face occlusion due to hair style (**d**) gas mask used in the movie "The Dark Knight Rises," and (**e**) football helmet. In the *bottom row from left* (**a**) motor bike helmet (**b**) cricket wicket keeper helmet (**c**)–(**d**) skiing mask, and (**e**) traditional face Hijab

extreme pose changes when only one eye is visible. In these cases, the single visible periocular region can be used or mirrored and matched to the opposite side periocular region in the database.

Periocular biometrics also offer many advantages over the iris biometric [12, 22, 46]. Periocular biometrics can be acquired from a larger distance by surveillance cameras even for the non-cooperative subjects whereas iris scanning requires the subjects to open their eyes fully and look into a scanner from a very close range. The latter may not be socially acceptable to users performing daily life routine activities [40, 41]. In many practical scenarios, the subject may not be cooperative or willing, or even aware that he/she is being imaged. For example, in banks and shopping malls, valued customers may need to be identified without causing inconvenience to them.

In this chapter we present a comprehensive study for the feasibility of the periocular region as a biometric for person identification for non-cooperative subjects. We consider realistic scenarios such as automatic detection of periocular regions, the availability of videos rather than a single handpicked frame, and propose a two stage inverse Error Weighted Fusion scheme to achieve the state-of-the-art results.

Many research efforts on periocular biometrics are mostly based on single image matching [1, 20, 38, 54–56] where a single best frame per subject is handpicked to form a gallery database. Each image in the query sequence is matched with the gallery database to find the best match. Such techniques inherently suffer from the lack of information since only one image cannot contain all possible variations in the periocular region of an individual (Fig. 10.2). For example, if the gallery

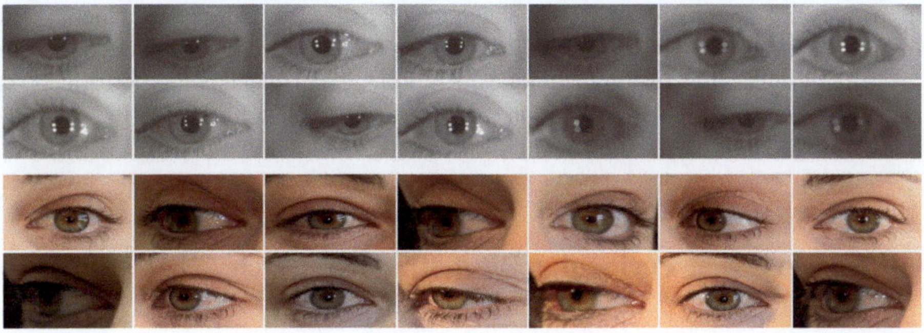

Fig. 10.2 Appearance variations of the periocular region of the same subject in the MBGC NIR video dataset (*top row*) and UBIPr dataset (*bottom row*)

image is taken from a frontal view with centered eyeballs and the query images have pose variations with eyeballs at the extreme ends, recognition performance will significantly degrade.

In contrast, in our work we formulated periocular region based identification as an image-set classification problem [48, 49]. Each set contains multiple periocular images of the same person and represents a wide range of variations including different eyelid positions, illuminations, eyeball movements, and poses. We observe that the left and the right periocular regions of the same subject possess more similarities to each other than to other subjects. Therefore, the reflection of one region is combined with the other to form a single large set per identity. We believe that this strategy better suits linear modeling techniques. We construct a gallery database from image-sets of different identities. The query set also contains multiple periocular images of the same person and is assigned the label of the nearest gallery set. Compared to single image matching, set-to-set matching offers significantly more information because multiple images in the same set model the appearance variations [16, 49].

We recommend four types of features to be extracted from each periocular region. We combined six state-of-the-art image-set classification techniques including the Affine Hull based image-set Distance (AHISD) [14], Convex Hull based image-set Distance (CHISD) [14], Discriminative Canonical Correlation (DCC) [26], Manifold-Manifold Distance (MMD) [52], Manifold Discriminant Analysis (MDA) [51], and Sparse Approximated Nearest Point (SANP) distance [21]. A brief overview of these techniques is given in Sect. 10.2. Note that our objective is to show the feasibility of image-set based classification for periocular biometrics as opposed to a performance comparison of the image-set classification algorithms.

Our preliminary results were initially published in [48], and later in [49]. In the current chapter we have extended our work in a number of directions. We present a detailed literature survey of the periocular recognition techniques, non-cooperative face recognition, and image-set based face recognition techniques. We have tested different imaging modalities beyond the visible spectrum, such as infra-red and hyperspectral image cubes. We have also developed a fully automatic video based periocular region detection system for all three modalities and investigated the use of different classifier fusion techniques. Rigorous experiments were performed on more datasets and higher recognition rates were observed compared to the state-of-the-art methods.

10.2 Review of Periocular Biometrics Research

Periocular region includes the iris [40], eyes, eyelids, eye lashes, and part of the eyebrows [55]. Recognition using periocular biometrics is an emerging research area. Initial feasibility studies were done by Park et al. [40, 41] who detected the iris in the visible spectrum images and extracted features at fixed grid positions around the center of the limbus circle. Experiments on a subset of the FRGC v2.0 [43] dataset showed that their algorithm's performance depends on the accurate iris detection and is sensitive to the eyeball movements.

The use of local binary patterns (LBP) [2] to represent the texture of the periocular region has been investigated by several researchers. Miller et al. [33] used city block distance to classify LBP features. Their work was extended by Adams et al. [1] using genetic algorithms to select the optimal subset of LBP features. They experimented with manually generated periocular data from a subset of high-resolution frontal images of FRGC [43] with neutral expression, less appearance variations, and controlled illumination. Woodard et al. [55] used the LBP features and color histograms to represent the local appearance of the periocular region. They achieved better performance by using city block distance for matching LBP features and Bhattacharya coefficient for matching color histograms. They performed experiments on periocular data generated from two databases: FRGC v2 face [43] and MBGC v1 NIR video [36]. In the case of MBGC NIR dataset, the complete videos were not used. Manually extracted periocular regions from the first two frames of the video were used as gallery and periocular region from a random frame was chosen as a probe. The periocular regions had the same scale and no appearance variations which made the analysis unrealistic.

Bharadwaj et al. [9] performed score level fusion of global GIST features and local circular LBP features of the periocular regions. They used the visible spectrum eye images of UBIIRIS v2 [44] dataset which contained slight appearance variations. Xu et al. [56] proposed Walsh Transform based local binary patterns (WLBP). Periocular region containing both eyes were cropped using the detected iris centers. They performed experiments on the FRGC v2 [43] database and achieved a verification rate of 17.0 at 0.1 % FAR. Woodard et al. [54] simultaneously utilized the iris and periocular biometrics by performing score level fusion. LBP features were extracted from the periocular region while the iris texture was encoded using the Gabor filters. Their experiments showed that the periocular recognition performed better than iris recognition for the MBGC v1 NIR portal videos.

Hollingsworth et al. [20] investigated human performance on the periocular recognition task. By presenting unlabeled pairs of NIR periocular images to different humans, they found that humans can recognize the periocular region with 92 % accuracy. They also calculated the performance of three computer algorithms on the periocular recognition task and found that the performance of humans and computer was similar.

Boddeti et al. [11] used probabilistic matching based on optimal trade-off synthetic discriminant function correlation filter to classify periocular regions. They performed experiments on the face and FOCS database [37] which contains periocular region appearance variations in the form of illumination, blur, and off-angle iris. They reported better recognition rates of the ocular regions than the iris on the FOCS database.

Juefei-Xu and Savvides [24] employed different schemes to correct the illumination and pose variations of the periocular regions. Class dependent feature analysis was used to classify the WLBP features [56]. Pauca et al. [42] used SIFT features for classification of the periocular regions and also introduced the COIR database.

Padole and Proenca [39] introduced the UBIPr database which contains scale and pose variations as well as non-uniform illumination variations. They observed that the eye corners offer a better reference for grid construction compared to the iris center. Alonso-Fernandez and Bigün [3] used retinoscopic sampling to generate grids of various geometry centering the pupil. Gabor features calculated from the grids were used for periocular region classification. Oh et al. [38] combined periocular and sclera biometrics using score level fusion. They used structured random projections for feature extraction from periocular regions and a modified LBP operator for feature extraction from sclera region. They obtained improved equal error rates on UBIRIS v1 dataset.

Tan and Kumar [47] proposed an online iris and periocular recognition system using images captured at a distance in less constrained environment. An improved iris segmentation method was proposed using a random walk algorithm. The segmented iris was then refined using improved post-processing techniques. A hybrid iris and periocular recognition was then performed using log-Gabor feature for the iris and local features (SIFT, LBP, and HOG) for the periocular region. Experiments were performed using publicly available UBIRIS.v2, FRGC, and CASIA.v4-distance databases and improved recognition rates were achieved.

Cao and Schmid [13] used Gabor LBP and Generalized LBP combined with Gabor Weber descriptors for matching cross spectral images of periocular regions in the short wave infra-red (SWIR), midwave infra-red (MWIR), and NIR spectrum to visible spectrum periocular regions at short (1.5 m) and long (50 and 106 m) standoff distances. Experiments were performed on the Pre-TINDERS (Tactical Imager for Night/Day Extended-Range Surveillance) and TINDERS databases collected by Pinellas County Sheriff's Office (PCSO).

Mahalingam et al. [30] proposed a system for recognizing the medically altered faces due to gender transformation. Their experiments showed that the periocular region represented with simple texture based features such as LBP outperforms full face recognition. Their periocular-fused patch based face matcher outperformed

PittPatt SDK v5.2.2 [23] by 76.8 % and Cognetic FaceVACS v8.5 [15] by 56.23 % for rank-1 accuracy on a challenging database collected from Youtube videos.

Nie and Kumar [35] performed periocular recognition using automatic feature learning with unsupervised Convolutional Restricted Boltzmann Machine (CRBM). They showed that automatic feature learning has improved the recognition accuracy compared to the hand crafted features. Fusion of hand crafted features with CRBM further improved the recognition rate.

Alonso-Fernandez and Bigun [5] proposed a periocular recognition system using near infra-red imagery. Gabor filters with five frequencies and six orientations equally spaced in the log-polar frequency plane were used to represent the periocular region. Experiments were performed on the BioSec (1200 NIR images) database and a recognition rate of 98 % was achieved. In their following work [7] they compared the performance of different periocular recognition algorithms on near infra-red periocular and iris images and showed that the recognition performance of the periocular region is better than the iris, mostly due to the small size of the iris or the use of visible illumination.

Table 10.1 summarizes the major research efforts in the area of periocular biometrics. Recently, Alonso-Fernandez and Bigun [4, 6] have also presented an extensive review on the state of the art in periocular recognition and a comprehensive list of databases employed in periocular research. Existing research has investigated the use of texture or point features calculated from a single handpicked periocular image. Therefore, these methods do not model the periocular region variations including partial occlusion, eyeball and eyelid movements, pose, and illumination conditions. In order to mitigate some of these challenges, we propose an image-set based approach for periocular biometric recognition. For each subject, the gallery may contain one or more image-sets per subject. An image-set will contain multiple periocular region images of the same subject thereby modeling a wide range of variations.

10.3 Review of Non-cooperative Face Recognition Techniques

Biometric recognition technologies has seen rapid growth in recent years, both in research and in commercial applications [15, 23]. Highly distinctive and permanent biometrics such as fingerprints, iris, or retinal scans require a cooperative subject. Such biometrics become unfeasible for a non-cooperative subject which is observed unobtrusively at a distance. Facial images can be captured at a distance; however, occlusion still remains a major challenge to non-cooperative face recognition.

Ao et al. [8] presented a review of the common issues in the design of face recognition in non-cooperative and at a distance scenarios and present effective solutions for making FRAD systems for practical deployments. Issues such as face occlusion, low resolution, and out of focus images are addressed in detail and solutions and strategies are also suggested.

Table 10.1 Recognition rates (RR) of the existing state- of-the-art periocular biometric recognition techniques

Author	Year	Database	#Subj	RR
Park et al. [40]	2009	Proprietary	30	80.80
Miller et al. [33]	2010	FRGC, FERET	464	89.70
Adams et al. [1]	2010	FRGC, FERET	464	92.16
Woodard et al. [54]	2010	MBGC v1 NIR	85	96.50
Woodard et al. [55]	2010	MBGCv1NIR, FRGCv2	85, 410	87.00, 91.20
Xu et al. [56]	2010	FRCG	466	53.20
Bharadwaj et al. [9]	2010	UBIRIS v2 [44]	261	73.60
Park et al. [41]	2011	FRGCv2	568	87.32
Boddeti et al. [11]	2011	FOCS [37]	136	94.20
Pauca et al. [42]	2011	COIR [42]	285	95.00
Padole and Proenca [39]	2012	UBIPr [39]	342	38.00[a]
Juefei-Xu and Savvides [24]	2012	Compass [24]	342	60.00[a]
Alonso-Fernandez and Bigün [3]	2012	CASIAv3 [10], BioSec [19]	249, 200	5.66, 13.08[b]
Tan and Kumar [47]	2013	UBIRISv2,FRGC, CASIAv4	261,466, 142	39.6, 59.4, 94
Oh et al. [38]	2014	UBIRIS v1	131	6.89[b]
Cao and Schmid [13]	2014	Pre-TINDERS	48	0.75[a]
Nie and Kumar [35]	2014	UBIPr	300	0.11[b]
Mahalingam et al. [30]	2014	Youtube videos	38	57.79
Fernandez et al [5, 7]	2015	BioSec NIR	200	98 %

[a] Verification rate at 0.01 false accept rate
[b] Equal error rate

Medioni et al. [32] presented a non-cooperative person identification at a distance using a hybrid 3D+2D approach. As the natural head and body motion provides enough viewpoint variation to perform stereo-motion for 3D face reconstruction, they generated a 3D face model from 2D sequence of images captured a distance. However, the quality of 3D reconstruction in this approach depends on the quality of 2D images, therefore for noisy and very low resolution images the performance of the system may by degraded severely.

Maeng et al. [29] presented a night time person identification system based on Near Infra-red face images captured at a distance. For this purpose they collected a new NIR Face Recognition at a Distance Database (NFRAD-DB) of 50 subjects. The probe NIR images were captured at a distance of 60 m and compared to gallery visible spectrum images captured at 1 m. They evaluated the performance of the commercial FaceVACS [15], SIFT key point based, and Multi-scale Local Binary Pattern based face recognition algorithms on their database. A rank-1 recognition rate of only 28 % has been achieved from these methods that shows the algorithms included in the study were not effective for night time person identification. They showed that NIR modality is more suitable for face recognition for low illumination conditions.

Wheeler et al. [53] proposed a face recognition at a distance for non-cooperative users using an active vision system. Subjects were detected and tracked with wide-field-of-view cameras and near-field-of-view pan-tilt-zoom cameras which are automatically controlled to collect high-resolution facial images. They performed face recognition experiments in large transportation hubs on facial images captured at a distances of 25–50 m and achieved good recognition rate.

10.4 Review of Image-Set Classification Algorithms

Although image-set classification has been well studied for face recognition, to the best of our knowledge, periocular region recognition has not been formulated before as an image-set classification problem. We argue that a set based approach is more suitable for periocular biometrics because it can model a number of variations such as pose, position of eyeball, and partially open eyelids that are unavoidable in real scenarios. We use six state-of-the-art image-set classification algorithms broadly divided into sample and structure based.

10.4.1 Sample Based Image-Set Classification

Sample based techniques measure the distance between nearest neighbor samples of two image-sets. Let $X = \{x_i\}_{i=1}^{n} \in \mathscr{R}^{m \times n}$ be an image-set, where $x_i \in \mathscr{R}^m$ is a feature vector and n be the number of feature vectors in a set which may vary across the image-sets. The feature vectors can simply be the image pixel values or some features calculated from the pixels such as the PCA coefficients or LBP features [2]. Each image-set may be considered as a point cloud in $\mathscr{R}^m$. All points in a probe image-set X_p are compared with all points in each gallery set X_g to find the nearest pair of points (x_i, x_j) such that $x_i \in X_p$ and $x_j \in X_g$. If x_i and x_j have zero mean and unit magnitude, the nearest neighbor pair (x_i, x_j) is the one that maximizes the cosine of the angular distance:

$$\max_{g}\left(\max_{x_i, x_j} X_g^t X_p\right). \tag{10.1}$$

The probe image-set label is predicted as the label of x_j.

Cevikalp and Triggs [14] considered each image-set as a convex geometric region in $\mathscr{R}^m$. Set dissimilarity was measured by the distance of closest approach between the regions represented by the affine (AHISD) or convex hulls (CHISD). The region contained all the affine combinations $x_p = \sum_{i=1}^{n} X_{pi}\alpha_{pi}$ and $x_g = \sum_{j=1}^{n} X_{gj}\alpha_{gj}$, where $\alpha_p, \alpha_g \in \mathscr{R}^m$. For the case of affine hull $\sum_{i=1}^{n} \alpha_{pi} = \sum_{j=1}^{n} \alpha_{gj} = 1$ and for the convex hull $0 \leq (\alpha_{pi}, \alpha_{gj}) \leq 1$. The minimum distance was computed as

$$\min_{g}\Big(\min_{\alpha_p,\alpha_g} ||X_g\alpha_g - X_p\alpha_p||^2\Big). \tag{10.2}$$

For the case of affine hull, the minimum distance was computed using least squares while for the case of convex hull, an SVM was used.

Instead of searching the nearest points with dense combinations of samples in the corresponding image-set, Hu et al. [21] proposed that each of the two points should be approximated by a sparse combination from the samples of the respective set. They argued that the sparse approximated nearest points (SANP) will lie close to some facet of the affine hull, and hence, implicitly incorporate structural information of the sets as well. By restricting the SANPs to be close to the facets, this approach can reject outliers resulting in higher accuracy.

10.4.2 Structure Based Image-Set Classification

Structure based techniques represent the underlying structure of an image-set with one or more linear subspaces. Structural similarity of the sets is usually measured using subspace to subspace distance. Kim et al. [26] proposed Discriminative Canonical Correlation (DCC) which performs discriminative learning on canonical correlations between the structures of image-sets. More specifically, a discriminant function is learned that maximized the within-class and minimized the between-class canonical correlations. Image-sets are compared after transforming them optimally by the learned discriminant function.

Wang et al. [52] proposed Manifold-Manifold Distance (MMD) which clustered each image-set into multiple linear local models and represented each model by a linear subspace. The similarity between two sets was defined as the canonical correlation between the nearest local models. In addition, the nearest point distance was also combined with the structural similarity to calculate the final similarity between two sets.

Wang and Chen [51] proposed Manifold Discriminant Analysis (MDA) that uses a hierarchical divisive clustering approach to represent each image-set by multiple local linear models. The local models are transformed by a linear discriminant function where different classes are better separable. The similarity between two sets is calculated as the pair-wise local model distances in the learned embedding space.

10.5 Detection and Extraction of Periocular Region

Distance variations between a subject and a camera may induce significant scale changes while head pose variations may induce rotational effects in the video. These scale and rotation variations do not provide any discriminative information. We will discuss two algorithms for automatic periocular region extraction and normalization in videos and hyperspectral cubes.

10.5.1 Periocular Region Detection in RGB Videos

Most algorithms [54, 55] select only the best frames with a fixed scale and rotation for matching purpose, though performing a rotation and scale normalization would improve the results. In a video frame (NIR or visible), an eye pair can be detected by using a Haar feature based ensemble of weak classifiers as proposed by Viola and Jones [50]. Then we crop a relatively large area containing both eyes and the nose bridge. For each eye, we automatically detect the pupil center using an improved version of Masek's algorithm [28, 31]. We first apply adaptive thresholds on the eye image to obtain a binary image. Following this, the largest local number of contour points is searched by contour processing the binary image. A circle is fitted to the detected largest contour points and its center is then selected as the pupil's center. The pupil centers in each eye are denoted by P_1 and P_2 in Fig. 10.3.

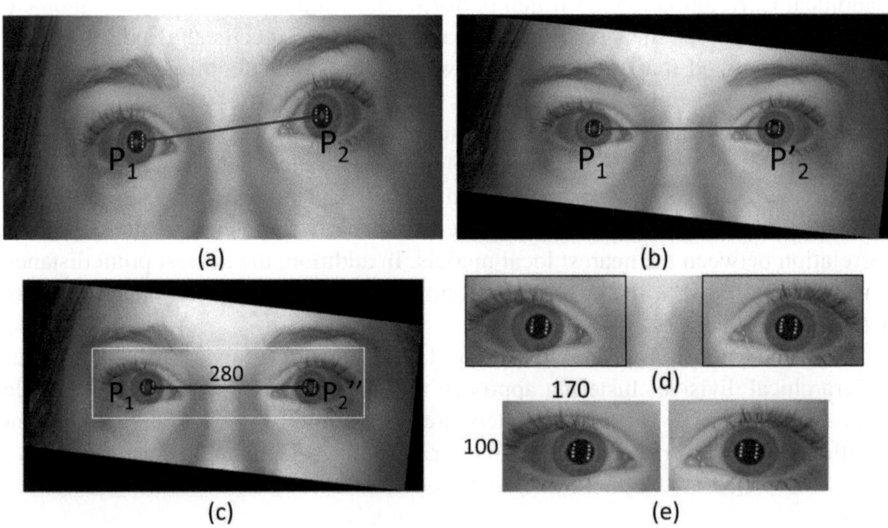

Fig. 10.3 Periocular region normalization in MBGC NIR and visible spectrum videos. (**a**) Detected pupil centers. (**b**) Rotation normalization. (**c**) Scale normalization. (**d**) Nose bridge removal. (**e**) Normalized periocular regions

For in-plane rotation normalization, we rotate the cropped images such that the line through P_1 and P_2 becomes horizontal. For scale normalization, we fix the distance between P_1 and P_2 and we scale the full periocular region equally in both dimensions. By using fixed distance constraints, we divide the normalized periocular region into left and right periocular regions. The nose bridge is removed and both regions simulate the situations when only one eye is visible. Thus we only utilize the eye region for recognition because in many practical situations, the other regions may not be available or may be changed artificially.

To enable single eye based recognition, we flip all the periocular regions to one side. Linear combinations of the original and flipped periocular regions can generate novel appearances and hence account for more intra-person variations. Note that the effect of flipping from left to right or right to left is the same.

10.5.2 Periocular Region Detection in Hyperspectral Image Cubes

We extend our analysis beyond the visible spectrum towards hyperspectral periocular biometric recognition. A hyperspectral image cube consists of images captured at multiple wavelengths of the electromagnetic spectrum. The signal to noise ratio of the bands near the blue wavelength region is very low and in general, no periocular region detector may ensure 100 % detection rate in these images. However we assume that the location of the pupil centers in all bands are aligned. Thus we solve the problem of miss detections by estimating collective pupil center positions over all bands in a hyperspectral cube.

In each image of a hyperspectral cube, we automatically detect the eye pair using Viola and Jones detector [50] and for each eye, we automatically detect the pupil centers as follows. First, we apply thresholding on the eye image to obtain a segmented binary image. Next, we apply the morphological opening operation on the binary image. The opening operation enhances the dense region of the binary image corresponding to the pupil center and smooths the sparsely occupied regions (Fig. 10.4). We apply the opening operation n times and a value of $n = 3$ gave good results in our experiments. The location of the global minimum in the resultant opened image is considered as the pupil center. Here, we do not require circle

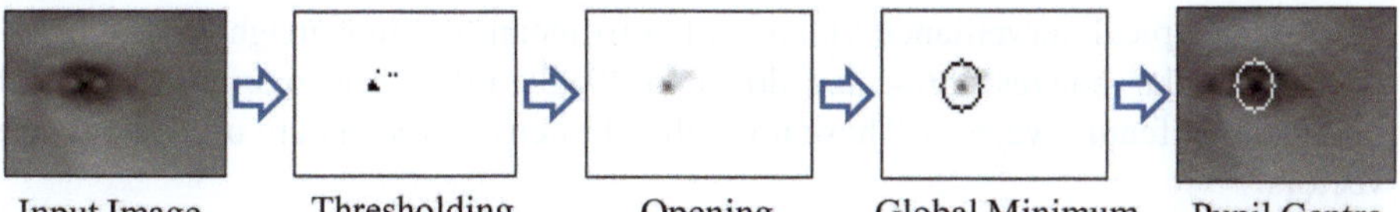

Fig. 10.4 Pupil center detection in hyperspectral databases. Detection errors are corrected by using the proposed algorithm

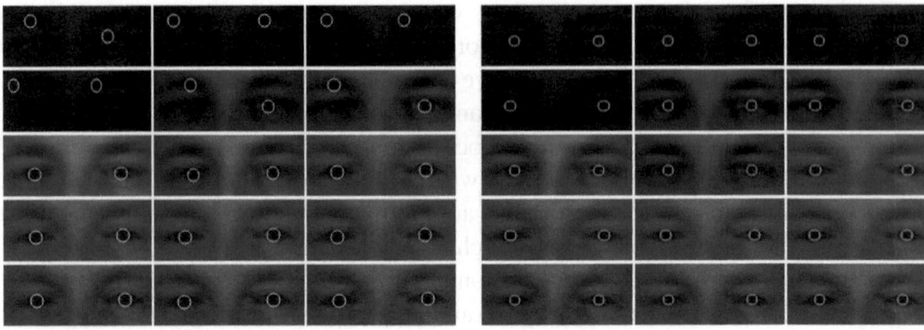

Fig. 10.5 Detected pupil center (*left*) and corrected pupil centers (*right*) in some of the bands of CMU hyperspectral database

fitting because of the availability of multiple bands to assist accurate pupil center localization. Figure 10.4 shows the proposed pupil center detection process.

We independently detect the pupil center in each band (Fig. 10.5). Then, each detection votes in a circle of radius r around the detected position, using Gaussian weights, in an accumulation array. The votes are accumulated over all bands of the same cube. The average of the coordinates in a small vicinity getting maximum votes over all bands is selected as the pupil center for the whole hyperspectral cube and used for rotation and scale normalization. The same rotation and scale normalization is applied to all bands of a cube. Cropping and flipping of the periocular regions is similar to the video databases.

10.6 Feature Extraction

Each algorithm discussed in the previous sections proposed a different set of features to be extracted from the periocular region and used for classification. In this section we study only following four simple and more commonly used features:

10.6.1 Normalized Intensity Vectors

To mimic typical surveillance videos, all zero mean and unit magnitude normalized periocular images are scaled down to 20×30 pixels and rearranged as 600 dimensional feature vectors. These normalized intensity vectors are used as feature vectors.

10.6.2 PCA Coefficients of Image Vectors

Instead of simply scaling down the images, a better way is to reduce the dimensionality of the normalized images is to project them on most significant PCA basis. Images are rearranged as vectors. Let $X \in R^{p \times q}$ be the matrix containing image vectors $x \in R^{p}$ are as columns. Each image vector x has dimensionality p and there are q images in X. Using SVD X is decomposed as: $X = USV^{\top}$, where $U \in R^{p \times q}$ and $V \in R^{q \times q}$ are orthogonal matrices and $S \in R^{q \times q}$ is a positive diagonal matrix such that $S(i,i) \geq S(i+1, i+1)$. First $r < q$ vectors of V are selected as new basis $\hat{V} \in R^{p \times r}$. Projection of an image vector is computed as $\hat{x} = \hat{V}^{\top} x \in R^{r}$, the dimensionality of $\hat{x}$ is r. In our experiments we have used $r = 400$. The resulting 400 dimensional coefficient vectors are used as feature vectors. Note that for learning the PCA basis, only the gallery sets are used as training data. Using PCA coefficients as features has two main advantages. By dropping the least significant principal components, PCA suppresses the effect of noise such as partial and unaligned periocular regions present in the image-sets. By reducing the dimensionality of the normalized images to 400, the image-set classification algorithms achieved significant speedup.

10.6.3 LBP Features

From the 20×30 images, LBP features are extracted using circular (8,1) neighborhoods [2]. The LBP operator labels a pixel with a binary number computed by thresholding the gray-scale differences between the pixel and its neighborhood. The LBP of the pixel z in an image I can be represented as follows:

$$
\begin{aligned}
\mathrm{LBP}_{P,R}(z) &= \{\mathrm{LBP}^{(p)}_{P,R}(z)\}_{p=1,\ldots,P}, \\
\mathrm{LBP}^{(p)}_{P,R}(z) &= s(I^{g}(v_p) - I^{g}(z)), \\
s(z) &= \begin{cases} 1 & \text{if } z \geq 0, \\ 0 & \text{if } z < 0, \end{cases}
\end{aligned}
\tag{10.3}
$$

where $I^{g}(z)$ is the gray-scale value of the pixel z in the image I and $\{v_p\}_{p=1,\ldots,P}$ as a set of P equally spaced pixels located on a circle of radius R and center z. We choose LBP features for their robustness to monotonic gray-scale changes resulting from global or local illumination variations [57]. Figure 10.6 shows the process of LBP filtering. The resulting LBP coded images are used as feature vectors.

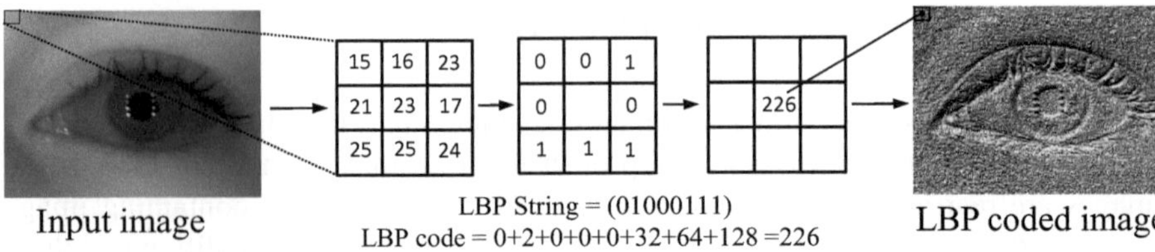

Fig. 10.6 LBP filtering process to generate features of type 2

10.6.4 PCA Coefficients of LBP Features

LBP features are computed over the normalized images as discussed above. LBP encoded images are rearranged as vectors and the dimensionality of the LBP code vectors is reduced to 400 using PCA. The PCA process is similar the one discussed above.

10.7 Score Level Fusion of Classifiers

We consider video based periocular biometric as a set based classification problem. The label of a probe image-set p is to be predicted by computing its similarity with g gallery image-sets. We have c different classifiers, each generating a different similarity vector $\{S_k\}_{k=1}^{c}$ where $S_k = \{s_{k,j}\}_{j=1}^{g}$ encodes the similarity scores for each gallery set for a particular classifier. To obtain better accuracy, these similarity vectors generated by different classifiers are to be fused into one similarity vector $\mathscr{S} = \{\mathsf{s}_j\}_{j=1}^{g}$.

For this purpose, we investigate the efficiency of different classifier fusions techniques [27, 45] for the periocular biometric. We consider fusion schemes including the sum, product, min, max, and median rules, and majority voting. We will also discuss our own proposed a two stage fusion scheme using classifier error weighted summation technique (see Fig. 10.7) [49]. Let L_p^{Rule} be the predicted label by a specific rule, where Rule$\in$ { sum, prod, max, min, med, vote, SMF}.

$$\text{Sum rule}: L_p^{\text{sum}} \triangleq \max_j \left(\sum_{k=1}^{c} s_{k,j} \right).$$

$$\text{Product rule}: L_p^{\text{prod}} \triangleq \max_j \left(\prod_{k=1}^{c} s_{k,j} \right).$$

$$\text{Max rule}: L_p^{\text{max}} \triangleq \max_j \left(\max_k (s_{k,j}) \right).$$

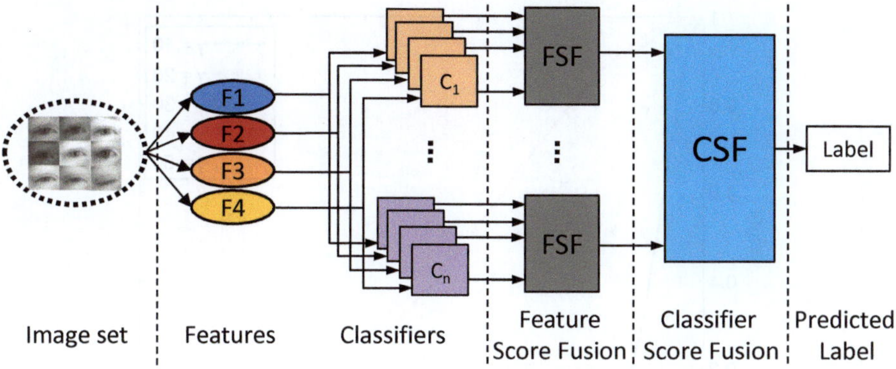

Fig. 10.7 Block diagram of the two stage fusion: Feature Score Fusion (FSF) and Classifier Score Fusion (CSF). Different fusion strategies are used at both stages (Table 10.3)

$$\text{Min rule}: L_p^{\min} \triangleq \max_j \left(\min_k (s_{k,j}) \right).$$

$$\text{Median rule}: L_p^{\text{med}} \triangleq \max_j \left(Med_k(s_{k,j}) \right).$$

$$\text{Majority voting rule}: L_p^{\text{vote}} \triangleq \text{Mode}_k(\{l_k\}_{k=1}^{c}), \tag{10.4}$$

where $l_k \triangleq \max_j(s_{k,j})$ is the label predicted by the kth classifier. In these schemes, all classifiers are given equal weight. In contrast, we propose to weight each classifier inversely proportional to the error of that classifier on the validation dataset. We refer to this technique as *Error Weighted Fusion* (EWF). Specifically, the following weight function is learned during the training phase:

$$w_k = \exp\left(-\frac{\xi_k^2}{\gamma^2} \right), \tag{10.5}$$

where ξ_k is the error rate of the kth classifier on the training data. The parameter γ controls the decay of the weight function with the increasing error rate: $1 \geq w_k \geq 0$. Figure 10.8 shows the variation of w_k for different combinations of ξ_k and γ. An appropriate value of γ is selected that minimizes the error on training data after fusion of classifiers. The fused similarity vector is computed as:

$$\mathscr{S} = \sum_{k=1}^{c} \exp\left\{ \ln\left(\frac{S_k - \delta_k}{\lambda_k} \right) - \left(\frac{\xi_k^2}{\gamma^2} \right) \right\}, \tag{10.6}$$

where δ_k is used to shift the values in the S_k to start from zero and λ_k is the range of values in S_k. After fusion, the label L_p of the probe is assigned by the index containing maximum value over all gallery image-sets: $L_p^{\text{SMF}} = \max_j(\mathscr{S}_j)$. Note that

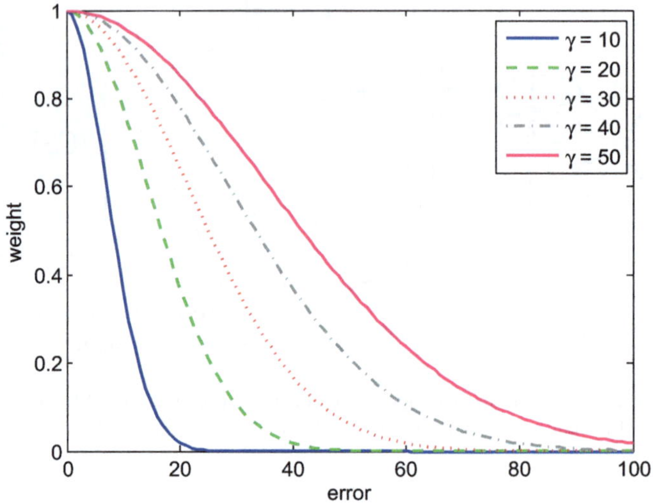

Fig. 10.8 Plot of the proposed weight function w_k for different combinations of ξ_k and γ

Table 10.2 Database details

Database	Subjects	Image-sets/subject	Images/set
UBIPr [39]	342	6	5
MGBC v2 Visible [43]	91	1–6	23–352
MGBC v2 NIR [43]	114	1–12	6–48
CMU Hyperspectral [17]	48	1–5	130

if similarity is changed by distance, max function will be changed by min function. We perform score level fusion at two stages namely Feature Score Fusion (FSF) and Classifier Score Fusion (CSF) as shown in Fig. 10.7. We then experimentally study the performance of the fusion schemes defined above.

10.8 Datasets Used for Periocular Biometric

To make our study on periocular region based person identification more diverse, we performed experiments on UBIPr database [39] and three public face databases including MBGC Visible spectrum and NIR video databases [43] and the CMU hyperspectral database [17]. These databases are summarized in Table 10.2 and their details are given below.

10.8.1 UBIPr Visible Spectrum Database

The *UBIPr Visible Spectrum Database* [39] consists of visible spectrum periocular images with variations in scale, illumination, eyeball movements, partial occlusion, and head pose changes. The images were acquired at five different camera distances (4–8 m), three poses (0°, 30°, and −30°), three gazes (0°, 30°, and −30°), and three levels of pigmentation (light, medium, and heavy). The images in this database are converted to gray scale and normalized to reduce the effects of scale and pose variations.

10.8.2 MBGC Visible Spectrum and NIR Datasets

The *MBGC v2 Visible Spectrum* (VS) videos have significant scale and illumination variations (Fig. 10.9a). Many frames have low signal to noise ratio and motion blur resulting in significant intra-person variations. The NIR illuminator used during *MBGC v2 NIR* video acquisition flashed for only a brief moment resulting in significant illumination variations, defocusing, motion blur, sensor noise, specular reflections, partially occluded faces and off-angle eyes making biometric recognition very challenging (Fig. 10.9b). Moreover, only part of the face is visible in most videos.

Fig. 10.9 Sample video frames from the MBGC v2 (**a**) Visible and (**b**) NIR databases

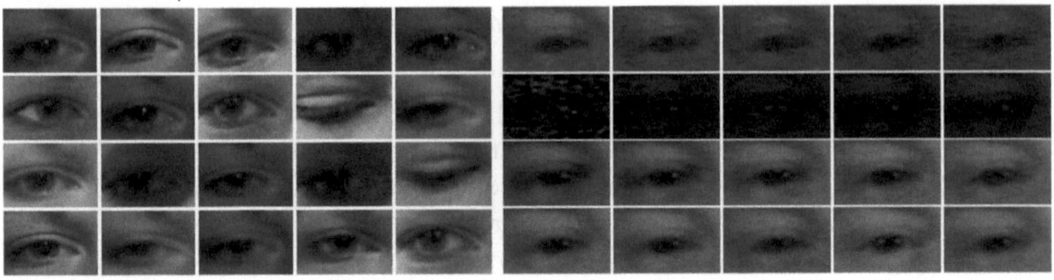

Fig. 10.10 Sample images from one image-set of MBGC visible spectrum video (*left*) and CMU Hyperspectral database (*right*). See Fig. 10.2 for sample images from MBGC NIR and UBIPr databases

In each video frame, the periocular regions are extracted as detailed in Sect. 10.5.1. The distance between pupil centers is fixed to 280 pixels and a normalized image of size 430 × 100 containing both periocular regions is cropped. This normalized image is divided into left and right periocular regions by removing the central nose bridge portion of 90 × 100 pixels. The final image size of the left and right periocular regions is 170 × 100 (Fig. 10.3). The right side periocular regions are mirrored and combined with the left side regions to form a single set (Fig. 10.10). Note that left to right mirroring gives the same results as right to left mirroring.

10.8.3 CMU Hyperspectral Dataset

The *CMU Hyperspectral Database* [17] contains face images in 65 bands covering the spectral range of 450–1100 nm at 10 nm step. The spatial resolution of the images is 640 × 480. In each image (band), we automatically detect the periocular region and normalize for rotation and scale variations. The distance between the eye centers is fixed to 90 pixels. A normalized image of size 160 × 40 is then cropped. The periocular region is divided into left and right regions by removing the nose bridge. The final image size of the periocular regions is 40 × 28 pixels (Fig. 10.10). The right periocular images are mirrored and combined with the left ones to form a set.

10.8.4 Experiments and Results

For each database, the gallery is constructed by randomly selecting one image-set per subject. The remaining image-sets are used as probes. Experiments are repeated tenfold by randomly selecting different gallery and probe combinations each time. For image-set classification, six algorithms are used including AHISD, CHISD

[14], SANP [21], DCC [26], MMD [52], and MDA [51]. For every algorithm, we performed tenfold experiments for each of the four feature vector types.

For the sample based algorithms, the default input parameter values are used. For the structure based algorithms, the parameters are carefully tuned so that the best performance can be obtained on training data. For DCC, the subspace dimensions are set to 10 which preserves 90 % energy and the corresponding 10 maximum canonical correlations are used to define set similarity. The embedding space is set to 100. For MMD and MDA, the ratio between Euclidean and Geodesic distance is varied from 2.0 to 0.01 with a step of 0.5. Maximum recognition rates were observed at 0.10 for MMD and 2.0 for MDA. The number of connected nearest neighbors for computing geodesic distance was tested at 14, 12, and 10. The best results were observed at the value of 10. The maximum canonical correlation was used in defining MMD. For MDA, the number of between-class NN local models was set to 5 and the dimension of MDA embedding space was set to 10 as recommended by Wang et al. [51, 52].

The error rate ξ_k in the weighted sum based fusion scheme is learned during training using a held-out validation set which is disjoint from the test set. For each database the validation set is generated by randomly choosing one image-set for each subject. The parameter ξ_k is found using the validation set. We observe that the accuracy of different algorithms on the validation set is almost the same as the test sets. After experimenting various values of γ, a fixed value $\gamma = 40$ is used for all the datasets.

10.8.5 Results and Discussion

The proposed image-set based periocular biometric recognition has exhibited recognition rates much higher than the existing state-of-the-art techniques. In the following sections, results of the six classification algorithms and different fusion schemes are discussed for each of the four databases. Results of existing algorithms in the literature along with brief experimental settings are summarized in Table 10.1.

UBIPr Database Results on this database are reported in Fig. 10.11. The performance of each algorithm on individual feature types is shown in Fig. 10.11a, where CHISD, AHISD, and SANP achieve better performance than DCC, MMD, and MDA. Thus sample based algorithms have outperformed structure based ones because only five images per set were not sufficient to estimate the structure. LBP features consistently achieve the highest performance among the sample based techniques. The last group of bars in Fig. 10.11a show the performance of *Error Weighted Fusion* (EWF) of all algorithms on different features. Here, again LBP features have achieved the highest recognition rate.

The performance of each algorithm is also studied after the feature score fusion (FSF) of all features using EWF. Figure 10.11b, c shows the CMC and ROC curves, respectively, for each algorithm after EWF feature score fusion. The curves of

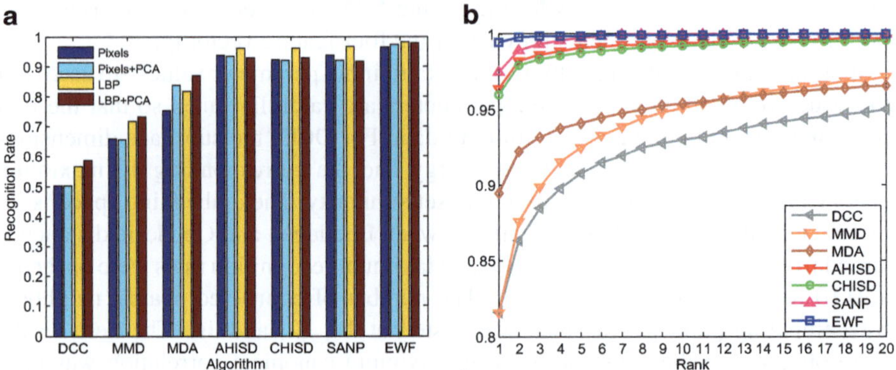

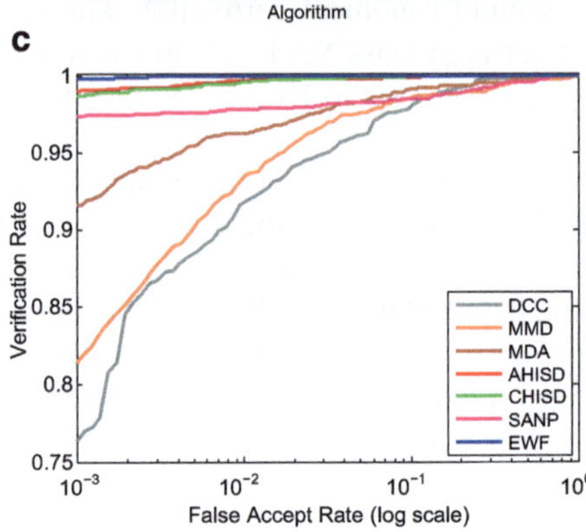

d

Algorithm	Rank-1 recognition rate (%)	Verification rate (%) at 0.001 FAR
DCC	81.8±2.8	76.3±1.6
MMD	81.6±1.9	81.6±2.5
MDA	89.5±2.4	91.6±2.1
AHISD	96.4±1.9	98.0±1.9
CHISD	96.7±1.8	98.7±1.1
SANP	97.4±1.2	97.4±1.1
EWF	**99.5±0.5**	**99.7±0.1**

Fig. 10.11 Results of tenfold experiments on the UBIPr database. (**a**) Average recognition rates of individual algorithms on individual features. (**b**) CMC and (**c**) ROC curves of individual algorithms after feature score fusion. (**d**) Summarized results

the sample based algorithms are again above the structure based algorithms. The best performance in Fig. 10.11b, c corresponds to the proposed two stage fusion algorithm (FSF + CSF as shown in Fig. 10.7). The proposed two stage fusion approach achieved the highest recognition rate of 99.5±0.5 % and the highest verification rate of 99.9±0.1 % at 0.001 FAR. The latter is significantly higher than the 38 % verification rate at 0.001 FAR reported by Padole and Proenca in [39] using fusion of LBP+HOG+SIFT features. The consistent best performance of EWF demonstrates the effectiveness of the proposed image-set based classification as well as feature (FSF) and classifier score fusion (CSF) techniques.

MBGC Visible Range Video Database Figure 10.12 shows the results on this database. The average recognition rates of each algorithm for each feature type are given in Fig. 10.12a. Here, the structure based algorithms have outperformed the sample based ones because the number of images per set is significantly higher than the UBIPr dataset leading to improved structure estimation. The last group of bars are for feature score fusion of all algorithms. Feature type 4 (LBP+PCA) has shown

a b

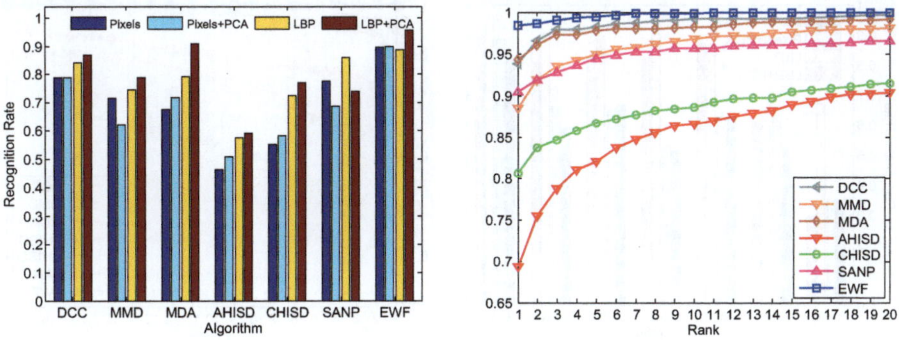

c

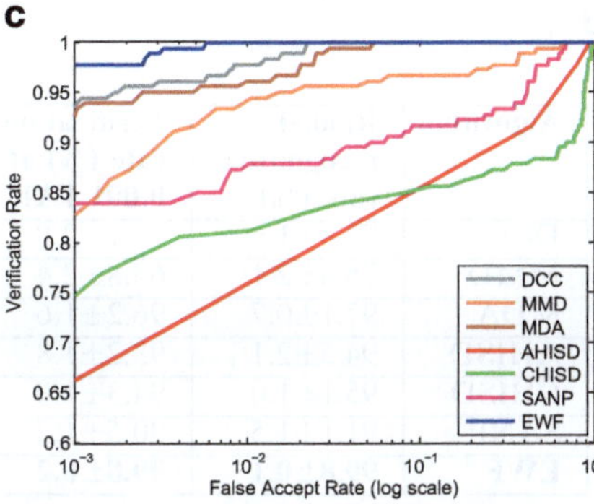

d

Algorithm	Rank-1 recognition rate(%)	Verification rate(%)at 0.001FAR
DCC	93.9±1.0	93.4±1.9
MMD	88.5±1.9	83.4±1.5
MDA	94.4±1.2	92.3±1.1
AHISD	69.4±2.9	66.5±3.3
CHISD	80.7±2.1	76.8±2.0
SANP	90.5±2.1	84.0±2.2
EWF	**98.5±1.0**	**97.8±0.5**

Fig. 10.12 Results of tenfold experiments on the visible range MBGC video database. (**a**) Average recognition rates of individual algorithms on individual features. (**b**) CMC and (**c**) ROC curves of individual algorithms after feature score fusion. (**d**) Summarized results

better performance than the other features for structure based techniques and the feature score fusion. The large number of samples per set not only helps in better structure estimation but also facilitates better PCA subspace estimation after LBP filtering. Since the images in this database have more noise and blur, PCA and LBP together perform better noise filtering than either one alone.

The performance of each algorithm after feature score fusion (FSF) of all features using EWF is given in Fig. 10.12b, c and summarized in Fig. 10.12d. Here, the plots of structure based techniques are above the sample based techniques. The proposed two stage fusion (FSF+CSF) using EWF has outperformed all other algorithms by achieving a rank-1 recognition rate of 98.5±1.0 % and verification rate of 97.8±0.5 % at 0.001 FAR.

MBGC NIR Database Results for this database are given in Fig. 10.13. The performance of individual algorithms on each feature type is shown in Fig. 10.13a. Due to fewer images per set, sample based algorithms have again performed better than structure based ones with the exception of MDA. Note that in MDA, a discriminative function is learned such that the geometric distance between different

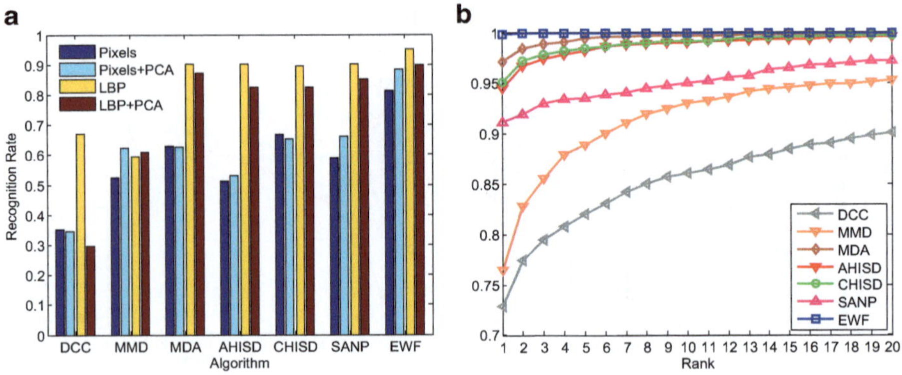

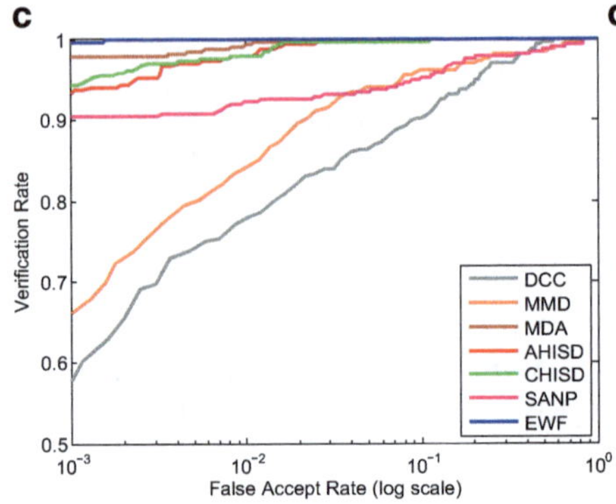

d

Algorithm	Rank-1 recognition rate (%)	Verification rate (%) at 0.001 FAR
DCC	72.8±1.6	55.1±2.0
MMD	76.4±2.1	65.8±2.4
MDA	97.1±0.7	96.2±1.6
AHISD	94.5±2.1	93.2±1.8
CHISD	95.1±1.0	94.3±1.0
SANP	91.1±1.5	90.5±1.2
EWF	**99.8± 0.1**	**99.8± 0.2**

Fig. 10.13 Results of tenfold experiments on MBGC NIR video database: (**a**) Average recognition rates of individual algorithms on individual features. (**b**) CMC and (**c**) ROC curves of individual algorithms after feature score fusion. (**d**) Summarized results

set centers is maximized. The average of the NIR imagery of the periocular region, which also includes the iris, is more discriminative than the visual spectrum images. This fact is also reflected by the higher mean and lower standard deviation of the NIR recognition rate of 99.9±0.1 %, compared to visual spectrum recognition 98.5±1.0 %. One may argue that the higher performance of NIR images is due to better image quality but this is balanced by the higher number of images per set in the visible spectrum image-sets. The NIR illuminator gives better illumination, but at the same time induces non-linear specularities. LBP filtering has remained more successful in removing the uneven illumination variations therefore, LBP features have shown better performance than raw pixels and raw pixels + PCA. The reason why LBP+PCA resulted in lower performance is because an accurate PCA subspace could not be estimated from fewer images.

The performance of each algorithm after feature score fusion using EWF fusion is shown in Fig. 10.13b, c. The CMC and ROC curves of the sample based techniques are above the structure based techniques. The proposed two stage fusion (FSF+CSF) using EWF achieved a rank-1 identification rate of 99.9±0.1 % and

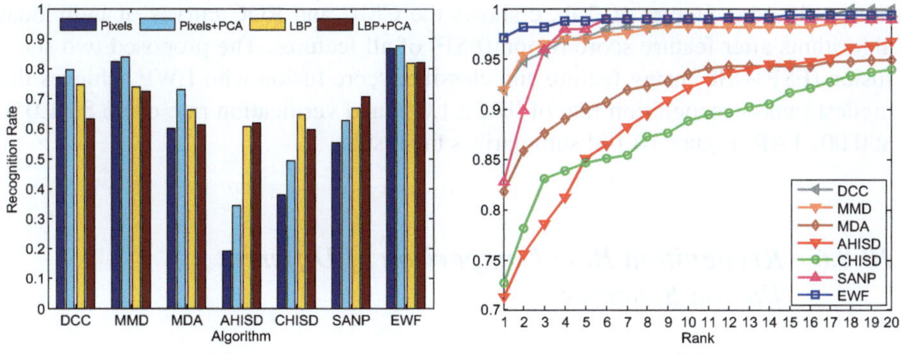

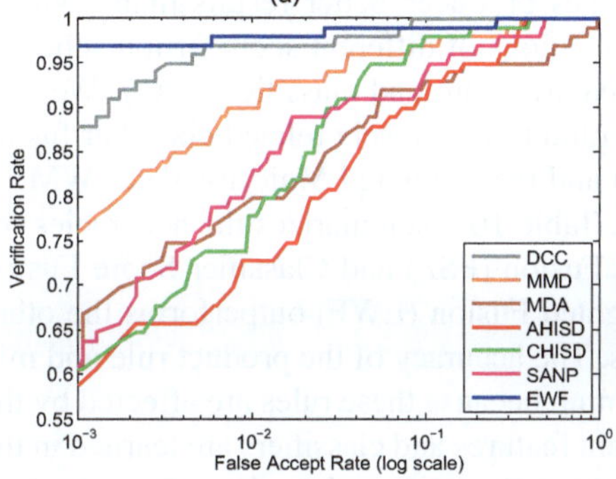

Algorithm	Rank-1 recognition rate (%)	Verification rate (%) at 0.001 FAR
DCC	91.9±2.5	78.8±3.3
MMD	91.6±2.1	74.7±2.3
MDA	81.8±1.6	66.7±1.7
AHISD	71.3±2.9	58.5±3.8
CHISD	72.7±3.6	60.6±4.1
SANP	82.8±2.1	63.6±1.8
EWF	97.2±1.0	96.8±1.0

(d)

Fig. 10.14 Results of tenfold experiments on CMU Hyperspectral database: (**a**) Average recognition rates of individual algorithms on individual features. (**b**) CMC and (**c**) ROC curves of individual algorithms after feature score fusion. (**d**) Summarized results.

a verification rate of 99.8±0.2 % at 0.001 FAR (Fig. 10.13d). The previous best identification rate on this database was reported by Woodard et al. [54] which is 96.5 % using only 85 subjects of the database (Table 10.1). We achieved higher identification rate while using 114 subjects from the database which is a more challenging scenario. We excluded 35 subjects because only four periocular regions per subject could be automatically detected. Moreover, Woodard et al. [54] manually extracted the periocular regions whereas we performed this task automatically.

CMU Hyperspectral Face Database For thoroughness, we also performed experiments on the hyperspectral face cubes. To the best of our knowledge, this is the first work on hyperspectral periocular biometrics that includes as many as 65 bands covering the visible and infra-red range (from 450 to 1100 nm at 10 nm step). The results on this database are compiled in Fig. 10.14. On individual features (Fig. 10.14a), structure based algorithms have performed better than sample based ones (Fig. 10.14a) because the number of images per set are 130 leading to a good structure estimation. Pixels+PCA gives the best performance for structure based techniques and when feature score fusion is performed (the last set of bars in 10.14a). Figure 10.14b, c shows the CMC and ROC curves of individual algorithms after feature score fusion (FSF) of all features. The proposed two stage fusion (FSF+CSF) using feature and classifier score fusion with EWF achieves the highest rank-1 recognition rate of $97.2 \pm 1.0\,\%$ and verification rate of $96.8 \pm 1.0\,\%$ at 0.001 FAR. Figure 10.14d summarizes the results.

10.8.6 Recognition Rate Comparison of Different Fusion Schemes

We observe that score fusion at two stages produces better results than a single stage fusion. We have explored all combinations of different score fusion schemes discussed in Sect. 10.7 at both stages (feature score and classifier score). We also compare the proposed EWF fusion algorithm to the closely related classifier fusion methods, Simple Weighted Vote (SWV) and the Weighted Majority Vote (WMV), presented by Moreno Seco et al. in [34]. Table 10.3 summarizes the accuracies for different combinations of Feature Score Fusion (FSF) and Classifier Score Fusion (CSF). The proposed inverse Error Weighted Fusion (EWF) outperforms the other fusion methods when used at both stages. The accuracy of the product rule and min rule when used at both levels is the minimum because these rules are affected by the lowest performers. The weights of different features and classifiers are learned in the training stage according to the performance on the training data. Therefore, the poor performers get lower weights in the test stage and the overall accuracy significantly improves over the sum rule.

10.8.7 Effect of Periocular Region Misdetection on Accuracy

The accuracy of periocular region detection may affect the performance of periocular recognition. The accuracy of the proposed algorithm will depend on the total images in a set and the ratio of misdetections to the correct detection. Figure 10.15 shows samples from an image-set of the visible range MBGC video database. Alongside correct detections, the set also contains around $14\,\%$ noisy samples in the form of misdetected periocular regions. In our experiments, this set was misclassified by the AHISD algorithm because the affine hull of the set was not able to accommodate for the outlier samples. On the other hand, the same image-set was classified correctly by the DCC algorithm as DCC models the image-set structure using the most dominant subspace basis. However, if most of the set samples contain large detection errors and there are not enough correct samples available for accurate set structure estimation, then the accuracy of the set based periocular recognition may slightly degrade. We observe that the proposed framework is robust to approximately $17\,\%$ outlier samples in image-sets without significant change in accuracy.

Table 10.3 Average identification Rates and std (%) of tenfold experiments using different combinations of Feature Score Fusion (FSF) and Classifier Score Fusion (CSF)

	CSF							
FSF	Prod	Max	Median	Voting	Sum	SWV	WMV	EWF
MBGC visible range video database								
Prod	33.6±1.2	94.7±1.1	97.5±0.6	78.5±1.8	95.6±1.0	96.3±1.1	97.5±0.6	97.1±1.1
Max	63.7±1.1	77.7±2.2	94.8±0.9	93.0±0.7	95.4±1.5	95.1±1.2	96.3±1.1	97.1±0.7
Median	73.7±3.7	91.7±2.5	97.5±0.8	96.5±0.7	97.1±1.1	97.0±1.8	97.2±1.4	97.2±1.4
Voting	–	–	–	97.8±0.4	–	–	–	
Sum	69.9±3.5	91.5±1.6	97.5±0.8	97.0±0.7	97.3±1.2	97.4±1.2	97.3±1.5	97.3±1.5
SWV	68.7±3.6	90.3±1.9	97.0±0.8	96.7±1.5	96.7±1.5	94.3±1.8	96.7±1.5	97.2±1.4
EWF	74.1±3.6	91.5±2.5	96.9±1.6	97.7±0.1	97.4±1.4	97.4±1.4	97.7±1.5	**98.5±1.0**
MBGC NIR video database								
Prod	18.4±2.5	91.8±1.3	97.5±0.9	73.6±2.0	97.4±1.0	97.4±1.0	99.1±0.5	99.0±0.7
Ma	65.2±1.7	83.3±2.5	96.9±1.3	97.2±1.0	97.9±0.8	97.7±0.8	98.7±0.8	99.0±0.5
Median	75.5±2.2	93.3±1.7	98.0±0.7	98.2±0.9	98.2±0.5	98.7±0.8	99.1±0.5	99.1±0.5
Voting	–	–	–	97.6±0.9	–	–	–	
Sum	73.0±2.0	91.5±2.0	98.1±0.7	98.2±0.9	98.1±0.5	97.7±0.8	99.6±0.4	99.6±0.4
SWV	71.8±1.7	90.7±1.3	97.9±0.8	97.7±0.8	98.4±0.4	96.8±0.9	98.4±0.4	97.7±0.8
WMV	73.2±2.1	93.1±1.7	98.4±0.6	97.7±0.8	98.1±0.5	98.7±0.8	98.4±0.4	99.6±0.4
EWF	73.9±2.9	93.7±2.0	98.4±0.6	98.8±0.8	98.3±0.4	98.7±0.8	99.6±0.4	**99.9±0.1**

Maximum values are shown bold

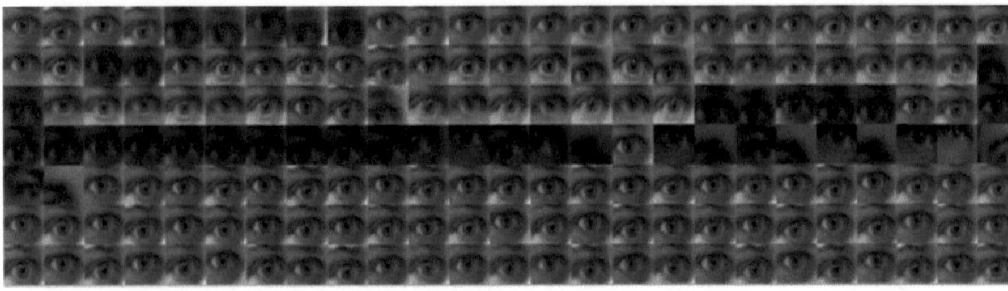

Fig. 10.15 Sample images from an image-set of the visible range MBGC video dataset. The image-set contains noise in the form of misdetected periocular regions

Table 10.4 Execution time in seconds for matching a single probe image-set with 114 gallery image-sets in the MBGC NIR database using 600-d LBP features

Algorithm	DCC	MMD	MDA	AHISD	CHISD	SANP	FSF+CSF
Train	13.11	6.16	21.95	N/A	N/A	N/A	164.91
Test	0.14	0.11	0.09	0.15	2.31	5.23	32.12

10.8.8 Execution Time

We compare the execution times of the six image-set classification algorithms on the five databases. Table 10.4 shows the average execution time for matching one probe set with 114 gallery sets, on a 3.4 GHz CPU with 8 GB RAM for the MBGC NIR database. The relative execution times for the other databases are similar. We observe that the sample based techniques are computationally more expensive than the structure based techniques. The overall execution time of the proposed two stage fusion approach is the sum of execution time for all algorithms. Note that the execution time for SANP is lower than reported by the original authors [21] and in our preliminary work [48]. This was achieved by normalizing all features to unit norm. This significantly improves the convergence speed of the Accelerated Proximal Gradient method used to find the sparse approximated nearest points in the SANP algorithm without affecting the accuracy.

10.9 Conclusion and Future Work

In this chapter an extensive study of periocular biometric is performed using the visible, NIR, and the hyperspectral imaging technologies. Application of this emerging biometric was studied for use in real world applications. Periocular biometric recognition was formulated as an image-set classification problem. Such a formulation provided robustness to intra-person variations such as pose, eyelid, and eyeball movements. Different methods for periocular data extraction were presented. Periocular data representation was also studied and four simple feature types were used for further experiments. Performance of six image-set classification algorithms was studied for periocular biometric. Seven fusion schemes were studied including our own proposed inverse Error Weighted Fusion (EWF) algorithm. In our experiments, EWF was found to yield better performance than other fusion schemes. Performance of different algorithms and fusion schemes was evaluated on four public databases, including MBGC Visible and NIR, CMU Hyperspectral, and UBIPr datasets. This study demonstrated feasibility of periocular biometrics for real world person identification in scenarios where the full face is not visible or the person is non-cooperative. In future, the performance of various data reduction and dimensionality reduction techniques need to be studied for finding an optimal trade-off between the execution time and the recognition performance of the system.

Acknowledgements This work was made possible by NPRP grant number NPRP 7-1711-1-312 from the Qatar National Research Fund (a member of Qatar Foundation). The statements made herein are solely the responsibility of the authors.

References

1. J. Adams, D. Woodard, G. Dozier, P. Miller, K. Bryant, G. Glenn, Genetic-based type ii feature extraction for periocular biometric recognition: less is more, in *International Conference on Pattern Recognition, 2010* (2010), pp. 205–208
2. T. Ahonen, A. Hadid, M. Pietikainen, Face description with local binary patterns: application to face recognition. IEEE Trans. Pattern Anal. Mach. Intell. **28**(12), 2037–2041 (2006)
3. F. Alonso-Fernandez, J. Bigün, Periocular recognition using retinotopic sampling and gabor decomposition, in *European Conference on Computer Vision, Workshops (2), 2012* (2012), pp. 309–318
4. F. Alonso-Fernandez, J. Bigun, A survey on periocular biometrics research. Pattern Recogn. Lett. (2015)
5. F. Alonso-Fernandez, J. Bigun, Near-infrared and visible-light periocular recognition with gabor features using frequency-adaptive automatic eye detection. IET Biom. **22**(4), 74–89 (2015)
6. F. Alonso-Fernandez, J. Bigun, Periocular biometrics: databases, algorithms and directions, in *International Conference on Biometrics and Forensics (IWBF)* (2016), pp. 1–6
7. F. Alonso-Fernandez, A. Mikaelyan, J. Bigun, Comparison and fusion of multiple iris and periocular matchers using near-infrared and visible images, in *Proceedings of the International Workshop Biometrics and Forensics* (2015)
8. M. Ao, D. Yi, Z. Lei, S.Z. Li, Face recognition at a distance: system issues, in *Handbook of Remote Biometrics: For Surveillance and Security* (Springer, London, 2009), pp. 155–167
9. S. Bharadwaj, H. Bhatt, M. Vatsa, R. Singh, Periocular biometrics: when iris recognition fails, in *IEEE International Conference on Biometrics: Theory, Applications, and Systems, 2010* (2010), pp. 1–6
10. BIT, CASIA Iris Image Database (2013), http://biometrics.idealtest.org
11. V. Boddeti, J. Smereka, B. Kumar, A comparative evaluation of iris and ocular recognition methods on challenging ocular images, in *International Joint Conference on Biometrics, 2011* (2011), pp. 1–8
12. K.W. Bowyer, K. Hollingsworth, P.J. Flynn, Image understanding for iris biometrics: a survey. Comput. Vis. Image Underst. **110**(2), 281–307 (2008)

13. Z. Cao, N. Schmid, Matching heterogeneous periocular regions: short and long standoff distances, in *International Conference on Image Processing (ICIP)* (2014), pp. 4967–4971
14. H. Cevikalp, B. Triggs, Face recognition based on image sets, in *IEEE International Conference on Computer Vision and Pattern Recognition, 2010* (2010), pp. 2567–2573
15. Cognitec, FaceVACS-VideoScan, *Face Recognition Commercial Software* (2016), http://www.cognitec.com
16. Z. Cui, H. Chang, S. Shan, B. Ma, X. Chen, Joint sparse representation for video-based face recognition. Neurocomputing **135**, 306–312 (2014)
17. L. Denes, P. Metes, Y. Liu, Hyperspectral face database. Technical Report CMU-RI-TR-02-25 (Robotics Institute, Pittsburgh, 2002)
18. J.X. Du, C.M. Zhai, Y.Q. Ye, Face aging simulation and recognition based on NMF algorithm with sparseness constraints. Neurocomputing **116**, 250–259 (2013)
19. J. Fierrez-aguilar, J. Ortega-garcia, D. Torre-toledano, J. Gonzalez-rodriguez, Biosec baseline corpus: a multimodal biometric database. Pattern Recogn. **40**(4), 1389–1392 (2007)
20. K. Hollingsworth, S. Darnell, P. Miller, D. Woodard, K. Bowyer, P. Flynn, Human and machine performance on periocular biometrics under near-infrared light and visible light. IEEE Trans. Inf. Forensics Secur. **7**(2), 588–601 (2012)
21. Y. Hu, A. Mian, R. Owens, Face recognition using sparse approximated nearest points between image sets. IEEE Trans. Pattern Anal. Mach. Intell. **34**(10), 1992–2004 (2012)
22. J. Huang, X. You, Y. Yuan, F. Yang, L. Lin, Rotation invariant iris feature extraction using gaussian Markov random fields with non-separable wavelet. Neurocomputing **73**(4–6), 883–894 (2010)
23. G. Inc., Pittsburgh pattern recognition. *Face Recognition Commercial Software* (2016), www.pittpatt.com
24. F. Juefei-Xu, M. Savvides, Unconstrained periocular biometric acquisition and recognition using COTS PTZ camera for uncooperative and non-cooperative subjects, in *IEEE Workshop on the Applications of Computer Vision, 2012* (2012), pp. 201–208
25. F. Juefei-Xu, K. Luu, M. Savvides, T. Bui, C. Suen, Investigating age invariant face recognition based on periocular biometrics, in *International Joint Conference on Biometrics, 2011* (2011), pp. 1–7
26. T.K. Kim, J. Kittler, R. Cipolla, Discriminative learning and recognition of image set classes using canonical correlations. IEEE Trans. Pattern Anal. Mach. Intell. **29**(6), 1005–1018 (2007)
27. J. Kittler, M. Hatef, R.P.W. Duin, J. Matas, On combining classifiers. IEEE Trans. Pattern Anal. Mach. Intell. **20**(3), 226–239 (1998)
28. Y. Lee, R. Micheals, P. Phillips, Improvements in video-based automated system for iris recognition, in *IEEE Workshop on Motion and Video Computing (WMVC), 2009* (2009), pp. 1–8
29. H. Maeng, H. Choi, U. Park, S. Lee, A. Jain, Nfrad: near-infrared face recognition at a distance, in *International Joint Conference on Biometrics (IJCB)* (2011), pp. 1–7
30. G. Mahalingam, K. Ricanek, A. Albert, Investigating the periocular-based face recognition across gender transformation. IEEE Trans. Inf. Forensics Secur. **9**(12), 2180–2192 (2014)
31. L. Masek, Recognition of human iris patterns for biometric identification. Technical Report, The University of Western Australia (2003)
32. G. Medioni, J. Choi, C. Kuo, A. Choudhury, L. Zhang, D. Fidaleo, Non-cooperative persons identification at a distance with 3d face modeling, in *IEEE International Conference on Biometrics: Theory, Applications, and Systems (BTAS)* (2007), pp. 1–6
33. P.E. Miller, A.W. Rawls, S.J. Pundlik, D.L. Woodard, Personal identification using periocular skin texture, in *Proceedings of the ACM Symposium on Applied Computing, 2010* (2010), pp. 1496–1500
34. F. Moreno Seco, J. Inesta, P. Ponce de Leon, L. Mico, Comparison of classifier fusion methods for classification in pattern recognition tasks, in *International Workshops on Structural and Syntactic Pattern Recognition* (2006), pp. 705–713
35. L. Nie, A. Kumar, Periocular recognition using unsupervised convolutional rbm feature learning, in *International Conference on Pattern Recognition (ICPR)* (2014), pp. 299–404

36. NIST, Multiple Biometric Grand Challenge (MBGC) dataset (2008), http://face.nist.gov/mbgc/
37. NIST, Face and Ocular Challenge Series (FOCS) dataset (2010), http://www.nist.gov/itl/iad/ig/focs.cfm
38. K. Oh, B.S. Oh, K.A. Toh, W.Y. Yau, H.L. Eng, Combining sclera and periocular features for multi-modal identity verification. Neurocomputing **128**, 185–198 (2014)
39. C. Padole, H. Proenca, Periocular recognition: analysis of performance degradation factors, in *International Conference on Biometrics, 2012* (2012), pp. 439–445
40. U. Park, A. Ross, A. Jain, Periocular biometrics in the visible spectrum: a feasibility study, in *IEEE International Conference on Biometrics: Theory, Applications, and Systems, 2009* (2009), pp. 1–6
41. U. Park, R. Jillela, A. Ross, A. Jain, Periocular biometrics in the visible spectrum. IEEE Trans. Inf. Forensics Secur. **6**(1), 96–106 (2011)
42. V.P. Pauca, M. Forkin, X. Xu, R. Plemmons, A.A. Ross, Challenging ocular image recognition, in *BTHI, SPIE*, vol. 8029, (2011), pp. 80291V-1–80291V-13
43. P. Phillips, P. Flynn, T. Scruggs, K. Bowyer, J. Chang, K. Hoffman, J. Marques, J. Min, W. Worek, Overview of the face recognition grand challenge, in *IEEE International Conference on Computer Vision and Pattern Recognition, 2005*, vol. 1 (2005), pp. 947–954
44. H. Proenca, S. Filipe, R. Santos, J. Oliveira, L. Alexandre, The ubiris.v2: a database of visible wavelength iris images captured on-the-move and at-a-distance. IEEE Trans. Pattern Anal. Mach. Intell. **32**(8), 1529–1535 (2010)
45. A. Sinha, H. Chen, D. Danu, T. Kirubarajan, M. Farooq, Estimation and decision fusion: a survey. Neurocomputing **71**(13–15), 2650–2656 (2008)
46. Y. Song, W. Cao, Z. He, Robust iris recognition using sparse error correction model and discriminative dictionary learning. Neurocomputing **137**, 198–204 (2014)
47. C. Tan, A. Kumar, Towards online iris and periocular recognition under relaxed imaging constraints. IEEE Trans. Image Process. **22**(10), 3751–3765 (2013)
48. M. Uzair, A. Mahmood, A. Mian, C. McDonald, Periocular biometric recognition using image sets, in *IEEE Workshop on the Applications of Computer Vision, 2013* (2013), pp. 246–251
49. M. Uzair, A. Mahmood, A. Mian, C. McDonald, Periocular region based person identification in the visible infra-red and hyperspectral imagery. Neurocomputing **149**, 854–867 (2014)
50. P. Viola, M. Jones, Robust real-time face detection. Int. J. Comput. Vis. **57**, 137–154 (2004)
51. R. Wang, X. Chen, Manifold discriminant analysis, in *IEEE International Conference on Computer Vision and Pattern Recognition, 2009* (2009), pp. 429–436
52. R. Wang, S. Shan, X. Chen, W. Gao, Manifold-manifold distance with application to face recognition based on image set, in *IEEE International Conference on Computer Vision and Pattern Recognition, 2008* (2008), pp. 1–8
53. F. Wheeler, R. Weiss, P. Tu, Face recognition at a distance system for surveillance applications, in *IEEE International Conference on Biometrics: Theory Applications and Systems (BTAS)*, (2010), pp. 1–8
54. D. Woodard, S. Pundlik, P. Miller, R. Jillela, A. Ross, On the fusion of periocular and iris biometrics in non-ideal imagery, in *International Conference on Pattern Recognition, 2010* (2010), pp. 201–204
55. D. Woodard, S. Pundlik, J. Lyle, P. Miller, Periocular region appearance cues for biometric identification, in *IEEE International Conference on Computer Vision and Pattern Recognition Workshops, 2010* (2010), pp. 162–169
56. J. Xu, M. Cha, J. Heyman, S. Venugopalan, R. Abiantun, M. Savvides, Robust local binary pattern feature sets for periocular biometric identification, in *IEEE International Conference on Biometrics: Theory, Applications, and Systems, 2010* (2010), pp. 1–8
57. B. Yang, S. Chen, A comparative study on local binary pattern LBP based face recognition: LBP histogram versus LBP image. Neurocomputing **120**, 365–379 (2013)

Chapter 11
Robust Face Recognition Using Kernel Collaborative Representation and Multi-scale Local Binary Patterns

Muhammad Khurram Shaikh, Muhammad Atif Tahir, and Ahmed Bouridane

11.1 Introduction

Human face recognition is a very active research area comprising of variety of face descriptors [1] and classifiers to tackle the problems associated with it namely, pose, expression, illumination, occlusion, etc. But, due to in-built non-linear effects attached with face, it poses a great challenge to researchers for developing a universal technique [2, 3]. High dimensionality, image localization or face detection, stable features and need of strong classifiers are some of the great challenges faced by research communities. The majority of face recognition (FR) methods can hardly be regarded robust to deal with all variations in imaging conditions [4, 5].

As opposed to the normal division of FR methods [6], we consider its division into two broad categories as either intensity or facial feature based. Intensity based methods use the raw pixels as face features input to the classifier to make the decision. The recent methods which use intensity features and produce some good recognition rates are FR with sparse representation (SRC) [7], linear regression for face recognition (LRC) [8], two-dimensional PCA (2DPCA) [9] and Collaborative Representation for FR (CRC-RLS) [10]. However, the performance of almost all methods depends exclusively on correct detection, alignment and registration of images. Therefore, in these methods, image normalization plays a very important role in getting higher recognition rates.

M.K. Shaikh (✉) • A. Bouridane
Northumbria University at Newcastle, NE1 8ST Newcastle upon Tyne, UK
e-mail: mkshaikh75@yahoo.com; ahmed.bouridane@northumbria.ac.uk

M.A. Tahir
Northumbria University at Newcastle, NE1 8ST Newcastle upon Tyne, UK

National University of Computer and Emerging Sciences, Karachi, Pakistan
e-mail: muhammad.tahir@northumbria.ac.uk; atif.tahir@nu.edu.pk

R. Jiang et al. (eds.), *Biometric Security and Privacy*, Signal Processing for Security Technologies, DOI 10.1007/978-3-319-47301-7_11

Facial feature based methods transform the raw pixels into new feature subspace to get more discriminative and strong representation of face to assist in the accurate classification of images. Eigenface [11] and FisherFace [12] are one of the prominent facial feature based methods which revolutionize the research in the area of face recognition. LBP features proposed in [13] are used in a lot of FR methods and they normally outperform the Gabor feature based counterparts. The multi-resolution LBP [14] has resulted in improved performance as compared to uniscale LBP. But, the problem associated with LBP features applied at different radii, i.e., multi-scale LBP involve high dimensional feature vectors which lead to computationally expensive algorithms.

To solve this problem, our chapter tackles the issue of the high dimensionality firstly by downsampling the images as the feature space choice is no longer critical as reported in [7, 8]. Secondly, we further handle dimensionality issue resulting from multi-scale LBP representation and non-linear behaviour of face manifold by projecting our features to kernel subspace.

A new efficient regularized least square classifier is proposed for recognizing faces by Zhang et al. [10] and this classifier convincingly outperforms SRC [7] using $l2$ norm. The collaborative representation of a query sample in contrast to sparsity for classification as used in most previous works [15–17] is applied with low computational burden in CRC-RLS method. This approach yields high recognition accuracies. The limitation of this method is that it cannot able to generate high recognition rates due to severer illumination problem as it maps image coordinates directly from the feature space. But recently, a powerful texture descriptor, LBP [18] for face recognition that can withstand extreme lighting problem has gained popularity.

The multi-scale representation of images shows more accuracy and higher recognition rates compared to its single scale counterpart and tackles the problem of blur and illumination [14]. Utilizing this advantage, we represent faces with multi-scale LBP histograms [19] by extending the single scale representation.

Getting inspirations of the recent successful application of kernel based approaches [20], we propose a new kernel based CRC-RLS method using multi-scale LBP features as feature descriptors. We term this method as KCRC-RLS. The high dimensional feature vector resulting from the multi-scale representation of images using LBP feature descriptors maps onto kernel space using the kernel trick. We use four challenging face databases namely, AR, ORL, FERET and Extended Yale B to evaluate our proposed approach. The significant improvement in recognition rates is achieved compared to other famous approaches in this domain including LRC [8], SRC [7] and CRC-RLS [10].

Building on our previous work [21], the main contributions of this work are summarized as follows:

1. A new kernel based CRC-RLS method (KCRC-RLS) is proposed in this chapter to solve the non-linearity issues in FR.

2. Multi-scale LBP features are integrated with proposed KCRC-RLS approach to handling the problems associated with illumination, occlusion, etc.[1]

This chapter is organized as follows. Multi-scale local binary patterns (MLBP) are discussed in Sect. 11.2. The proposed KCRC-RLS method in conjunction with multi-scale LBP features is described in Sect. 11.3. The proposed method is evaluated using different parameter setting/protocols and results are presented in Sect. 11.4. Further, discussion on this proposed KCRC-RLS method is given in Sect. 11.5 and finally, conclusion is provided in Sect. 11.6.

11.2 Multi-scale Local Binary Patterns

The LBP method, shown in Eq. (11.1), is a highly powerful discriminative texture operator and it records a variety of patterns in different size neighbourhood of each pixel in the Q size histogram. The histogram features of LBP are robust to translation and rotation of image and invariant to local grey-scale variations. For generating LBP code, the centre pixel grey-scale value u_c is compared with its neighbour $u_j(0, 1, \ldots, Q-1)$ and if the pixel difference is greater than or equal to zero, then 1 is assigned otherwise 0. Figure 11.1 depicts the working of LBP operator. The values of neighbours which do not fall in the centre of the pixel are estimated by interpolation.

$$\mathrm{LBP}_{Q,R} = \sum_{q=0}^{Q-1} s(u_j - u_c)2^q \tag{11.1}$$

$$s(x) = \begin{cases} 1 & x \geq 0 \\ 0 & \text{otherwise} \end{cases}$$

The use of single scale in real world scenarios has shown substantial limitation due to small support area as 3×3 neighbourhood will not be able to capture

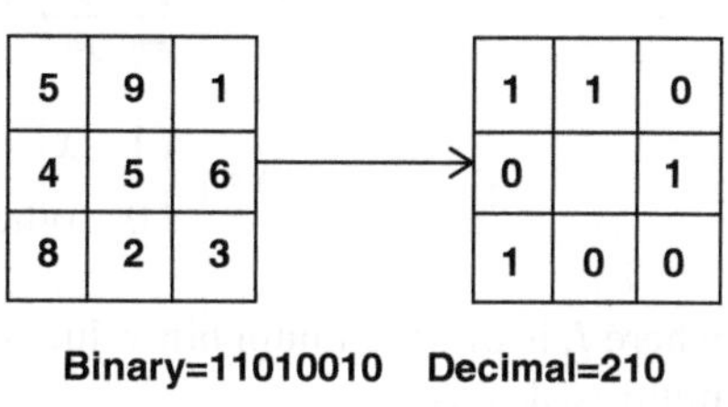

Fig. 11.1 Calculation of the LBP

[1]Another kernel based approach for CRC-RLS is also proposed recently by Yang et al. [22]. MLBPH features can be integrated with this kernel based approach.

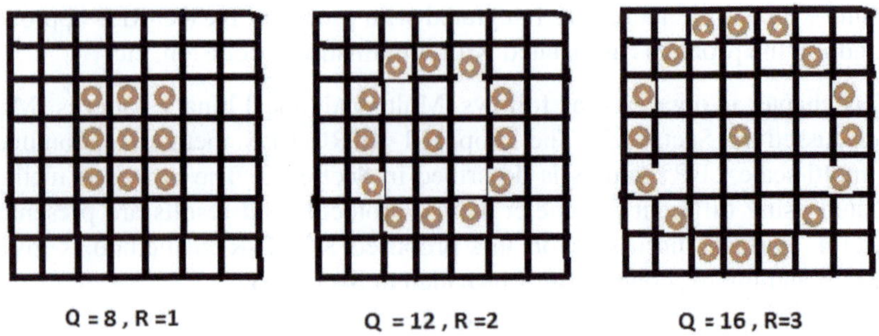

Fig. 11.2 Multi-scale LBP

large scale structures which are necessary for acquiring dominant features for some textures. Multi-scale LBP has been employed and used successfully [14, 19] to capture structures existed in different radii of LBP operator. The original LBP feature descriptor can be extended to multi-scale by using radii of different sizes. Multi-scale representation is achieved by sliding set of different radii LBP operators over an image and combining their results to capture non-local information. But, adding features from different radius LBP operator results in a problem of high dimensional feature vectors that can be minimized by feature selection technique to curb redundant information. Figure 11.2 provides a pictorial representation of the selection of different size Q and R for circular neighbourhoods.

By selecting different values of radii depending on the neighbour pixel distance from centre pixel, a multi-scale representation is created by concatenating LBP histograms from each scale with specific radius value. It has been reported that multi-scale analysis has been more accurate compared to its counterpart single-scale and it also tackles the problem of blur and illumination. The face descriptor for multi-scale analysis is achieved by applying LBP feature descriptor on different R scales for images. The resulting LBP histograms for each scale of image size $M \times N$ are computed by

$$H_{Q,r}(l) = \sum_{i=0}^{M-1} \sum_{j=0}^{N-1} g(\mathrm{LBP}_{Q,r}(i,j), l), l \in [0, L-1] \tag{11.2}$$

$$g(x,y) = \begin{cases} 1 & x = y \\ 0 & \text{otherwise} \end{cases}$$

where L is the maximum bin value of LBP and R is the maximum radius applied to multi-scale case.

The features from each scale with specific radius capture local information for an image and one long observation has been made by combining information/features from different scales or radii values.

The final multi-scale representation of face descriptor is given by

$$F_{Q,r} = [H_{Q,1}, H_{Q,2}, \ldots, H_{Q,R}] \tag{11.3}$$

Multi-scale approaches result in the problem of overfitting due to the high dimensional feature vector. In our technique, we handle this issue by combining features from different kernels. This kernel fusion is discussed in Sect. 11.3.3 which gives an effective solution for the problem of integrating features of local and multi-scale face representation.

11.3 Kernel Collaborative Representation Based Classification

This section first provides an overview of original CRC-RLS method [10] and later on our proposed kernel based approach is discussed with some other variants.

11.3.1 Collaborative Representation Using Regularized Least Square

Collaborative representation proposed by Zhang et al. [10] makes full use of the collaboration among classes in defining the query sample. Considering, there are k different classes, $i = 1, 2, \ldots, k$ with n training images for each class. Each training image from ith class is transformed into LBP histogram features $\mathbf{v}_i^m \in \mathbf{R}^{\mathbf{f}}$, $m = 1, 2, \ldots, p_i$. The total number of features obtained using LBP is f. Zhang et al. [10] represents y over x in Eq. (11.4) having low computational cost and collaborative representation of the query sample by

$$\hat{c} = (X^T X + \lambda I)^{-1} X^T y \tag{11.4}$$

In above Eq. (11.4), regularization parameter λ is applied to provide stability to the least square problem. The identity matrix is defined by I. All training images from all classes collaboratively representing training data are given by X.

Equation (11.4) can be re-written as

$$\hat{c} = Cy \tag{11.5}$$

Equation (11.5) shows that C contains only X and it is independent of y. This can be computed before solving the entire equation as a projection matrix. In order to calculate query vector y, we just need to project y onto already pre-computed C.

For this CRC-RLS method, they calculate residuals from every class as:

$$r_i = ||y - X_i \times \widehat{c_i}||_2 / ||\widehat{c_i}||_2 \tag{11.6}$$

where $\mathbf{X}_i = [\mathbf{v}_i^1, \mathbf{v}_i^2, \ldots \mathbf{v}_i^{p_i}] \in \mathbf{R}^{\mathbf{f} \times \mathbf{p_i}}$.

In Eq. (11.6) for every class i, the associated coefficient vector is given by c_i. To get discrimination among classes to assist in classification, it can be achieved by using l_2-norm $||\widehat{c_i}||_2$. The class having minimum residual gets the decision in its favour and it is represented by $\underset{i}{\operatorname{argmin}}\{r_i\}$.

11.3.2 Kernel Collaborative Representation with Regularized Least Square Classifier

Although the method proposed by Zhang et al. [10] is very computationally effective and has linear property. But, it is well established that linear methods performance degrades in the presence of non-linearities, e.g., expression, illumination and occlusion, etc., associated with face images. To tackle these non-linear behaviours by face images, kernel based approaches provide a handy solution by taking high dimensional feature vector into kernel space. We propose the kernelized version of CRC-RLS, a non-linear extension.

A *kernel* is a function K, such that for all $x_i, x_j \in X$

$$K(x_i, x_j) =< \phi(x_i).\phi(x_j) > \tag{11.7}$$

where ϕ is a mapping from X to an (inner product) feature space F.

Several variants of kernel functions can be applied. In this chapter, I have used RBF kernel with chi-squared distance metric: $K(x_i, x_j) = e^{-\frac{1}{A}\text{dist}_{\chi^2}(x_i, x_j)}$ where A is a scalar which normalizes the distances [23].

The original Eq. (11.4) of CRC-RLS [10] can be re-written in kernel space as:

$$\phi(\hat{c}) = (K + \lambda I)^{-1} K_y \tag{11.8}$$

where $(K + \lambda I)$ is positive definite and thus an inverse solution is possible.

As Eq. (11.4) represents the feature vectors of all training class samples using linear approach, the same equation can be extended to handle the high dimensionality of features by using different kernel approaches for subsiding any remaining non-linearities of the manifold.

Equation (11.8) can be re-written as

$$\phi(\hat{c}) = K^* K_y \tag{11.9}$$

Equation (11.9) shows that K^* contains only K and it is independent of K_y. This can be computed before solving the entire equation as a projection matrix. In order to calculate query vector y, we just need to project y onto already pre-computed K^*.

For our KCRC-RLS method, we calculate residuals from every class as given in Eq. (11.10)

$$r_i = ||K_y - K^* \times \phi(\widehat{c_i})||_2 / ||\phi(\widehat{c_i})||_2 \tag{11.10}$$

11.3.3 Kernel Level Fusion

In this chapter, we consider RBF Gaussian kernel for CRC-RLS method. But in reality, there are a variety of kernels can be applied, e.g., polynomial, spline, ANOVA, Bag of words, tree and graph kernels [24]. The features from different kernels can be combined to represent local and multi-scale face features of an image. The fusion of kernels can be mathematically represented by Cartesian product of the kernels (see Eq. (11.11)). After concatenation of kernel features, KCRC-RLS can be applied as a classifier.

$$K = \sum_{i=1}^{(r \times R)} K_i \tag{11.11}$$

11.4 Experiments and Results

In order to evaluate the effectiveness of the proposed method, extensive experiments were carried out on four standard databases: FERET [25], ORL [26], AR [27] and Extended Yale B [28]. In all experiments, the regularization parameter $\lambda = 0.001$ is used for "face" features as suggested in [10] while λ is set to 0.5 for "histogram" features. It is worthwhile to mention that we have selected these parameters after extensive simulation.

11.4.1 FERET Database

The FERET database is the largest publicly available database [25] and it is considered as a standard benchmark in reporting recognition rates of major face recognition algorithms.

11.4.1.1 Protocol 1

In our experiment, standard evaluation protocol similar to [8, 29, 30] is used. We use four images of given subject. Out of four images, galleries are formed containing two frontal "fa" and "fb" and two non-frontal "ql" and "qr" images of each individual. A total of 128 subjects is used in the first protocol. In the training phase, we have used 512 images while the remaining images are used in the testing phase. To achieve reliable collaboration among the classes, more than one training sample is required.

All the experiments are conducted using cross-validation approach in which combination of three samples are used for training while the left over is used for testing. Figure 11.3 shows a typical subject from the FERET database.

For face features, experiments are conducted on downsampled 7×6 feature subspace. In multi-scale LBP histogram features, an original image is processed with a radius varying from 1 to 8 with 8 neighbours. The final dimension of LBP is 2048. Table 11.1 compared the results with state-of-the-art methods. However, the face features applied on CRC-RLS approach performs marginally better than PCA, ICA but LRC approach shows improved results when compared to CRC-RLS (Face). The efficacy of using MLBP features in conjunction with RBF kernel version is quite evident as it outperforms all methods considerably. The increase of 50 and 39 % in performance as compared to original CRC-RLS is reported with KCRC-RLS. Furthermore, the results on frontal and non-frontal samples are quite promising as all recognition rates are close to each other except on ql.

The selection of different values of radii in multi-scale local binary pattern definitely makes a major impact on the recognition rates. The effect of selecting different radius values and its respective recognition rates is shown in Fig. 11.4.

11.4.1.2 Protocol 2

The performance measure of any state-of-the-art algorithm requires testing using a large number of samples. In order to validate the consistent performance of

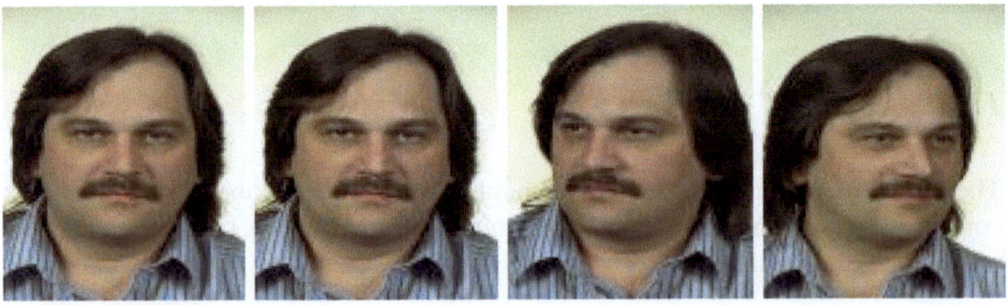

Fig. 11.3 A typical subject from the FERET database

Table 11.1 Results for the FERET evaluation protocol 1

Method	fa	fb	ql	qr
PCA [8]	74.22	73.44	65.63	72.66
ICA I [8]	73.44	71.09	65.63	68.15
LRC [8]	91.41	94.50	78.13	84.38
CRC-RLS$_{\mathbf{Image}}$ [10]	78.91	78.91	46.09	60.16
KCRC-RLS$_{\mathbf{Image}}$(RBF$_{\chi^2}$)	90.63	92.97	75.00	83.59
KCRC-RLS$_{\mathbf{MLBP}}$(Linear)	89.84	89.84	71.09	80.47
KCRC-RLS$_{\mathbf{MLBP}}$(RBF$_{\chi^2}$)	**99.22**	**100**	**92.97**	**98.44**

The reason of making our results bold is to show that our method has produced superior results when compared with other state-of-art methods

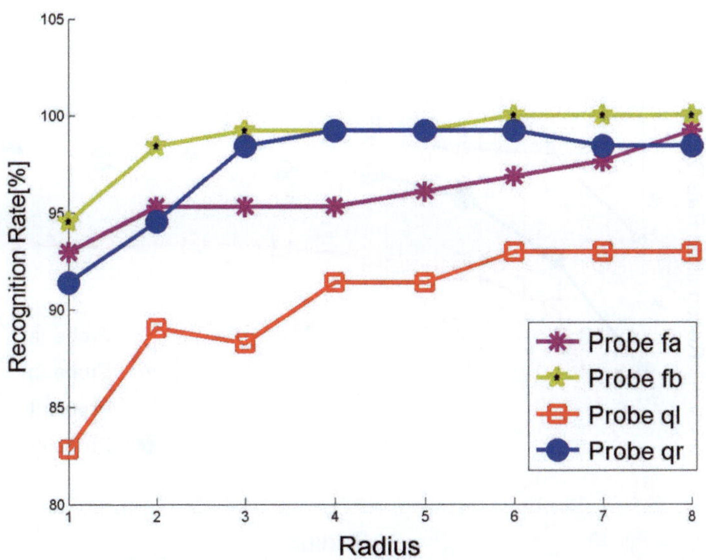

Fig. 11.4 Multi-scale LBP with varying radii vs. recognition rate protocol 1 FERET

our methodology, we have tested KCRC-RLS with 400 subjects of the FERET database selected randomly as compared to 128 subjects in protocol 1. We again get consistent results in comparison with protocol 1 as shown in Table 11.2. The results on frontal samples "fa" and "fb", i.e., 99.50 and 100 % and non-frontal samples

Table 11.2 Results for the FERET evaluation protocol 2

Method	fa	fb	ql	qr
PCA [8]	80.00	78.75	67.50	71.75
ICA I [8]	93.25	93.50	75.25	76.00
LRC [8]	91.41	94.50	78.13	84.38
CRC-RLS$_{\mathbf{Image}}$ [10]	71.50	69.25	38.00	39.00
KCRC-RLS$_{\mathbf{Image}}$(RBF$_{\chi^2}$)	93.25	92.25	69.50	76.75
KCRC-RLS$_{\mathbf{MLBP}}$(Linear)	90.00	89.25	72.25	77.75
KCRC-RLS$_{\mathbf{MLBP}}$(RBF$_{\chi^2}$)	**99.50**	**100**	**93.00**	**97.50**

The reason of making our results bold is to show that our method has produced superior results when compared with other state-of-art methods

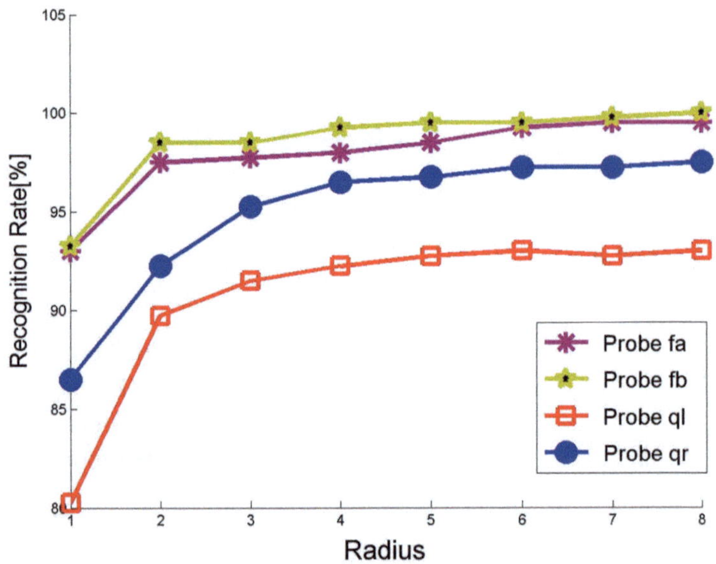

Fig. 11.5 Multi-scale LBP with varying radii vs. recognition rate protocol 2 FERET

"ql" and "qr", i.e., 93 and 97.5 % are remarkably outstanding and outperforming all reported methods in both protocols. The graph showing the effect of multi-scale LBP histogram application in our methodology is presented again.

The effect of selecting different radius values and its respective recognition rates is shown in Fig. 11.5.

11.4.2 ORL Database

AT&T Laboratories has maintained the ORL database which contains total 40 distinct individuals having ten different samples of each individual. Some images

Fig. 11.6 A typical subject from the ORL database

Table 11.3 Results for the ORL database

Method	RR (%)
Fisherfaces [9]	94.5
ICA [9]	85.00
Kernel Eigenfaces [9]	94.00
2DPCA [9]	96.00
LRC [8]	93.50
CRC-RLS$_{\mathbf{Image}}$ [10]	94.50
KCRC-RLS$_{\mathbf{Image}}$(RBF$_{\chi^2}$)	95.00
KCRC-RLS$_{\mathbf{MLBP}}$(Linear)	99.00
KCRC-RLS$_{\mathbf{MLBP}}$(RBF$_{\chi^2}$)	**100**

The reason of making our results bold is to show that our method has produced superior results when compared with other state-of-art methods

of individuals are collected over the period of time with varying lighting conditions. This database also incorporates facial expressions (open/closed eyes, smiling/not smiling) and facial details (glasses/no glasses) and some of the images from this database are presented in Fig. 11.6.

The standard evaluation protocol followed by Yang et al. [9] and Naseem et al. [8] is also adopted for our method. In this protocol, the training set contains first five images of each subject while the rest of the images are being used as probes. The long observation feature vector, multi-scale LBP has been made using LBP feature vector and all features from individual scale $R = [1, \ldots, 8]$ then merged. We have played around with image regions to find the most optimal region for a higher degree of classification accuracy and found that high recognition rates are achieved when an image is not divided. We follow the downsampling technique used by Naseem et al. [8] and in our experiments, all images are downsampled to 10×5 from the original size of 112×92.

The recognition rates of different state-of-the-art methods are reported and compared in Table 11.3. It is evident from the table that multi-scale LBP features in combination with kernel approach have produced outstanding results compared to other face recognition methods and it clearly enhances the performance on this database by 4.5 %.

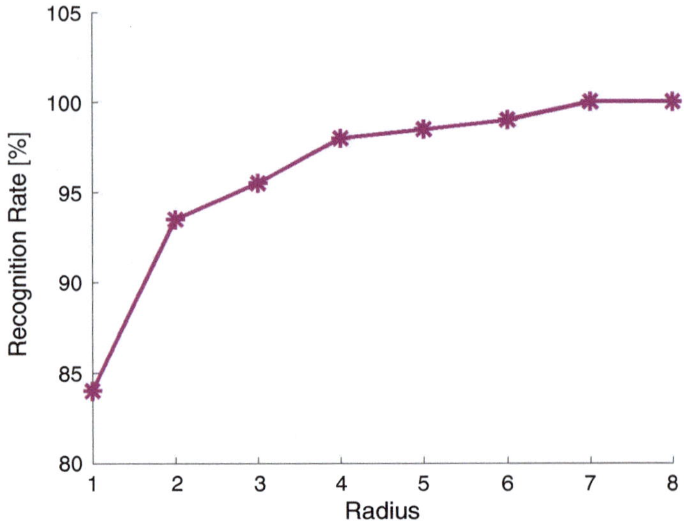

Fig. 11.7 Effect of radius selection on recognition rate of ORL database

Figure 11.7 shows the relationship of radii in LBP operator with recognition rates. It is clear from graph that multi-scale LBP has brought significant improvement in recognition rates as compared to the uniscale approach.

11.4.3 AR Database

The AR database [27] was developed at the Computer Vision Centre at the University of Alabama. It contains 4000 images of 126 different peoples out of whom 70 are male and 50 are female. This database captures variations in illumination, expression and occlusion. The expression variations consist of neutral, smile, anger and scream while occlusion contains sunglasses and scarf problems.

11.4.3.1 Experiments Related to Variations in Gesture

The gesture variation is considered to be one of the challenging tasks in face recognition algorithms as it may involve the case where the subject face localization may become difficult due to closure of the eyes. In recent face recognition techniques, the localization of face or image preprocessing plays an important role for the success of any method dependent on the correct face alignment. The need of methods less dependent on preprocessing stage may definitely subside the risk of getting any stable and practical face recognition methods in future.

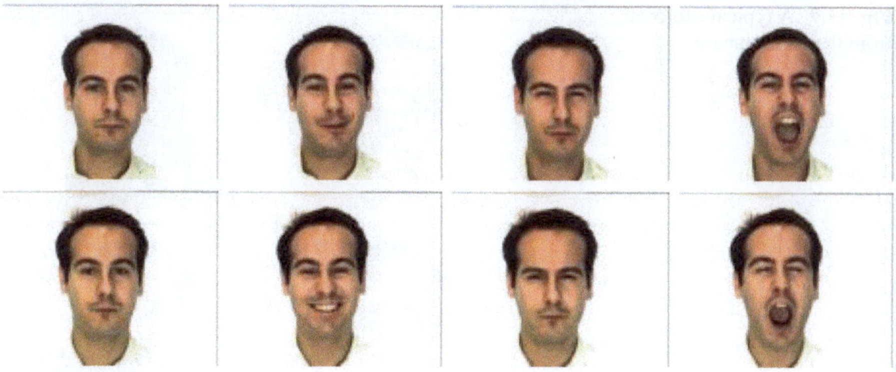

Fig. 11.8 A typical subjects from the AR database

Table 11.4 Results for the AR gesture variation protocol

Method	Neutral	Smile	Anger	Scream
LRC [8]	99.00	98.50	98.50	99.50
CRC-RLS$_{\mathbf{Image}}$ [10]	97.50	98.50	98.50	97.50
KCRC-RLS$_{\mathbf{Image}}$(RBF$_{\chi^2}$)	98.50	99.00	99.00	99.50
KCRC-RLS$_{\mathbf{MLBP}}$(Linear)	74.00	67.00	70.50	58.00
KCRC-RLS$_{\mathbf{MLBP}}$(RBF$_{\chi^2}$)	**100**	**100**	**100**	**100**

The reason of making our results bold is to show that our method has produced superior results when compared with other state-of-art methods

We have followed the standard evaluation protocol of [8, 9] as shown in Fig. 11.8. In our experiment, we have used the same 100 subjects (50 males and 50 females) out of total 128 subjects from AR face database. The gallery for conducting experiments is made up of 100 subjects with four different expressions from two different sessions resulting in total 800 subjects. While, the 600 individuals are used for training and rest for testing. The experiments are carried out again employing the same strategy as in the case of FERET database evaluation. The process is to use three expressions for building the training feature vector while the left over is utilized in making test feature vector.

When conducting experiments on intensity features, all images are downsampled to 10×10 resulting in 100 Dimensional (100D) feature vectors. For MLBP based feature vectors, the same strategy as applied in FERET case is employed with $R = [1, \ldots, 8]$. The resultant feature vector of MLBP is 2048D. The results are displayed in a table showing the KCRC-RLS$_{\mathbf{MLBP}}$ robustness and its superiority over other reported methods. It clearly outperforms CRC-RLS$_{\mathbf{Face}}$ and LRC by getting 100 % recognition rates on all expressions as evident in Table 11.4.

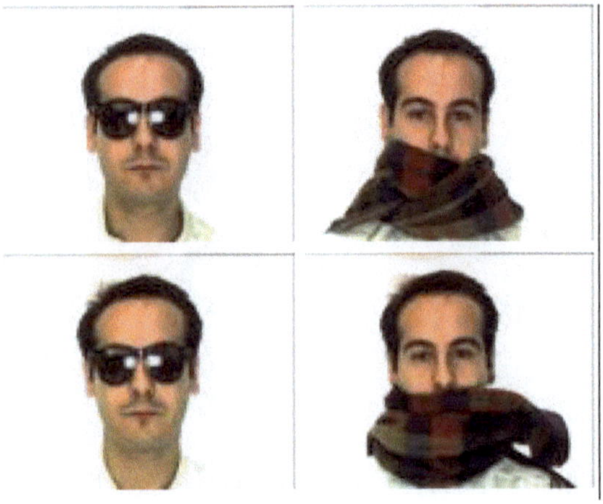

Fig. 11.9 A typical subjects from the AR database

11.4.3.2 Experiments Related to Variations in Occlusion

The challenge of occlusion resulting from the presence of caps, sunglasses, scarves, etc., poses a major threat for accurate classification of images. The problems associated with occlusion is to accurately localize the face from the image and inaccurate alignment may result in the incorrect classification of faces. Furthermore, the consequence of occlusion requires a major adjustment for face recognition methods requiring far more dependency on applying manual cropping and alignment of faces.

All our experiments do not make use of any preprocessing of images before applying the classification. In this experiment, two different occlusions problems are considered namely: sunglasses and scarfs. Two samples of 100 subjects containing 50 males and females are used. We have built the training feature vector consisting of all samples from the previous case, i.e., neutral, anger, smile and scream resulting in total of 800 individuals. Secondly, the query feature vectors consist of two sets of occlusion problems as mentioned in Fig. 11.9.

The same set of feature vectors for face and MLBP have employed again as in the case of expression variation. Table 11.5 shows the performance using AR Occlusion Protocol. For the sunglasses, our proposed KCRC-RLS outperforms the LRC and other methods as we achieve 100 % accuracy. Our method further outperforms all methods except SRC in the case of the very difficult scenario of the scarf and getting 50 % rise in comparison to LRC and CRC-RLS_{Face}.

11.4.4 Extended Yale B Database

The total face images in Extended Yale B database are 2414. Altogether, there are 38 different individuals having varying illumination conditions. The whole database is comprised of 5 subsets of images where subset 1 has images under normal

Table 11.5 Results for the AR occlusion protocol

Method	Sunglasses	Scarf
SRC [8]	87.00	**59.50**
LRC [8]	96.00	26.00
CRC-RLS$_{\mathbf{Image}}$ [10]	87.00	23.50
KCRC-RLS$_{\mathbf{Image}}$(RBF$_{\chi^2}$)	96.00	17.00
KCRC-RLS$_{\mathbf{MLBP}}$(Linear)	97.50	34.00
KCRC-RLS$_{\mathbf{MLBP}}$(RBF$_{\chi^2}$)	**100**	53.00

The reason of making our results bold is to show that our method has produced superior results when compared with other state-of-art methods

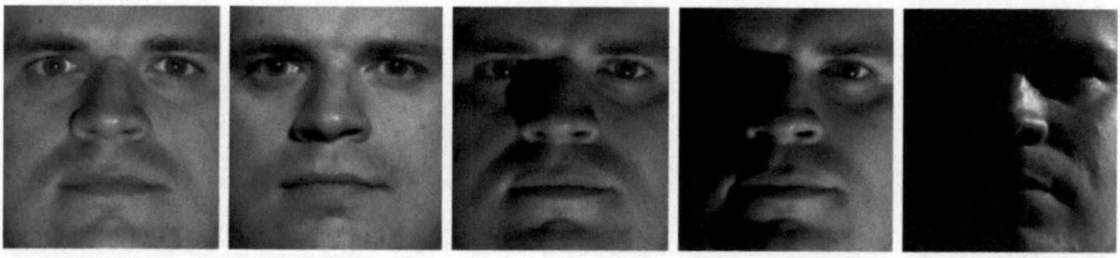

Fig. 11.10 Extended Yale B database subjects with various lighting conditions

illumination conditions and subset 2–5 contains images with extreme illumination variations. Figure 11.10 represents images from 5 subsets. The number of images per individual in subsets 1–5 are 7, 12, 12, 14 and 14, respectively. We follow the downsampling technique used by Naseem et al. [8] and in our experiments, all images are downsampled to 20×20 from the original size of 142×120. The long observation vector for our multi-scale LBP is created by applying 16 different radii on LBP operator $r = [1, \ldots, 16]$. In this experiment, we divided the image into 4×4 non-overlapping regions and concatenate multi-scale histograms of each region.

Table 11.6 provides a comparison of recognition rates of various state-of-the-art methods. It is evident from the table that CRC method fares well in normal lighting conditions for subsets 2 and 3 but its accuracy decreases in extreme lighting variations. The efficacy of multi-scale LBP features are also established when it is integrated with original CRC-RLS method as improvement of 57 % is obtained over subset 5. Our claim that kernel based approaches superiority over linear methods is also being established. We have received improvement of 15 and 60 % for subsets 4 and 5, respectively, when KCRC-RLS method is combined with multi-scale LBP features. For our method, we record marginal drop of 0.2 % for subset 2. The results also indicated that the proposed technique compares favourably with other face recognition methods especially dealing the case of illumination [31].

Table 11.6 Results for the Extended Yale B database

Method	S2	S3	S4	S5
PCA [8]	98.5	80.0	15.6	24.4
ICA I [8]	98.0	80.7	17.0	22.0
LRC [8]	**100**	**100**	83.3	33.6
CRC-RLS$_{\mathbf{Image}}$ [10]	**100**	**100**	86.5	36.9
CRC-RLS$_{\mathbf{MLBP}}$	99.8	**100**	92.4	81.1
KCRC-RLS$_{\mathbf{MLBP}}$(RBF$_{\chi^2}$)	99.8	**100**	**97.7**	**93.2**

The reason of making our results bold is to show that our method has produced superior results when compared with other state-of-art methods

Fig. 11.11 *First row* showing images successfully classified and *second row* displaying unsuccessfully identified images in both protocols of FERET database

11.5 Discussion

The results show the usefulness of kernel collaboration at the learning stage for face recognition. The proposed approach successfully captures discriminative information among the different classes by picking up collaboration between classes. This has led to a significant increase in the recognition rates when compared to other state-of-the-art methods. In the case of FERET database (which is mostly used in face recognition methods), we have achieved near perfect recognition rates in frontal matching. Figure 11.11 is showing some of the match and non-match cases in the cross-identification of four selected scenarios in FERET database considering both protocols.

We further report the efficacy of our proposed approach in a very tough and challenging problem of scarf occlusion in AR database. The results on this particular problem are very promising as compared to other methods.

In our matching, we use two samples of each subject by creating a test vector in occlusion protocol. It is noticed in our experiment that actually only 30 subjects

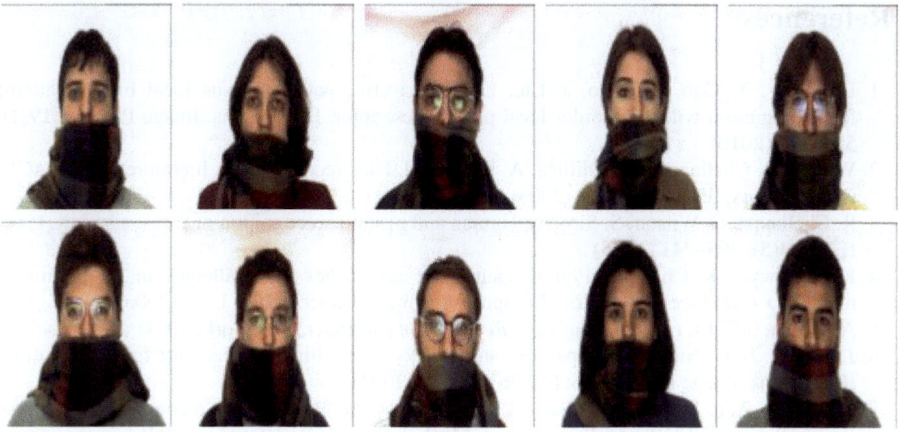

Fig. 11.12 Scarf occlusion protocol AR database: *first row* showing images successfully classified and *second row* displaying unsuccessfully identified images

out of 100 are not classified at all while at least one subject is identified correctly from rest of the subjects. The result on this particular problem can be increased using the approach of partially visible face matching. Some of the faces from scarf occlusion protocol are shown in Fig. 11.12 representing correct and incorrect classification. In summary, the presented approach is the best choice among the compared approaches. It provides an efficient and effective learning solution for face recognition.

11.6 Conclusion

CRC-RLS method is a very effective linear approach that gives outstanding results when combined with multi-scale LBP features. But, it has a major drawback that it cannot handle non-linear structures exist in images due to illumination, expression and occlusion. In order to overcome this deficiency, we propose the kernelized version of CRC-RLS method termed as KCRC-RLS utilizing multi-scale LBP features which have some built-in capability of tackling blur and illumination issues. The proposed KCRC-RLS method has produced some outstanding results on four challenging databases and produced better recognition rates in all parameter and protocol settings. The robustness of KCRC-RLS on different variations exist in face recognition is comprehensively tackled by our proposed approach. Our method does not require any kind of preprocessing of images or face any alignment issues in comparison to other state of the art. In future, the proposed method will be applied to more generic face recognition problems e.g. heterogeneous face recognition and modality identification for faces.

References

1. B. Zhang, Y. Gao, S. Zhao, J. Liu, Local derivative pattern versus local binary pattern: face recognition with high-order local pattern descriptor. IEEE Trans. Image Process. **19**(2), 533–544 (2010)
2. W. Zha, R. Chellappa, P.J. Phillips, A. Rosenfeld, Face recognition: a literature survey. ACM Comput. Surv. **35**(4), 399–458 (2003)
3. R. Chellappa, C. Wilson, S. Sirohey, Human and machine recognition of faces: a survey. Proc. IEEE **83**(5), 705–741 (1995)
4. K.W. Bowyer, K. Chang, P. Flynn, A survey of approaches and challenges in 3D and multi-modal 3D + 2D face recognition. Comput. Vis. Image Underst. **101**, 1–15 (2006)
5. A.K. Jain, S.Z. Li, *Handbook of Face Recognition* (Springer, New York, 2004)
6. J. Zou, Q. Ji, G. Nagy, A comparative study of local matching approach for face recognition. IEEE Trans. Image Process. **16**(10), 2617–2628 (2007)
7. J. Wright, A.Y. Yang, A. Ganesh, S.S. Sastry, Y. Ma, Robust face recognition via sparse representation. IEEE Trans. Pattern Anal. Mach. Intell. **31**(2), 210–227 (2009)
8. I. Naseem, R. Togneri, M. Bennamoun, Linear regression for face recognition. IEEE Trans. Pattern Anal. Mach. Intell. **32**(11), 2106–2112 (2010)
9. J. Yang, D. Zhang, A.F. Frangi, Y.J. Yang, Two dimensional PCA: a new approach to appearance-based face representation and recognition. IEEE Trans. Pattern Anal. Mach. Intell. **26**(1), 131–137 (2004)
10. L. Zhang, M. Yang, X. Feng, Sparse representation or collaborative representation: which helps face recognition? in *IEEE International Conference In Computer Vision* (2011), pp. 471–478
11. M. Turk, A. Pentland, Eigenfaces for recognition. J. Cogn. Neurosci. **3**(1), 71–86 (1991)
12. P.N. Belhumeur, J.P. Hespanha, D.J. Kriegman, Eigenfaces vs fisherfaces: recognition using class specific linear projection. IEEE Trans. Pattern Anal. Mach. Intell. **19**(7), 711–720 (1997)
13. T. Ahonen, A. Hadid, M. Pietikainen, Face description with local binary patterns: application to face recognition. IEEE Trans. Pattern Anal. Mach. Intell. **28**(12), 2037–2041 (2006)
14. T. Ojala, M. Pietikäinen, T. Maenpaa, Multiresolution gray-scale and rotation invariant texture classification with local binary patterns. IEEE Trans. Pattern Anal. Mach. Intell. **24**(7), 971–987 (2002)
15. S.H. Gao, I. Tsang, L. Chia, Kernel sparse representation for image classification and face recognition, in *Computer Vision - ECCV* (Springer, Berlin/Heidelberg, 2010), pp. 1–14
16. B. Cheng, J. Yang, S. Yan, Y. Fu, T. Huang, Learning with l1-graph for image analysis. IEEE Trans. Image Process. **19**(4), 858–866 (2010)
17. J. Wright, K. Yu, Y. Gong, T. Huang, Sparse representation for computer vision and pattern recognition. Proc. IEEE **98**(6), 1031–1044 (2010)
18. M. Pietikäinen, T. Ojala, Texture analysis in industrial applications, in *Image Technology - Advances in Image Processing, Multimedia and Machine Vision* (Springer, Berlin, 1996), pp. 337–359
19. C.H. Chan, J. Kittler, K. Messer, *Multi-Scale Local Binary Pattern Histograms for Face Recognition* (Springer, Berlin/Heidelberg, 2007), pp. 809–818
20. C. Chan, M.A. Tahir, J. Kittler, M. Pietikäinen, Multiscale local phase quantization for robust component-based face recognition using kernel fusion of multiple descriptors. IEEE Trans. Pattern Anal. Mach. Intell. **35**(5), 1164–1177 (2013)
21. M. Tahir, A. Bouridane, Face recognition using kernel collaborative representation and multiscale local binary patterns, in *IET Conference on Image Processing* (2012)
22. M. Yang, Z. Lei, C.S. Simon, Z. David, Robust kernel representation with statistical local features for face recognition. IEEE Trans. Neural Netw. Learn. Syst. **24**(6), 900–912 (2013)
23. J. Zhang, M. Marszalek, S. Lazebnik, C. Schmid, Local features and kernels for classification of texture and object categories: a comprehensive study. Int. J. Comput. Vis. **73**(2), 213–238 (2007)

24. T. Hofmann, S. Bernhard, J.S. Alexander, Kernel methods in machine learning. Ann. Stat. **36**(3), 1171–1220 (2008)
25. P.J. Phillips, H. Moon, S.A. Rizvi, P.J. Rauss, The FERET evaluation methodology for face-recognition algorithms. IEEE Trans. Pattern Anal. Mach. Intell. **22**(10), 1090–1104 (2000)
26. S. Ferdinando, A. Harter, Parameterisation of a stochastic model for human face identification, in *2nd IEEE Workshop on Applications of Computer Vision* (1994), pp. 138–142
27. A. Martínez, R. Benavente, The AR face database, Computer Vision Center Technical Report 3 (1998)
28. A.S. Georghiades, P.N. Belhumeur, D.J. Kriegman, From few to many: illumination cone models for face recognition under variable lighting and pose. IEEE Trans. Pattern Anal. Mach. Intell. **23**(6), 643–660 (2001)
29. X. Jiang, B. Mandal, A. Kot, Eigenfeature regularization and extraction in face recognition. IEEE Trans. Pattern Anal. Mach. Intell. **30**(3), 383–394 (2008)
30. J. Lu, K.N. Plataniotis, A.N. Venetsanopoulos, A. Kot, Ensemble-based discriminant learning with boosting for face recognition. IEEE Trans. Pattern Anal. Mach. Intell. **17**(1), 166–178 (2006)
31. T. Zhang, Y. Tang, B. Fang, Z. Shang, X. Liu, Face recognition under varying illumination using gradientfaces. IEEE Trans. Image Process. **18**(11), 2599–2606 (2009)

Chapter 12
Recognition of 3D Faces with Missing Parts Based on SIFT and LBP Methods

Narimen Saad and NourEddine Djedi

12.1 Introduction

In recent years, approaches of recognition faces in 2D still working with images and videos have got a very great precision that can even surpass face recognition of human capabilities. Though, face recognition is ever a hard task, particularly in computer vision if the lighting conditions or pose issues are not verified. Not long ago, the accessibility of 3D facial data acquired with scanners has increased interest for 3D facial testing solutions using 3D facial form to improve recognition accuracy. In fact, it is expected the 3D geometry of the face to have less sensitivity to variations in illumination or pose changes. Since the last Face Recognition Grand Challenge (FRGC 2005) [1], a lot of 3D face recognition approaches have been proposed and tested (for further discussion, see the investigation in [2] and survey of the literature) [3].

In summary, 3D face recognition approaches can be classed into two broad classes: global (or holistic) perform face matching from representations extracted from the total face; and the partition of the local region (or base-region), the face in the regions, and the extract and match suitable descriptors for each of them. There are also hybrid solutions that combine global and local descriptors and multimodal solutions incorporating 2D and 3D information to improve recognition accuracy [3].

In general, the overall results are obtained from the whole surface of the face, which usually makes the compact and therefore efficient calculation. Holistic representations are robust to noise, although they are sensitive to face alignment and missing parts, and the precision of recognition is gravely concerned if the 3D model of the face is obstructed by the hair, ears, and neck. In inclusion, precision significantly decreases in the presence of non-neutral facial expressions.

N. Saad (✉) • N. Djedi
LESIA Laboratory, Department of Computer Science, University of Biskra, Biskra, Algeria
e-mail: saad_narimen@live.fr

R. Jiang et al. (eds.), *Biometric Security and Privacy*, Signal Processing for Security Technologies, DOI 10.1007/978-3-319-47301-7_12

Nevertheless, the local representatives are from small parcels on the surface of the face, which can even reduce to small regions throughout identified keypoints. The characterizing feature of local approaches is their potential to deal with missing parts, facial expressions and occlusions: The description of the face results from the combination of local descriptors computed for many parts of the face, missing parts or the deformation of some parts due to non-neutral or occlusions facial expressions does not compromise the global match. Subsequently, many new approaches to 3D face recognition have applied local facial features reporting highly precisions on benchmark databases such as the FRGC version 2.0 dataset [3]. In particular, new researches have exposed that the local descriptors based around salient keypoints can be usefully applied to represent 3D objects and faces. Within [20], a 3D keypoint detector and descriptor inspired SIFT (Scale Invariant Feature Transform) [3], was created and used to implement 3D face recognition using a hybrid 2D and 3D approach using also SIFT descriptor index detector 2D texture of face images [3].

12.1.1 Previous Works

However, to date, benchmark databases were designed to emphasize, especially facial recognition capacity 3D approaches to manage changes due to changes in facial expressions. A second problem that can significantly affect the quality of recognition and has not been widely discussed is the effect of the missing parts and limited occlusions of the face: In general, the difficulty of relevant recognition of a topic when just one part of the face scan is available. Missing parts may arise from self-occlusions of the face owing to pose variations, while the face occlusions are probably happening in real applications due to scarves, glasses, the hair, or caps. Perceptively, these problems may be increased with the advent of scanners 4D (3D plus time) that are able to get 3D sequences of body or facial scans. In this instance, due to the natural motion of the subjects, the existence of occlusions and missing parts can greatly augment. The effects of facial occlusions were first elaborated in 2D face recognition applications, where local matching methods have been used according to the adjustment of misclassification of occluded local regions through exact classification provided by non-occluded parts. In 3D, only a few solutions have explicitly examined the difficulties constituted by missing parts and occlusions in the conception and experimentation of recognition methods: usually, holistic approaches cannot effectively control these conditions. Alternatively, local methods have the potential to approach partial face matching, but only some solutions have been specifically created to this end.

In [4], a 3D face detection approach is proposed which is able to detect faces even in presence of face occlusions induced by objects like scarves, eyeglasses, caps, or by the hand. The approach is based on detection of the occluded area and restoration of the missing region, employing the information of the non-occluded parts [3]. Occlusions aredetected through comparing a face with a generic facial model, and by utilizing a threshold on the facial surface distances. The experiments were performed on a training set of 132 images with various non-occluded facial expressions and a test set of 76 acquisitions.

In [5], a 3D face recognition method which is capable to operate in the existence of both expression variations and occlusions is suggested. The method is founded on the use of the Average Region Models (ARMs) for registration: The facial area is separated (manually) into many meaningful components like cheeks, mouth, eyes, and chin regions, and registration of faces is performed by dividing dense alignments to relative ARMs [3].

The differences between the gallery and probe faces generated for individual regions are then united to define the final dissimilarity score. Under changes, such as those created by occlusions, the combination method can determine noisy areas and discard them. The performance of this method is tested on the Bosphorus 3D face database [6] that is planned for face analysis in the presence of variations facial expressions, poses differences and occlusions.

In [7], several overlapping regions around the nose are segmented and the scores of Iterative Closest Point (ICP) matching in these regions are incorporated with each other. This notion is prolonged in [8] by applying an ensemble of 38 regions which densely cover the face, and choosing the best performing sub-set of 28 regions to implement matching using the ICP algorithm. A recognition test accounting for missing parts in the probe faces is mentioned. However, as well in this case regions segmentation across several facial scans effectively depends on the exact identification of the nose tip [9]. Local approaches founded on regions are restricted by the necessity to identify few facial landmarks used to identify regions of interest in matching faces. In inclusion, since parts of these regions may be missing or occluded, the extraction of region descriptors is generally avoided, and regions comparison is attached to the computationally costly ICP matching. Methods that employ keypoints of the face guarantee to address a few of these limitations. Especially, some recent studies have demonstrated that local descriptors calculated throughout salient keypoints can be usefully used to demonstrate 3D objects and faces.

In [10], a 3D keypoint detector and descriptor inspired by the Scale Invariant Feature) Transform (SIFT [11] has been created and used to achieve 3D face recognition using a hybrid 2D + 3D approach that uses the SIFT descriptor, detector to index 2D texture face images [9]. In [12], SIFT is applied to find and perform salient points in multiple 2D range images extracted from 3D face models for the aim of 3D face recognition. The same idea is applied in [13] to perform the recovery of 3D objects by visual similarity, but in this case the points of a sampling grid used, and SIFT descriptors are calculated for them. In [14], SIFT feature descriptors calculated in correlation with facial landmarks of depth images are applied to classify 3D facial expressions.

At last, SIFT descriptors have been as well used in [15] to implement 2D expression recognition from non-frontal face images.

In [3], propose an approach to 3D face recognition based on local properties of the face. The approach is motivated that the information captured by the SIFT descriptor in connection to dispersed set of keypoints of the face is also local to capture those characterizing features of the 3D face that support specific face recognition. So as to become rather discriminate, the SIFT descriptors should be complemented with data that model the morphological changes of the face through regions that are wider than the neighborhood used to extract the SIFT descriptors. For this goal, they introduce facial profile to model the depth of the face scan along a segment connecting two SIFT keypoints. Given a face model in the form of a range image, the SIFT model is used for the detection and the description of keypoints. In the differentiation of two faces, SIFT descriptors are matched to calculate the similarity between pairs of keypoints identified in the two range images. Then, the distance between the two faces is derived by constituting the individual distances between facial profiles that originate by equal keypoints. The proposed solution is tested using the FRGC v1.0 dataset.

Tang et al. [16] developed a 3D facial recognition algorithm using LBP (Local Binary Pattern) varieties of expression, operator extension LBP, which is widely used in the analysis of 2D face. First, to describe the human face more accurately and reduce the effect of its local distortion, a 3D facial division system is proposed. Then, for each region of the face, the statistical histogram is used to summarize the details of the face. Finally, 3D face recognition algorithm proposed is tested on BJUT-3D database and FRGCv2. The authors have obtained promising results and concluded that it is possible to apply the representation of LBP on 3D face recognition. Because of its discriminatory capacity and simplicity of calculation, the LBP descriptor (Local Binary Pattern) has become a popular approach in various computer vision applications. This is true, for example, the detection and analysis of faces, biometrics, medical image analysis, motion and activity analysis, and retrieval of image databases or video [17, 18]. Many researchers on face recognition became interested with LBP.

The authors in their paper IEEE PAMI in 2006 [19] showed his great success in face recognition. According to [18], LBP is not considered only as a simple texture operator, but is the foundation of a major new research direction for local descriptors binary on image and video. In recent years, various LBP variants have been proposed to improve its robustness and increase its discriminatory capacity and its applicability to different types of problems [19]. Motivated by the success use of LBP and its variants in 3D face recognition, we offer this work as an important extension of our recent work [20], where a general framework for 3D face recognition is based on the fusion multi descriptors binary local. The summary of this work is presented in Sect. 12.5.

In this work, following the researches in [3] we first propose a novel 3D face recognition method based on local properties of the face. The contribution is founded by the fact that the information captured by the SIFT descriptor in correlation to a selection of keypoints of the face. Being enough discriminating, we introduce the profile face to model the depth of the face analysis along a segment connecting two SIFT keypoints, then measuring the dissimilarity between the profiles using the SVM method. The suggested solution is proven using the dataset FRGC v1.0. To assess the accuracy of recognition in the existence of missing parts, we applied an evaluation protocol where only the profiles that come from a part of the testing of the probe are estimated with gallery profiles scans. Following the researches in [16] we propose a second contribution of 3D face recognition based on local properties of the face. Where the first contribution is improved by the fact that we use SIFT descriptor on LBP image for feature extraction 3D range images. For range images we use the LBP operator, that the fusion of the four values of the radius LBP improves the performance of 2D and 3D information.

For the finest of our knowledge, no one of these approaches which treat 3D data of the face using the fusion of SIFT model and LBP of feature extraction were experienced throughout FRGC V 1.0 dataset.

12.1.2 Overall Description of the Proposed System FRS3D

Our automatic face recognition system includes three main subsystems where we use SIFT descriptor on LBP image for feature extraction 3D range images. Figure 12.1 illustrates these steps.

This section explains the different steps of the final proposed system. With our verification system proposed: we examine how the local characteristics of

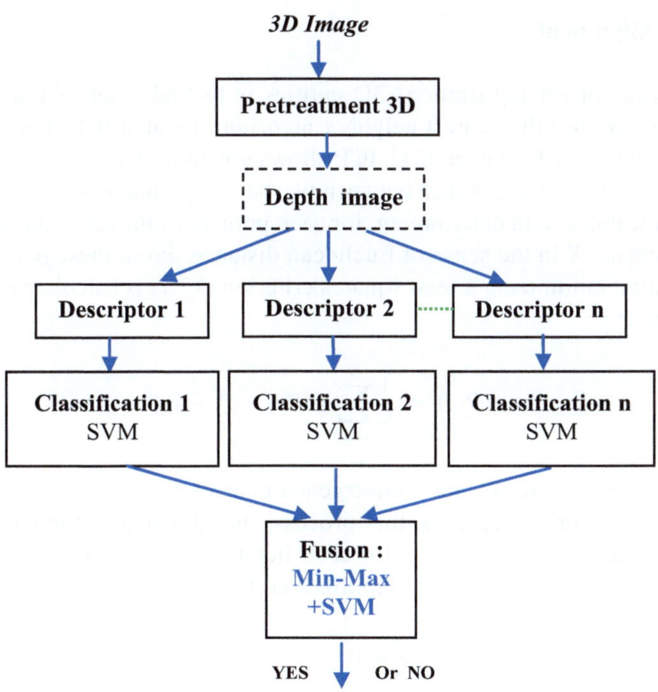

Fig. 12.1 Automatic face recognition system using SIFT KEYPOINTS + LBP

3D information contribute to facial recognition when the lighting, the expression changes, and the combination of expression under illumination are taken into account. All processes included in the training and test phase are fully automatic. Our system, as shown in Fig. 12.1, includes four major steps:

12.1.2.1 Pretreatment

By the alignment (translation and rotation) of the input 3D image relative to the reference image (of neutral expression face image) using ICP algorithm (Iterative Closest Point). According to the aligned images, we get 3D depth images.

Assume that a face is described by a 3D point cloud captured by a 3D laser scanner. Each point cloud is made up of thousands of points in 3D space. These discrete points approximately describe the surface of the face. We use the face database FRGC v1.0 3D. Each point is described with 3D spatial coordinates and the coordinates of respective RGB colors. In this section, we explain how the original 3D data are pretreated. We will align the data, and so we will get the depth of images. This part prepares feature extraction in the next section. This pretreatment comprises two main steps, the alignment of 3D face surfaces and the acquisition of the depth.

12.1.2.2 Alignment

The alignment of (or registration) 3D entities is to find a set of transformations to reconcile. We use the nearest neighbor algorithm iterated ICP (Iterative Closest Point) introduced by Besl et al. [21]. ICP allows you to find the rigid transformation (R, t) that minimizes the distance between two sets of points: $P=\{p_i\}$ and $X=\{y_i\}$. The principle consists in determining, for each point p_i of the set P, the closest point in the second set X in the sense of Euclidean distance. From these pairings, a rigid transformation minimizing a least squares criterion (12.1) is calculated and applied to all the set P:

$$e(R,t) = \frac{1}{N}\sum_{i=0}^{N} \|(Rp_i + t) - y_i\|^2 \tag{12.1}$$

This process is iterated until convergence characterized by the stability of the minimum error, for example. In this process, the global transformation (R, t) is updated incrementally as follows: for each iteration k of the algorithm, $R_k{:}R=R$ and $t=t+t_k$. The criterion to be minimized at iteration k becomes (12.2):

$$e(R_k, t_k) = \frac{1}{N}\sum_{i=0}^{N} \left\| \left(R_k (Rp_i + t) + t_k - y_i \right\|^2 \tag{12.2}$$

12.1.2.3 Depth and Intensity Image

The depth images are obtained from the aligned 3D data. The data is converted into a 3D depth image (see Fig. 12.2a) and a color image (see Fig. 12.3a). Inmost images, the nose is the part of the face closest to the 3D scanner. For each pixel, the average is calculated using a window size of 9×9 around it. The use of a window of size 3×3 which calculates the sum of the depth of its corresponding pixels, the nose is detected as the pixel coordinates of the center window that returns the minimum value (see Fig. 12.2b). After detection of the nose, extracted a sub-image centered on the nose of the center, with a size of 57×47. For images in RGB colors, we used the intensity images (see Fig. 12.3b). However, due to the original quality 3D data, the depth images and intensity usually contain a lot of noise, such as holes and outliers. We can get improved images by the following methods: For each pixel, the average is calculated using a window (5×5) (see Figs. 12.2c and 12.3c). If the pixel value is less than a given threshold, that pixel is replaced by the mean pixel. The change in

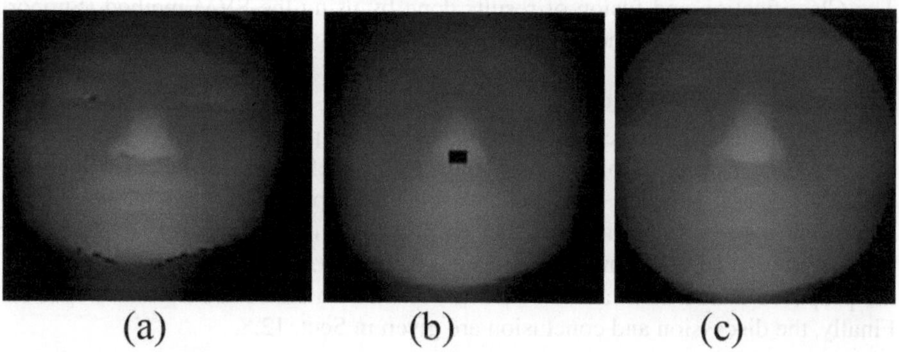

Fig. 12.2 Pretreatment of depth image. (**a**) Depth image, (**b**) detecting the tip of the nose, (**c**) average image 5×5

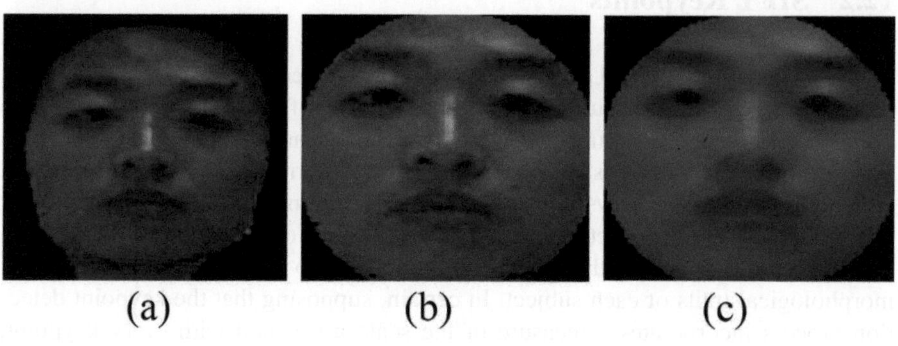

Fig. 12.3 Pretreatment of intensity image. (**a**) The color image, (**b**) the intensity image, (**c**) average image 5×5

light strongly influences the presentation intensity images. To address this problem, the histogram equalization is used to reduce the influence of illumination variations.

12.1.2.4 Feature Extraction

The method of extracting features is very important for the whole system. It is expected that these features are invariant to rotation, scale, expression changes, and lighting. In our system, we use the following local methods:

- Local Binary Patterns (LBP),
- Transformation of visual features invariant to scale (Scale Invariant Feature Transform (SIFT)).

12.1.2.5 Classification

The Classification and fusion of results done by using the SVM method (Support Vector Machines) and standardization of scores using the Min-Max method that is chosen for its simplicity.

This paper is organized as follows: In Sect. 12.2, the important characteristics of the SIFT descriptor are summarized, and its adaptation to our contribution is detailed. The extraction of the face and profiles defining a distance to measure the dissimilarity between the profiles using the support vector machine (SVM) algorithm, a second contribution, that we use SIFT descriptor on LBP image for the feature extraction from 3D range images in Sect. 12.5. The experiments with the proposed contribution and the results obtained on the FRGC v1.0 are reported. Finally, the discussion and conclusion are given in Sect. 12.8.

12.2 SIFT Keypoints

The representation of the 3D model concerned face detection based on a number of keypoints on the 3D face surface and the description of the surface of the 3D face corresponding to the keypoints and along the paths linear linking pairs of keypoints. As opposed to possibilities that keypoints should correspond to face significant landmarks, such as eyebrows, eyes, mouth, cheeks, and nose, we stick to a particular supposition about the expected position of the dots key on the facial surface.

Relatively, we include the position of keypoints to influence through specific morphological traits of each subject. In certain, supposing that the keypoint detection process incorporates a measure of the scale associated with every keypoint, we suppose that detected keypoints correspond to markers mean-full and operate the assumption so usually below the repeatability object: The most position—stable keypoints detected at coarse scales—do not change considerably in the models face the identical issue [3].

In our contribution, we used the SIFT algorithm to identify keypoints and description. SIFT was neither for 2D images in gray levels and cannot be immediately applied to the 3D face scans. Although, the 3D information of scanned faces can be captured through range images that use the gray scale of each pixel to represent the depth of a face scan.

These must be very distinctive, have a low probability of mismatch, be tolerant to image noise and changes in lighting, and they must also be in the presence of practically invariant to range, rotation, unimportant changes in the management and visualization of local distortions. Although a lot of likely keypoints at different locations in an image can be detected, only the most distinctive and useful invariants for compatibility must be kept. These frequently fall on the edges and corners of the image, and can be of many different sizes and orientations as well. To detect keypoints, SIFT descriptors are calculated. In short, a SIFT descriptor for a little region of the image, e.g., 4×4, size is calculated from the vector gradient histograms of the pixels in the patch. Eight likely gradient directions, and therefore the total size of SIFT descriptor is $4 \times 4 \times 8 = 128$ elements. This descriptor is unified to ameliorate invariance to changes in illumination (not relevant in the case of the range of images), and performed by other ways to ensure scale invariance and the rotation as well. These properties make the SIFT descriptor can issue a solid and powerful local representative of the range of the image and, consequently, of the face surface. In particular, SIFT descriptors were extracted using the following settings:

- SIFT keypoints are extracted and preserved their angles of range and orientation;
- 4×4 Orientation histograms regions of each keypoint four sampling are used to calculate the SIFT descriptor [3].

As in Fig. 12.4, detectedby SIFT keypoints are presented for the range of images of three various analyses of the face. The prime two images represent the same subject, while the third is a different topic. In wide, the position of the keypoints

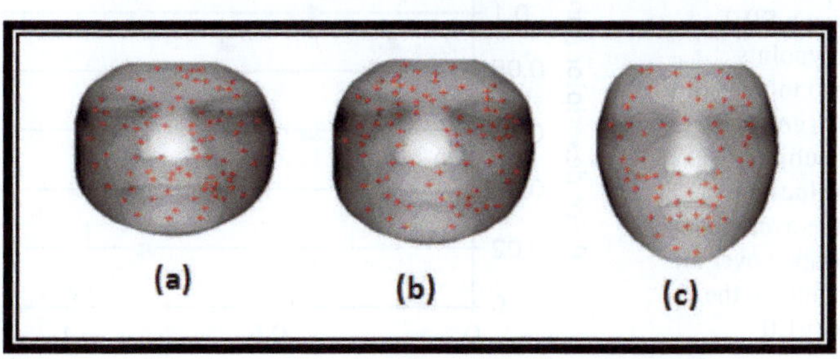

Fig. 12.4 Range of images from three 3D face scans show SIFT points detected with 3D depth information

on a detected face scanning depends on morphological features of the face. This is confirmed by the fact that the spatial arrangement of keypoints detected on different scans of the same subject—along a neutral expression—is much comparable. Though, the information that is captured by combining local descriptors (e.g., SIFT descriptors) with keypoints detected in a facial analysis is not sufficiently discriminating to support accurate recognition of the identity of the object. This is due to several reasons, which relate most significantly to the salience of local descriptors and inter-subject iterated. The latter refers to the absence of any warranty that the position of the keypoints on the face model should be separate on the subject. In fact, since the position of keypoints has a dependence on the morphology of the face, the detection of keypoints in models of different subjects understand a different layout on the face. Around the salience of local descriptors, calculated at detected keypoints it should be observed that these descriptors capture the morphology of the image surface in a small neighborhood (typically 16×16 pixels) centered at the keypoint. The report that is retained is just enough to allow recognition of the part of the face that is the keypoint: the nose, one of the points' extreme mouth or eyes, etc. [3].

However, this information is not sufficient to highlight the recognition of a specific person. For example, Fig. 12.5 indicatesthe repartition of the minimum distance values between the SIFT keypoints descriptors detected in two models as matching. The green curve is the repartition of minimum within-subject distance and represents the distribution of the minimum distance values between keypoints detected on different facial scans of the same subject. Differently, the red curve is the repartition of the minimum distance and inter represents the distribution of values of the minimum distance among keypoints detected on face scans with different subjects. As anticipated, the within-subject distribution gets a minimum distance of one form (peak) at a much decreased range value that the distribution of the minimum distance between subjects. However, the two distributions are not separated enough to allow for accurate recognition based only on statistical analysis with the minimum distance values between keypoints detected on two facial

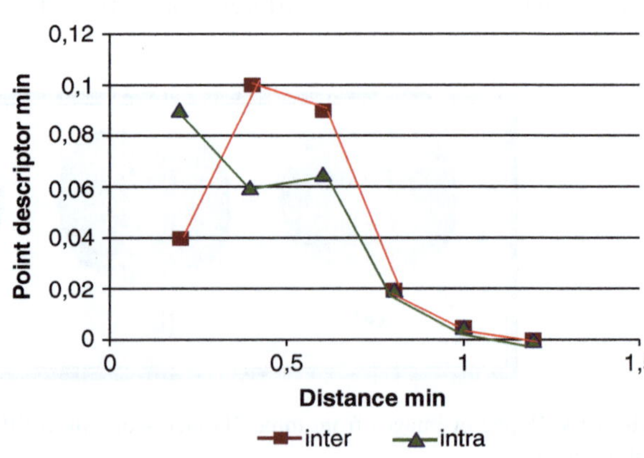

Fig. 12.5 Distribution of the values of the minimum distance between SIFT descriptors keypoints detected in two models of the same subjects (*green curve*) and different subjects (*red curve*). The values are obtained by the average minimum distance over all pairs of keypoints in the dataset FRGC v1.0

scans. This is principally incited by the event that the distinctive facial features are not concentrated in restrictive regions of some points of the face. Distinguishing facial features are distributed all over the face and, rather than being related to the morphology of the face in proximity to a sparse set of keypoints of the face changes progress since a keypoint to another. This resume the definition of face profiles that is discussed in detail in the next section [3].

12.3 Face Profiles with Sift Keypoints

Two SIFT keypoints detected on a range image to identify a face profile, its 1D function corresponding to the depth values corresponding along the points of the segment that connects the two keypoints. More formally, let:

- $I(x)$ where $x \in \Re^2$ be the image of a range scanning of the face; x_1, x_2 two keypoints;

Then the profile identified by the ordered pair (x_1, x_2) is defined as:

$$P_{i,x_1,x_2}(t) = I((1-t)\,x_1 + tx_2) \qquad t \in [0,1] \tag{12.3}$$

In describing the proposed model face, distinctive facial features are captured retaining SIFT keypoint descriptors detected on the range image and the profiles identified by pairs of these keypoints. The result data organized in a graph structure with nodes correlated to the keypoints and edges correlated to profiles.

SVM (Support Victor Machine) Classification derived from the theory of statistical learning. The essential idea is to project data belonging to different classes not linearly separable, with the input space, in a larger space called feature space, so that the data becomes linearly separable [22]. In this space, the optimal hyper-plane built-in technique is used to compute the ranking function separating the classes.

Case of two classes:

$$\{(x_1, y_1), \ldots (x_n, y_n)\} \quad \text{where } y_i \in \{-1, +1\} \quad \text{and} \quad x_i \text{ input} \tag{12.4}$$

Given the graphics of the two faces, the dissimilarity is measured by the first attribute to every node of the graph its first nearest node present in the latter graph, proximity is calculated as the Euclidean distance between the 128-dimensional SIFT descriptors related to the keypoints [3]. Then, for each pair of corresponding nodes in the two graphs whole profiles built from the keypoints are differentiated to associate the two profiles with the minimum distance. Determined the generic profile P1 extract from the couple of keypoints (x_1, x_2) of the face analysis I1 and the profile P2 extract from the couple of keypoints (x_3, x_4) of the face analysis I2 the distance between the two profiles is measured:

$$D(P1(t), P2(t)) = \int_0^1 |P1(t) - P2(t)|\, dt \tag{12.5}$$

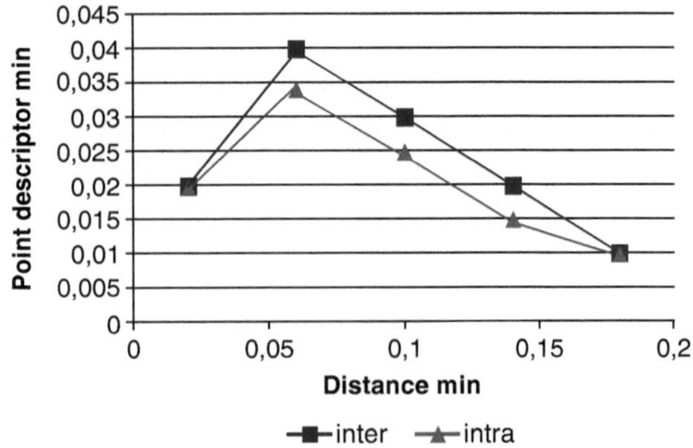

Fig. 12.6 Distribution of the minimum distance values between the detected keypoints in different scans of the same subject (*green curve*) and different subjects (*red curve*). These curves are calculated by averaging the values of the minimum distance over all pairs of scans of the series of data FRGC v1.0

At last, the distance among the two scans' faces is measured by the average minimum distance values profiles over every pair of observed nodes. We look for the linear classifier that separates the data with the lowest generation error. This classifier is a hyper-plane maximizes the margin of error, which is the sum of the distances between the hyper-plane and (positive, negative) examples of the closest hyper-plane [22]; requires classification of these two classes applied the SVM method that was presented previously. In the case where the data cannot be separated with a linear function, a non-linearity can be introduced by using a non-linear function.

Finally, the distance between the two face scans is calculated by averaging the profile minimum distance values beyond all pairs of matching nodes.

Consequently, the scheme can be used without any change to support matching of just a partial scan of a face against a complete scan of the face. In this reason, the proposed solution enables facial recognition even in the presence of missing parts and/or occlusions. The case in which the availability of only a part of a face affects the accuracy of the recognition is discussed in the next. In Fig. 12.6, the repartitions of the minimum distance values between the profiles faces that generated from various keypoints corresponding models of the same subject (within-subject distance) and various subjects (inter-subject distance) are described. These curves are calculated by averaging the values of the minimum distance of all pairs of scans of the database FRGC v1.0 [3].

12.4 Experimental Results (1)

In what follows, we present the results of the use of representation and matching contribution proposed to support the recognition of 3D face scans in the presence of the missing parts. For this experiment, we used the dataset FRGC v1.0 which includes 943 3D faces scanning of 275 people presenting a neutral facial expression (see Fig. 12.7).

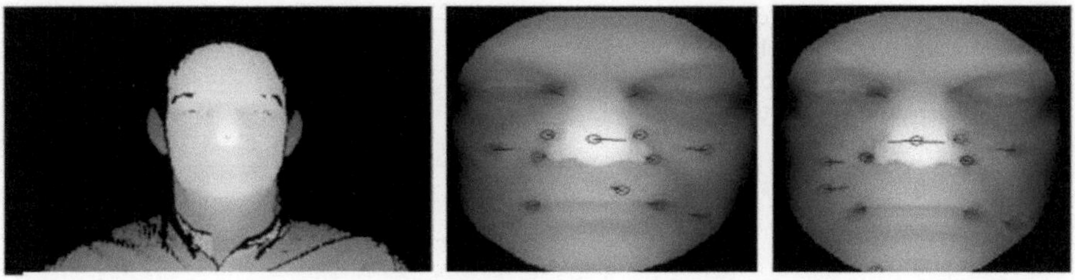

Fig. 12.7 Depth images corresponding to the original face image with FRGCV1.0 dataset

Facial scans are given as 480×640 matrix size of 3D points along a bit mask showing the available points of the scan (i.e., prominent points corresponding typically to the shoulders and head). Since as dissimilar topics distance from the sensor during the acquisition, the actual number of effective's points in a scanning can differ. Persons were acquired with front view of the shoulder, with very small variations in pose. Some analyses include occlusions due to facial hair. More details on FRGC dataset can be found in [23].

Depth images are obtained with the 3D face models according to the procedural changes and reported in Sect. 12.2 pretreatment [3]. According to the proposed contribution, first SIFT keypoints are extracted range of images from all faces. This translates into a variable number of keypoints for the image, depending on the specific characteristics of the surface of the face. To every depth image, just the first 15 keypoints (selected after ordering all the keypoints of the coarser the finest scale value) are preserved, and facial profiles are calculated between each pair of these keypoints.

The validity of the proposed contribution with respect to the missing parts of the face was tested applying the next experimental settings. The pattern (model) for every topic is selected to form the gallery (i.e., we chose the initial model of each topic as gallery pattern).

This showed in a gallery with reference models 275, while all models have been used as probes and compared with those of the gallery. To assess the accuracy of the proposed solution differentiated to parts of the face, every survey face model is divided into two parts (parts of the left and right side with observing to the perfect vertical plane with symmetry since the tip) and the profiles which are derived from a single part of the face differentiated to the gallery scans.

Exactly, two separated matching experiments were followed out, first exercising since a probe the left portion of every probe face model, the second by using the right portion of each probe face model. The efficiency of recognition has been measured by the rank of recognition rate k and presented cumulative similar characteristics curves (CMC) [3].

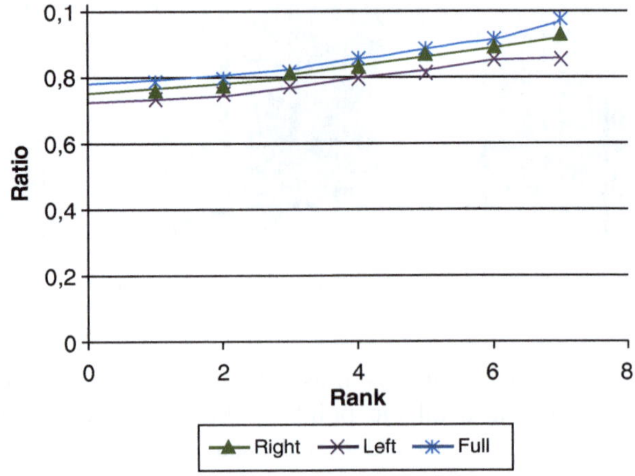

Fig. 12.8 Plots CMC curves obtained using probe models from profiles of keypoints on the face (*blue*), left side of the face (*purple*), and the right side of the face (*green*)

In special, recognition of rank *k* experience is successful if the face of the gallery constituting the identical person in the present probe is ranked in first position *k* of the ranking list. CMC curves are evaluated to every value of *k*, the rate of success rank *k* experiments.

Figure 12.8 reports CMC curves, for example, where just the profiles that come from a region of the face model used for recognition (purple plot refers to the right side with face scanning and green plot refers to the left).

To estimate the robustness of the proposed contribution, Fig. 12.8 also noted with a CMC blue curves plot for the case where all the profiles are used for recognition. One can observe that no pertinent motion emerges profiles employing the left or right side of the face. The rank-1 recognition rate is adjacent to 80 % in the two cases. It is equally important to observe that there is just a little decrease in performance with respect to the case in which the face profiles of the global face are used in the match [3].

However, the ROC curves (Receiver Operating Characteristic) measures the performance of a system of authentication (also uses the term recognition 1:1). The ROC curve (Fig. 12.9) provides a False Rejection Rate (FRR) vs. False Acceptance Rate (FAR). CMC curve is just another way to show data and can be calculated from the FAR and FRR. A comparative study specifying the relation between CMC and ROC curves can be found in [24].

12.5 Fusion of SIFT and LBP for Feature Extraction

We propose a second contribution 3D face recognition based on local properties of the face. The first contribution is improved by the fact that we use SIFT descriptor on LBP image for feature extraction 3D range images [8]. For range images we use the LBP operator, the fusion of the four values of the radius LBP improves performance

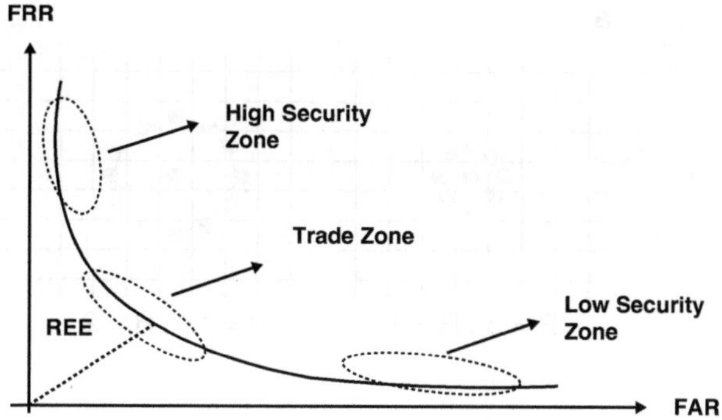

Fig. 12.9 ROC curve y

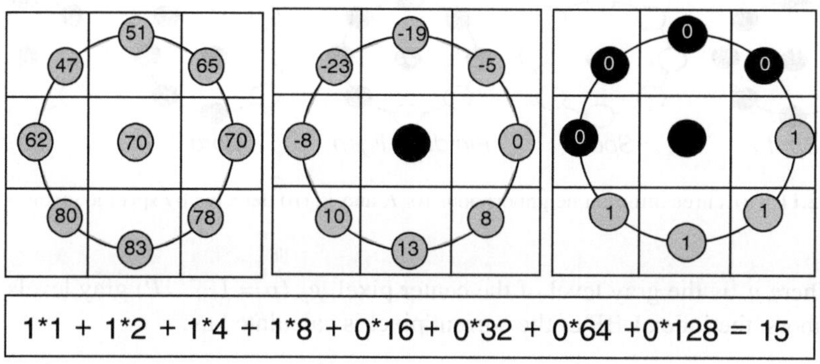

Fig. 12.10 The LBP operator

2D and 3D information. The LBP operator was at first suggested by Ojala et al. [35] in order to characterize the texture of an image. Calculating the LBP value for each pixel is "threshold" its eight neighbors with a threshold whose value is the gray level of the current pixel. All the neighbors will take a while value 1 if the value is greater than or equal to the current pixel to 0 if the value is less than (Fig. 12.10). The LBP code of the current pixel is then produced by concatenating the 8 values to form a binary code. One thus obtains, as for an image to gray scale, image LBP values containing pixels whose intensity is between 0 and 255.

LBP was subsequently expanded using different sized neighborhoods. In this case, a circle of radius R around the central pixel is considered. The values of sampled points on the P edge of the circle are taken and compared with the value of the central pixel. For values P sample points in the vicinity for any radius R, an interpolation is necessary.

We adopt the notation (P, R) for defining the vicinity of points P radius R of a pixel. Figure 12.11a illustrates three different neighborhoods to R and P values.

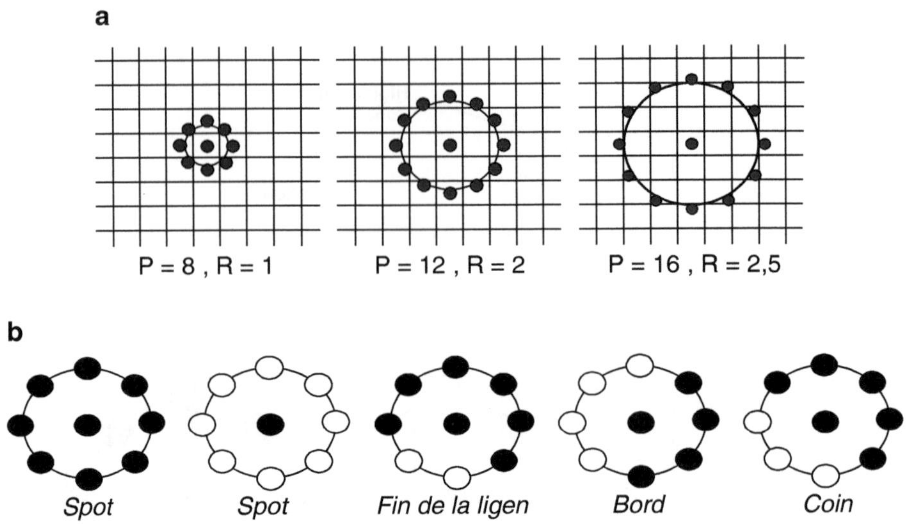

Fig. 12.11 (**a**) Three different neighborhoods for *R* and *P*, (**b**) detected by specific textures LBP

Where g_c is the gray level of the center pixel, g_p ($p = 1 \ldots P$) gray levels of its neighbors, the index LBP of the current pixel is calculated as:

$$\mathrm{LBP}_{P,R}\left(x_c, y_c\right) = \sum_{p=1}^{P} s\left(g_p + g_c\right) 2^{p-1}$$

Where

$$s(x) = \begin{cases} 1 \text{ if } x \geq 0 \\ 0 \text{ if } x < 0 \end{cases} \tag{12.6}$$

where (x_c, y_c) are the coordinates of the current pixel, *P* LBP, LBP *R* is the code for the radius *R* and the number of neighboring *P*. The LBP operator obtained with $P = 8$ and $R = 1$ ($\mathrm{LBP}_{8,1}$) is extremely adjacent to the LBP operator original. The main difference is that the pixels must first be interpolated to get the values of the points on the circle (circular instead of rectangular neighborhood). The important property of the LBP code is that this code is invariant to uniform changes to Global illumination because the LBP of a pixel depends solely on the differences between the level of gray and its neighbors. Given the depth image of the face, we can generate a set of LBP multi-scale representations of the face. Some examples are illustrated in Fig. 12.12. In this figure, the number of sampling points varies from 8 points to 24 points and the value of radius ranges from 1 pixel to 4 pixels. We used the Lowes code [25] to calculate the descriptor. The LBP images are calculated from the depth image.

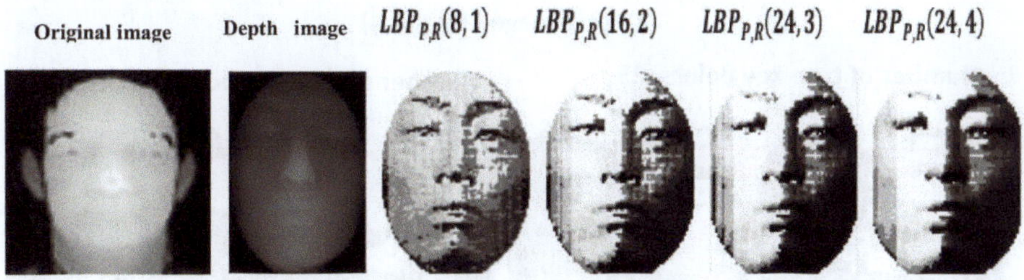

Fig. 12.12 Multi-scale LBP of the depth image of face

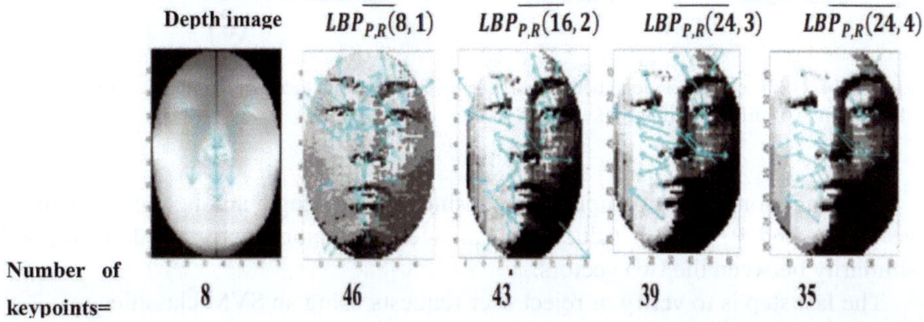

Fig. 12.13 Keypoints detected with the SIFT descriptor of the depth image original and four $\text{LBP}_{P,R}$ images associated

12.6 Experimental Results (2)

We applied the operator SIFT on $\text{LBP}_{P,R}$ images separately. Because, the $\text{LBP}_{P,R}$ highlights smooth local characteristics of the depth image and the intensity of the face. We can better detect a number of keypoints by using SIFT on $\text{LBP}_{P,R}$ image, that using of SIFT in the original images. We conducted a statistical work based on data FRGC v1.0. The average number of extracted keypoints of each LBP image P, R, is equal to 52 for a depth images and equal to 162 for intensity images. In contrast, the average number of keypoints extracted from each depth image of face is limited to 14. In the intensity, the number of keypoints is limited to 63. We used the code Lowes [25] to calculate the descriptor. Figure 12.13 shows the keypoints extracted from the image of the original face and its four images $\text{LBP}_{P,R}$ associates. To calculate the similarity between the learning face and to test the keypoints of the descriptor SIFT was matched using the Euclidean distance (see Fig. 12.14). Although the Euclidean distance is optimal in theory, the diverse experiences found that Euclidean distance is surpassed by other distances. One is the cosine [26] which is defined by:

$$\mathrm{S}\left(\mathrm{A},\mathrm{B}\right)=\frac{\mathrm{A}^{\mathrm{T}}\,\mathrm{B}}{||\mathrm{A}||\,||\mathrm{B}||} \tag{12.7}$$

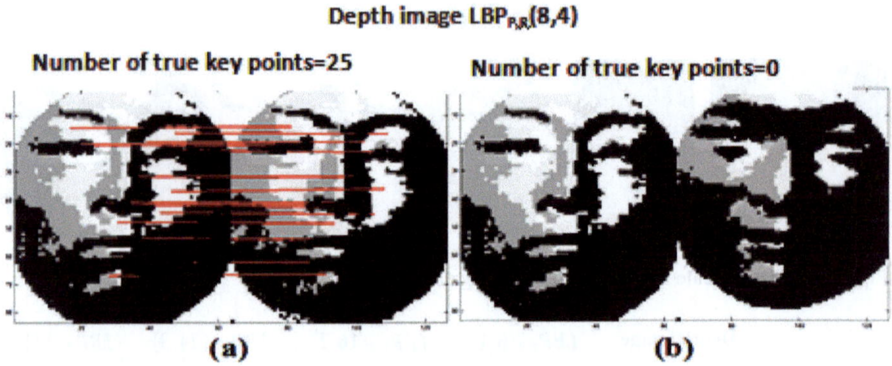

Fig. 12.14 SIFT correspondent between learning and testing faces belonging to (**a**) the same identity and (**b**) different identities

This function simply computes the cosine with the angle among the two feature vectors A and B. A high value of normalized correlation corresponds to a good similarity between the two vectors.

The last step is to verify or reject user requests using an SVM classifier.

SVM, also called separators Wide Margins, introduced by Vapnik [27] and used in fusion data in Refs. [23, 28–32]. They aim to define a hyper-plane of separation between the two classes; hyper-plane minimizes the classification error on a training set.

This method of supervised training can learn a separating more or less complex depending on the nature of the selected core. The simplest is the linear core nucleus of seeking a linear separator in the N-dimensional space of scores.

The purpose of the SVM is to find a separator that minimizes the classification error on the training set, but can be performed also in general on non-used training data. For this, the concept used is the margin (hence the name to wide margin separators). The margin is the mean square distance between the splitter and the midwife apprentices elements closest to it called support vector (Fig. 12.15). Those elements are called support vectors since the partition is only optimized on these elements of the training set.

Any classifier aims to clarify an element x, here $x = (s_1, \ldots s_N)$ is a vector of dimension N scores in one of the possible classes.

In our problem, there are two classes, client or imposter, whose label will be noted with $y = -1$ and 1. A value of -1 corresponds to the class of impostors and 1 to class Guest. The classifier thus aims to determine f such that:

$$y = f(x) \tag{12.8}$$

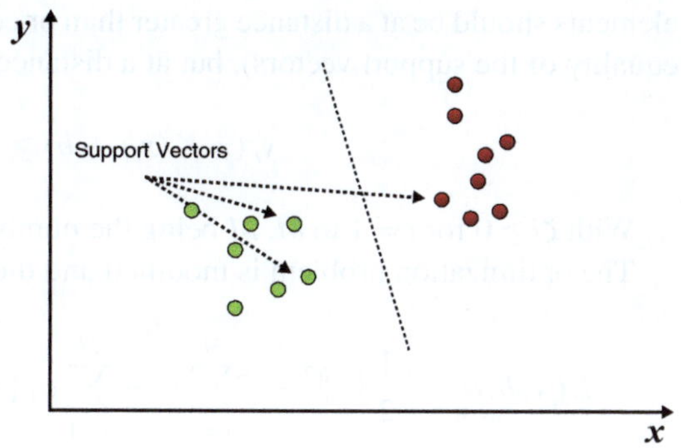

Fig. 12.15 Linear separation in a two-dimensional space

SVM aims to find the best linear separation (in terms of maximum margin, i.e., the best generalization) in a space transformed by K. This core function is to determine the vector w and the constant b such as separator has the equation [33]:

$$w.K(x) + b = 0 \tag{12.9}$$

The distance between a point in space x_i and the hyper-plane equation $y.K(x) + b = 0$ is equal to [33]:

$$h(x_i) = \frac{w.K(x_i) + b}{\|w\|} \tag{12.10}$$

To maximize the margin, it is necessary to minimize $||w||$ while maximizing for x_i defined as support vector. These support vectors are x_i for $i = 1$: m from the learning base such as: $w.K(x) + b = \pm 1$.

Solving this optimization problem can be done by using Lagrange multipliers where the Lagrangian is given by:

$$L(w, b, \alpha) = \frac{1}{2}\|w\|^2 - \sum_{i=1}^{M} \alpha_i \Big(y_i \big(w.K(x_i) + b \big) - 1 \Big) \tag{12.11}$$

With coefficients α_i called Lagrange multipliers [33].

To solve this optimization problem, we must minimize the Lagrangian with respect to w and b, and maximize compared to α. In practice, it is often impossible to find a linear separator (even in space transformed by the kernel function) because there is always misclassification. Vapnik [27] has introduced the technique of "flexible margin." This principle of flexible margin tolerates poor rankings by introducing variables resilience's ζi that allow to release the constraints on learning elements should be at a distance greater than or equal to 1 margin (the corresponding equality of the support vectors), but at a distance greater or equal to $1-\zeta i$, i.e.:

$$y_i\,(w.K\,(x_i)+b)\geq 1-\zeta i \tag{12.12}$$

With $\zeta i\geq 0$ for $i=1$ to M, M being the number of the learning basics [33]. The optimization problem is modified and the Lagrangian becomes [23]:

$$L\,(w,b,\alpha)=\frac{1}{2}\|w\|^2-C\sum_{i=1}^{M}\zeta i-\sum_{i=1}^{M}\alpha_i\left(y_i\left(w.k\,(x_i)+b\right)-1\right) \tag{12.13}$$

where C is a positive constant that adjusts the balance between the number of classification errors and the margin width.

A fusion system consists of two modules: a fusion module and a decision module (see Fig. 12.16). The problem becomes a problem of classification into 2 classes (YES or NO, Client or Impostor) from a real number vector whose dimension is equal to the number of subsystems.

There are two approaches which combine the scores of different systems:

1. The first approach is to treat the subject as a combination of problems.
2. The second approach is to see this as a classification problem.

In our work we use the scores of standardization from the Min-Max method (Eq. 12.5) for simplification.

– Standardization by the Min-Max method [33]:

With s_{ij} is the jth output rating of the ith system, where $i=1, 2, \ldots, N$ and $j=1, 2, \ldots, M$ (N is the number of systems and M the number of scores available in the entire training data). The standard scores Max Min-s_{ik} for the test score is given by:

$$S'_{ik}=\frac{s_{ik}-\min\,(\{s_{i.}\})}{\max\,(\{s_{i.}\})-\min\,(\{s_{i.}\})'} \tag{12.14}$$

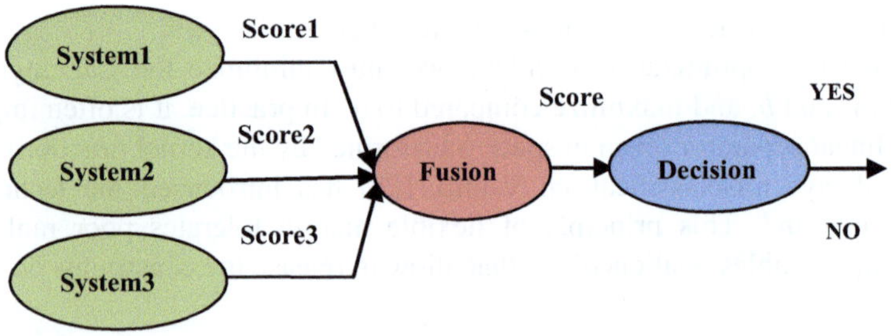

Fig. 12.16 Schema of the fusion scores

where $\{s_{i.}\} = \{s_{i1}, s_{i2}, \ldots, s_{iM}\}$. The Min-Max normalization retains the original distribution of scores on a scale factor and transforms all scores in the interval [0, 1].

For the fusion, we used a classification method which is SVM, because it is an efficient well-known method in the field of classification in two classes.

We use a SVM with Radial Basis Function (RBF) kernel (see Eq. 12.9). SVM has been implemented using the LibSVM which is a free available library [34]. The RBF kernel used is of the form:

$$K_{\text{RBF}}(u, v) = e^{-\gamma\|u-v\|^2} \tag{12.15}$$

where γ is a parameter that adjusts the width of the margin. SVM also requires the setting of parameter C which adjusts the balance between the margin width and the number of allowable errors (as described in Eq. 12.13). The different error rates using the merger by a SVM with RBF kernel without and with standardization methods [33].

The table above shows the equal error rate (REE) and the verification rate (VR) in all evaluations and testing by this method feature extraction. We conducted a statistical work based on data FRGC v1.0.

Table 12.1 shows the error rate in the total evaluation and test for features extraction by LBP + SIFT. The number of neighboring points (P) varies from 8 to 24 points and the value the radius (R) varies from one to four pixels. The table shows that the fusion of four LBP ($R = 1, 2, 3, 4$ and $P = 8, 16, 24$) plus SIFT works best with an REE = 4.67 % and VR = 94. 98 %.

LBP method for 3D information provides more performance than the 2D information (see Sect. 12.1 depth and intensity image), for four values of the radius (R).

The following parameters are used to measure the standard of a biometric system performance with a verification scenario:

1. False Rejection Rate (FRR): This rate represents the percentage of people expected to be recognized, but which is rejected by the system.
2. False Acceptance Rate (FAR): This rate represents the percentage of persons believed not to be recognized but which is nevertheless accepted by the system.
3. Equal Error Rate (REE): The rate is calculated from the two principal criteria and constitutes a point of measuring current performance. This point where FRR = FAR, i.e., the best compromise between false rejection and false acceptance.

Table 12.1 Shows the error rate in the total evaluation and test for features extraction by LBP + SIFT

(*P*,*R*)	Intensity image (2D)		Depth image (3D)	
	Eval REE	Test VR	Eval REE	Test VR
(8,1)	8.97	81.67	6.73	88.44
(16,2)	6.09	87.47	6.30	89.89
(24,3)	4.18	91.66	7.25	88.7
(24,4)	3.7	89.49	6.34	90.51
1 + 2 + 3 + 4	2.91	92.94	2.67	94.98

4. Recognition rate (Rank-one or RR): This measures the percentage of entries that are correctly identified.
5. Verification rate (VR): VR = 100 - FRR - FAR.

12.7 Comparison with Other Methods and Other Works

Table 12.2 summarizes the performance of the proposed method and other published results of benchmark, so we can note that our results are better than the other methods, yielding satisfactory results.

The single performances of work records on 3D face recognition with missing parts are that of Faltemier et al. [36], until the work in [5] of Aly UZ, G Okberk. They reported the rank-1 recognition rate using dataset FRGC v2.0 of 88 and 89.2 %, separately for regions in the left and right side of the face. However, the experimental setting applied in these evaluations presumes completely that the regions are not affected by a corresponding missing part (i.e., parts can fail, but not their parts) and it is difficult to fusion the results.

Berretti et al. [3] proposed an approach for 3D face recognition based on face local properties. The approach is motivated by the fact that the information captured by the SIFT descriptor in connection to dispersed keypoints set of the face is also local to capture those characterizing features of the 3D face that support specific face recognition. In the differentiation of two faces, SIFT descriptors are matched to calculate the similarity among pairs of keypoints identified on the two range images. Then, the distance among the two faces is derived by constituting the individual distances between facial profiles that originate by equal keypoints. The proposed method is tested using the FRGC v1.0 dataset.

To estimate the precision of recognition in presence of missing parts, they adopted an evaluation protocol that just the profiles that originate from one part of the probe scan are compared to the profiles of gallery scans. The rank-1 recognition rate is adjacent to 70 %.

Following the work in [3, 5, 36] we have proposed an approach to 3D face recognition based on local properties of the face, we also adopted the evaluation

Table 12.2 Shows the performance of the proposed method and other published results of the benchmark

Approach	Comparison
1. Faltemier et al.	– Sensitive except missing parts (difficult fusion of the results).
2. Aly UZ, G Okberk	– Treatment of occlusion only.
3. Berritti, Del Bimbo	– Rate 70 % for a rank $k = 1$ without classification data of each part of the face.
4. Our approach	– Rate 80 % for a rank $k = 1$ with classification data of each face part + Treatments occlusion and missing parts; – Using four LBP + SIFT improves our results with VR = 94.98 %.

Table 12.3 Classification results for the FRGC v1.0 dataset with SIFT + LBP method and comparison with our other methods used

Feature extraction methods	Recognition rate (%)
1. PCA and Hu + SVM	77.36
2. SIFT + SVM	79.98
3. SIFT and LBP + SVM	94.98

protocol that just the profiles that originate from one part of the probe scan are compared to the profiles of gallery scans with classification data of each face part using SVM method. The rank-1 recognition rate is adjacent to 80 %. Then we have improved our results by using for feature extraction four LBP + SIFT with VR = 94.98 %.

Table 12.3 summarizes the performance of the proposed method and comparison with other methods that we had implemented, then we have tested and noted that the results of the operator SIFT on $LBP_{P,R}$ images are superior to other methods, yielding satisfactory results.

12.8 Conclusion, Discussion, and Future Work

In this work, we proposed a new contribution for 3D face recognition which is capable to recognize a 3D face model probe also in cases that only a part of the probe model is vacant. The first contribution uses the detection of keypoints and identification/characterization of the keypoints with SIFT description model for the representation of the range of a face image. Detected keypoints for network profiles identification, each variation profile representative of the surface with line segments attaching pairs of keypoints. Our contribution makes no hypothesis about matching detected keypoints to particular landmarks on the face. Consequently, it may support comparison network profiles extracted from a probe model versus extraction of the network profiles from a model of the same gallery if the model of the probe is only part of the face by employing the SVM algorithm.

First results conducted on the dataset FRGC v1.0 show the viability of the contribution. Comparison with the only alternative approach proposed in the literature by Faltemier et al. [36] suggests firstly that the suggested solution is less sensitive than [1] of missing parts, but, on the other hand, it needs to be improved to make it comparable (if not better) precision values of recognition. The proposed solution get a rate of 80 % for a rank $k = 1$ with classification data of each face part. However, the approach of Berretti et al. [3] get a rate of 70 % for a rank $k = 1$ without classification data of each face part. For the latter, the plots reported in Fig. 12.2 make evidence that the curves for the case where only the profiles that come from a part of the face model used for recognition. This constitutes a probable loss of precision which highlights the presence of keypoints and/or profiles that are much equal, while to be extracted from models of different subjects. Using four LBP + SIFT improves our result with VR = 94.98 %.

Future work will investigate the use of a model to estimate the mutual information between the issues and keypoints/profiles so as to identify the keypoints/profiles that can be reliably associated with models a single subject, and we are looking to use FRGC V2.0 database. Multimodality will be approached from the angle appearance, more as a form in the first manner; and various scores on various parts of the face (e.g., eyes, mouth) combining together in another manner. Our research is not specifically facial recognition therefore my knowledge of the specific approaches to this and the current areas of interest are extensive, but we have knowledge of SIFT and LBP in other approaches of face works.

References

1. P.J. Phillips, P.J. Flynn, T. Scruggs, K.W. Bowyer, J. Chang, K. Hoffman, J. Marques, J. Min, W. Worek. Overview of the face recognition grand challenge, in *IEEE Workshop on Face Recognition Grand Challenge Experiments*, San Diego, CA, June 2005, pp. 947–954
2. K.W. Bowyer, K.I. Chang, P.J. Flynn, A survey of approaches and challenges in 3d and multimodal 3D + 2D face recognition, January (2006)
3. S. Berretti, A. Del Bimbo, P. Pala. Recognition of 3D faces with missing parts based on profile networks, in *Proceedings of the ACM Workshop on 3D Object Retrieval (3DOR '10)*, ACM, USA, 2010, pp. 81–86
4. A. Colombo, C. Cusano, R. Schettini, Gappy PCA classification for occlusion tolerant 3D face detection. J. Math. Imaging Vis. **35**, 193–207 (2009)
5. N. Alyuz, B.G. Okberk, L. Akarun, 3d face recognition system for expression and occlusion invariance, in *IEEE 2nd International Conference on Biometrics*, September 2008
6. A. Savran, N. Alyuz, H. Dibeklioglu, O. Celiktutan, B. Gdkberk, B. Sankur, L. Akarun, Bosphorus database for 3D face analysis, in *Proceedings of First COST 2101 Workshop on Biometrics and Identity Management*, May 2008
7. K.I. Chang, K.W. Bowyer, P.J. Flynn, Multiple nose region matching for 3d face recognition under varying facial expression. IEEE Trans. Pattern Anal. Mach. Intell. **28**(6), 1695–1700 (2006)
8. S. Berretti, A. Del Bimbo, P. Pala, 3d face recognition using iso-geodesic stripes, in *IEEE Transactions on Pattern Analysis and Machine Intelligence*, 2010
9. S. Berretti, A. Del Bimbo, P. Pala, Facial curves between keypoints for recognition of 3D faces with missing parts, in *IEEE Computer Society Conference on Computer Vision and Pattern Recognition Workshops (CVPRW)*, 20–25 June 2011, pp. 46–51
10. A.S. Mian, M. Bennamoun, R. Owens, Keypoint detection and local feature matching for textured 3d face recognition. Int. J. Comput. Vision **79**(1), 1–12 (2008)
11. D. Lowe, Distinctive image features from scale-invariant key points. International Journal of Computer Vision, **60**(2), 91–110, (2004)
12. M. Mayo, E. Zhang, 3d face recognition using multiview key point matching, in *IEEE International Conference on Advanced Video and Signal Based Surveillance*, Genoa, Italy, September 2009, pp. 290–295
13. R. Ohbuchi, T. Furuya, Scale-weighted dense bag of visual features for 3d model retrieval from a partial view 3d model, in *Proceedings of Workshop on Search in 3D and Video*, Kyoto, Japan, September 2009
14. S. Berretti, A. Del Bimbo, P. Pala, B. Ben Amor, M. Daoudi, Selected sift features for 3d facial expression recognition, Turkey, August 2010

15. W. Zheng, H. Tang, Z. Lin, T.S. Huang, A novel approach to expression recognition from non-frontal face images, in *Proceedings of IEEE International Conference on Computer Vision*, Kyoto, Japan, September 2009, pp. 1901–1908
16. H. Tang, B. Yin, Y. Sun, Y. Hu, 3D face recognition using local binary patterns. Signal Process. **93**, 2190–2198 (2013)
17. M. Pietikinen, A. Hadid, G. Zhao, T. Ahonen, Computer vision using local binary patterns. Comput. Imaging Vision **40** (2011)
18. S. Brahnam, L.C. Jain, L. Nanni, A. Lumini, *Local Binary Patterns: New Variants and Applications*, Studies in Computational Intelligence, vol 506 (Springer, 2014)
19. T. Ahonen, A. Hadid, M. Pietikainen, Face description with local binary patterns: application to face recognition. IEEE Trans. Pattern Anal. Mach. Intell. **28**(12), 2037–2041 (2006)
20. X. Tan, B. Triggs, Enhanced local texture feature sets for face recognition under difficult lighting conditions. IEEE Trans. Image Process. **19** (2010)
21. P.J. Besl, N.D. McKay, A method for registration of 3-d shapes. IEEE Trans. Pattern Anal. Mach. Intell. **14**(2), 239–256 (1992)
22. A. Atia, *Face Recognition in Computer Biometric Authentication, for Academic Master in Computer Science* (Biskra University, 2011)
23. L. Allano, Scores of merger strategies and dependence measures applied to the bases of virtual people. Thesis of Doctor of the National Institute of Telecommunications, January 12, 2009
24. R.M. Bolle, J.H. Connell, S. Pankanti, A.W. Senior, The relation between the ROC curve and the CMC. IBM Research Report (2006)
25. D. Lowe, Demo Software: SIFT Keypoint Detector, http://www.cs.ubc.edu.ca/lowe/ (2006)
26. W. Hwang, H. Wang, H. Kim, S. Kee, S.-C. Kee, J. Kim, Face recognition system using multiple face model of hybrid fourier. IEEE Trans. Image Process. **20**(4), 1152–1165 (2011)
27. V.N. Vapnik, The Nature of Statistical Learning Theory (1995)
28. F. Wang, J. Han, Multimodal biometric authentication based on score level fusion using support vector machine. Opto-Electron. **17**(1), (2009)
29. D.A. Ramli, S.A. Samad, A. Hussain, Score information decision fusion using support vector machine for a correlation filter based speaker authentication system, in *Proceedings of the International Workshop on Computational Intelligence in Security for Information Systems CISIS'08*, vol. 53, pp. 235–242 (2009)
30. P. Ejarque, J. Hernado, D. Hernando, D. Gómez, Eigenfeatures and supervectors in feature and score fusion for SVM face and speaker verification. Biomet. ID Manage. Multimodal Commun. **5707**, 81–88 (2009)
31. D.R. Kisku, P. Gupta, J.K. Sing, Fusion of multiple matchers using SVM for offline signature identification. Secur. Technol. **58**, 201–208 (2009)
32. M. Farrús, P. Ejarque, A. Temko, J. Hernando, Histogram equalization in SVM multimodal person verification. Adv. Biomet. **4642**, 819–827 (2007)
33. A. Ouamene, Biometric recognition multimodal fusion of 2D and 3D Face, for Doctor of Science in Electronics, June (2015)
34. The LibSVM library, http://www.csie.ntu.edu.tw/cjlin/libsvm/
35. D. Harwood, T. Ojala, M. Pietikäinen, S. Kelman, S. Davis, *Texture Classification by Center-Symmetric Auto-correlation, Using Kullback Discrimination of Distributions, Computer Vision Laboratory* University of Maryland, College Park, Maryland (1993)
36. T.C. Faltemier, K.W. Bowyer, P.J. Flynn, A region ensemble for 3d face recognition. IEEE Trans. Inf. Forensics Secur. **3**(1), 62–73 (2008)

Chapter 13
Face Anti-spoofing in Biometric Systems

Zinelabidine Boulkenafet, Zahid Akhtar, Xiaoyi Feng, and Abdenour Hadid

13.1 Introduction

Over the past decade, automated face recognition systems have been adopted in various applications because of face's rich features that offer a strong biometric cue for recognizing individuals [8, 53]. In fact, facial recognition systems are already being used at large scale. For instance, UIDAI program provides identity to all persons resident in India using face, and Microsoft Kinect employs face recognition to access dashboard and automatic sign-in to Xbox Live profile. Likewise, face biometrics based access control is a ubiquitous feature available now on mobile devices as an alternative to passwords, e.g., Android KitKat mobile OS, Lenovo VeriFace, Asus Smart-Logon, and Toshiba SmartFace. Since deployment of face recognition

---

The original version of this chapter was revised: All the figures are updated. The correction to this chapter is available at https://doi.org/10.1007/978-3-319-47301-7_17

Z. Boulkenafet (✉)
University of Oulu, Oulu, Finland
e-mail: zboulken@ee.oulu.fi

Z. Akhtar
INRS-EMT, University of Quebec, Varennes, QC, Canada
e-mail: zahid.eltc@gmail.com

X. Feng
Northwestern Polytechnical University, Xi'an, China
e-mail: fengxiao@nwpu.edu.cn

A. Hadid
University of Oulu, Oulu, Finland

Northwestern Polytechnical University, Xi'an China
e-mail: hadid@ee.oulu.fi

R. Jiang et al. (eds.), *Biometric Security and Privacy*, Signal Processing
for Security Technologies, DOI 10.1007/978-3-319-47301-7_13

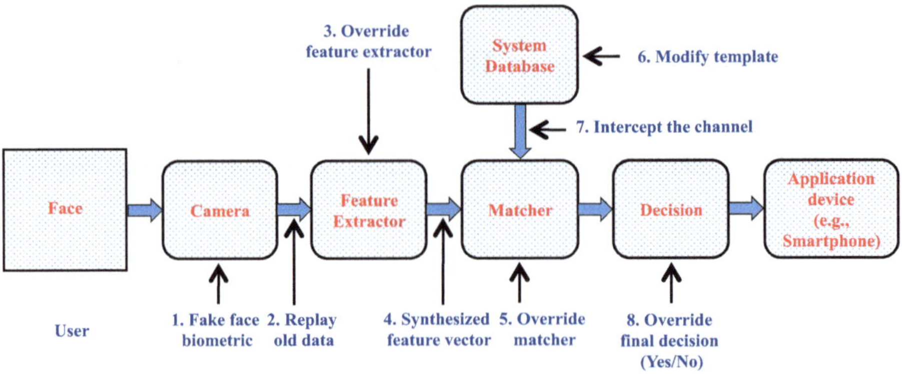

Fig. 13.1 Points of attack in a face verification system

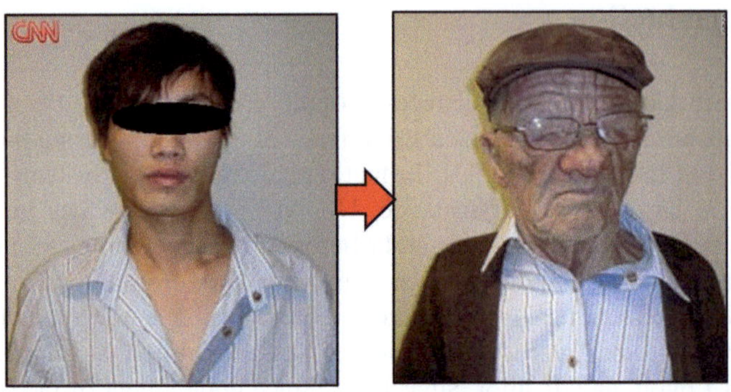

Fig. 13.2 In 2010, a passenger boarded a plane in Hong Kong with an old man mask and arrived in Canada to claim asylum

systems are growing year after year, people are also becoming more familiar to their use in daily life. Consequently, security weakness of face recognition systems is getting better known to the general public. However, vulnerabilities of face systems to attacks are mainly overlooked [7, 17, 18], even when it is not difficult nowadays to find web sites or even tutorial videos giving detailed guidance on how to attack face systems to gain unauthorized access. Eight possible different points where security of face recognition systems can be compromised are shown in Fig. 13.1.

In particular, existing systems are vulnerable to facial spoofing attacks [26]. Facial spoof attack is a process in which a fraudulent user can subvert or attack a face recognition system by masquerading as registered user and thereby gaining illegitimate access and advantages [3, 6, 10, 15, 34, 42, 53]. Face spoofing attack is also known as "direct attack" or "presentation attack". Face spoofing is also a major issue for companies selling face biometric-based identity management solutions [10]. For example, in Hong Kong, a young passenger boarded a plane while wearing an old-man mask and arrived in Canada to claim asylum in 2010 as also shown in Fig. 13.2. It is worth mentioning that face spoofing does not require advanced technical skills, which increases thereby potential number of attackers. Moreover,

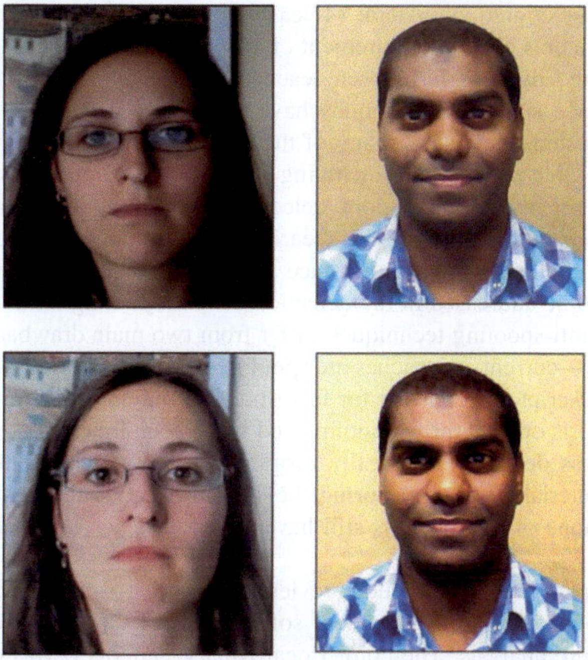

Fig. 13.3 Example of images captured from real faces (*upper row*) and from photo spoofing attacks (*lower row*)

as illustrated in Fig. 13.3, face images captured from spoofing attacks can look very similar to images captured from real ones, which makes very difficult to detect face spoofing attacks.

In [53] authors reported experimental results to point out that the probability of spoofed faces being successfully accepted as genuine user might be up to 70 %, even when a state-of-the-art Commercial Off-The-Shelf (COTS) face recognition system is employed. Hence, one can extrapolate that current COTS face systems are predominantly not designed to effectively counteract against spoofed faces. This vulnerability is now listed in the National Vulnerability Database of the National Institute of Standards and Technology (NIST) in the USA.

Quintessential face anti-spoofing technique is face liveness detection, which aims at disambiguating human live face samples from spoof artifacts [10, 45]. A large number of face liveness detection methods have been proposed in the literature [10, 12, 13, 36, 39, 40, 46, 49, 56]. Besides liveness detection, multibiometrics is also considered as natural countermeasure against face spoofing attacks [2, 4, 5, 22].

Face liveness detection is an onerous task that demands certain difficult requirements [10, 19] such as: (1) non-invasive, the method should not be harmful for the user or require an excessive contact with the user; (2) speed, the outcome has to be generated in a very short interval; (3) performance, beside having good spoof detection rate, the anti-spoofing must not decrease the recognition performance of the main face recognition system; (4) easy to embed in already functional face recognition systems plus no requirement of new piece of hardware.

Though, big advances have been reached in face anti-spoofing over the last decade, the face spoofing techniques have also evolved and become more and more sophisticated. Inevitably, many of the existing face anti-spoofing techniques are still vulnerable to spoofing, including various commercial systems that claim to have some degree of face spoof detection embedded. A coherent survey on face spoofing and anti-spoofing [33] clearly points out the face spoofing being yet to be a huge challenge for existing face recognition systems. Namely, there are many issues to be addressed in the detection of face spoofing attacks. Specifically, existing face anti-spoofing techniques suffer from two main drawbacks: (1) lack of generalization—current approaches are spoof material- and/or trait-dependent, such that feature descriptors proposed for face spoofing may not function effectively if employed for iris or fingerprint spoofing and vice versa. Likewise, the performance of face liveness detectors drastically drops when they are presented with novel fabrication materials (not used during the system design/training stage); (2) high error rates—none of the methods still have shown to reach a very low acceptable error rates.

In this chapter, after an assiduous review of the state-of-the-art in face spoofing and anti-spoofing, we present a novel software-based face anti-spoofing method via color texture analysis. The color Local Binary Patterns (LBP) descriptor [27] is utilized to extract the joint color-texture information from the face images. Specifically, the uniform LBP histograms are extracted from the individual image bands. These histograms are subsequently concatenated to form the final descriptor. To attain insight into which color space is more discriminative to distinguish real face from fake ones, three color spaces, namely *RGB*, *HSV*, and *YCbCr* are considered independently. Extensive experiments on two challenging benchmark datasets, namely CASIA face anti-spoofing and Replay-Attack databases, clearly indicate that color texture based method outperforms gray-scale counterparts in detecting various types of spoofing attacks. Moreover, inter-database evaluation shows very promising generalization capabilities of the proposed method compared to state-of-the-art techniques.

The rest of the chapter is organized as follows. In Sect. 13.2, an exhaustive overview of published works in the field of face spoofing and anti-spoofing is outlined. A brief description of large and publicly available face spoofing databases is presented in Sect. 13.2.3. A case study based on the use of color texture analysis as face anti-spoofing tool is introduced in Sect. 13.3. Future research directions and conclusions are described in Sects. 13.4 and 13.5, respectively.

13.2 State-of-the-Art in Face Spoofing and Anti-spoofing

In this section, we present an overview of face spoofing with their liveness detection methods.

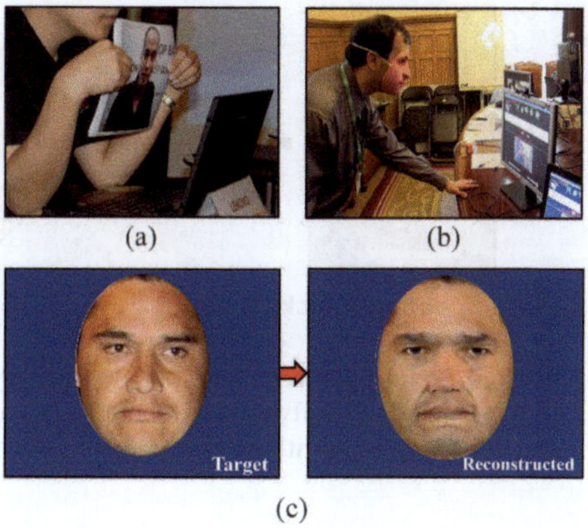

Fig. 13.4 Examples of face spoofing using (**a**) photograph, (**b**) 3Dmask, (**c**) reverse-engineered face image

13.2.1 Face Spoofing

Despite great deal of progress in face recognition systems, face spoofing still poses a serious threat. Most of the existing academic and commercial facial recognition systems may be spoofed by Akhtar [2] and Akhtar et al. [10] (see Fig. 13.4): (1) a photo of a genuine user; (2) a video of a genuine user; (3) a 3D face model (mask) of a genuine user; (4) a reverse-engineered face image from the template of a genuine user; (5) a sketch of a genuine user; (6) an impostor wearing specific make-up to look like a genuine user; (7) an impostor who underwent plastic surgery to look like a genuine user; (8) a photo or a video, generated using computer graphics, of a genuine user. The easiest and most common face spoofing attack is to submit a photograph or video of a legitimate user to the face recognition systems.

13.2.2 Face Anti-spoofing

The pioneer studies on biometric vulnerabilities, such as [47, 50], highlighted the necessity of developing efficient protection schemes against face spoofing attacks. Over the last two decades, a number of face anti-spoofing or liveness detection methods have been proposed [26, 33, 53]. The published face anti-spoofing techniques can be broadly grouped into four categories: (1) motion analysis based methods, (2) texture analysis based methods, (3) image quality analysis based methods, (4) hardware based methods (Fig. 13.5).

13.2.2.1 Motion Analysis Based Methods

Methods in this category are mainly based on the spontaneous movement clues generated by two-dimensional spoofing attacks such as photographs and videos.

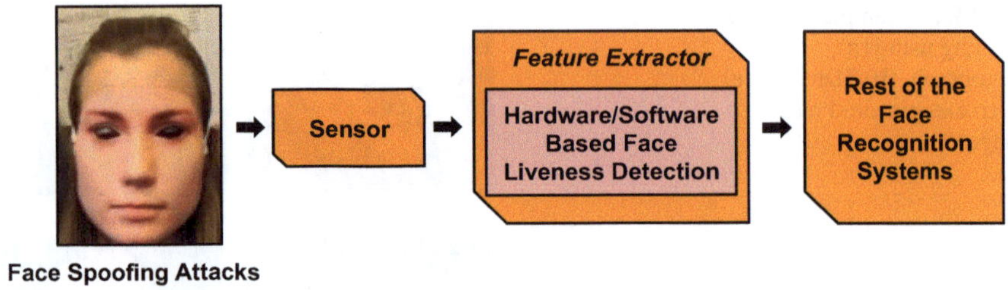

Fig. 13.5 Spoofing attacks might be detected by hardware- or software-based presentation attack detection methods. Software-based methods are inexpensive and non-invasive because they use signal-processing algorithms

They analyze the motion features of the input to determine the realism of the face sample to the system. For instance, it is well known that human eye-blink occur once every 2–4 s (Pan et al.'s). Therefore, [45] exploited such spontaneous eye-blinking factor to devise a liveness detection for photo-spoofing. In particular, the proposed algorithm utilizes an undirected conditional random field framework to model the eye blinking, while relaxing the independence assumption of generative modelling and state dependence limitations from hidden Markov models. Likewise, Tan et al. [49] exploited another object movement fact that the real human 3D faces significantly move differently from planer objects, and such deformation patterns can be used for face anti-spoofing. The features are extracted using a variational retinex-based and difference-of-Gaussians (DoG) [39] based approaches. The extracted features are then used for live or spoof classification. However, the above-mentioned methods are very vulnerable to spoofing attacks using videos. To overcome this limitation, authors in [36] designed an algorithm heavily based on a short sequence of images using a binary detector. The method captures and tracts the subtle movements of different selected facial parts using a simplified optical flow analysis followed by a heuristic classifier. The same authors also presented another technique to fuse different experts systems introduced in the former works as liveness attributes, e.g., eye-blinks and mouth movements. Bao et al. [13] proposed a systems to estimate face motion via optical-flow system to detect attacks produced with planar media such as prints or screens.

Despite their advantages and accuracy, motion-based face anti-spoofing suffers from computational complexity, speed, and face video attacks. In addition, they are not applicable for face recognition systems, where only a single image of the user is available.

13.2.2.2 Texture Analysis Based Methods

The methods in this category assume that surface properties (e.g., pigments) of real faces are different from spoof prints, thereby examining the skin properties such as skin texture and skin reflectance be helpful for spoof detection. Most common detectable texture patterns due to artifacts are printing failures or blurring effects. Contrary to motion analysis based methods, techniques in this category need only a single static image sample rather than video data. Consequently, these algorithms are generally faster and user-friendly. Li et al. [40] described a method for print-attack face spoofing by exploiting differences in the 2-D Fourier spectra of live and spoof images. The method presumes that photographs are normally smaller in size and contain fewer high frequency components compared to real faces. The method only works well for down-sampled photos of the attacked identity, but likely fails for higher-quality samples. In [9, 11, 12, 35, 37, 55] authors developed micro-texture analysis based methods to detect printed photo attacks using "bidirectional reflectance distribution function (BRDF)," "local binary pattern (LBP)," "modified census transform (MCT)," "local phase quantization (LPQ)," "vector quantization (VQ)," and "CENsus TRansform hISTogram (CENTRIST)," respectively . One drawback of the presented methods is the requirement of reasonably sharp input images.

13.2.2.3 Image Quality Analysis Based Methods

Recently, authors in [34] proposed a biometric liveness detection method for iris, fingerprint, and face images utilizing 25 image quality measures (21 full-reference measures and 4 non-reference measures). Nevertheless, all 25-image quality measures are required to attain acceptable results. In addition, in the proposed algorithm no face-specific information has been considered in devising informative features for face spoof detection. On the contrary, an image quality analysis based technique via four face specific features was proposed in [53], which uses image distortion analysis method. This method compared to procedure in [34] aims to improve generalization ability under cross-database scenarios.

13.2.2.4 Hardware Based Methods

Besides software face anti-spoofing methods, also several hardware-based face liveness detection techniques have been developed that are grounded in imaging technologies. Majority of the solutions in this category employ light spectrum that is outside of the visual spectrum (e.g., 3D depth [52], complementary infrared (CIR) or near infrared (NIR) images [56]) to compare the reflectance information of real faces and spoof materials. To this aim, a specific setup of LEDs and photo-diodes at two different wavelengths is used. Lately, authors in [30] carried out a study on thermal imaging for face anti-spoofing by collecting large database of thermal face images for real and spoofed access attempts. Also, it is commonly believed that multimodal biometric systems are natural anti-spoofing technique [2]. To this end, Chetty et al. [22] combined face and voice, and analyzed the correlation between the lips movement and the speech being produced. In particular, the system used a microphone and a speech analyzer for anti-spoofing. On the whole,

Table 13.1 A summary of the most representative face anti-spoofing techniques

Category	Features used	Attack type	Database	Error (%)
Motion analysis based	Motion detection using retinotopic grid [36]	Photo	Private DB	0.5
	Eye blink detection using Conditional Random Fields (CRF) [45]	Photo	Public NUAA DB	2.0
	Face motion detection using Optical Flow Lines (OFL) [13]	Photo	Private DB	1.0
	Context-based using correlation between face motion vs. background motion [25]	Photo, video	Public REPLAY-ATTACK DB	5.0
Texture analysis based	Face texture using Lambertian model [49]	Photo	Public NUAA DB	15.0
	Face texture using LBPs [42]	Photo, video	Public NUAA, PRINT-ATTACK, and YALE-RECAPTURED DBs	0.5
	Multi-scale LBP for 3D mask [31]	3D mask	3DMAD DB	2.00
Image quality analysis based	25 general image quality measures [34]	Photo, video	Public REPLAY-ATTACK DB	15.2
	Image distortion based quality measures [53]	Photo, video	Public MSU MFSD DB	7.41
Hardware based	Multimodality: face and voice [22]	Video	Public VidTIMIT and UCBN DBs	2.0
	Thermal images [30]	3D mask	Public I^2BVSD DB	13.0
	Reflectance in 3D [38]	3D mask	Private DB	5.50

though hardware-based solutions tend to provide better results and performances, they require extra piece of hardware thereby increasing the cost of the system. A summary with relevant features of the most representative works in face liveness detection is presented in Table 13.1.

13.2.3 Face Spoofing Databases

To assess the effectiveness of our proposed anti-spoofing technique, we considered two public face anti-spoofing databases: CASIA Face Anti-Spoofing Database (CASIA FASD) and Replay-Attack Database. These two datasets are the most

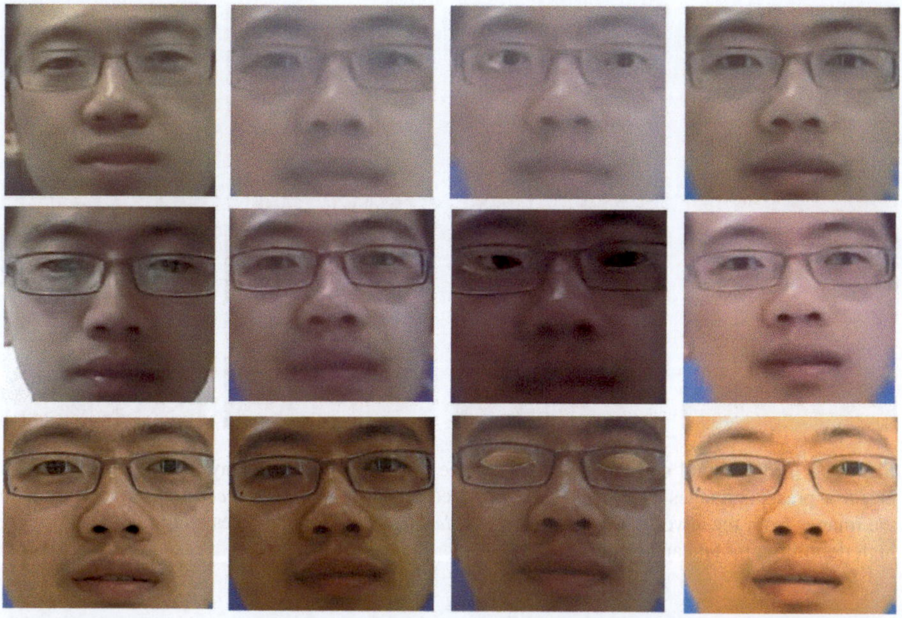

Fig. 13.6 Cropped and normalized example face images from the CASIA FASD. *From top to bottom*: low, normal, and high quality images. *From the left to the right*: real faces and the corresponding warped photo, cut photo, and video replay attacks

challenging face anti-spoofing benchmark databases that consist of recordings of real client accesses and various spoofing attack attempts captured with different imaging qualities, including mobile phones, webcams, and digital system cameras. In the following, a brief description of the two databases, is given.

13.2.3.1 CASIA Face-Anti-spoofing Database

The CASIA Face Anti-Spoofing Database (CASIA-FASD) [57] contains video recordings of genuine and fake faces (see, Fig. 13.6, for cropped and normalized example images). The real faces were recorded from 50 genuine subjects, whereas the fake faces were made from the high quality recordings of the genuine faces. Three fake face attacks were designed: *warped photo attacks*, i.e. facial motion simulated by bending (warping) a photograph, *cut photo attacks* (photographic masks), i.e. the eye regions were cut off and the attacker hides behind the mask and exhibits eye-blinking through the holes, and *video attacks*. Both of real accesses and attack attempts were recorded using three imaging qualities: *low*, *normal*, and *high*. The 50 subjects were divided into two subject-disjoint subsets for training and testing (20 and 30, respectively).

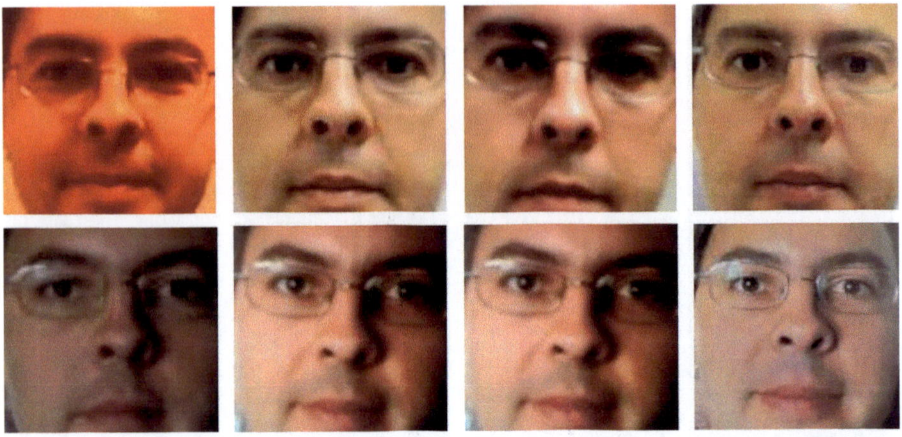

Fig. 13.7 Cropped and normalized example face images from the Replay-Attack Database. The *first row* presents images taken from the controlled scenario, while the *second row* corresponds to the images from the adverse scenario. *From the left to the right*: real faces and the corresponding high definition, mobile and print attacks

13.2.3.2 Replay-Attack Database

The Replay-Attack Database [23] consists of video recordings of real accesses and attack attempts to 50 clients (see, Fig. 13.7, for cropped and normalized example images). Using a built-in camera of a MacBook Air 13-in. laptop, a number of videos were recorded of each person in the database under two illumination conditions: *controlled*, i.e. uniform background and a fluorescent lamp was used to illuminate the scene, and *adverse*, i.e. nonuniform background and the day-light was the only source of illumination. Under the same conditions, a high resolution pictures and videos were taken for each person using a Canon PowerShot SX150 IS camera and an iPhone 3GS camera. These recordings were used to generate the fake face attacks.

Three types of attacks were designed: (1) *print attacks*, i.e. high resolution pictures were printed on A4 paper and displayed to the camera); (2) *mobile attacks*, i.e. high resolution pictures and videos were displayed on the iPhone 3GS screen; and (3) *high definition attacks*, i.e. the pictures and the videos were displayed on an iPad screen with resolution of 1024 by 768 pixels. According to the support used in presenting the fake face devices in front of the camera, two types of attacks were defined: *hand based attacks*, i.e. the attack devices were held by the operator and *fixed-support attacks*, i.e. the attack devices were set on a fixed support. For the evaluation, the 50 subjects were divided on three subject-disjoint subsets for training, development, and testing.

13.3 Case Study: Face Anti-spoofing Based on Color Texture Analysis

Texture analysis of gray-scale face images can provide sufficient means to reveal the recapturing artifacts of fake faces if the image resolution (quality) is good enough to capture the fine details of the observed face. However, if we take a close look at the cropped facial images of a genuine human face and corresponding fake ones in Fig. 13.8, it is basically impossible to explicitly name any textural differences between them because the input image resolution is not high enough.

To emulate the color perception properties of the human visual system, color mapping algorithms give a huge importance to the preservation of the spatially local luminance variations at the cost of the chroma information [16]. Human eye is indeed more sensitive to luminance than to chroma, thus fake faces still look very similar to the genuine ones when the same facial images are shown in color (see, Fig. 13.8). However, if only the corresponding chroma component is considered, some characteristic differences can be already noticed. While the gamut mapping and other artifacts cannot be observed clearly in the gray-scale or color images, they are very distinctive in the chrominance channels. Thereby, color texture analysis of the chroma images can be used for detecting these gamut mapping and other (color) reproduction artifacts.

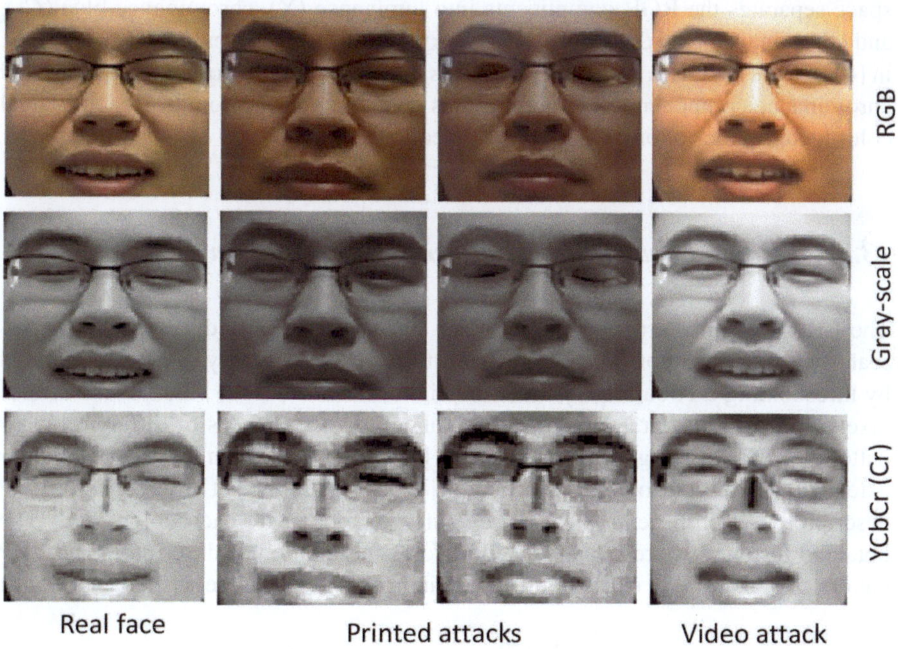

Fig. 13.8 Example of a genuine face and corresponding print and video attacks in RGB, gray-scale and YCbCr colour space

Inspired by the aforementioned observations, we propose, in this work, a new face anti-spoofing method based on color texture analysis. The color Local Binary Patterns (LBP) descriptor proposed in [27] is used to extract the joint color-texture information from the face images. In this descriptor, the uniform LBP histograms are extracted from the individual image bands. Subsequently, these histograms are concatenated to form the final descriptor. To gain insight into which color space is more discriminative to distinguish real face from fake ones, we considered three color spaces, namely *RGB*, *HSV*, and *YCbCr*.

13.3.1 Color Spaces

RGB is the most used color space for sensing, representation, and displaying of color images. However, its application in image analysis is quite limited due to the high correlation between the three color components (red, green, and blue) and the imperfect separation of the luminance and chrominance information.

In this work, we considered two other color spaces to explore the color texture information in addition to RGB: the *HSV* and the *YCbCr*. Both of these color spaces are based on the separation of the luminance and the chrominance information. In the *HSV* color space, the hue and the saturation dimensions define the chrominance of the image while the value dimension corresponds to the luminance. The *YCbCr* space separates the RGB components into luminance (Y), chrominance blue (C_b), and chrominance red (C_r). More details about these color spaces can be found, e.g., in [41]. In the *HSV* and *YCbCr* color spaces, the representation of the luminance and chrominance components is different, thus they can provide complementary facial color texture descriptions for spoofing detection.

13.3.2 Color Local Binary Pattern Representation

The LBP descriptor proposed by Ojala et al. [44] is a highly discriminative gray-scale texture descriptor. For each pixel in an image, a binary code is computed by thresholding a circularly symmetric neighborhood with the value of the central pixel. Finally, a histogram is created to collect the occurrences of different binary patterns. LBP was originally intended to handle gray-scale images but was later extended to exploit also color information. In [27], a simple yet efficient color LBP descriptor was proposed. The LBP operator is applied on each color band. The obtained histograms are concatenated to form the final color descriptor. The LBP pattern of a pixel (x, y) extracted from the image band (i) can be written as follows:

$$\mathrm{LBP}^{(i)}_{P,R}(x,y) = \begin{cases} \sum_{n=0}^{P-1} \delta(r_n^{(i)} - r_c^{(i)}) \times 2^n \text{ if } U^{(i)} <= 2 \\ P(P-1) + 2 \text{ otherwise} \end{cases} \tag{13.1}$$

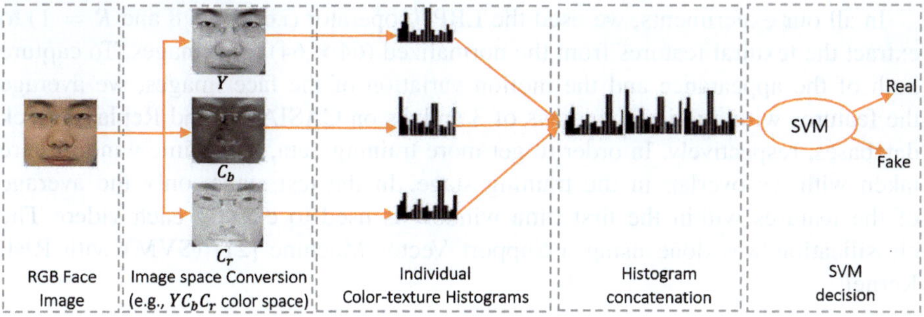

Fig. 13.9 Illustration of the proposed face anti-spoofing approach

where

$$U^{(i)} = \left|\delta(r_{p-1}^{(i)} - r_c^{(i)}) - \delta(r_0^{(i)} - r_c^{(i)})\right| + \sum_{n=1}^{P} \left|\delta(r_n^{(i)} - r_c^{(i)}) - \delta(r_{n-1}^{(i)} - r_c^{(i)})\right| \tag{13.2}$$

and $\delta(x) = 1$ if $x >= 0$, otherwise 0. r_c and $r_n (n = 0, \ldots, P-1)$ denote the intensity values of the central pixel (x, y) and its P neighborhood pixels located at the circle of radius R $(R > 0)$.

To summarize, let I be a face image represented in a color space S ($S \in \{RGB, HSV, YCbCr\}$) and, let $H_s^{(i)}, \{i = 1 : M\}$ be its uniform LBP histogram extracted from the M channel of the space S. The color LBP features of the image I represented in the space S can be defined by

$$H_s = [H_s^{(1)} \ldots H_s^{(M)}] \tag{13.3}$$

To detect spoofing attacks, the color LBP features extracted from face images are fed into a Support Vector Machine (SVM) classifier as shown in Fig. 13.9.

13.3.3 Experimental Setup

In our experiments, we followed the defined protocols of the two databases which allows a fair comparison against the state-of-the-art. On CASIA-FA database, the model parameters are trained and tuned using fourfold subject-disjoint cross-validation on the training set and the results are reported in terms of Equal Error Rate (EER) on the test set. Replay-Attack database provides also a separate validation set for tuning the model parameters. Thus, the results are given in terms of EER on the development set and the Half Total Error Rate (HTER) on the test set.

In all our experiments, we used the $LBP^{(i)}_{8,1}$ operator (i.e., $P = 8$ and $R = 1$) to extract the textural features from the normalized (64 × 64) face images. To capture both of the appearance and the motion variation of the face images, we average the features within a time windows of 3 and 4 s on CASIA-FA and Replay-Attack databases, respectively. In order to get more training data, these time windows are taken with 2 s overlap in the training stage. In the test stage, only the average of the features within the first time window is used to classify each video. The classification was done using a Support Vector Machine [21] (SVM) with RBF Kernel.

13.3.4 Results and Discussions

In Fig. 13.10, we demonstrate the effect of the color information by measuring the similarity between the local binary pattern (LBP) descriptions extracted from a genuine face (Real face 1), another genuine face (Real face 2), and two fake face

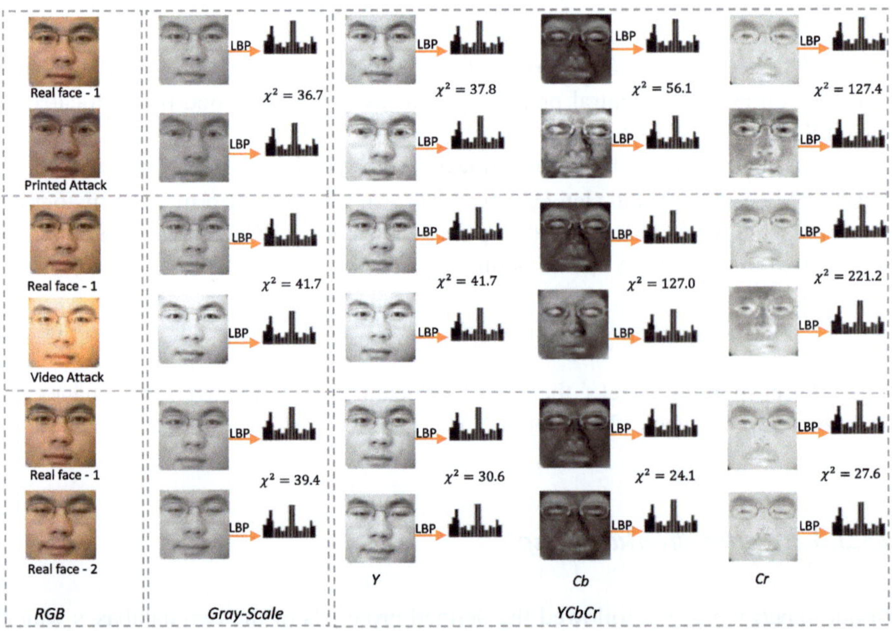

Fig. 13.10 The similarity between the holistic LBP descriptions extracted from genuine faces and fake ones. The original RGB images are shown on the *left*. In the *middle*, the similarity between the LBP descriptions extracted from the gray-scale images is presented. The similarity between the LBP descriptions extracted from the different color channels in the YCbCr color space is presented on the *right*

images (Printed Attack and Video Attack) of the same person. The similarity is measured using the chi-square distance:

$$d_{\chi^2}(H_x, H_y) = \sum_{i=1}^{N} \frac{(H_x(i) - H_y(i))^2}{H_x(i) + H_y(i)}, \tag{13.4}$$

where H_x and H_y are two LBP histograms with N bins. In addition to its simplicity, the chi-square distance is shown to be effective to measure the similarity between two LBP histograms. From Fig. 13.10, we can observe that the chi-square distance between gray-scale LBP histograms of the genuine face and the printed fake face is smaller than the one between two genuine face images. Moreover, the difference in similarity between the texture descriptions of genuine faces and the chi-square distance between the genuine face and the video attack is not significant. It is worth noting, however, that similarity measured with pure chi-square distance does not necessarily indicate that there are no intrinsic disparities in the gray-scale texture representation that could be exploited for face spoofing detection.

Tables 13.2 and 13.3 present the results of different LBP based color texture descriptions and their gray-scale counterparts. From these results, we can clearly see that the color texture features significantly improve the performance compared to the gray-scale LBP-based countermeasure. When comparing the different color spaces, *YCbCr* based representation yields to the best overall performance. The color LBP features extracted from the *YCbCr* space improves the performance on CASIA-FA and Replay-Attack databases by 64.5 and 81.4 %, respectively, compared to the gray-scale LBP features.

Table 13.2 The performance in terms of EER% of the gray-scale LBP and the color LBP descriptors on CASIA FA database

Method	1	2	3	4	5	6	7
Gray-scale-LBP	16.5	17.2	23.0	24.7	16.7	27.0	24.8
RGB-LBP	17.9	17.6	11.1	18.0	12.5	14.3	16.1
HSV-LBP	12.2	10.3	14.4	11.0	11.1	**7.8**	13.4
YCbCr-LBP	9.0	**9.5**	**6.0**	9.9	8.8	12.1	8.8
YCbCr+HSV-LBP	**7.8**	10.1	6.4	**7.5**	**5.4**	8.1	**6.2**

The bold values means the best permanences

Table 13.3 The performance of the gray-scale LBP and the color LBP descriptors on Replay-Attack database

Method	EER%	HTER%
Gray-scale-LBP	15.3	15.6
RGB-LBP	5.0	6.6
HSV-LBP	6.4	7.0
YCbCr-LBP	0.7	3.3
YCbCr+HSV-LBP	**0.4**	**2.9**

The bold values means the best permanences

From Table 13.2, we can also observe that the features extracted from the *HSV* color space seem to be more effective against video attacks than those extracted from the *YCbCr* color space. Thus, we studied the benefits of combining the two color texture representations by fusing them at feature level. The color LBP descriptions from the two color spaces were concatenated, thus the size of the resulting histogram is $59 \times 3 \times 2$. The results in Tables 13.2 and 13.3 indicate that a significant performance enhancement is obtained, thus confirming the benefits of combining the different facial color texture representations.

Table 13.4 compares the performance of our proposed countermeasure against the state-of-the-art face anti-spoofing methods. From this table, we can notice that our method outperforms the state-of-the-art results on the challenging CASIA-FA database, and yields in very competitive results on the Replay-Attack database.

To gain insight into the generalization capabilities of our proposed method, we conducted a cross-database evaluation. In these experiments, the countermeasure was trained and tuned with one database (CASIA-FA or Replay-Attack) and then tested on the other database. The results of these experiments are summarized in Table 13.5.

Table 13.4 Comparison between the proposed countermeasure and state-of-the-art methods

Method	Replay-Attack		CASIA
	EER%	HTER%	EER%
IQA based [32]	–	–	32.4
CDD [55]	–	–	11.8
DOG (baseline)[57]	–	–	17.0
Motion+LBP [37]	4.5	5.1	–
Motion [29]	11.6	11.7	26.6
LBP [23]	13.9	13.8	18.2
LBP-TOP [28]	7.8	7.6	10.6
Motion mag [14]	**0.2**	**0.0**	14.4
Deep learning[43]	6.1	2.1	7.3
Proposed method	0.4	2.9	**6.2**

The bold values means the best permanences

Table 13.5 Inter-Test results in terms of HTER% on CASIA and Replay-Attack databases

Method	Replay-Attack		CASIA	
	Dev	Test	Train	Test
Motion [29]	50.2	50.2	47.7	48.2
LBP [29]	44.9	47.0	57.3	57.9
LBP-TOP [29]	48.9	50.6	60.0	61.3
Motion-Mag [14]	50.0	50.20	43.8	50.3
Deep-learning [43]	48.2	48.8	45.7	45.4
Our method (SVM-RBF)	22.5	20.6	47.5	43.9
Our method (SVM-linear)	**17.7**	**16.7**	**38.6**	**37.6**

The bold values means the best permanences

In the first experiment, we evaluated the performance on CASIA-FA database while training and tuning the countermeasure on the Replay-Attack database. Table 13.5 reports a HTER values of 47.5 and 43.9 % on the training and the testing sets, respectively. While, in the second experiment, when the countermeasure is trained and tuned on CASIA-FA database and then tested on Replay-Attack database, the HTER values on the development and the test sets are 22.5 and 20.6 %, respectively. Although these results are very competitive to those of state-of-the-art methods, specially on Replay-Attack database, they are still degraded compared to the intra-test results (when the countermeasure is trained and tested on the same database).

Complex classifiers, like SVM-RBF, might be more sensitive to over-fitting than simpler classification schemes. The two face anti-spoofing benchmark datasets are rather small and the variations in the provided data are also limited which increases the chance of over-fitting with powerful texture features and complex classification schemes. Inspired by the observations, e.g., in [23, 29], we proposed to mitigate this problem by using linear SVM instead of SVM-RBF

The experiments using linear SVM models show very interesting results compared to those of the SVM-RBF models. On the CASIA-FA database, the H TER values on the training and the testing sets have been reduced to 38.6 and 37.6 %, respectively. On Replay-Attack database, the HTER values have been reduced to 17.7 and 16.7 % (on the development and the test sets, respectively), that are comparable to those obtained with the gray-scale LBP descriptor in the intra-test evaluation (15.3 and 15.6 %).

The model optimized on the Replay-Attack dataset is not able to generalize as well as the model based on the CASIA FA. The reason behind this is that the CASIA FA dataset contains more variations in the collected data (e.g., imaging quality and proximity between the camera and face) compared to the Replay-Attack database. Therefore, the model optimized for Replay-Attack database has difficulty to perform well in the new environmental conditions. One way to deal with this problem is to train countermeasure with a joint training set by combining the train set of both databases, as described in [29].

13.4 Research Opportunities

In this section, we will discuss some open issues and some research directions for face anti-spoofing.

13.4.1 *Generalization to Unknown Attacks*

Many visual cues for non-intrusive spoofing detection have been already explored and impressive results have been reported on individual databases. However, the varying nature of spoofing attacks and acquisition conditions makes it impossible to predict how single anti-spoofing techniques, e.g. facial texture analysis, can generalize the problem in real-world applications. Moreover, we cannot foresee all possible attack scenarios and cover them in databases because the imagination of the human mind always finds out new tricks to fool existing systems. As one obviously cannot foresee all possible types of fake faces, one-class approach modeling only the genuine facial texture distribution could be a promising direction. This has been successfully applied in voice anti-spoofing [1], for instance.

13.4.2 Fusion of Countermeasures

It is reasonable to assume that no single superior technique is able to detect all known, let alone unseen, spoofing attacks. Therefore, the problem of spoofing attacks should be broken down into attack-specific subproblems that are solvable if a proper combination of complementary countermeasures is used. In this manner, a network of attack-specific spoofing detectors could be used to construct a flexible anti-spoofing framework in which new techniques can be easily integrated to patch the existing vulnerabilities in no time when new countermeasures appear. This obviously raises the problem of fusing different spoofing countermeasures which has not been studied much besides the algorithms [48, 51, 54] proposed within the context of the IJCB 2011 competition on countermeasures to 2D facial spoofing attacks [20].

13.4.3 Biometric System and Countermeasures

A spoofing countermeasure is usually not designated to operate as a stand-alone procedure but in a joint operation with a recognition system. However, most works on anti-spoofing tend to focus only on the spoofing detection part hence omitting to integrate the countermeasure into a recognition system. In practice, integrating the countermeasure will affect the performance of the recognition system. While it will reduce its vulnerability to spoofing attacks, it may also decrease the recognition performance. The open issue is how to combine the spoofing countermeasure and the biometric recognition so that the combined biometric recognition system is robust to spoofing and does not suffer from reduced recognition accuracy [24].

13.4.4 Contextual Information

Face images captured from face spoofs may visually look very similar to the images captured from live faces. Thus, face spoofing detection may be difficult to perform based on only single face image or a relatively short video sequence. Depending on

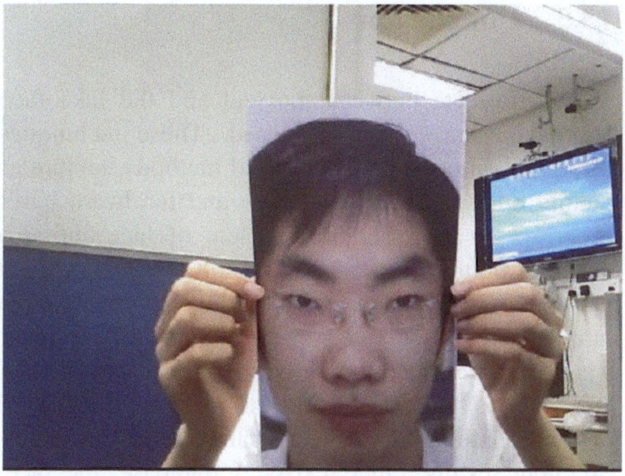

Fig. 13.11 Sample showing the possibility of using the context information in face anti-spoofing

the imaging and fake face quality, it is nearly impossible, even for humans, to tell the difference between a genuine face and a fake one without any scene information or unnatural motion or facial texture patterns. However, we can immediately notice if there is something suspicious in the view, e.g. if someone is holding a video display or a photograph in front of the camera. Therefore, scenic cues can be exploited for determining whether display medium is present in the observed scene as shown in the Fig. 13.11.

13.4.5 Challenge-Response Approach

Liveness and motion analysis based spoofing detection is rather difficult to perform by observing only spontaneous facial motion during short video sequences. This problem can be simplified by prompting the user to do some specific random action or challenge (such as a smiling and moving the head to the right). The user's response (if any) will provide liveness evidences. This is called challenge-response approach for spoofing detection. The drawback of such an approach is that it requires user cooperation, thus making the authentication process a time-consuming. Another advantage of non-intrusive techniques is that from challenge-response based countermeasures it is rather easy to deduce which liveness cues need to be fooled. For instance, the request for uttering words suggests that analysis of synchronized lip movement and lip reading is utilized, whereas rotating head in a certain direction reveals that the 3D geometry of the head is measured. For non-intrusive approaches, it is usually not known which countermeasures are used, thus the system might be harder to deceive [46].

13.5 Conclusions

To discriminate between the real and the fake face images many methods and approaches have been proposed. These techniques can be grouped into four categories: motion analysis based methods, texture analysis based methods, image quality analysis based methods, and hardware based methods. In this work, we proposed to approach the problem of face anti-spoofing from the color texture analysis point of view. Instead of the gray-scale texture features used in the previous works, we investigated the importance of the joint color texture features in discriminating between the real and the fake images. The LBP texture features were extracted from the individual image channels of the: *RGB*, *HSV*, and *YCbCr* color spaces than concatenated to form the final descriptors. Extensive experiments on two challenging spoofing databases, CASIA-FA and Replay-Attack, showed excellent results. On CASIA-FA database, the face representation based on the combination of *HSV* and *YCbCr* color spaces beat the state-of-the-art. Furthermore, in our inter-database evaluation, the proposed approach showed very promising generalization capabilities. As future work, more experiments should be conducted in order to get more insight into the color texture based face anti-spoofing and to derive problem-specific facial color representations. In addition to further improve the performance of the color texture features, we have also discussed some open issues and some research directions that should be investigated in the future to enhance the robustness of the biometric systems against the spoofing attacks.

References

1. F. Alegre, A. Amehraye, N. Evans, A one-class classification approach to generalised speaker verification spoofing countermeasures using local binary patterns, in *Sixth International Conference Biometrics: Theory, Applications and Systems (BTAS)* (2013), pp. 1–8
2. Z. Akhtar, Security of multimodal biometric systems against spoof attacks. PhD Thesis, University of Cagliari, Italy, 2012
3. Z. Akhtar, G.L. Luca Foresti, Face spoof attack recognition using discriminative image patches. J. Electr. Comput. Eng. **2016**, Article ID 4721849 (2016), p. 14
4. Z. Akhtar, G. Fumera, G.L. Marcialis, F. Roli, Robustness analysis of likelihood ratio score fusion rule for multimodal biometric systems under spoof attacks, in *IEEE International Carnahan Conference on Security Technology (ICCST)* (2011), pp. 1–8
5. Z. Akhtar, G. Fumera, G.L. Marcialis, F. Roli, Robustness evaluation of biometric systems under spoof attacks, in *16th International Conference on Image Analysis and Processing (ICIAP 2011)* (2011), pp. 159–168
6. Z. Akhtar, S. Kale, N. Alfarid, Spoof attacks on multimodal biometric systems, in *International Conference on Information and Network Technology (ICINT)*, Chennai, India, April 29–30, 2011, pp. 46–51
7. Z. Akhtar, G. Fumera, G.L. Marcialis, F. Roli, Evaluation of multimodal biometric score fusion rules under spoof attacks, in *5th IAPR International Conference on Biometrics (ICB)*, New Delhi (2012), pp. 402–407
8. Z. Akhtar, G. Fumera, G.L. Marcialis, F. Roli, Evaluation of serial and parallel multibiometric systems under spoofing attacks, in *IEEE Fifth International Conference on Biometrics: Theory, Applications and Systems (BTAS)* (2012), pp. 283–288

9. Z. Akhtar, C. Micheloni, C. Piciarelli, G.L. Foresti, MoBio_LivDet: mobile biometric liveness detection, in *Proceedings of 11th IEEE International Conference on Advanced Video and Signal Based Surveillance (AVSS)* (2014), pp. 187–192
10. Z. Akhtar, C. Micheloni, G.L. Foresti, Biometric liveness detection: challenges and research opportunities. IEEE Secur. Priv. **13**(5), 63–72 (2015)
11. A. Anjos, S. Marcel, Counter-measures to photo attacks in face recognition: a public database and a baseline, in *Proceedings of IEEE International Joint Conference on Biometrics (IJCB)* (2011), pp. 1–6
12. J. Bai, T. Ng, X. Gao, Y. Shi, Is physics-based liveness detection truly possible with a single image? in *IEEE International Symposium on Circuits and Systems* (2010), pp. 3425–3428
13. W. Bao, H. Li, N. Li, W. Jiang, A liveness detection method for face recognition based on optical-flow field, in *IEEE International Conference on Image Analysis and Signal Processing* (2009), pp. 233–236
14. S. Bharadwaj, T.I. Dhamecha, M. Vatsa, R. Singh, Computationally efficient face spoofing detection with motion magnification, in *2013 IEEE Conference on Computer Vision and Pattern Recognition Workshops* (2013), pp. 105–110. ISSN: 2160-7508
15. B. Biggio, Z. Akthar, G. Fumera, G.L. Marcialis, F. Roli, Robustness of multi-modal biometric verification systems under realistic spoofing attacks, in *2011 International Joint Conference on Biometrics (IJCB), Washington, DC* (2011), pp. 1–6
16. N. Bonnier, Contribution to spatial gamut mapping algorithms. Doctoral dissertation, Télécom ParisTech, 2008
17. Z. Boulkenafet, J. Komulainen, A. Hadid, Face anti-spoofing based on color texture analysis, in *IEEE International Conference on Image Processing (ICIP)* (2015), pp. 2636–2640
18. Z. Boulkenafet, J. Komulainen, A. Hadid, Face spoofing detection using colour texture analysis. IEEE Trans. Inf. Forensics Secur. **11**(8), 1818–1830 (2016)
19. J.D. Bustard, J.N. Carter, M.S. Nixon, A. Hadid, Measuring and mitigating targeted biometric impersonation. IET Biom. **3**(2), 55–61 (2014)
20. M. Chakka et al., Competition on counter measures to 2-d facial spoofing attacks, in *2011 International Joint Conference on Biometrics (IJCB)* (IEEE, New York, 2011), pp. 1–6
21. C.C. Chang, C.J. Lin, LIBSVM: a library for support vector machines. ACM Trans. Intell. Syst. Technol. (TIST) **2**(3), 27 (2011)
22. G. Chetty, M. Wagner, Liveness detection using cross-modal correlations in face-voice person authentication, in *Proceedings of the Annual Conference of the International Speech Communication Association (Interspeech)* (2005), pp. 2181–2184
23. I. Chingovska, A. Anjos, S. Marcel, On the effectiveness of local binary patterns in face anti-spoofing, in *International Conference of the Biometrics Special Interest Group (BIOSIG)* (2012), pp. 1–7
24. I. Chingovska, A. Anjos, S. Marcel, Anti-spoofing in action: joint operation with a verification system, in *Proceedings of the IEEE Conference on Computer Vision and Pattern Recognition Workshops* (2013), pp. 98–104
25. I. Chingovska et al., The 2nd competition on counter measures to 2D face spoofing attacks, in *Proceedings of International Conference on Biometrics (ICB)* (2013), pp. 1–6
26. I. Chingovska, A.R. dos Anjos, S. Marcel, Biometrics evaluation under spoofing attacks. IEEE Trans. Inf. Forensics Secur. **9**(12), 2264–2276 (2014)
27. J.Y. Choi, K.N. Plataniotis, Y.M. Ro, Using colour local binary pattern features for face recognition, in *Proceedings of IEEE International Conference on Image Processing (ICIP)* (2010), pp. 4541–4544
28. T. de Freitas Pereira, A. Anjos, J.M. De Martino, S. Marcel, LBP-TOP based countermeasure against face spoofing attacks, in *Asian Conference on Computer Vision Workshops* (2012), pp. 121–132
29. T. de Freitas Pereira, A. Anjos, J.M. De Martino, S. Marcel, Can face anti-spoofing countermeasures work in a real world scenario? in *International Conference on Biometrics (ICB), 2013* (2013), pp. 1–8

30. T.I. Dhamecha, A. Nigam, R. Singh, M. Vatsa, Disguise detection and face recognition in visible and thermal spectrums, in *Proceedings of IEEE International Conference on Biometrics (ICB)*, (2013), pp. 1–6
31. N. Erdogmus, S. Marcel, Spoofing face recognition with 3D masks. IEEE Trans. Inf. Forensics Secur. **9**(7), 1084–1097 (2014)
32. J. Galbally, S. Marcel, Face anti-spoofing based on general image quality assessment, in *International Conference on Pattern Recognition (ICPR)* (2014), pp. 1173–1178
33. J. Galbally, S. Marcel, J. Fierrez, Biometric antispoofing methods: a survey in face recognition. IEEE Access **2**, 530–1552 (2014)
34. J. Galbally, S. Marcel, J. Fierrez, Image quality assessment for fake biometric detection: application to iris, fingerprint and face recognition. IEEE Trans. Image Process. **23**, 710–724 (2014)
35. D. Gragnaniello, G. Poggi, C. Sansone, L. Verdoliva, An investigation of local descriptors for biometric spoofing detection. IEEE Trans. Inf. Forensics Secur. **10**(4), 849–863 (2015)
36. K. Kollreider, H. Fronthaler, J. Bigun, Non-intrusive liveness detection by face images. Image Vis. Comput. **27**(3), 233–244 (2009)
37. J. Komulainen, A. Hadid, M. Pietikainen, A. Anjos, S. Marcel, Complementary countermeasures for detecting scenic face spoofing attacks, in *Proceedings of International Conference on Biometrics* (2013), pp. 1–7
38. N. Kose, J.L. Dugelay, Mask spoofing in face recognition and countermeasures. Image Vis. Comput. **32**(10), 779–789 (2014)
39. Y. Li, X. Tan, An anti-photo spoof method in face recognition based on the analysis of fourier spectra with sparse logistic regression, in *Chinese Conference on Pattern Recognition (CCPR)* (2009)
40. J. Li, Y. Wang, T. Tan, A.K. Jain, Live face detection based on the analysis of fourier spectra, in *Biometric Technology for Human Identification* (2004), pp. 296–303
41. R. Lukac, K.N. Plataniotis, *Color Image Processing: Methods and Applications* (CRC Press, Boca Raton, 2006)
42. J. Maatta, A. Hadid, M. Pietikainen, Face spoofing detection from single images using texture and local shape analysis. IET Biom. **1**(1), 3–10 (2012)
43. D. Menotti et al., Deep representations for iris, face, and fingerprint spoofing detection. IEEE Trans. Inf. Forensics Secur. **10**(4), 864–879 (2015)
44. T. Ojala, M. Pietikäinen, T. Mäenpää, Multiresolution gray-scale and rotation invariant texture classification with local binary patterns. IEEE Trans. Pattern Anal. Mach. Intell. **24**(7), 971–987 (2002)
45. G. Pan, L. Sun, Z. Wu, S. Lao, Eyeblink-based anti-spoofing in face recognition from a generic webcamera, in *International Conference on Computer Vision (ICCV)* (2007), pp. 1–8
46. G. Pan, L. Sun, Z. Wu, Y. Wang, Monocular camera-based face liveness detection by combining eye-blink and scene context. J. Telecommun. Syst. **47**, 215–225 (2009)
47. N.K. Ratha, J.H. Connell, M.R. Bolle, An analysis of minutiae matching strength, in *International Conference on Audio- and Video-based Biometric Person Authentication (AVBPA)* (2001), pp. 223–228
48. W.R. Schwartz, A. Rocha, H.P. Edrini, Face spoofing detection through partial least squares and low-level descriptors, in *International Joint Conference on Biometrics (IJCB)* (2011), pp. 1–8
49. X. Tan, Y. Li, J. Liu, L. Jiang, Face liveness detection from a single image with sparse low rank bilinear discriminative model, in *11th European Conference on Computer Vision (ECCV)*, vol. 6316 (2010), pp. 504–517
50. L. Thalheim, J. Krissler, P.M. Ziegler, Biometric access protection devices and their programs put to the test. C'T (2002)
51. R. Tronci, D. Muntoni, G. Fadda, M. Pili, N. Sirena, G. Murgia, F. Roli, Fusion of multiple clues for photo-attack detection in face recognition systems, in *International Joint Conference on Biometrics (IJCB)* (2011), pp. 1–6
52. T. Wang, J. Yang, Z. Lei, S. Liao, S.Z. Li, Face liveness detection using 3d structure recovered from a single camera, in *International Conference on Biometrics (ICB)* (2013), pp. 1–6

53. D. Wen, H. Han, A.K. Jain, Face spoof detection with image distortion analysis. IEEE Trans. Inf. Forensics Secur. **10**(4), 746–761 (2015)
54. J. Yan, Z. Zhang, Z. Lei, D. Yi, S.Z. Li, Face liveness detection by exploring multiple scenic clues, in *12th International Conference on Control Automation Robotics and Vision (ICARCV)* (2012), pp. 188–193
55. J. Yang, Z. Lei, S. Liao, S.Z. Li, Face liveness detection with component dependent descriptor, in *Proceedings of International Conference on Biometrics (ICB)*, (2013) pp. 1–6
56. Z. Zhang, D. Yi, Z. Lei, S.Z. Li, Face liveness detection by learning multispectral reflectance distributions, in International Conference on Face and Gesture (2011), pp. 436–441
57. Z. Zhang, J. Yan, S. Liu, Z. Lei, D. Yi, S.Z. Li, A face anti-spoofing database with diverse attacks, in *International Conference on Biometrics (ICB)* (2012), pp. 26–31

Chapter 14
Biometric Template Protection: A Systematic Literature Review of Approaches and Modalities

Mulagala Sandhya and Munaga V.N.K. Prasad

14.1 Introduction

The increased requirements of security in authentication applications made exponential growth in biometric based systems. Conventional authentication systems like token or password could not meet these security requirements because they can be lost, forgotten, or misplaced. Hence, recent years have seen an epidemic growth in identity theft [1]. Biometric recognition refers to the automated recognition of individuals based on their anatomical and/or behavioral characteristics. The most commonly used anatomical biometric traits are a fingerprint, face, iris, palmprint, hand geometry, hand vein, finger vein, voice, etc. Behavioral biometric characteristics include keystroke dynamics, gait, etc. [2, 3]. Typically, the biometric characteristics are obtained by using sensors. During enrollment stage, these characteristics are passed to the feature extraction module and stored as templates. During verification/authentication stage, the query biometric is processed in the same way. Now, the query template is compared with the enrolled template that yields in acceptance or rejection. Exposure of biometric traits in diverse applications makes a serious compromise on user's privacy. In near future, biometric authentication is applied to networks for remote user authentication [4]. In such applications, the server stores the biometric information of users called templates. The distinctiveness and permanence of biometrics made it

---

M. Sandhya (✉)
Institute for Development and Research in Banking Technology, Hyderabad, India

School of Computer and Information Sciences, University of Hyderabad, Hyderabad, India
e-mail: msandhya.phd@gmail.com

M.V.N.K. Prasad
Institute for Development and Research in Banking Technology, Hyderabad, India
e-mail: mvnkprasad@idrbt.ac.in

R. Jiang et al. (eds.), *Biometric Security and Privacy*, Signal Processing for Security Technologies, DOI 10.1007/978-3-319-47301-7_14

advantageous over conventional authentication systems. But, the same artifact made the permanent loss of identity in biometric systems [5]. The model of a biometric system obliterates many hazards that lead to several security risks and privacy threats. To address this issue, biometric templates are protected using various schemes that are classified as cancelable biometrics, biometric cryptosystems, hybrid methods, and Homomorphic Encryption based methods [6].

14.1.1 Evolution of Biometric Template Protection Schemes

The evolution of BTP schemes is shown in Fig. 14.1. In 1996, Soutar et al. [7] built a biometric cryptosystem named it as Mytec1. Later, authors enhanced its version to Mytec2 in 1998 [8]. Juels and Wattenberg [9] initiated a novel method named **"Fuzzy Commitment Scheme"** that combines error correction codes with cryptography for protecting biometric template. Later in the early 2000s, Ratha et al. [10] provided the importance of **"non-invertible transforms."** The native biometric is transformed and cannot be recovered thereby increasing security. In 2002, **"Fuzzy Vault Scheme"** was presented by Juels and Sudan [11]. A key is locked using the biometric features resulting in a vault. In the same year 2002, Feng et al. [12] proposed the key generating cryptosystem using **"Quantization Scheme."** In 2003, Goh et al. [13] proposed a salting method called **"Biohashing"** which transforms the biometric features. The transform selected need not necessarily be non-invertible. **"Fuzzy Extractors"** were developed in 2004 by Dodis et al. [14]. This was termed as **"Secure Sketch"** by Sutcu et al. [15]. The secure sketch approach is similar to robust hashing and similarity-preserving transformations. In 2006, Boult et al. [16, 17] developed revocable biotokens that combine both cryptosystems and features transformation, hence called **"Hybrid Method."** In 2009, **"Homomorphic Encryption"** is introduced into biometrics by Ye et al. [18]. A detailed discussion of the aforementioned techniques using a single biometric trait or multiple biometric traits (multi-modal BTP) is given in Sects. 14.5–14.8.

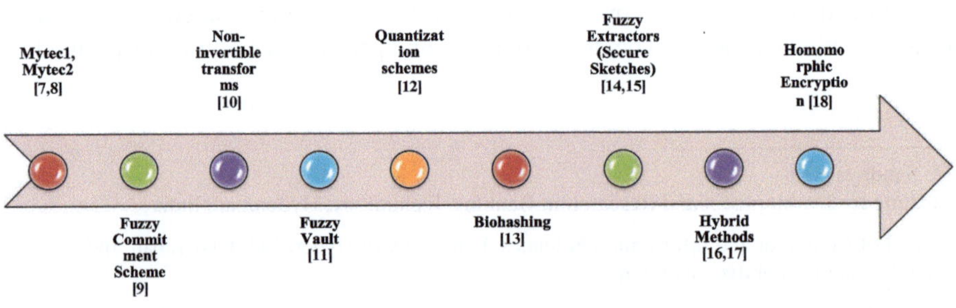

Fig. 14.1 Evolution of biometric template protection schemes

14.1.2 Motivation and Contributions

The contributions for this chapter are presented below:

- BTP provides methods and solutions for protection of various biometric templates without degrading the accuracy of the recognition system. Consequently, this study emphasizes various techniques for biometric template protection based on different categories viz. Biometric Cryptosystems, Cancelable Biometrics, Hybrid methods, and Homomorphic Encryption based methods .
- By considering ongoing research in biometric template protection techniques, we noticed the necessity of systematic survey in this area. Hence, the available research based on a broad and methodical search in the existing databases is summarized.
- A detailed investigation has been carried out to study various techniques for biometric template protection.
- Extraction outcomes, research implications, and future research directions in the area of BTP are presented.

14.1.3 Chapter Organization

This chapter is structured as follows: Earlier and related surveys on BTP methods are presented in Sect. 14.2. Our review technique used to find available research, research questions, and search criteria is given in Sect. 14.3. The classification of approaches for BTP resulted from the methodical search is described in Sect. 14.4. Section 14.5 gives the taxonomy of Biometric Cryptosystems. Section 14.6 describes the taxonomy of Cancelable Biometrics. Section 14.7 presents the Hybrid methods. Section 14.8 presents the BTP methods based on Homomorphic Encryption. Section 14.9 presents the results and analysis of our Systematic Literature Review (SLR). Section 14.10 gives the research implications and future research directions. Finally, the conclusions are furnished in Sect. 14.11.

14.2 Earlier Reviews

Uludag et al. [19] presented various methods for biometric cryptosystems. This paper focused on giving details about the existing research in the "key generation" and "key binding" schemes. The existing work is classified and compared regarding privacy protection, practicality, sensitivity to invariance, and security. The authors also discussed how naive attacks can happen by presenting biometric sample at biometric matcher. Authors suggested the need for devising effective multi-biometric strategies without performance degradation.

Jain et al. [1] provided a classification of vulnerabilities of a biometric system. The authors classified the approaches to BTP into "salting," "non-invertible transforms," "key binding biometric cryptosystem," and "key generating biometric cryptosystem." The review is made on what entities are stored? and how the intra-user variations are handled? The implementation of each of the template protection approach is discussed along with advantages and limitation of each of the approach. Also, a discussion of the countermeasures to be taken to address the vulnerabilities is presented. This paper gives a rigorous review of 92 articles by covering categories of template security approaches, the implementation, and pitfalls in the methods proposed by existing work. The authors conclude that "there is no best approach for BTP." The selection of an approach depends on application scenario, requirements, biometric trait, and its representation by features, the extent of intra-user variations.

Rathgeb et al. [20] provided a survey on "state-of-the-art" approaches of biometric cryptosystems and cancelable biometrics. This paper discusses the experimental results of "fuzzy commitment schemes," "fuzzy vault schemes" in the literature. The biometric cryptosystems designed using password hardening, biohashing, and their operation modes were presented. As per the author's perspective, the cancelable biometrics were classified into "non-invertible transforms" and "biometric salting." The experimental results of the existing cancelable biometric design are summarized. A brief comparison between the cancelable biometrics and biometric cryptosystems was made, and an outlook for futures perspective is given. The review also discussed the possible attacks on biometric cryptosystems and cancelable biometric systems.

Rane et al. [21] presented a tutorial on secure biometric architectures. A high-level comparison is made on four secure biometric architectures namely "secure sketch, fuzzy commitment, secure computation, and cancelable biometrics." An illustration of the "design tension between maximizing security and privacy is presented."

Patel et al. [22] provided an overview of available cancelable schemes. The classification is made based on registration based or registration-free, signal domain or feature domain. The techniques involving tokens, bio convolving, salting, palmhash, and palmphasor that need a special matcher are discussed. The techniques that work with conventional matcher such as minimum distance graphs, image warping, non-invertible transforms, block-remapping, and curtailed circular convolution are presented. The performance of various schemes is summarized using "Genuine Acceptance Rate (GAR), False Acceptance Rate (FAR), False Rejection Rate (FRR) and Equal Error Rate (EER)." The possible attacks against cancelable biometric templates are discussed.

Natgunanathan et al. [23] presented a comprehensive overview of "privacy-preserving biometric schemes (PPBSs)" in the last decade. This paper presented various drawbacks of existing PPBSs and gives guidance for "future privacy-preserving biometric research." The functional mechanisms of popular PPBSs are considered for presenting the state-of-the-art literature on these mechanisms.

14.3 Our Systematic Literature Review Technique

In contrast with the conventional expert literature review, a "Systematic Literature Review (or simply a systematic review, structured literature review or SLR) is a literature review focused on a research question that tries to identify, classify, select and synthesize all high-quality research evidence relevant to that question" [24]. An SLR is a structured process that identifies, evaluates, and interprets existing evidence of research that is pertinent to a specific research query. Hence, it follows a predefined search category. An SLR provides a framework for locating new research tasks. Figure 14.2 shows the steps of systematic reviews. SLR discovers gaps in ongoing research and recommends areas for further examination. Although the initial void for future work was filled in earlier reviews, this survey differs from them in the following ways:

- This survey paper included emerging approach for novices called homomorphic encryption.
- There is no SLR on BTP till date making research difficult in identifying gaps and latest research trends. For this, we conduct an SLR on BTP and identify the approaches, modalities, fusion of modalities (multi-modal), and hybrid methods for protecting biometric templates.
- We answered the research questions shown in Table 14.1 according to the guidelines of SLR [24].
- We included recently presented methods for protecting biometric templates and percentage of BTP categories, the percentage of BTP schemes developed for each biometric trait, year-wise distribution of BTP schemes.
- We listed out active research communities in developing BTP schemes.
- All online available datasets on which BTP schemes are evaluated are listed.

.

14.3.1 Sources of Information for Collecting Studies on BTP

The following electronic databases are used for searching studies for this SLR on BTP as guided in [24]:

1. IEEE Xplore Digital Library (ieeexplore.ieee.org)
2. ACM Digital Library (dl.acm.org)
3. ScienceDirect (www.sciencedirect.com)
4. SpringerLink (www.spingerlink.com)
5. Taylor & Francis Online (www.tandfonline.com)
6. IET Digital Library (digital-library.theiet.org)
7. Google Scholar (scholar.google.co.in)
8. IEICE Knowledge Discovery (https://search.ieice.org/bin/search.php?lang=E)
9. WorldScientific (www.worldscientific.com)
10. Hindawi Publishing Corporation (www.hindawi.com).

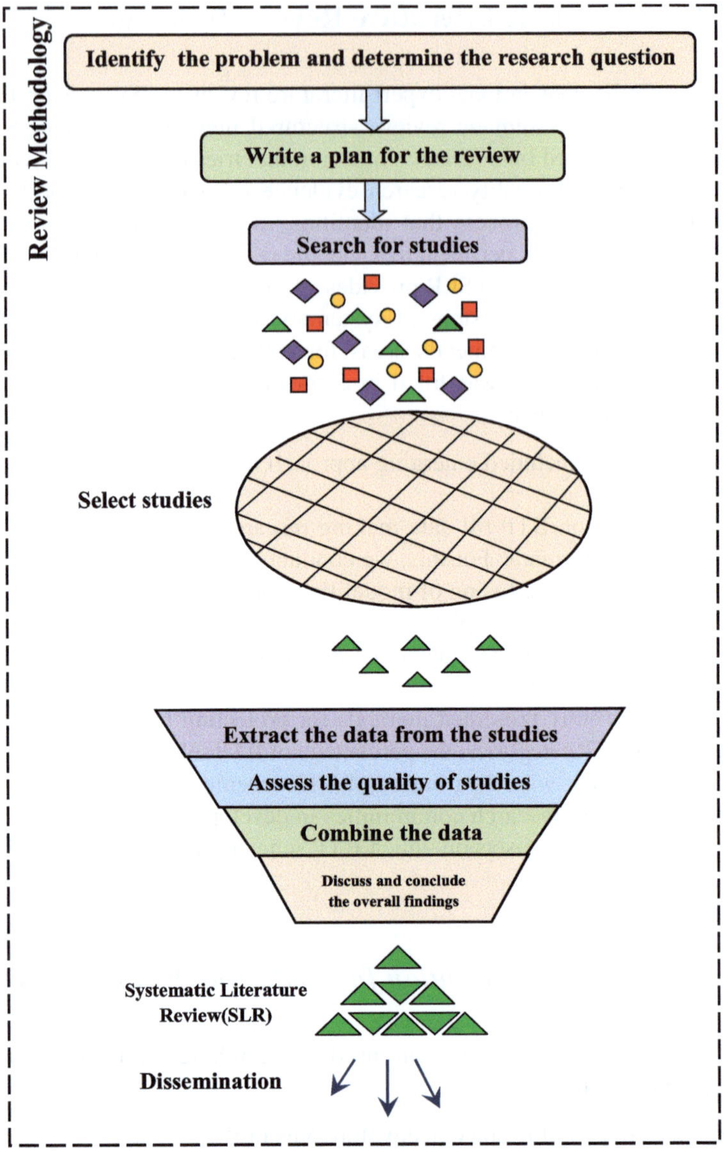

Fig. 14.2 Review methodology based on [24, 25]

14.3.2 Search Criteria

We improve the review scope of the SLR by inclusion and exclusion criteria given in Table 14.2. The various search strings used in this SLR are given in Table 14.3. This SLR included quantitative and qualitative research articles written in English

Table 14.1 Research questions and motivation

S. No	Research questions	Motivation
1	What is the current state-of-the-art of BTP methods?	The main aim of this review is: (i) To getinsight on the state of art of biometric template protection methods. This study presented and compared the different types and subtypes of BTP. (ii) To make future research through relative investigation of existing research (iii) To understand the research gap that needs to be addressed and to find the future directions in this area of research
2	What are the main research motivations for BTP?	
3	How to identify and classify approaches to BTP?	
4	Which modality's templates are protected using BTP schemes?	
5	What are the potential attacks against BTP schemes?	
6	How to understand the existing protection methods for better security and performance?	
7	What are the online available datasets for evaluating BTP methods?	
8	What are the open issues and challenges for protecting biometric templates	

Table 14.2 Inclusion and exclusion criteria

	Criteria	Prenominal
Inclusion	Peer reviewed research articles that propose methods, solution, experiments and evaluates different biometric template protection methods	These papers contain significant content and quality through peer review. Different biometric template protection solutions are studied
Exclusion	Books, book chapters, thesis, editorials, short papers, editorials, abstracts, non-English scripts, white papers	The relevant papers of authors of the thesis, books, thesis are searched in journals, conferences and incorporated. The white papers are situational and the studies that do not provide any reasonable solutions are excluded

from 2005–2016. We presented the research articles in journals, conferences, workshops, symposium, magazines, and technical reports from industry.

14.3.3 Initial and Final Selection of Studies

The primary research on BTP was started in 1998. But extensive work on BTP was carried out after 2005. Hence, our SLR reports research work after 2005. We searched the studies based on inclusion and exclusion criteria described in Table 14.2. Further, we also filtered the papers at different stages as shown in

Table 14.3 Search strings used for SLR

S. No	Keywords	Synonyms	Year	Content type
1	Biometric template protection	Biometric template security, secure biometrics, template protection, biometric information protection, biometric security	2005–2016	Journal, conference, workshop, symposium, transactions, magazine, symposium, technical reports from industry
2	Cancelable biometrics	Cancelable templates, non-invertible templates, changeable biometrics, biohashing, cancelable biometrics, secure hashing salting feature transformation		
3	Biometric cryptosystem	Multi-biometric cryptosystem, biometric encryption, helper data scheme, bio cryptosystem		
4	Biometric key binding	Fuzzy commitment, fuzzy vault, key binding cryptosystem		
5	Revocable biometrics	Biotokens		
6	Secure sketch	Fuzzy extractors		
7	Biometric key generation	Private key generation, key generating cryptosystem		
8	Homomorphic encryption	Privacy-preserving encryption		

Fig. 14.3. Our methodical search fetched 358 articles at stage-1, filtering articles by title returned 280 articles at stage-2, by abstract and conclusion returned 210 articles at stage-3, by full-text returned 148 articles at stage-4, and finally filtering by common challenges and investigation returned 119 articles.

14.4 Classification of Approaches for BTP

14.4.1 Architecture Overview

The architecture of BTP as given in ISO/IEC 24745: (2011) [26] is provided in Fig. 14.4. In enrollment phase, a "Pseudonymous Identifier (PI)" and "Auxiliary Data (AD)" are obtained from a biometric sample by encoding it with "Pseudonymous Identifier Encoder (PIE)." The pair {PI, AD} is called "Renewable

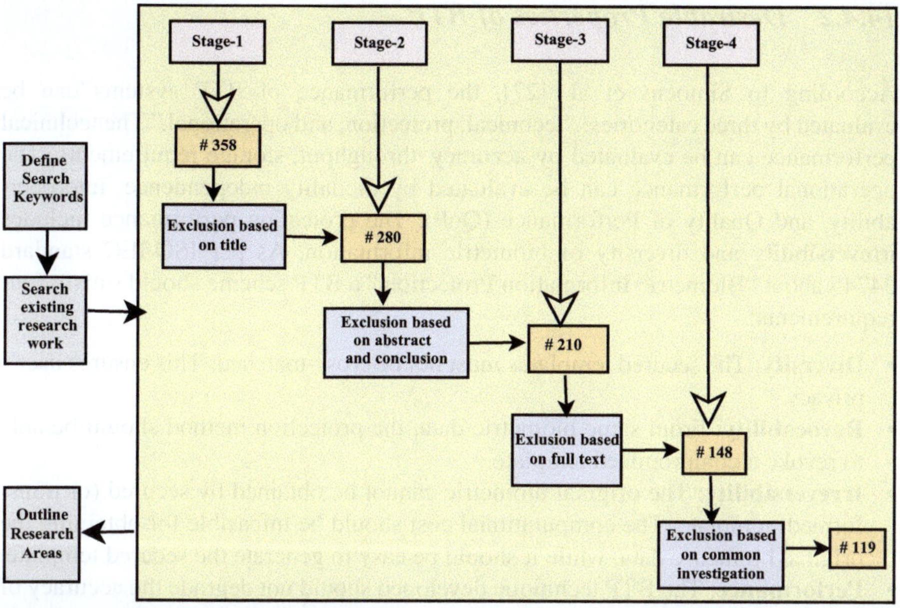

Fig. 14.3 Stages in selection of studies

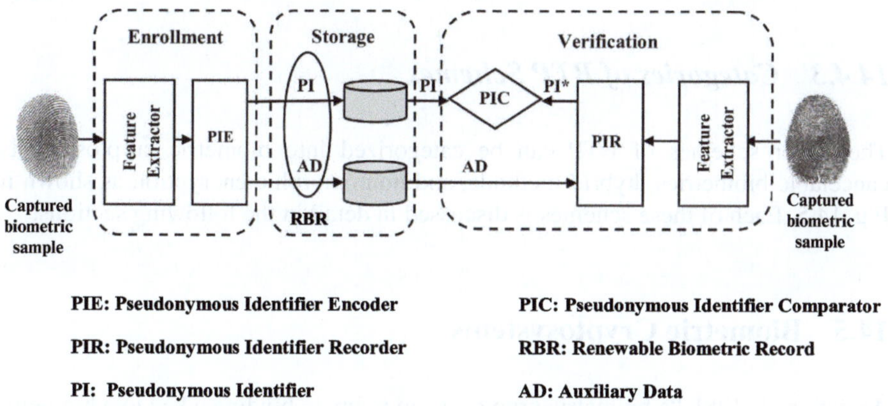

Fig. 14.4 The architecture of BTP based on ISO/IEC 24745: (2011) [26]

Biometric Record (RBR)." During verification phase, the extracted biometric feature and AD are given to "Pseudonymous Identifier Recorder (PIR)." Hence, a "Pseudonymous Identifier (PI*)" is produced following the "Pseudonymous Identity Comparator (PIC)" that compares PI and PI*, which results in a similarity score.

14.4.2 Desirable Properties of BTP

According to Simoens et al. [27], the performance of BTP systems can be evaluated by three categories: "technical, protection, and operational." The technical performance can be evaluated by accuracy, throughput, storage requirements. The operational performance can be evaluated by modality independence, interoperability, and Quality of Performance (QoP). The protection performance includes irreversibility and diversity of biometric information. As per ISO/IEC standard 24745 about "Biometric Information Protection," a BTP scheme should satisfy four requirements:

- **Diversity**: The secured templates must not be cross-matched. This ensures user's privacy.
- **Revocability**: From same biometric data, the protection method should be able to revoke a compromised template.
- **Irreversibility**: The original biometric cannot be obtained by secured (or transformed) template. The computational cost should be infeasible for obtaining the original biometric data, while it should be easy to generate the secured template.
- **Performance**: The BTP technique developed should not degrade the accuracy of the recognition system.

14.4.3 Categories of BTP Schemes

The major schemes of BTP can be categorized into biometric cryptosystems, cancelable biometrics, hybrid methods, and homomorphic encryption as shown in Fig. 14.5. Each of these schemes is discussed in detail in the following sections.

14.5 Biometric Cryptosystems

As defined in [28], "a biometric cryptosystem refers to binding a key to a biometric feature or generating a key from the biometric feature." In a biometric cryptosystem, "helper data" is used to bind/generate keys [1]. The biometric cryptosystems are classified into the "key binding" and "key generation" systems based on how the "helper data" is derived.

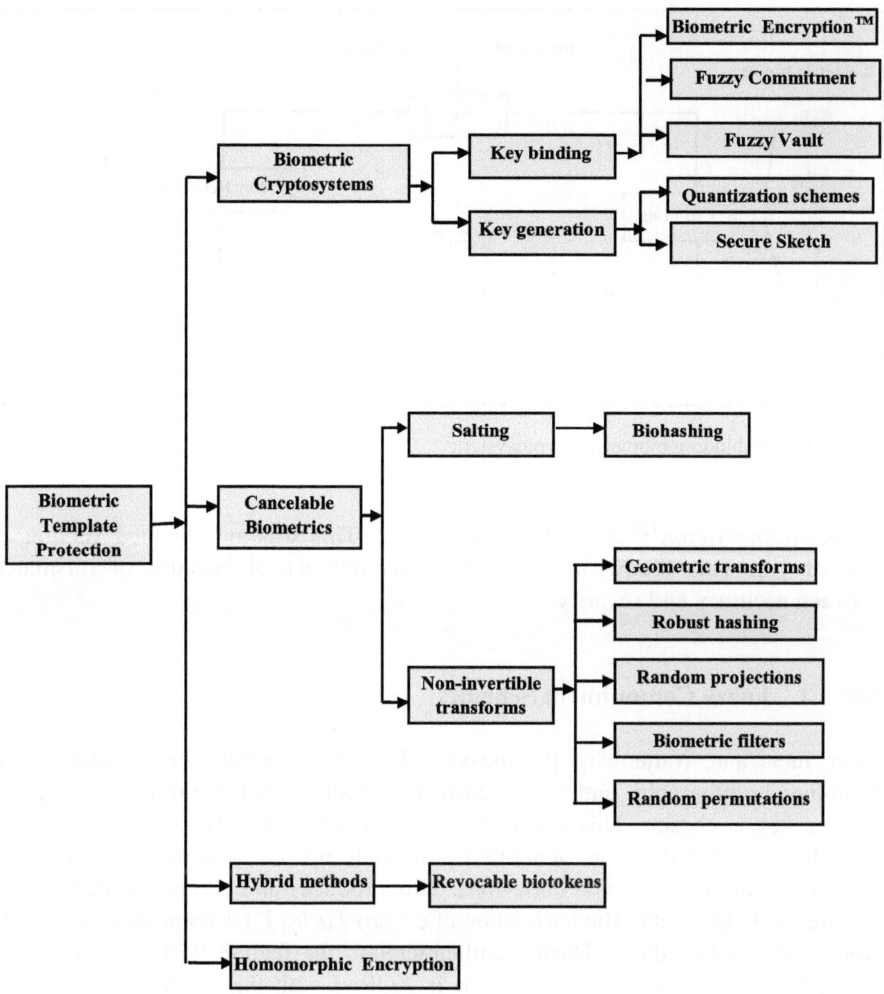

Fig. 14.5 A hierarchical taxonomy of biometric template protection schemes

14.5.1 Key Binding Biometric Cryptosystems

In a "key binding" cryptosystem, a user-specific chosen key is bound to a biometric template to obtain helper data. The combination of key and the biometric template bound is stored as a secure template, often termed as "helper data." By a suitable decoding attempt, the keys are acquired from the helper data [19]. An overview of "key binding" cryptosystem is shown in Fig. 14.6. The first "key binding" biometric cryptosystem was developed for fingerprints by Soutar et al. [8], named

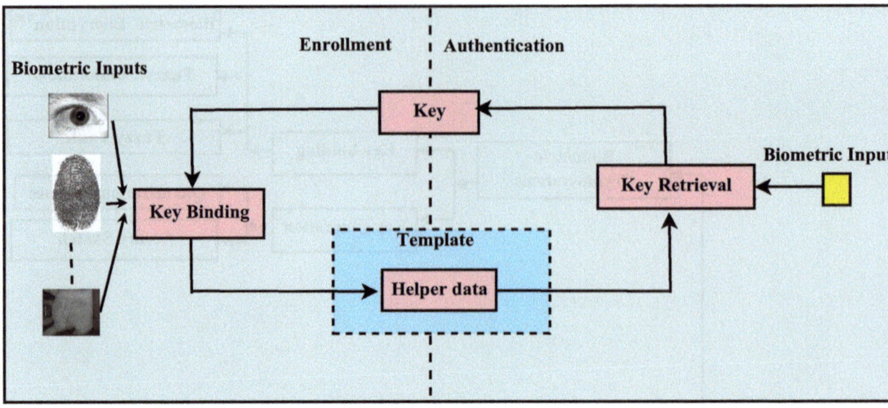

Fig. 14.6 Key binding biometric cryptosystem

BiometricEncryption^{TM} and called as Mytec2. This was an extended version of Mytec1 [7]. This biometric encryption was impractical because of mismatch between accuracy and security.

14.5.1.1 Fuzzy Commitment Schemes

Later Juels and Wattenberg [9] introduced **"fuzzy commitment scheme"** that combines cryptography and "error correcting codes." In the enrollment stage, a random key is chosen. This key is encoded using "error correcting codes." Now, a random codeword "c" is generated. An XOR operation is done between the biometric feature vector and codeword. This produces an encrypted template and is stored as helper data. The hash value of c ("*say hash(c)*") is computed and stored along with "helper data." During authentication, the feature vector of the query image is presented to the decoder and is XORed with the stored helper data to get the codeword c^1. If $hash(c) = hash(c^1)$, then $c = c^1$ and hence the user is considered to be genuine and accepted else treated as an imposter and rejected. The generic framework of "Fuzzy Commitment Scheme" is shown in Fig. 14.7. The summary of "key binding" cryptosystems using "fuzzy commitment scheme" is listed in Table 14.4.

Hao et al. [29] developed a technique to combine cryptography to the iris biometric. A binary string is generated from iris codes. The authors coined a "two-layer error correction" method that combines "Hadamard" and "Reed–Solomon" codes. Veen et al. [30] integrated helper data system with face biometric. In Zheng et al. [31] other than using error correcting codes in "fuzzy commitment scheme," error-tolerant lattice functions are used. Simulation is done on iris plants. Bringer et al. [32] introduced a min-sum decoding of "fuzzy commitment scheme" for iris codes. Reed–Muller codes are used to build a matrix from the iris codes.

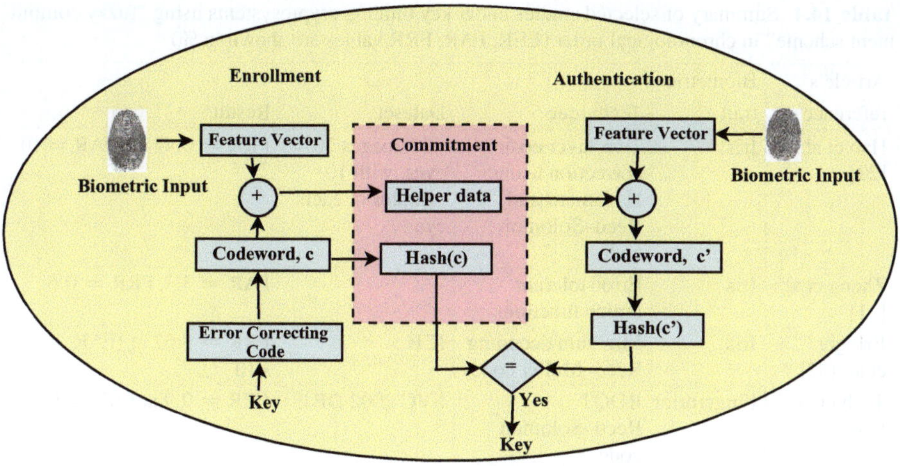

Fig. 14.7 "Fuzzy commitment scheme"

The authors extended their work in Bringer et al. [33] by detecting the suitable error correcting code regarding a given database of biometric data. Teoh et al. [34] built a "Randomized Dynamic Quantization Transformation Technique (RDQT)" for fingerprint data using the "fuzzy commitment scheme." Mairorana et al. [35] built a user-adaptive error correction method for online signatures. In Meng et al. [36] authors built a "near-infrared (NIR)" face biometric by enhancing BioHash algorithm. An NXOR mask is inputted to the "error correcting code." Rathgeb et al. [37] provided a methodical construction of using "fuzzy commitment schemes" for iris by making an analysis of error distributions.

In Lu et al. [38], the face features based on "Principal Component Analysis (PCA)" are integrated with "fuzzy commitment scheme." A binarization algorithm for online signatures based on "fuzzy commitment scheme" was applied in Mairorana et al. [39]. This method is based on function based signature verification and achieved better accuracy than feature based representation methods. Nandakumar et al. [40] proposed "Binarized Phase Spectrum (BiPS)" for fingerprints. A binary string is obtained from minutiae points by quantizing Fourier phase spectrum. The BiPS representation is secured by "fuzzy commitment scheme" employing turbo codes. Ignatenko et al. [48] investigated the privacy leakage in "fuzzy commitment schemes" in four scenarios, namely "memoryless totally symmetric, memoryless input-symmetric, memoryless, and stationary ergodic of biometric data statistics". Rathegeb et al. [41] used intra-class error analysis to rearrange iris codes. Rathgeb et al. [42] presented a feature level fusion method for "fuzzy commitment schemes" on iris data. The binary data is rearranged and combined in a way that error correction abilities are utilized properly. The authors presented a statistical attack for iris "fuzzy commitment schemes" in Rathgeb et al. [49, 50]. The authors examined the effect of iris blur and noise to "fuzzy commitment schemes" in Rathgeb et al. [51]. Zhou et al. [52, 53] show that secrets from iris codes can be

Table 14.4 Summary of selected studies under key binding cryptosystems using "fuzzy commitment scheme" in chronological order (EER, FAR, FRR values are shown in %)

Article's reference	Biometric trait	Technique	Dataset	Result
Hao et al. [29]	Iris	Two-layer error correction using Hadamard and Reed–Solomon codes	700 images 70 eyes, with 10 samples of each eye	FRR = 0.47 at (FAR = 0)
Zheng et al. [31]	Iris	Error-tolerant lattice functions	–	FAR = 3.3, FRR = 0.6
Bringer et al. [32]	Iris	Min-sum decoding Reed–Muller codes	ICE	FRR = 5.62 at (FAR $<10^{-5}$)
Teoh et al. [34]	Fingerprint	RDQT Reed–Solomon code	FVC 2002 DB1	FRR = 0.9 at FAR = 0
Mairona et al. [35]	Online signature	Adaptive selection of error correction	MCYT	FRR = 13.07 at FAR = 4
Bringer et al. [33]	Iris fingerprint	Min-sum decoding Reed–Muller codes	ICE, CASIA, FVC 2000	For ICE, FRR = 5.62 at (FAR $<10^{-5}$), For CASIA, FRR = 6.65 at FAR = 0, for FVC 2000 FRR = 2.73 at FAR = 5.53
Meng et al. [36]	Face	Enhance BioHash NXOR mask	NIR face	FRR = 7.99 at FAR = 0.11
Rathgeb et al. [37]	Iris	Block level ECC	CASIAv3	FRR = 4.64 at FAR = 0
Lu et al. [38]	Face	PCA gray codes	CMU PIE	FRR = 0.47 at FAR = 0
Mairo et al. [39]	Online signature	Signature binarization algorithm	MCYT	EER = 1.43
Nandakumar et al. [40]	Fingerprint	BiPS turbo codes	FVC 2002 DB1, DB2	DB1 FNMR = 16.2 at FMR = 0, EER = 2.1, DB2 FNMR = 12.6 at FMR = 0, EER = 2.1
Rathgeb et al. [41]	Iris	Intra-class error analysis Hadamarad codes	CASIA v3	FRR = 4.92 at FAR = 0
Rathgeb et al. [42]	Iris	Reliability balanced feature level fusion Hadamard codes	CASIA v3	FRR = 4.58 at FAR<0.01
Li et al. [43]	Fingerprint	Minutia triplets combining ECC	FVC 2002 DB2	FRR = 4.85 at FAR = 0

(continued)

Table 14.4 (continued)

Article's reference	Biometric trait	Technique	Dataset	Result
Imamverdiyev et al. [44]	Fingerprint	Minutia texture descriptors	FVC 2002 DB2	FRR = 95.3 at FAR = 0
Yuan et al. [45]	Face and ear	Feature level fusion using PCA BCH encoding	ORL face USTB ear3	EER = 6.1
Billeb et al. [46]	Voice	Gaussian mixture universal background Hadamard code	Text independent digit corpus	FNMR=0.9764 at FMR=0
Wang et al. [47]	Face and iris	Score level fusion Aczel–Alsina T-norm	CASIA iris NVIE face	EER = 1.163×10^{-3}

retrieved by Markov model in an iris fuzzy commitment. The authors measure the privacy and secrecy of iris "fuzzy commitment schemes" using "information theoretical metrics" in Zhou et al. [54].

Kelkboom et al. [55] provided an explanation for large deviation in key sizes for "fuzzy commitment schemes." Li et al. [43] developed a minutiae triplet binary feature method for fingerprints. Using the binary string as input to "fuzzy commitment scheme" various "error correction codes" are combined to design a fingerprint cryptosystem. Imamverdiyev et al. [44] designed a fingerprint cryptosystem using texture descriptors such as "Gabor filter, local binary pattern (LBP), and a local direction pattern (LDP), and their combinations." The texture descriptors are binarized and given as input to the "fuzzy commitment scheme." Yuan et al. [45] built a multi-modal cryptosystem by fusing human face and human ear at feature level using principal component analysis. In Billeb et al. [46], the authors used binarization technique to extract voice reference data from speaker models like Gaussian mixture models and universal background models. Wang et al. [47] presented a multi-biometric template protection technique using face and iris data. Aczel–Alsina triangular norm is used to perform a score level fusion of the proposed method. A Delaunay Neighbor Structure (DNS) was constructed for fingerprints in Sandhya and Prasad [56]. The bit string representation of DNS was encoded with BCH coding in the "fuzzy commitment scheme."

14.5.1.2 Fuzzy Vault Schemes

Another popular "key binding" cryptosystems are **"Fuzzy Vault Schemes"** shown in Fig. 14.8. Juels and Sudan [11] is the pioneer in building a "fuzzy vault scheme." The key idea is to lock a key k by an unordered set A yielding a vault V_A. During enrollment, a polynomial p encodes key k. A is projected onto p. Chaff points are added to secure the genuine points of p. During authentication, if another set B overlaps A, key k is reconstructed. Following Juels and Sudan [11] framework, many

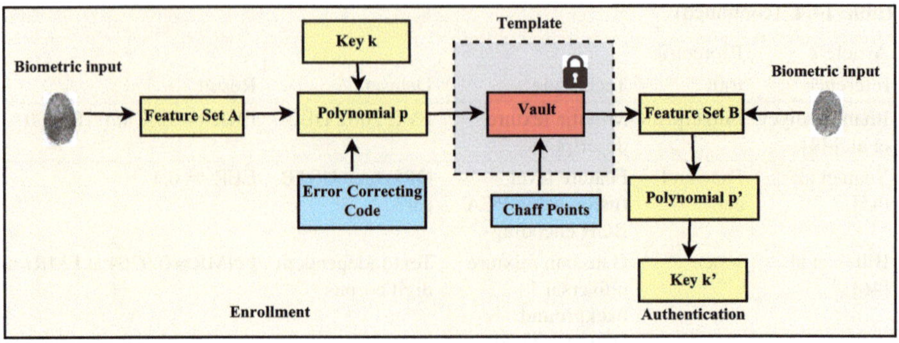

Fig. 14.8 "Fuzzy vault scheme"

trials to implement fuzzy vaults for fingerprints are proposed. The summary of "key binding" cryptosystems using "fuzzy vault scheme" is listed in Table 14.5.

Uludag et al. [57] used minutiae line based representation scheme for constructing the fuzzy vault for fingerprints by securing 128-bit AES keys in minutiae points of a fingerprint. This method derived experimental results and analysis for FVC 2002 DB2 databases in Uludag et al. [58]. Nandakumar et al. [59] built a fully automatic fingerprint cryptosystem using "fuzzy vault scheme." High curvature points obtained from orientation field of minutiae points are used as helper data to align the template and query minutiae. The authors extended their work in Nandakumar et al. [60] by using password for hardening the fuzzy vault. Nagar et al. [61] used minutiae descriptors that record orientation and ridge frequency data in a minutia's neighborhood for locking the polynomial evaluations in fuzzy vault.

Nyang et al. [62] built a fuzzy vault for face biometric using weighted features. Lee et al. [63] proposed a "fuzzy vault" system using local iris features. Numerous iris features from local regions are extracted and a clustering method is applied to get the exact values. A shift matching method is used for aligning the iris templates with iris data. Kelkboom et al. [64] presented multi-algorithm fusion at "feature, score, and decision-level" for 3D face images. Zhou et al. [65] generate artificial minutiae from a PIN or password for fingerprints. Now, the feature sets of genuine minutiae and artificial minutiae are fused to generate a protected template. Nagar et al. [66] developed a multi-modal cryptosystem that fuses iris, fingerprint, and face at feature level. This work used both "fuzzy vault" and "fuzzy commitment" methods for designing the cryptosystem. Berlekamp Massive algorithm is used to decode the vault. Liu et al. [67] developed a fuzzy vault for fingerprints using lorez chaotic system. Nguyen et al. [75] proposed a chaff point generation method for fingerprint "fuzzy vault" by dividing fingerprint image into cells. Each cell generates a random chaff point. Eskander et al. [76, 77] designed a machine learning approach that adapts user key size is presented for offline signature verification. Bhateja et al. [68] described a online signature based cryptosystem based on artificial neural network classifier. Firstly, the signature is broke into time slices, then a "weighted back propagation algorithm" is employed to train each slice. "AdaBoost algorithm" is employed to combine the decisions of different networks. Liu et al. [69] proposed

Table 14.5 Summary of selected studies under "key binding" cryptosystems using "fuzzy vault scheme" in chronological order (EER, FAR, FRR values are shown in %)

Article's reference	Biometric trait	Technique	Dataset	Result
Uludag et al. [57]	Fingerprint	CRC encoding and decoding	IBM-GTDB	FRR = 27 at FAR = 0
Uludag et al. [58]	Fingerprint	CRC encoding and decoding	FVC 2002 DB2	GAR=72.6 at FAR=0
Nandakumar et al. [59]	Fingerprint	High curvature points orientation field	FVC 2002 DB2, MSU-DBI	For FVC 2002 DB2, GAR = 97 at FAR = 0.24, for MSU-DBI, GAR=96.9 at FAR=0.16
Nandakumar et al. [60]	Fingerprint	Password hardening orientation field	FVC 2002 DB2, MSU-DBI	For FVC 2002 DB2, GAR = 90 at FAR = 0, for MSU-DBI, GAR = 80.6 at FAR = 0
Nyang et al. [62]	Face	Weighted features	–	–
Nagar et al. [61]	Fingerprint	Minutiae descriptors	FVC 2002 DB2, FVC 2002 DB2	For FVC 2002 DB2, GAR = 95 at FAR = 0.01
Lee et al. [63]	Iris	Multiple features from multiple local regions shift matching	BERC v1, CASIA v1, CASIA v3	For BERC v1, GAR = 99.925 at FAR = 0.035, for CASIA v1, GAR=83.408 at FAR=0.0235, for CASIA v3, GAR=91.171 at FAR=0.0275
Kelkboom et al. [64]	3D face	Multi-algorithm fusion	FRGC v2	EER = 2.58
Zhou et al. [65]	Fingerprint	Artificial minutiae	NIST SD14	FNMR=0.051625 at FMR=0
Nagar et al. [66]	Iris, fingerprint, face	Feature level fusion secure sketch	CASIA v1, FVC 2002 DB2, XM2VTS	GAR = 75
Liu et al. [67]	Fingerprint	Lorez chaotic system	FVC 2004 DB2	–
Liu et al. [69]	Palmprint	Multidimensional fuzzy vault	HA-BJTU	GAR = 80
Bhateja et al. [68]	Online signature	Artificial neural network based classifier AdaBoost algorithm	SVC 2004	GAR=17.78 at FAR=2.24

(continued)

Table 14.5 (continued)

Article's reference	Biometric trait	Technique	Dataset	Result
Rathgeb et al. [70]	Iris	Ridge features	IITD Iris	GAR = 97.21 at FAR = 0
Nguyen et al. [71]	Fingerprint	Ridge features	FVC 2002 DB1, DB2, FVC 2004 DB1, DB3	For FVC 2002 DB1, GAR = 86 at FAR = 0, for FVC 2002 DB2, GAR = 89 at FAR = 0
Tams et al. [72]	Fingerprint	Minutia orientation, frequency, local minutia structures	FVC 2002 DB1	GAR = 82 at FAR = 0
Bansal et al. [73]	Fingerprint	Hadamard transformation	FVC 2002 DB1	GAR = 77 at FAR = 0
Caili et al. [74]	Fingerprint	Pair-polar (P-P) structures	FVC 2000 DB1, FVC 2002, FVC 2004 DB2, FVC 2006 DB2, DB3	For FVC 2002 DB2, FRR = 11.50 at FAR = 0, EER = 3.37, for FVC 2006 DB2, FRR = 5.78 at FAR = 0, EER = 1.59

a multidimensional "fuzzy vault" that can handle intra-class variances. A subspace error-tolerant mechanism is embedded into "fuzzy vault scheme."

Rathgeb et al. [70] presented a "fuzzy vault" system by transforming bloom filter based representation of biometric templates into unordered set of integer values. Nguyen et al. [71] developed a fingerprint "fuzzy vault" based on ridge features information and noise generation technique to improve the security of the system. Tams et al. [78] redesigned a minutiae based "fuzzy vault" execution, which is resistant to attacks via record multiplicity. The authors extended their work in Tams et al. [72] by improving "fuzzy vault scheme" that employs minutiae orientation, frequency descriptors, and local minutia structures. Bansal et al. [73] applied Hadamard Transformation for fingerprint "fuzzy vault." Laftih et al. [79] presented an attack by altering the original data in fuzzy vault biometric cryptosystems. Leng et al. [80] built a hybrid cancelable palmprint cryptosystem by combining palmprint texture code and fuzzy vault. The 2DPalmHash code and fuzzy vault are mixed to protect palmprint templates. Li et al. [74] proposed fingerprint cryptosystem using pair-polar (P-P) minutiae structures.

14.5.2 Key Generating Biometric Cryptosystems

The "key generating" cryptosystems generate keys directly through biometric templates. The framework of "key generating" cryptosystems is shown in Fig. 14.9.

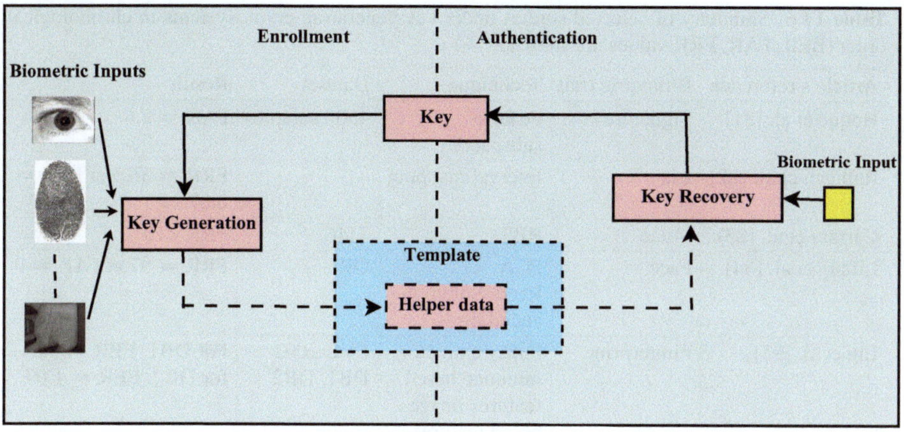

Fig. 14.9 Key generating biometric cryptosystem

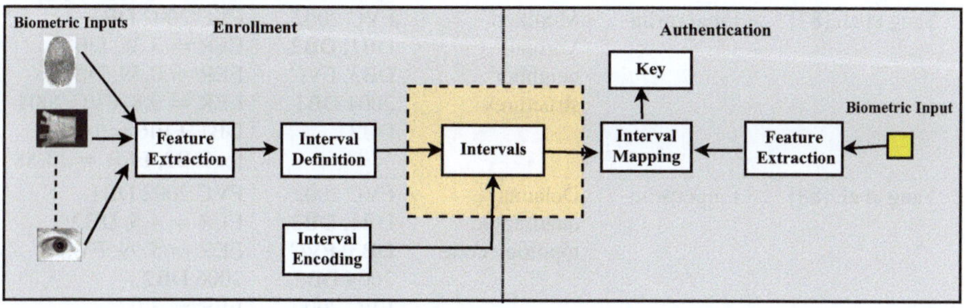

Fig. 14.10 Quantization scheme (adapted from [20])

One of the key generating cryptosystems is **Quantization scheme**. In this method, helper data are quantized to obtain stable keys. These schemes take feature vectors of several biometric samples and obtain intervals of the feature elements. The intervals are encoded and then stored in the form of helper data. During authentication, biometric features are calculated and mapped to the determined intervals. This results in a key as shown in Fig. 14.10. The summary of "key generating" cryptosystems is listed in Table 14.6.

Hoque et al. [81] generated keys from signatures by dividing feature space into subspaces and in turn into cells. Rathgeb et al. [82] applied interval mapping technique to generate cryptographic keys from iris characteristics. Carrara et al. [83] applied Randomized Biometric Templates (RBT) algorithm to voice biometrics. The voice signal is split into time quanta. Now, features are extracted from each quantum for evaluation. "Self-Organizing Map (SOM)" is used to model the speech signal. Lifang et al. [84] generated keys using face features. "Principal Component Analysis (PCA)" is applied to face image which generates a binary vector. Distinguishable bits are selected to form a key. The optimal bit order number

Table 14.6 Summary of selected studies under key generating cryptosystems in chronological order (EER, FAR, FRR values are shown in %)

Article's reference	Biometric trait	Technique	Dataset	Result
Hoque et al. [81]	Signature	Feature subspaces	4500 samples	FAR = 7.6
Rathgeb et al. [82]	Iris	Interval mapping	–	FRR = 36.5 at FAR = 0.07
Carrara et al. [83]	Voice	RBTs	TI46	FRR <10
Lifang et al. [84]	Face	PCA Reed–Solomon encoding	ORL	FRR = 97 at FAR = 0
Liu et al. [85]	Fingerprint	Fuzzy extractor minutiae based features image based features	FVC 2002 DB1, DB2	For DB1, EER = 3.81, for DB2, EER = 1.07
Ranjan et al. [86]	Fingerprint	Divide and conquer	–	–
Yang et al. [87]	Fingerprint	Modified Voronoi neighbor structures	FVC 2002 DB1, DB2, DB3, FVC 2004 DB1, DB2	FVC 2002 DB1 EER = 3.38, DB2 EER = 0.59, DB3 EER = 9.8, FVC 2004 DB1, EER = 16.52, for DB2, EER = 14.88
Yang et al. [88]	Fingerprint	Delaunay quadrangle topology code	FVC 2002 DB1, DB2, DB3, FVC 2004 DB2, FVC 2006 DB2, DB3	FVC 2002 DB1 EER = 4.5, DB2 EER = 5.99, FVC 2006 DB2, EER = 3.07

is saved in a lookup table. Liu et al. [85] developed a "key generating" system from fingerprints using fuzzy extractors and combining minutiae based and image based features. Ranjan et al. [86] used Divide and Conquer method for fingerprint images, and a bio key is generated using the minutia set. Yang et al. [87] presented a fingerprint cryptosystem by "modifying Voronoi Neighbor Structures (VNS)." Yang et al. [88] focused on using "Delaunay Quadrangle-based fingerprint authentication system using topology code for local registration and security enhancement."

14.6 Cancelable Biometrics

Cancelable biometric systems uses a transformation function that is dependent on a factor, called as key. Figure 14.11 represents the generic framework of a cancelable biometric system.

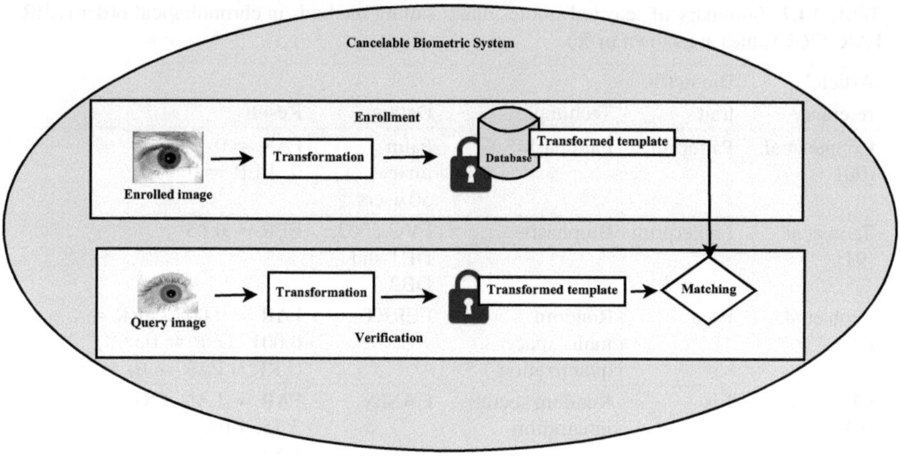

Fig. 14.11 Framework of cancelable biometric system

14.6.1 Salting

Salting is a BTP approach, where biometric features are transformed using an invertible function. In these systems, the key should be stored securely or recalled by the user for authentication as the transformation used is invertible. Since keys are introduced, a low FAR results for salting methods. It is easy to revoke the compromised templates by changing user-specific keys. In case such keys are used, they need to be presented at authentication [89]. The summary of salting methods is listed in Table 14.7.

Connie et al. [90] proposed a salting method called palmhash. The palmprint templates are hashed with pseudo-random keys. This palmhash code is stored in portable devices. In Jeong et al. [95], authors proposed an appearance based method for generating cancelable face templates. A normalization is applied between "Principal Component Analysis (PCA)" and "Independent Component Analysis (ICA)" coefficient vectors of face templates. A biometric hash is generated from the online signature in Yip et al. [94]. The comparison and analysis of palmhash code and palmphasor code were extensively discussed in Leng et al. [107, 108]. Teoh et al. [92] presented a salting method for face templates. Fisher discriminant analysis is used to extract discriminative projections of a face template and projected randomly in orthogonal directions. This random projection is applied with a binarization scheme to reduce intra-user variations. This maximizes the entropy of the template. A detailed description on analysis of Biohashing methods and its variants was given in Kong et al. [109]. Teoh et al. [91] introduced biophasor technique for fingerprints. This method is iterated mixing of pseudo-random number with fingerprint feature. This work considered stolen key scenario and was extended in Teoh et al. [96] for face templates and Teoh et al. [103] for speech recognition.

Chin et al. [93] proposed a salting method called S-iris encoding. The encoding method uses iterated "inner product" of iris feature and the pseudo-random number.

Table 14.7 Summary of selected studies under salting methods in chronological order (EER, FAR, FRR values are shown in %)

Article's reference	Biometric trait	Technique	Dataset	Result
Connie et al. [90]	Palmprint	Palmhash	Palm images of 50 users	FAR = 0, FRR = 0, EER = 0
Teoh et al. [91]	Fingerprint	Biophasor	FVC 2002, DB1 and DB2	EER = 0.65
Teoh et al. [92]	Face	Random multi-space quantization	FERET	FAR = 0.002, FRR = 0.001, EER = 0.002, (FRR at FAR = 0) = 0.43
Chin et al. [93]	Iris	Random secret integration	CASIA	FAR = 1.43, FRR = 3.75, EER = 2.59, d^1 = 6.85
Yip et al. [94]	Hand signature	Replaceable online signature hash	SVC	EER = 0, mean = 0.422, std = 0.009
Jeong et al. [95]	Face	Appearance based approach PCA and ICA	AR face	–
Teoh et al. [96]	Face	Multi-space random projections (MRP)	FERET	EER = 16
Kuan et al. [97]	Hand signature	Mixing of BioPhasor and 2^N discretization	SVC	EER = 9.38, mean = 0.051, std = 0.026
Him et al. [98]	Face	Extended random projection	AR face and BERC visual face	EER = 10.915
Wang et al. [99]	Face	Discretized random orthonormal transformation	ORL and GT	FAR = 0, FRR = 0, EER = 0
Teoh et al. [100]	Face	BioHash	Eigen-face	EER = 2.11 for genuine token EER = 0 for stolen token
Zuo et al. [101]	Iris	GRAY-SALT, BIN-SALT	MMU	–
Hirata et al. [102]	Finger vein	Number theoretic transform	Own dataset	–

(continued)

Table 14.7 (continued)

Article's reference	Biometric trait	Technique	Dataset	Result
Teoh et al. [103]	Speech	Probabilistic random projection (PRP)	YOHO speech corpus	FAR = 3.92, FRR = 3.97, EER = 3.95
Ouda et al. [104]	Iris	BioEncoding	CASIA v1.0	EER = 1.3, $d^1 = 6.35$
Ouda et al. [105]	Iris	One-factor authentication	CASIA v3.0	EER = 2.3, $d^1 = 5.52$
Pillai et al. [106]	Iris	Sectored random projections (SRP)	MMU	mean = 0.4997, std = 0.0136

An extended random projection method was designed in Kim et al. [98] for face databases by using better feature extraction transformation rather than template transformation. A discretized random orthonormal projection was proposed by Wang et al. [99] for face databases. In Teoh et al. [100], the quantized random projection using Johnson–Lindenstrauss Lemma was adopted for BioHash. This paper also discusses stolen-token problem. In Zuo et al. [101], authors proposed two methods: GRAY-SALT and BIN-SALT for generating cancelable iris templates. The Gabor features of iris are added with synthetic iris patterns. The problem of outlier amplification does not effect this method. Ouda et al. [104, 105] developed cancelable iris codes based on tokenless BioEncoding methods. In contrast with Biohashing, BioEncoding is a one-factor authentication scheme where tokens are not used but solely dependent on iris codes. The security of BioEncoding based salting methods is discussed in detail in Osama et al. [110].

14.6.2 Non-invertible Transforms

Unlike salting methods, non-invertible transforms apply the irreversible transformation. The parameters of the transformation function, termed as key, should be produced during authentication as shown in Fig. 14.11. The original biometric cannot be recovered, hence increases security [10]. For example, the loss is due to the artifact that the biometric templates are difficult to align such as fingerprints. To reduce this difficulty, later alignment-free methods were proposed for fingerprints. Further, non-invertible transforms achieve better revocability and diversity compared to salting approach [1]. An effort should be made in the design of non-invertible transforms to maintain the trade-off between discriminability and non-invertibility. The summary of non-invertible transforms is listed in Tables 14.8 and 14.9.

Table 14.8 Summary of selected studies under non-invertible transforms in chronological order (EER, FAR, FRR values are shown in %)

Article's reference	Biometric trait	Technique	Dataset	Result
Ratha et al. [111]	Fingerprint	Cartesian, polar, functional transformation	IBM-99	GAR = 0.90
Chikkerur et al. [112]	Fingerprint	Localized texture based representation	IBM-99	GAR = 1.0
Zuo et al. [101]	Iris	GRAY-COMBO, BIN-COMBO	MMU	ROC curve, histogram of scores
Mairorana et al. [113]	Online signature	HMM	MCYT online signature corpus	EER = 13.3
Mairorana et al. [114]	Online signature	HMM, stemming, function mixing, shifting function	MCYT online signature corpus	EER = 13.8
Xu et al. [115]	Voice	Knowledge signatures	Collection of 900 samples	GAR = 99.5
Hammerle et al. [116]	Iris	Block-remapping, image warping	CASIA v3	EER = 0.2
Yang et al. [117]	Fingerprint	Perpendicular projection of pair minutiae to a circle	FVC 2002 DB2	EER = 13
Rathgeb et al. [118]	Iris	Local intensity variations in iris texture	CASIA v3	(FRR at FAR = 0) = 1.978, EER = 1.016
Farberbock et al. [119]	Iris	Rectangular and polar block-remapping, mesh warping	CASIA v3	Block-remapping, EER = 3.1, mesh warping EER = 2.0
Mairorana et al. [120]	Online signature	Baseline approach, mixing approach, shifting approach, HMM	MCYT online signature corpus	EER = 9.12
Pillai et al. [106]	Iris	Sectored random projections (SRP)	MMU	mean = 0.4997, std = 0.0136
Lee et al. [121]	Fingerprint	Minutia based bit strings	FVC 2004 DB1, DB2, DB3	DB1 EER = 10.3, mean = 0.117 DB2 EER = 9.5, mean = 0.133 DB3 EER = 6.8, mean = 0.168

(continued)

Table 14.8 (continued)

Article's reference	Biometric trait	Technique	Dataset	Result
Nanni et al. [122]	Online signature	Combining local, global, regional matchers	MCYT	EER = 3
Jin et al. [123]	Fingerprint	Minutiae vicinity decomposition (MVD)	FVC 2002 DB2	EER = 3.90
Pillai et al. [124]	Iris	Random projections sparse representation	ICE2005, ND-IRIS-0405MBGC videos	–
Ahmad et al. [125]	Fingerprint	Pair-polar coordinates	FVC 2002 DB1, DB2, DB3	DB1 EER = 9 DB2 EER = 6 DB3 EER = 27
Takahashi et al. [126]	Fingerprint	Correlation invariant random filtering (CIRF)	Veridicom 5th sense	–
Wang et al. [127]	Fingerprint	Densely infinite to one mapping (DITOM)	FVC 2002 DB1,DB2, DB3	DB1 EER = 3.5 DB2 EER = 5 DB3 = 7.5
Das et al. [128]	Fingerprint	Minimum distance graphs	FVC 2002 DB1, DB2	FVC 2002 DB1, EER = 2.27 FVC 2002 DB2, EER = 3.79
Moujahdi et al. [129]	Face	Random projection logistic map	YALE,UMIST	EER = 3.34
Ferrara et al. [130]	Fingerprint	Non-invertible minutiae cylinder code	FVC 2002 DB1, DB2, DB3, DB4 FVC 2006 DB2	For FVC 2006 DB2, FNMR = 1.42 atFMR = 0 EER = 0.12 $FNMR_{1000}$ = 0.14

Ratha et al. [10] introduced cancelable biometrics using non-invertible transforms. The idea of cancelable biometrics is to alter the raw biometric templates by using either feature or signal domain transformations. Templates that are in a uniform space (either feature or signal) even after transformation are called cancelable templates [149]. Three such transformations namely "Cartesian, polar and functional transformation" for fingerprints were proposed in Ratha et al. [111]. Before applying these transformations, the image has to be pre-aligned by estimat-

Table 14.9 Summary of selected studies under non-invertible transforms in chronological order (EER, FAR, FRR values are shown in %)

Article's reference	Biometric trait	Technique	Dataset	Result
Wong et al. [131]	Fingerprint	Multi-line code	FVC 2002 DB1, DB2, FVC 2004 DB1	FVC 2002 DB1,EER = 4.69, DB2, EER = 5.03, FVC 2004 DB1, EER = 10.36
Wong et al. [132]	Fingerprint	Enhanced multi-line code	FVC 2002 DB1, DB2, FVC 2004 DB1, DB2	FVC 2002 DB1, EER = 1.97, DB2, EER = 2.54, FVC 2004 DB1, EER = 6.53, DB2, EER = 9.2
Yang et al. [133]	Fingerprint	Delaunay triangle local structure	FVC 2002 DB1, DB2	FVC 2002 DB1, EER = 5.93, for DB2, EER = 4
Rathgeb et al. [134, 135]	Iris	Adaptive bloom filters	CASIA-v3	1-FNMR = 97.95 at FMR = 0.01, EER = 1.14
Ferrara et al. [136]	Fingerprint	Non-invertible minutiae cylinder code	FVC 2006 DB2	FNMR = 1.42 at FMR = 0, EER = 0.13
Rathgeb et al. [137]	Iris	Mixing iris codes adaptive bloom filters	IITD v1.0	EER = 0.5, Compression = 10
Jin et al. [138]	Fingerprint	Randomized graph based Hamming embedding (RGHE)	FVC 2002 DB1, DB2, FVC 2004 DB1, DB2	FVC 2002 DB1 EER = 4.36, DB2 EER = 1.77, FVC 2004 DB1, EER = 24.71, for DB2, EER = 21.82
Prasad et al. [139]	Fingerprint	Multi-line neighboring relation	FVC 2002 DB1, DB2, DB3	DB1 EER = 0.62, DB2 EER = 1.33, DB3 EER = 2.64
Jin et al. [140]	Fingerprint	2D random projected MVD	FVC 2002 DB1, DB2	DB1 EER = 3.07, DB2 EER = 1.02
Prasad et al. [141]	Fingerprint	Minimum spanning tree	FVC 2002 DB1, DB2, DB3, FVC 2004 DB1, DB2	FVC 2002 DB1 EER = 0.8, DB2 EER = 0.8, DB3 EER = 2.4, FVC 2004 DB1 EER = 4, DB2 EER = 4
Moujahdi et al. [142]	Fingerprint	Fingerprint shell	FVC 2002 DB1, DB2	EER for DB1 = 2.03, EER for DB2 = 1.01, DB1 d^1 2.4703, DB1 K-S test = 0.7812
Leng et al. [143]	Palmprint	Multi-directional 2D palmphasor	PolyU v2	EER = 0.1938
Escobar et al. [144]	Fingerprint	Chaotic encryption using logistic map	–	–

(continued)

Table 14.9 (continued)

Article's reference	Biometric trait	Technique	Dataset	Result
Sandhya and Prasad [145]	Fingerprint	k-Nearest neighborhood structure (k-NNS)	FVC 2002 DB1, DB2, DB3	DB1 EER = 4.71, DB2 EER = 3.44, DB3 EER = 8.79, DB1 d^1 3.14, K-S test = 0.91, DB2 d^1 2.95, K-S test = 0.93 DB3 d^1 1.66, K-S test = 0.75
Sandhya et al. [146]	Fingerprint	Delaunay triangle feature sets FS-INCIR, FS-AVGLO	FVC 2002, 2004, DB1, DB2 and DB3	FVC 2002 DB1 EER = 3.96, DB2 EER = 2.98, DB3 EER = 6.89, DB1 d^1 3.96, K-S test = 0.92, DB2 d^1 3.65, K-S test = 0.94, DB3 d^1 1.66, K-S test = 0.86
Prasad et al. [147]	Fingerprint	Multiple spiral curves	FVC 2002, DB1, DB2 and DB3	FVC 2002 DB1 EER = 7.85, DB2 EER = 5.29, DB3 EER = 17.55
Wong et al. [5]	Fingerprint	Kernel PCA enabled multi-line code	FVC 2002 DB1, DB2, FVC 2004 DB1, DB2	FVC 2002 DB1, EER = 1.61, DB2, EER = 1.69, FVC 2004 DB1, EER = 3.73, DB2, EER = 3.74
Wang et al. [148]	Fingerprint	Blind system approach	FVC 2002 DB1, DB2 DB3	EER for DB1 = 4, DB2 = 3, DB3 = 8.5

ing the singular point location and orientation. In "Cartesian transformation," the minutiae points are evaluated after alignment of x-axis with minutia's orientation. The Cartesian system is divided into fixed size cells. The transformation consists of changing the cell positions of the coordinate system [149]. In "polar transformation," the minutiae points are computed in polar coordinates on the singular point. The "Cartesian and polar transformation" had a limitation that "a small change in minutia position prior to transformation may result in a large change after transformation." This was reduced using "functional transformation" that employs locally smooth functions [111, 150]. This work was extended in Chikkerur et al. [112] by generating registration-free cancelable templates for fingerprints that result in better verification rates compared to Ratha et al. [111]. In Hammerle et al. [116], iris texture is remapped by using a distorted grid mesh.

Zuo et al. [101] suggested GRAY-COMBO, BIN-COMBO transforms to generate cancelable iris biometric. In their work, a small performance fall was reported.

Farberbock et al. [119] applied block and mesh deformation in two variants to iris images, i.e., rectangular and polar iris images by extracting the iris texture from images, but before feature extraction. Row-permutations were applied to iris codes in Rathgeb et al. [118] to secure iris templates. Mairorana et al. [113, 114, 120] developed a secure recognition of biometrics whose templates are represented as a set of sequences. The transformed templates are generated by linearly convolving the set of sequences. In this work, the authors used Hidden Markov Models (HMM) for matching the templates. For experimental evaluation, online signature database is used. A group signature scheme called as knowledge signatures was implemented for generating cancelable voice templates in Xu et al. [115]. This technique makes the users of a group to sign so that it does not reveal their identity as described in Camenisch et al. [151]. Yang et al. [117] developed a "non-invertible" transform for fingerprints by considering local and global features of minutia points. The distances between minutiae pair are perpendicularly projected to a circle. Another method for non-invertible transformations is using random projections as described in Pillai et al. [106]. Sectored Random Projections for iris codes were proposed which performed better compared to random projections and robust to the occlusions of eyelids and eyelashes. The authors extended their work in [124] by using sparse representations.

A new representation of minutiae points of a fingerprint image is done using bit strings in Lee et al. [121]. The minutia points of a fingerprint image are mapped to a 3D array that is divided into small cells. A bit string is generated by finding which cells include the minutiae points. In Nanni et al. [122] the local, global, and regional matchers are merged to protect online signature templates. In Ahmad et al. [125], a pair-polar relationship of minutiae is obtained. A correlation based filter method using chip matching was proposed in Takahashi et al. [126]. The use of "Correlation Invariant Random Filtering (CIRF)" results in the same accuracy as of Fingerprint Verification Competition. Cancelable fingerprint templates were designed based on "Densely Infinite To One Mapping (DITOM)" approach in Wang et al. [127]. Das et al. [128] designed a method based on "Minimum Distance graphs." A hashing algorithm is built using these graphs and a corresponding search algorithm was employed to match the hashes generated. Moujahdhi et al. [129] developed a non-invertible transform by building a spiral cube for face biometric using chaotic logistic map for random projection of feature vectors. A non-invertible transform is built by the intersection of hyperplanes. Ferrara et al. [130] built a non-invertible Minutiae Cylinder Code (pMCC) for fingerprints as an enhancement to Cappelli et al. [152]. Wong et al. [131] generated cancelable fingerprint templates using a multi-line code of minutiae points. The authors extended their work in Wong et al. [132] by building an enhanced multi-line code for fingerprints. Yang et al. [133] developed a cancelable fingerprint template based on local structures of Delaunay triangulation in "polar-coordinate space." An adaptive bloom filter application to the iris codes was proposed in Rathgeb et al. [134, 135]. The application of this bloom filter based approach enabled template protection, compression, and acceleration of biometric identification. The authors extended their work in Rathgeb et al. [137] by mixing binary iris templates of both eyes of a single subject at a feature level.

This method reported error rates below 0.5 %. The unlinkability and irreversibility analysis for adaptive bloom filter based method was given in Rathgeb et al. [153].

In Prasad et al. [141], authors used minimum spanning tree based methods for generating cancelable templates for fingerprints. Two methods namely Cartesian space method and boundary based method were used to generate chain codes from fingerprint minutiae. Bit strings are produced using plane based quantization method. A cancelable fingerprint template was obtained using random projected MVD features in Jin et al. [123, 140]. Moujahdi et al. [142] proposed "spiral curve representation for securing fingerprint templates." Jin et al. [138] developed a "Randomized Graph-based Hamming Embedding (RGHE)" method for protecting fingerprint templates. A multi-neighboring relation was proposed in Prasad and Kumar [139] where plane based quantization technique followed by DFT and user-specific random matrix multiplication was used to generate renewable templates for fingerprints. The authors extended their work using k-Nearest Neighborhood Structure in Sandhya and Prasad [145]. Leng et al. [143] presented a cancelable palmprint method based on "multi-directional two-dimensional PalmPhasor fusion." In Sandhya et al. [146], a feature set construction method was presented using the properties of Delaunay triangles. The Delaunay triangles are built using the minutiae points of the fingerprint image. By constructing multiple spiral curves, the authors in Prasad et al. [147] designed a cancelable template using contiguous right angled triangles drawn from fingerprint minutia points. Escobar et al. [144] proposed a non-invertible transform for fingerprints using chaotic encryption based on the logistic map and Murillo-Escobar's algorithm. Wong et al. [5] built a binary cancelable template using minutia descriptor called multi-line code (MLC). The MLC template is transformed to a fixed-length bit string using kernel principal components analysis (KPCA). Wang et al. [148] presented a blind system identification approach for cancelable fingerprint template generation. A binary string derived from quantized pair-minutiae vectors is protected using its frequency samples.

14.7 Hybrid Methods

Hybrid methods of BTP can be designed by combining Cancelable Biometrics and Cryptosystems. Various schemes discussed so far are combined by using different traits. The summary of hybrid methods for BTP is listed in Table 14.10.

Boult et al. [16] introduced revocable biotokens for face biometrics. This technique divides the data into two parts "fractional part" and "integer part." The "fractional part" does the transformation and "integer part" is encrypted. The authors extended their work for fingerprints in Boult et al. [17]. Feng et al. [154] developed a "three-step hybrid algorithm based on random projection, discriminability-preserving (DP) transform, and fuzzy commitment scheme." Nagar et al. [155] built a hybrid cryptosystem with minutiae descriptors for fingerprints. This work used both "fuzzy vault" and "fuzzy commitment" schemes for building the cryptosystem. The helper data extraction involves "fuzzy vault" encoding

Table 14.10 Summary of selected studies under hybrid methods in chronological order (EER, FAR, FRR values are shown in %)

Article's reference	Biometric trait	Technique	Dataset	Result
Boult et al. [16]	Face	Revocable biotoken	FERET	GAR = 100 at FAR = 0
Boult et al. [17]	Fingerprint	Revocable biotoken	FVC 2000, FVC 2002, FVC 2004 DB1, DB2	FVC 2000 DB1, EER = 0.029, FVC 2000 DB2, EER = 0.025, FVC 2002 DB1, EER = 0.021, FVC 2002 DB2, EER = 0.012, FVC 2004 DB1, EER = 0.086, FVC 2004 DB2, EER = 0.075
Feng et al. [154]	Fingerprint	Random projection DP transform fuzzy commitment	FERET, CMU-PIE, FRGC	–
Nagar et al. [155]	Fingerprint	Minutiae descriptors	FVC 2002 DB2	GAR = 97 at FAR = 0
Liu et al. [156]	Palmprint	Random multi-space projection, heterogeneous fuzzy vault	HA-BJTU	EER = 0.08
Chin et al. [157]	Fingerprint, palmprint	Random tilling 2^N discretization	FVC 2002 DB1, FVC 2004 DB1, PolyU	EER = 0 at $d^1 = 10.51$
Kumar et al. [158]	Iris, palmprint	Cell-array	PolyU, IITD palm, IITD iris	FRR = 0.91 at FAR = 0
Sandhya and Prasad [159] (2016)	Fingerprint	Delaunay triangulation convolution code	FVC 2002 DB1, DB2, DB3	FVC 2002 DB1, EER = 1.66, FVC 2002 DB2, EER = 1.89, FVC 2002 DB3, EER = 6.87
Jin et al. [160] (2016)	Fingerprint	ECC-free key binding modified RGHE	FVC 2002 DB1 DB2, DB3, DB4 FVC 2004 DB2	–

and the ordinate values are secured using "fuzzy commitment." Liu et al. [156] combined multi-space random projection for cancelable and heterogeneous fuzzy vault technique for enhancing security of palmprints. Chin et al. [157] built a hybrid system using fingerprint and palmprint features. This method combines random tilling and 2^N discretization scheme. Kumar et al. [158] present a framework using two functions called parity code and hash code. One cell-array is used to scatter the hash code. Another cell-array is filled with chaff points. Now, parity code is tangled by a regenerative XorCoding. Sandhya and Prasad [159] constructed Delaunay triangles from fingerprint minutiae. The features are now transformed and then a cryptosystem was built using "fuzzy commitment scheme." Convolution code was used to generate the error correction code. Jin et al. [160] proposed an "ECC-free key binding scheme along with cancelable transforms for minutiae based fingerprint biometrics. The minutiae information is secured by a strong non-invertible cancelable transform called modified Randomized Graph based Hamming Embedding (RGHE)."

14.8 Homomorphic Encryption

As an alternative to the aforementioned techniques, other approach called "Homomorphic Encryption (HE)" from the traditional cryptography is introduced into biometrics by Ye et al. [18]. HE schemes allow a "limited subset of computation on the encrypted data." Combining HE with biometric recognition systems would meet the BTP requirements without degrading the accuracy. The summary of selected studies under "Homomorphic Encryption" schemes developed in biometrics is listed in Table 14.11.

Table 14.11 Summary of selected studies under "homomorphic encryption" in chronological order

Article's reference	Biometric trait	Technique	Dataset
Ye et al. [18]	Iris	k-ABAC kAQ	CASIA
Erkin et al. [161]	Face	Paillier homomorphic encryption	ORL face
Sadeghi et al. [162]	Face	Oblivious transfer (OT)	–
Rane et al. [163]	Fingerprint	Secure Hamming distance	Proprietary database of 1035 fingers
Barni et al. [164, 165]	Fingerprint	Fingercode	408 images acquired by a CrossMatch verifier 300 sensor
Osadchy et al. [166]	Iris	SciFi	CMU-PIE
Kulkarni et al. [167]	Iris	Secure Hamming distance	ICE 2005
Karabat et al. [168]	Any biometric	THRIVE	–
Barrero et al. [169]	Online signature	Dynamic time warping Paillier cryptosystem	Signature subset from BioSecureID multi-modal database sensor

Ye et al. [18] presented "Anonymous Biometric Access Control (ABAC)" which uses "k-Anonymous Quantization (kAQ) framework." kAQ uses a lookup table to recognize k candidates. HE-based matching protocol is applied on these k candidates. Erkin et al. [161] proposed a privacy-preserving face recognition system for eigen-faces by using the "Paillier homomorphic encryption scheme". Later, Sadeghi et al. [162] improve the efficiency of this system. Rane et al. [163] presented Hamming distance calculation for fingerprint applications. Barni et al. [164, 165] demonstrated a distributed biometric system by exploiting "cryptosystems, homomorphic encryption on Fingercode templates in a semi-honest model." Osadchy et al. [166] proposed a "secure hamming distance based HE for face biometrics. The system is called SCiFI". Kulkarni et al. [167] proposed a HE method by calculating values stored on server by performing XOR operation between biometric template vector and corresponding user's key. Karabat et al. [168] introduced "THRIVE: threshold homomorphic encryption based secure and privacy-preserving biometric verification system" that is applicable to any biometric. Barrero et al. [169] presented a "Paillier's homomorphic probabilistic encryption" on online signature systems.

14.9 Analysis and Discussion of SLR Results

The analysis of the state-of-the-art literature on BTP schemes is made addressing the research questions discussed in Table 14.1. The percentage distribution of broad BTP categories is shown in Fig. 14.12a. The percentage of BTP schemes developed on each biometric trait is shown in Fig. 14.12b. From Fig. 14.12a, it is observed that 49 % of the published literature is on cancelable biometrics, 35 % is on biometric

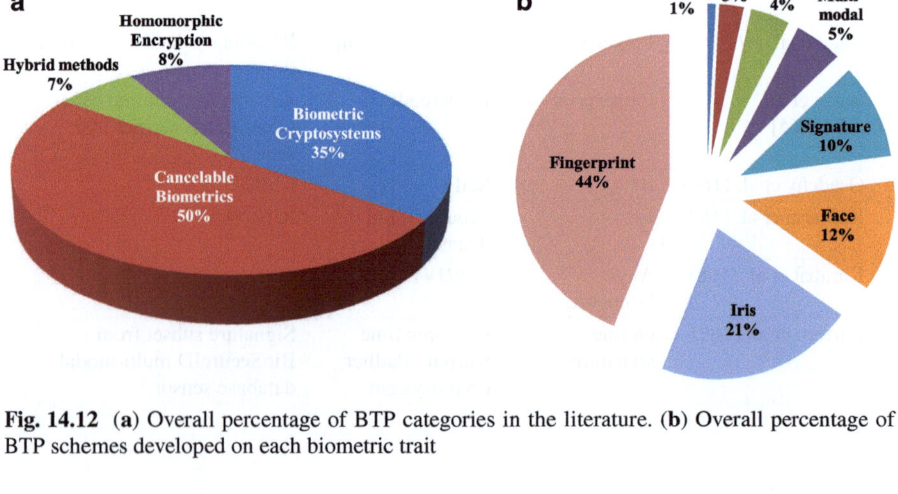

Fig. 14.12 (**a**) Overall percentage of BTP categories in the literature. (**b**) Overall percentage of BTP schemes developed on each biometric trait

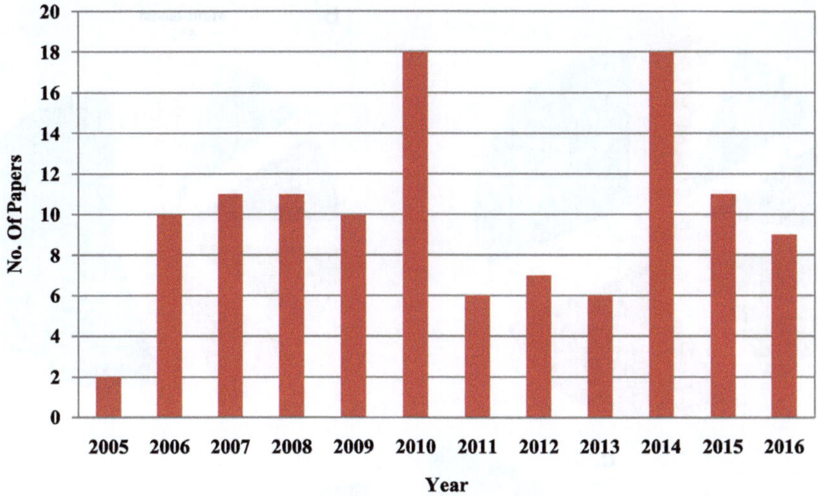

Fig. 14.13 Count of papers published in each year selected for study (upto June, 2016)

cryptosystems, 8 % on hybrid methods, and 8 % on Homomorphic Encryption. Hence, the hybrid methods, and methods using Homomorphic Encryption, which have been paid little attention so far, need to be developed to make use of advantages of multiple schemes. From Fig. 14.12b, it is observed that BTP schemes developed so far are 44 % on fingerprints, 21 % on the iris, 12 % on the face, 10 % on the signature, 5 % on multi-modal, 4 % on palmprint, 3 % on voice, and 1 % on Finger vein traits. Hence, we can say multi-modal biometric template protection schemes need to be developed to make use of advantages of multi-biometric systems. The count of research papers published in each year is shown in Fig. 14.13. It is noticed from Fig. 14.13, that the first set of papers that were published in 2005 are few and significantly increased in 2010. Further, a consistent increase is seen in the past three years.

Figure 14.14a–c represents the percentage of BTP schemes developed on each biometric trait under subcategories of biometric cryptosystems. Figure 14.15a, b represents the percentage of BTP schemes developed on each biometric trait under subcategories of cancelable biometrics. Figure 14.16a, b represents the percentage of BTP schemes developed on each biometric trait under hybrid methods and homomorphic encryption based methods, respectively.

After the selection and analysis of studies, we had a look at the affiliation of authors and listed out the active research communities in developing BTP schemes in Table 14.12. The publicly available databases on which BTP schemes are evaluated in the literature are listed in Table 14.13.

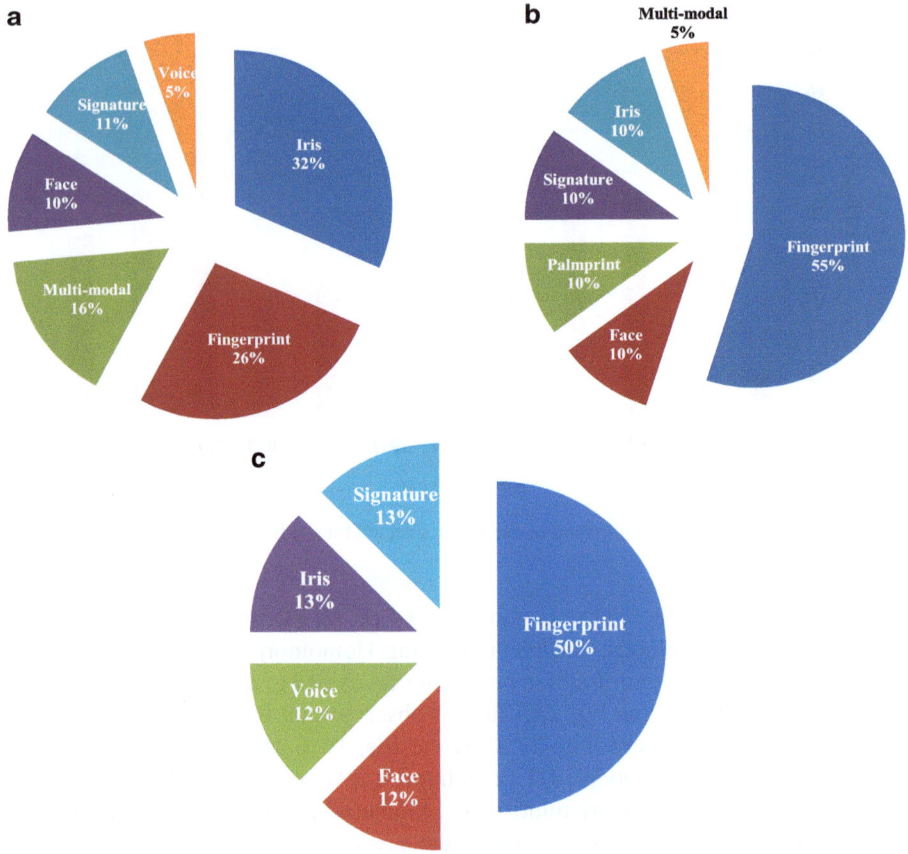

Fig. 14.14 Percentage of BTP schemes developed on each biometric trait under subcategories of biometric cryptosystems. (**a**) "Fuzzy commitment schemes". (**b**) "Fuzzy vault schemes". (**c**) "Key generating systems"

14.10 Research Implications and Future Directions

14.10.1 Benefits for Researchers and Practitioners

This SLR provides the classification of approaches and comparisons for BTP in a systematic way. This SLR makes a methodical and comprehensive study of 119 most relevant research articles. By using this SLR, researchers and practitioners can find relevant information of BTP by using search strings in Table 14.3 thereby reducing time and complexity of searching the studies.

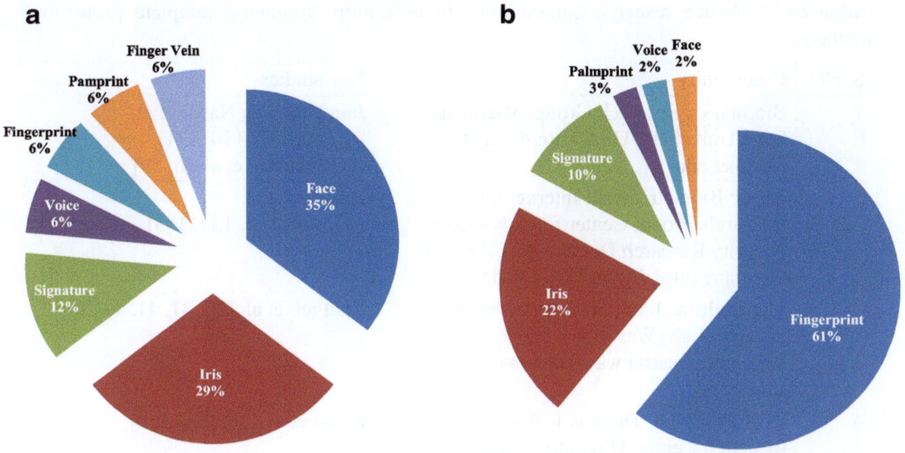

Fig. 14.15 Percentage of BTP schemes developed on each biometric trait under subcategories of cancelable biometrics. (**a**) "Salting methods". (**b**) "Non-invertible transforms"

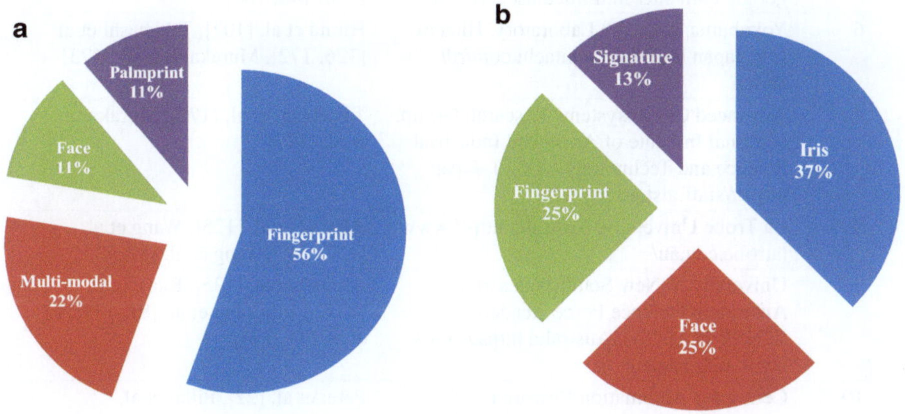

Fig. 14.16 Percentage of BTP schemes developed on each biometric trait under. (**a**) "Hybrid methods". (**b**) "Homomorphic encryption" based methods

14.10.2 Future Research Directions

After analyzing the studies in this SLR, we noticed the following issues and challenges that are not addressed by researchers working on BTP methods in the past decade.

- Most of the BTP techniques designed so far are analyzed on small and mid-size databases. However, these schemes have to be assessed on a large scale and comprehensive databases to prove their significance.

Table 14.12 Active research communities in designing "biometric template protection" methods

S. No	Community	Key studies
1	Biometrics Research Group, Michigan State University, USA http://biometrics.cse.msu.edu/	Jain et al. [1], Nagar et al. [61, 66, 155, 170, 171], Nandakumar et al. [6, 59]
2	da/sec Biometrics and Internet Security Research Group, Center for Advanced Security Research Darmstadt (CASED), Germany https://www.dasec.h-da.de/	Rathgeb et al. [70, 134, 135, 137, 153]
3	The Multimedia Signal Processing and Security Lab (WaveLab), University of Salzburg, Austria wavelab.at/member-uhl.shtml	Rahtgeb et al. [20, 37, 41, 49, 50]
4	IBM Research, Thomas J. Watson Research Center, USA https://www.research.ibm.com/labs/watson/	Ratha et al. [10, 111, 150]
5	Multimedia Security Lab, Yonsei University, South Korea https://sites.google.com/site/ multimediasecuritylab/	Teoh et al. [34, 92, 96, 100, 103], Wong et al. [132], Jin et al. [138, 140, 160]
6	Yokohama Research Laboratory, Hitachi Ltd, Japan http://www.hitachi.com/rd/about/	Hirata et al. [102], Takahashi et al. [126, 172], Murakami et al. [173]
7	Advanced Cryptosystems Research Group, National Institute of Advanced Industrial Science and Technology (AIST), Japan https://staff.aist.go.jp/	Takahashi et al. [172], Murakami et al. [173]
8	La Trobe University, Australia http://www.latrobe.edu.au/	Ahmad et al. [125], Wang et al. [127, 148], Yang et al. [87, 133]
9	University of New South Wales at the Australian Defence Force Academy (UNSW@ADFA), Australia https://www.unsw.adfa.edu.au/	Ahmad et al. [125], Rang et al. [127, 148], Yang et al. [87, 133], Li et al. [74, 174]
10	Centre for Automation Research, University of Maryland, USA http://www.cfar.umd.edu/	Patel et al. [22], Pillai et al. [106, 124]
11	Biometric Systems and Multimedia Forensics LAB, University of "Roma TRE", Italy http://biomedia4n6.uniroma3.it/index.html	Mairorana et al. [35, 39, 113, 114, 120], Nanni et al. [122]
12	Biometrics Systems Laboratory, University of Bologna, Italy http://biolab.csr.unibo.it/home.asp	Adler et al. [175], Cappelli et al. [152], Nanni et al. [122], Ferrara et al. [130, 136]

(continued)

Table 14.12 (continued)

S. No	Community	Key studies
13	CyLab, Biometrics Center, Carnegie Mellon University, USA https://www.cylab.cmu.edu/ research/center-biometrics.html	Savvides et al. [176]
14	Universiti tunku abdul rahman Kuala Lumpur, Malaysia www.utar.edu.my	Jin et al. [138, 140, 160]
15	Institute for Development and Research in Banking Technology (IDRBT), India www.idrbt.ac.in	Prasad et al. [139, 141, 147], Sandhya et al. [56, 145, 146, 159, 178, 179]

- There is a rampant growth in the use of mobile devices in the past decade. But, till date the authentication of the mobile device is under the user control until it is active. The development of template protected active authentication systems make mobile devices secure from impersonation.
- From this SLR, it is noticed that 49 % of the published literature is on cancelable biometrics, 35 % is on biometric cryptosystems, 8 % on hybrid methods, and 8 % on Homomorphic Encryption. The hybrid template protection approaches and methods using Homomorphic Encryption, which have been paid little attention so far need to be developed to make use of advantages of multiple schemes.
- From this SLR, it is also observed that Homomorphic Encryption schemes developed so far are not evaluated on benchmark databases, which is needed to prove their significance.
- It is also observed from the SLR, that only 5 % of BTP methods are on multi-modal biometrics, which is to be developed to make use of advantages of multi-biometrics.
- There is no common platform to host datasets and to evaluate benchmark systems. Hence, common "Institutional Review Board (IRB) protocols" and "consent forms" that can be used internationally by both academia and industry are needed to evaluate BTP schemes. The count of research papers published in each year is shown in Fig. 14.13. It is noticed from Fig.14.13, that the first set of papers that were published in 2005 are few and significantly increased in 2010. Further, a consistent increase is seen in the past three years.
- The definition of a common methodology to systematically evaluate the security level of a BTP method is needed.

14.10.3 Threats to Validity

- **Threats to completeness:** The key strand to be considered in the designing an SLR is the search string used for retrieving the related studies. The construction of search string allows researchers not to waste time on the refinement of

Table 14.13 Publicly available databases on which BTP schemes are evaluated in the literature and their source

S.No	Database	Biometric trait	Abbreviation	Source
1	FVC 2000	Fingerprint	Fingerprint Verification Competition	http://bias.csr.unibo.it/fvc2000/
2	FVC 2002			http://bias.csr.unibo.it/fvc2002/
3	FVC 2004			http://bias.csr.unibo.it/fvc2004/
4	FVC 2006			http://bias.csr.unibo.it/fvc2006/
5	FVC Ongoing			http://bias.csr.unibo.it/fvcongoing/
6	NIST-SD14		National Institute of Standards and Technology-Special Database	http://www.nist.gov/srd/nistsd14.cfm
6	ICE	Iris	Iris Challenge Evaluation	http://www.nist.gov/itl/iad/ig/ice.cfm
7	IITD iris v1		IIT Delhi Iris version 1	http://www4.comp.polyu.edu.hk/~csajaykr/IITD/Database_Iris.htm
8	CASIA Iris v1		National Laboratory of Pattern Recognition Institute of Automation, Chinese Academy of Sciences	http://biometrics.idealtest.org/dbDetailForUser.do?id=1
9	CASIA Iris v3			http://biometrics.idealtest.org/dbDetailForUser.do?id=3
10	CASIA Iris v4			http://biometrics.idealtest.org/dbDetailForUser.do?id=4
11	FERET	Face	Facial Recognition Technology	http://www.nist.gov/itl/iad/ig/colorferet.cfm
12	CALTECH		California Institute of Technology.	http://www.vision.caltech.edu/html-files/archive.html
13	CMU-PIE		Carnegie Mellon University-Pose, Illumination and Expression	http://vasc.ri.cmu.edu/idb/html/face/
14	NIR face		Hong Kong Polytechnic University, Near-Infrared	http://www4.comp.polyu.edu.hk/~biometrics/polyudb_face.htm
15	ORL face		AT& T Laboratories, Cambridge	http://www.cl.cam.ac.uk/research/dtg/attarchive/facedatabase.html
16	NVIE		Natural Visible and Infrared Facial Expression	http://nvie.ustc.edu.cn/
17	FRGC		Face Recognition Grand Challenge	http://www.nist.gov/itl/iad/ig/frgc.cfm
17	XM2VTS		Multi Modal Verification for Teleservices and Security applications	http://www.ee.surrey.ac.uk/CVSSP/xm2vtsdb/
18	AR face		–	http://www2.ece.ohio-state.edu/~aleix/ARdatabase.html

(continued)

Table 14.13 (continued)

S.No	Database	Biometric trait	Abbreviation	Source
19	Poly U	Palmprint	Hong Kong Polytechnic University Palmprint	http://www4.comp.polyu.edu.hk/~biometrics/MultispectralPalmprint/MSP.htm
20	SVC 2004	Signature	Signature Verification Competition	http://www.cse.ust.hk/svc2004/
21	MCYT		(Ministerio de Cienciay Tecnología, Spanish Ministry of Science and Technology	Ortega-Garcia et al. [177]
22	TI 46	Voice	Texas Instruments 46-Word Speaker-Dependent Isolated Word Corpus	http://catalog.ldc.upenn.edu/LDC93S9
23	YOHO speech		YOHO Speaker Verification	https://catalog.ldc.upenn.edu/LDC93S9

irrelevant studies [24]. In this SLR, we used ten search databases and executed the search strings in Table 14.3 and filtered the articles based on inclusion and exclusion criteria defined in Table 14.2 and Fig. 14.3. But, this may result in the threat to completeness like missing some articles due to linguistic barriers.

- **Threats to the selection of studies:** Due to different views of researchers on inclusion and exclusion criteria defined in Table 14.2 and Fig. 14.3, the selection of primary studies may be different for each researcher. So, two researchers worked together to reduce the bias. If still, a disagreement persists, other experts in conducting SLRs were called to make a conclusion in the selection of studies.
- **Threats to data extraction:** From executing the search strings on ten search databases, 358 articles were retrieved. Finally, we found 119 articles relevant to our study. We cross-checked our review results by searching individual journals.

14.11 Concluding Remarks

Existing surveys like [1, 20–23] reflect various research issues in BTP methods. They filled the initial void for future work in BTP. In contrast, this SLR provides a methodical and comprehensive survey on BTP methods. On the research questions, this SLR answered that 49 % of literature was on cancelable biometrics approach, 35 % was on the biometric systems approach, 8 % on hybrid approaches, and 8 % on Homomorphic Encryption schemes. Also, BTP schemes developed so far are 44 % on fingerprints, 21 % on the iris, 12 % on the face, 10 % on the signature, 5 % on multi-modal, 4 % on palmprint, 3 % on voice, and 1 % on Finger vein traits.

We addressed the future directions in designing BTP methods. Extraction outcomes are discussed in detail thereby strengthening the evidence and use of this SLR. We believe that researchers from academia and industry need to collaborate, regardless of they are geographically dispersed, for exploring and progressing in designing BTP techniques further for larger databases, by developing a joint research agenda.

References

1. A.K. Jain, K. Nandakumar, A. Nagar, Biometric template security. EURASIP J. Adv. Signal Process. **2008**, 579416 (2008)
2. A. Jain, A. Ross, S. Prabhakar, An introduction to biometric recognition. IEEE Trans. Circuits Syst. Video Technol. **14**, 4–20 (2004)
3. A Jain, A.A. Ross, K. Nandakumar, *Introduction to Biometrics* (Springer, Berlin, 2011)
4. D.C.L. Ngo, A.B.J. Teoh, J. Hu, *Biometric Security* (Cambridge Scholars Publishing, Newcastle upon Tyne, 2015)
5. W.J. Wong, A.B. Teoh, Y.H. Kho, M.D. Wong, Kernel PCA enabled bit-string representation for minutiae-based cancelable fingerprint template. Pattern Recogn. **51**, 197–208 (2016)
6. K. Nandakumar, A. Jain, Biometric template protection: bridging the performance gap between theory and practice. IEEE Signal Process. Mag. **32**(5), 88–100 (2015)
7. G.J. Tomko, C. Soutar, G.J. Schmidt, Fingerprint controlled public key cryptographic system. US Patent 5832091 A (1996). http://www.google.co.in/patents/US5832091
8. C. Soutar, D. Roberge, A. Stoianov, R. Gilroy, B. Vijaya Kumar, Biometric encryption: enrollment and verification procedures, in *Proceedings of SPIE: Optical Pattern Recognition* (1998), pp. 24–35
9. A. Juels, M. Wattenberg, A fuzzy commitment scheme, in *Proceedings of the 6th ACM Conference on Computer and Communications Security* (1999), pp. 28–36
10. N.K. Ratha, J.H. Connell, R.M. Bolle, Enhancing security and privacy in biometrics-based authentication systems. IBM Syst. J. **40**(3), 614–634 (2001)
11. A. Juels, M. Sudan, A fuzzy vault scheme, in *Proceedings of IEEE International Symposium on Information Theory* (2002)
12. H. Feng, C.C. Wah, Private key generation from online handwritten signatures. Inf. Manage. Comput. Secur. **10**(4), 159–164 (2002)
13. A. Goh, D. Ngo, Computation of cryptographic keys from face biometrics, in *Communications and Multimedia Security. Advanced Techniques for Network and Data Protection* (Springer, Berlin, 2003), pp. 1–13
14. Y. Dodis, L. Reyzin, A. Smith, Fuzzy extractors: how to generate strong keys from biometrics and other noisy data, in *International Conference on the Theory and Applications of Cryptographic Techniques - EUROCRYPT* (2004), pp. 523–540
15. Y. Sutcu, Q. Li, N. Memon, Protecting biometric templates with sketch: theory and practice. IEEE Trans. Inf. Forensics Secur. **2**(3), 503–512 (2007)
16. T. Boult, Robust distance measures for face-recognition supporting revocable biometric tokens, in *7th International Conference on Automatic Face and Gesture Recognition (FGR)* (2006), pp. 560–566
17. T. Boult, W. Scheirer, R. Woodworth, Revocable fingerprint biotokens: accuracy and security analysis, in *Conference on Computer Vision and Pattern Recognition (CVPR)* (2007), pp. 1–8
18. S. Ye, Y. Luo, J. Zhao, S.C.S. Cheung, Anonymous biometric access control. EURASIP J. Inf. Secur. **2009**, 2:1–2:17 (2009)
19. U. Uludag, S. Pankanti, S. Prabhakar, A. Jain, Biometric cryptosystems: issues and challenges. Proc. IEEE **92**(6), 948–960 (2004)

20. C. Rathgeb, A. Uhl, A survey on biometric cryptosystems and cancelable biometrics. EURASIP J. Inf. Secur. **2011**(1), 3 (2011)
21. S. Rane, Y. Wang, S.C. Draper, P. Ishwar, Secure biometrics: concepts, authentication architectures, and challenges. IEEE Signal Process. Mag. **30**(5), 51–64 (2013)
22. V. Patel, N. Ratha, R. Chellappa, Cancelable biometrics: a review. IEEE Signal Process. Mag. **32**(5), 54–65 (2015)
23. I. Natgunanathan, A. Mehmood, Y. Xiang, G. Beliakov, J. Yearwood, Protection of privacy in biometric data. IEEE Access **4**, 880–892 (2016)
24. B. Kitchenham, S. Charters, Guidelines for performing systematic literature reviews in software engineering. Technical Report EBSE 2007–001, Keele University and Durham University Joint Report (2007)
25. B. Kitchenham, Procedures for performing systematic reviews, Joint Technical Report (Keele University Technical Report TR/SE-0401 and National ICT Australia Ltd. NICTA Technical Report 0400011T.1) (2004)
26. ISO/IEC JTC1 SC27 24745, Biometric information protection (2011)
27. K. Simoens, B. Yang, X. Zhou, F. Beato, C. Busch, E. Newton, B. Preneel Criteria towards metrics for benchmarking template protection algorithms, in *5th IAPR International Conference on Biometrics (ICB)* (2012), pp. 498–505
28. A. Cavoukian, A. Stoianov, in ed. by S. Li, A. Jain, *Encyclopedia of Biometrics* (Springer, New York, 2009), pp. 260–269
29. F. Hao, R. Anderson, J. Daugman, Combining cryptography with biometrics effectively. IEEE Trans. Comput. **55**(9), 1081–1088 (2006)
30. M. VanderVeen, T. Kevenaar, G.J. Schrijen, T.H. Akkermans, F. Zuo, Face biometrics with renewable templates, in *Proceedings of SPIE: Security, Steganography, and Watermarking of Multimedia Contents*, vol. 6072 (2006)
31. G. Zheng, W. Li, C. Zhan, Cryptographic key generation from biometric data using lattice mapping, in *18th International Conference on Pattern Recognition (ICPR)*, vol. 4 (2006), pp. 513–516
32. J. Bringer, H. Chabanne, G. Cohen, B. Kindarji, G. Zemor, Optimal iris fuzzy sketches, in *First International Conference on Biometrics: Theory, Applications, and Systems (BTAS)* (2007), pp. 1–6
33. J. Bringer, H. Chabanne, G. Cohen, B. Kindarji, G. Zemor, Theoretical and practical boundaries of binary secure sketches. IEEE Trans. Inf. Forensics Secur. **3**(4), 673–683 (2008)
34. A.B.J. Teoh, J. Kim, Secure biometric template protection in fuzzy commitment scheme. IEICE Electron. Expr. **4**(23), 724–730 (2007)
35. E. Maiorana, P. Campisi, A. Neri, User adaptive fuzzy commitment for signature template protection and renewability. J. Electron. Imaging **17**(1), 011011 (2008)
36. M. Ao, S. Li, Near infrared face based biometric key binding, in *Advances in Biometrics* (Springer, Berlin, 2009), pp. 376–385
37. C. Rathgeb, A. Uhl, Systematic construction of iris-based fuzzy commitment schemes, in *3rd International Conference on Biometrics (ICB)* (2009), pp. 940–949
38. H. Lu, K. Martin, F. Bui, K.N. Plataniotis, D. Hatzinakos, Face recognition with biometric encryption for privacy-enhancing self-exclusion, in *Proceedings of the 16th International Conference on Digital Signal Processing* (2009), pp. 625–632
39. E. Maiorana, P. Campisi, Fuzzy commitment for function based signature template protection. IEEE Signal Process Lett. **17**(3), 249–252 (2010)
40. K. Nandakumar, A fingerprint cryptosystem based on minutiae phase spectrum, in *IEEE International Workshop on Information Forensics and Security (WIFS)* (2010), pp. 1–6
41. C. Rathgeb, A. Uhl, Adaptive fuzzy commitment scheme based on iris-code error analysis, in *European Workshop on Visual Information Processing (EUVIP)* (2010), pp. 41–44
42. C. Rathgeb, A. Uhl, P. Wild, Reliability-balanced feature level fusion for fuzzy commitment scheme, in *International Joint Conference on Biometrics (IJCB)*, pp 1–7 (2011)
43. P. Li, X. Yang, H. Qiao, K. Cao, E. Liu, J. Tian, An effective biometric cryptosystem combining fingerprints with error correction codes. Expert Syst. Appl. **39**(7), 6562–6574 (2012)

44. Y. Imamverdiyev, A.B.J. Teoh, J. Kim, Biometric cryptosystem based on discretized fingerprint texture descriptors. Expert Syst. Appl. **40**(5), 1888–1901 (2013)
45. L. Yuan, Multimodal cryptosystem based on fuzzy commitment, in *17th International Conference on Computational Science and Engineering (CSE)* (2014), pp. 1545–1549
46. S. Billeb, C. Rathgeb, H. Reininger, K. Kasper, C. Busch, Biometric template protection for speaker recognition based on universal background models. IET Biom. **4**(2), 116–126 (2015)
47. N. Wang, Q. Li, A.A.A. El-Latif, J. Peng, X. Yan, X. Niu, A novel template protection scheme for multibiometrics based on fuzzy commitment and chaotic system. Signal Image Video Process. **9**, 1–11 (2015)
48. T. Ignatenko, F. Willems, Information leakage in fuzzy commitment schemes. IEEE Trans. Inf. Forensics Secur. 5(2), 337–348 (2010)
49. C. Rathgeb, A. Uhl, Statistical attack against iris-biometric fuzzy commitment schemes, in *Computer Vision and Pattern Recognition Workshops (CVPRW)* (2011), pp. 23–30
50. C. Rathgeb, A. Uhl, Statistical attack against fuzzy commitment scheme. IET Biom. **1**(2), 94–104 (2012)
51. C. Rathgeb, A. Uhl Iris-biometric fuzzy commitment schemes under signal degradation, in *International Conference on Image and Signal Processing (ICISP)* (2012), pp. 217–225
52. X. Zhou, A. Kuijper, R. Veldhuis, C. Busch, Quantifying privacy and security of biometric fuzzy commitment, in *International Joint Conference on Biometrics (IJCB)* (2011), pp. 1–8
53. X. Zhou, A. Kuijper, C. Busch, Retrieving secrets from iris fuzzy commitment, in *5th IAPR International Conference on Biometrics (ICB)* (2012), pp. 238–244
54. X. Zhou, C. Busch, Measuring privacy and security of iris fuzzy commitment, in *IEEE International Carnahan Conference on Security Technology (ICCST)* (2012), pp. 168–173
55. E. Kelkboom, J. Breebaart, I. Buhan, R. Veldhuis, Maximum key size and classification performance of fuzzy commitment for gaussian modeled biometric sources. IEEE Trans. Inf. Forensics Secur. **7**(4), 1225–1241 (2012)
56. M. Sandhya, M.V. Prasad, A bio-cryptosystem for fingerprints using delaunay neighbor structures (DNS) and fuzzy commitment scheme, in *Advances in Signal Processing and Intelligent Recognition Systems* (2016), pp. 159–171
57. U. Uludag, S. Pankanti, A.K. Jain, Fuzzy vault for fingerprints, in *Proceedings of 5th International Conference on Audio- and Video-Based Biometric Person Authentication, AVBPA 2005*, Hilton Rye Town, NY, USA, 20–22 July 2005
58. U. Uludag, A. Jain, Securing fingerprint template: fuzzy vault with helper data, in *Conference on Computer Vision and Pattern Recognition Workshop (CVPRW)* (2006), pp. 163–163
59. K. Nandakumar, A. Jain, S. Pankanti, Fingerprint-based fuzzy vault: implementation and performance. IEEE Trans. Inf. Forensics Secur. **2**(4), 744–757 (2007)
60. K. Nandakumar, A. Nagar, A.K. Jain, Hardening fingerprint fuzzy vault using password, in *International Conference on Biometrics (ICB)* (2007), pp. 927–937
61. A. Nagar, K. Nandakumar, A.K. Jain, Securing fingerprint template: fuzzy vault with minutiae descriptors, in *19th International Conference on Pattern Recognition (ICPR)* (2008), pp. 1–4
62. D. Nyang, K. Lee, Fuzzy face vault: how to implement fuzzy vault with weighted features, in *4th International Conference on Universal Access in Human-Computer Interaction, UAHCI* (2007), pp. 491–496
63. Y.J. Lee, K.R. Park, S.J. Lee, K. Bae, J. Kim, A new method for generating an invariant iris private key based on the fuzzy vault system. IEEE Trans. Syst. Man Cybern. B Cybern. **38**(5), 1302–1313 (2008)
64. E. Kelkboom, X. Zhou, J. Breebaart, R. Veldhuis, C. Busch, Multi-algorithm fusion with template protection, in *3rd International Conference on Biometrics: Theory, Applications, and Systems (BTAS)* (2009), pp. 1–8
65. X. Zhou, A. Opel, J. Merkle, U. Korte, C. Busch, Enhanced template protection with passwords for fingerprint recognition, in *Third International Workshop on Security and Communication Networks (IWSCN)* (2011), pp. 67–74
66. A, Nagar, K. Nandakumar, A. Jain, Multibiometric cryptosystems based on feature-level fusion. IEEE Trans. Inf. Forensics Secur. **7**(1), 255–268 (2012)

67. L. Hong-wei, W. Yao, A new fuzzy fingerprint vault using multivariable linear function based on lorenz chaotic system, in *International Conference on Computer Science and Automation Engineering (CSAE)*, vol. 1 (2012), pp. 531–534
68. A. Bhateja, S. Chaudhury, P. Saxena, A robust online signature based cryptosystem, in *14th International Conference on Frontiers in Handwriting Recognition (ICFHR)* (2014), pp. 79–84
69. H. Liu, D. Sun, K. Xiong, Z. Qiu, Palmprint based multidimensional fuzzy vault scheme. Sci. World J. **2014**, 8p. (2014)
70. C. Rathgeb, J. Wagner, B. Tams, C. Busch, Preventing the cross-matching attack in bloom filter-based cancelable biometrics, in *International Workshop on Biometrics and Forensics (IWBF)* (2015), pp. 1–6
71. T.H. Nguyen, Y. Wang, Y. Ha, R. Li, Performance and security-enhanced fuzzy vault scheme based on ridge features for distorted fingerprints. IET Biom. **4**(1), 29–39 (2015)
72. B. Tams, J. Merkle, C. Rathgeb, J. Wagner, U. Korte, C. Busch, Improved fuzzy vault scheme for alignment-free fingerprint features, in *International Conference of the Biometrics Special Interest Group (BIOSIG)* (2015), pp. 1–12
73. D. Bansal, S. Sofat, M. Kaur, Fingerprint fuzzy vault using Hadamard transformation, in *International Conference on Advances in Computing, Communications and Informatics (ICACCI)* (2015), pp. 1830–1834
74. C. Li, J. Hu, A security-enhanced alignment-free fuzzy vault-based fingerprint cryptosystem using pair-polar minutiae structures. IEEE Trans. Inf. Forensics Secur. **11**(3), 543–555 (2016)
75. T.H. Nguyen, Y. Wang, Y. Ha, R. Li, Improved chaff point generation for vault scheme in bio-cryptosystems. IET Biom. **2**(2), 48–55 (2013)
76. G. Eskander, R. Sabourin, E. Granger, Improving signature-based biometric cryptosystems using cascaded signature verification-fuzzy vault (SV-FV) approach, in *2014 14th International Conference on Frontiers in Handwriting Recognition (ICFHR)* (2014), pp. 187–192
77. G. Eskander, R. Sabourin, E. Granger, Offline signature-based fuzzy vault: a review and new results, in *IEEE Symposium on Computational Intelligence in Biometrics and Identity Management (CIBIM)* (2014), pp. 45–52
78. B. Tams, P. Mihailescu, A. Munk, Security considerations in minutiae-based fuzzy vaults. IEEE Trans. Inf. Forensics Secur. **10**(5), 985–998 (2015)
79. M. Lafkih, P. Lacharme, C. Rosenberger, M. Mikram, S. Ghouzali, M. El Haziti, D. Aboutajdine, Vulnerabilities of fuzzy vault schemes using biometric data with traces, in *International Wireless Communications and Mobile Computing Conference (IWCMC)* (2015), pp. 822–827
80. L. Leng, A.B.J. Teoh, Alignment-free row-co-occurrence cancelable palmprint fuzzy vault. Pattern Recogn. **48**(7), 2290–2303 (2015)
81. S. Hoque, M. Fairhurst, G. Howells, Evaluating biometric encryption key generation using handwritten signatures, in *Symposium on Bio-inspired Learning and Intelligent Systems for Security (BLISS'08)* (2008), pp. 17–22
82. C. Rathgeb, A. Uhl, An iris-based interval-mapping scheme for biometric key generation, in *Proceedings of 6th International Symposium on Image and Signal Processing and Analysis* (2009), pp. 511–516
83. B. Carrara, C. Adams, You are the key: generating cryptographic keys from voice biometrics, in *Eighth Annual International Conference on Privacy Security and Trust (PST)* (2010), pp. 213–222
84. L. Wu, X. Liu, S. Yuan, P. Xiao, A novel key generation cryptosystem based on face features, in *10th International Conference on Signal Processing (ICSP)* (2010), pp. 1675–1678
85. E. Liu, J. Liang, L. Pang, M. Xie, J. Tian, Minutiae and modified biocode fusion for fingerprint-based key generation. J. Netw. Comput. Appl. **33**(3), 221–235 (2010)
86. R. Ranjan, S. Singh, Improved and innovative key generation algorithms for biometric cryptosystems, in *3rd International Advance Computing Conference (IACC)* (2013), pp. 943–946
87. W. Yang, J. Hu, S. Wang, M. Stojmenovic, An alignment-free fingerprint bio-cryptosystem based on modified Voronoi neighbor structures. Pattern Recogn. **47**(3), 1309–1320 (2014)

88. W. Yang, J. Hu, S. Wang, A delaunay quadrangle-based fingerprint authentication system with template protection using topology code for local registration and security enhancement. IEEE Trans. Inf. Forensics Secur. **9**(7), 1179–1192 (2014)
89. A.B. Teoh, Y.W. Kuan, S. Lee, Cancellable biometrics and annotations on biohash. Pattern Recogn. **41**(6), 2034–2044 (2008)
90. T. Connie, A. Teoh, M. Goh, D. Ngo, Palmhashing: a novel approach for cancelable biometrics. Inf. Process. Lett. **93**(1), 1–5 (2005)
91. A. Teoh, D. Ngo, Biophasor: token supplemented cancellable biometrics, in *9th International Conference on Control, Automation, Robotics and Vision (ICARCV)* (2006), pp. 1–5
92. A. Teoh, A. Goh, D. Ngo, Random multispace quantization as an analytic mechanism for biohashing of biometric and random identity inputs. IEEE Trans. Pattern Anal. Mach. Intell. **28**(12), 1892–1901 (2006)
93. C.S. Chin, A.T.B. Jin, D.N.C. Ling, High security iris verification system based on random secret integration. Comput. Vis. Image Underst. **102**(2), 169–177 (2006)
94. W.K. Yip, A.B.J. Teoh, D.C.L. Ngo, Replaceable and securely hashed keys from online signatures. IEICE Electron. Expr. **3**(18), 410–416 (2006)
95. M. Jeong, C. Lee, J. Kim, J.Y. Choi, K.A. Toh, J. Kim, Changeable biometrics for appearance based face recognition, in *2006 Biometrics Symposium: Special Session on Research at the Biometric Consortium Conference* (2006), pp. 1–5
96. A. Teoh, C.T. Yuang, Cancelable biometrics realization with multispace random projections. IEEE Trans. Syst. Man Cybern. **37**(5)1096–1106 (2007)
97. Y.W. Kuan, A.B. Teoh, D.C. Ngo, Secure hashing of dynamic hand signatures using wavelet-fourier compression with biophasor mixing and 2^N discretization. EURASIP J. Appl. Signal Process. **2007**(1), 32 (2007)
98. Y. Kim, K.A. Toh, A method to enhance face biometric security, in *First International Conference on Biometrics: Theory, Applications, and Systems (BTAS)* (2007), pp. 1–6
99. Y. Wang, K. Plataniotis, Face based biometric authentication with changeable and privacy preservable templates, in *Biometrics Symposium* (2007), pp. 1–6
100. A.B. Teoh, Y.W. Kuan, S Lee, Cancellable biometrics and annotations on biohash. Pattern Recogn. **41**(6), 2034–2044 (2008)
101. J. Zuo, N. Ratha, J. Connell, Cancelable iris biometric, in *19th International Conference on Pattern Recognition (ICPR)* (2008), pp. 1–4
102. S. Hirata, K. Takahashi, Cancelable biometrics with perfect secrecy for correlation-based matching, in *Advances in Biometrics* (Springer, Berlin, 2009), pp. 868–878
103. A.B.J. Teoh, L.Y. Chong, Secure speech template protection in speaker verification system. Speech Comm. **52**(2), 150–163 (2010)
104. O. Osama, N. Tsumura, T. Nakaguchi, Bioencoding: a reliable tokenless cancelable biometrics scheme for protecting iris codes. IEICE Trans. Inf. Syst. **93**(7), 1878–1888 (2010)
105. O. Ouda, N. Tsumura, T. Nakaguchi, Tokenless cancelable biometrics scheme for protecting iris codes, in *20th International Conference on Pattern Recognition (ICPR)* (2010), pp. 882–885
106. J. Pillai, V. Patel, R. Chellappa, N. Ratha, Sectored random projections for cancelable iris biometrics, in *International Conference on Acoustics Speech and Signal Processing (ICASSP)* (2010), pp. 1838–1841
107. L. Leng, J. Zhang, Palmhash code vs. palmphasor code. Neurocomputing **108**, 1–12 (2013)
108. L. Leng, A.B.J. Teoh, M. Li, M.K. Khan, Analysis of correlation of 2d palmhash code and orientation range suitable for transposition. Neurocomputing **131**, 377–387 (2014)
109. A. Kong, K.H. Cheung, D. Zhang, M. Kamel, J. You, An analysis of biohashing and its variants. Pattern Recogn. **39**(7), 1359–1368 (2006)
110. O. Osama, N. Tsumura, T. Nakaguchi, On the security of bioencoding based cancelable biometrics. IEICE Trans. Inf. Syst. **94**(9), 1768–1777 (2011)
111. N.K. Ratha, S. Chikkerur, J.H. Connell, R.M. Bolle, Generating cancelable fingerprint templates. IEEE Trans. Pattern Anal. Mach. Intell. **29**(4), 561–572 (2007)

112. S. Chikkerur, N.K. Ratha, J.H. Connell, R.M. Bolle, Generating registration-free cancelable fingerprint templates, in *2nd International Conference on Biometrics: Theory, Applications and Systems (BTAS)* (2008), pp. 1–6
113. E. Maiorana, M. Martinez-Diaz, P. Campisi, J. Ortega-Garcia, A. Neri, Template protection for hmm-based on-line signature authentication, in *Computer Society Conference on Computer Vision and Pattern Recognition Workshops (CVPRW'08)* (2008), pp. 1–6
114. E. Maiorana, P. Campisi, J. Ortega-Garcia, A. Neri, Cancelable biometrics for hmm-based signature recognition, in *2nd International Conference on Biometrics: Theory, Applications and Systems (BTAS)* (2008), pp. 1–6
115. W. Xu, Q. He, Y. Li, T. Li, Cancelable voiceprint templates based on knowledge signatures, in *2008 International Symposium on Electronic Commerce and Security* (2008), pp. 412–415
116. J. Hämmerle-Uhl, E. Pschernig, A. Uhl, Cancelable iris biometrics using block re-mapping and image warping, in *Information Security* (Springer, Berlin, 2009), pp. 135–142
117. H. Yang, X. Jiang, A. Kot, Generating secure cancelable fingerprint templates using local and global features, in *2nd International Conference on Computer Science and Information Technology ICCSIT* (2009), pp. 645–649
118. C. Rathgeb, A. Uhl, Secure iris recognition based on local intensity variations, in *Image Analysis and Recognition* (Springer, Berlin, 2010), pp. 266–275
119. P. Farberbock, J. Hämmerle-Uhl, D. Kaaser, E. Pschernig, A. Uhl, Transforming rectangular and polar iris images to enable cancelable biometrics, in *Image Analysis and Recognition* (Springer, Berlin, 2010), pp. 276–286
120. E. Maiorana, P. Campisi, J. Fierrez, J. Ortega-Garcia, A. Neri, Cancelable templates for sequence-based biometrics with application to on-line signature recognition. IEEE Trans. Syst. Man Cybern. Part A Syst. Humans **40**(3), 525–538 (2010)
121. C. Lee, J. Kim, Cancelable fingerprint templates using minutiae-based bit-strings. J. Netw. Comput. Appl. **33**(3), 236–246 (2010)
122. L. Nanni, E. Maiorana, A. Lumini, P. Campisi, Combining local, regional and global matchers for a template protected on-line signature verification system. Expert Syst. Appl. **37**(5), 3676–3684 (2010)
123. J. Zhe, A. Jin, Fingerprint template protection with minutia vicinity decomposition, in *International Joint Conference on Biometrics (IJCB)* (2011), pp. 1–7
124. J. Pillai, V. Patel, R. Chellappa, N. Ratha, Secure and robust iris recognition using random projections and sparse representations. IEEE Trans. Pattern Anal. Mach. Intell. **33**(9), 1877–1893 (2011)
125. T. Ahmad, J. Hu, S. Wang, Pair-polar coordinate-based cancelable fingerprint templates. Pattern Recogn. **44**(10–11), 2555–2564 (2011)
126. K. Takahashi, S. Hirata, Cancelable biometrics with provable security and its application to fingerprint verification. IEICE Trans. Fundam. Electron. Commun. Comput. Sci. **94**(1), 233–244 (2011)
127. S. Wang, J. Hu, Alignment-free cancelable fingerprint template design: a densely infinite-to-one mapping (DITOM) approach. Pattern Recogn. **45**(12), 4129–4137 (2012)
128. P. Das, K. Karthik, B.C. Garai, A robust alignment-free fingerprint hashing algorithm based on minimum distance graphs. Pattern Recogn. **45**(9), 3373–3388 (2012)
129. C. Moujahdi, S. Ghouzali, M. Mikram, M. Rziza, G. Bebis, Spiral cube for biometric template protection, in *Image and Signal Processing* (Springer, Berlin, 2012), pp. 235–244
130. M. Ferrara, D. Maltoni, R. Cappelli, Noninvertible minutia cylinder-code representation. IEEE Trans. Inf. Forensics Secur. **7**(6), 1727–1737 (2012)
131. W.J. Wong, M.L.D. Wong, Y.H. Kho, Multi-line code: a low complexity revocable fingerprint template for cancelable biometrics. J. Cent. South Univ. **20**(5), 1292–1297 (2013)
132. W.J. Wong, A.B. Teoh, M.D. Wong, Y.H. Kho, Enhanced multi-line code for minutiae-based fingerprint template protection. Pattern Recogn. Lett. **34**(11), 1221–1229 (2013)
133. W. Yang, J. Hu, S. Wang, J. Yang, Cancelable fingerprint templates with delaunay triangle-based local structures, in *Proceedings of Cyberspace Safety and Security* (2013), pp. 81–91

134. C. Rathgeb, F. Breitinger, C. Busch, Alignment-free cancelable iris biometric templates based on adaptive bloom filters, in *International Conference on Biometrics (ICB)* (2013), pp. 1–8
135. C. Rathgeb, F. Breitinger, C Busch, H Baier, On application of bloom filters to iris biometrics. IET Biom. **3**(4), 207–218 (2014)
136. M. Ferrara, D. Maltoni, R. Cappelli, A two-factor protection scheme for MCC fingerprint templates, in *International Conference of the Biometrics Special Interest Group (BIOSIG)* (2014), pp. 1–8
137. C. Rathgeb, C. Busch, Cancelable multi-biometrics: mixing iris-codes based on adaptive bloom filters. Comput. Secur. **42**, 1–12 (2014)
138. Z. Jin, M.H. Lim, A.B.J. Teoh, B.M. Goi, A non-invertible randomized graph-based hamming embedding for generating cancelable fingerprint template. Pattern Recogn. Lett. **42**, 137–147 (2014)
139. M.V. Prasad, C.S. Kumar, Fingerprint template protection using multiline neighboring relation. Expert Syst. Appl. **41**(14), 6114–6122 (2014)
140. Z. Jin, B.M. Goi, A. Teoh, Y.H. Tay, A two-dimensional random projected minutiae vicinity decomposition-based cancellable fingerprint template. Secur. Commun. Netw. **7**(11), 1691–1701 (2014)
141. M.V. Prasad, P. Swathi, C. Rao, B. Deekshatulu, Minimum spanning tree (MST) based techniques for generation of cancelable fingerprint templates. Int. J. Pattern Recognit. Artif. Intell. **28**(06), 1456,013 (2014)
142. C. Moujahdi, G. Bebis, S. Ghouzali, M. Rziza, Fingerprint shell: secure representation of fingerprint template. Pattern Recogn. Lett. 45, 189–196 (2014)
143. L. Leng, A.B.J. Teoh, M. Li, M.K. Khan, A remote cancelable palmprint authentication protocol based on multi-directional two-dimensional palmphasor-fusion. Secur. Commun. Netw. **7**(11), 1860–1871 (2014)
144. Murillo-Escobar M, Cruz-Hernández C, Abundiz-Pérez F, López-Gutiérrez R (2015) A robust embedded biometric authentication system based on fingerprint and chaotic encryption. Expert Syst. Appl. **42**(21), 8198–8211
145. M. Sandhya, M.V. Prasad, k-nearest neighborhood structure (k-NNS) based alignment-free method for fingerprint template protection, in *International Conference on Biometrics (ICB)* (2015), pp. 386–393
146. M. Sandhya, M.V. Prasad, R.R. Chillarige, Generating cancellable fingerprint templates based on delaunay triangle feature set construction. IET Biom. **5**(2), 131–139 (2016)
147. M. Prasad, J. Anugu, C. Rao, Fingerprint template protection using multiple spiral curves, in *Proceedings of 3rd International Conference on Advanced Computing, Networking and Informatics* (2016), pp. 593–601
148. S. Wang, J. Hu, A blind system identification approach to cancelable fingerprint templates. Pattern Recogn. **54**, 14–22 (2016)
149. R.M. Bolle, J.H. Connell, N.K. Ratha, Biometric perils and patches. Pattern Recogn. **35**(12), 2727–2738 (2002)
150. N. Ratha, J. Connell, R.M. Bolle, S. Chikkerur, Cancelable biometrics: a case study in fingerprints, in *18th International Conference on Pattern Recognition*, vol. 4 (2006), pp. 370–373
151. J. Camenisch, M. Stadler, Efficient group signature schemes for large groups, in *Advances in Cryptology – CRYPTO'97* (1997), pp. 410–424
152. R. Cappelli, M. Ferrara, D. Maltoni, Minutia cylinder-code: a new representation and matching technique for fingerprint recognition. IEEE Trans. Pattern Anal. Mach. Intell. **32**(12), 2128–2141 (2010)
153. J. Bringer, C. Morel, C. Rathgeb, Security analysis of bloom filter-based iris biometric template protection, in *8th IAPR International Conference on Biometrics (ICB)* (2015), pp. 527–534
154. Y.C. Feng, P.C. Yuen, A.K. Jain, A hybrid approach for generating secure and discriminating face template. IEEE Trans. Inf. Forensics Secur. **5**(1), 103–117 (2010)

155. A. Nagar, K. Nandakumar, A.K. Jain, A hybrid biometric cryptosystem for securing fingerprint minutiae templates. Pattern Recogn. Lett. **31**(8), 733–741 (2010)
156. H. Liu, D. Sun, K. Xiong, Z. Qiu, A hybrid approach to protect palmprint templates. Sci. World J. **2014**, 686754 (2014)
157. Y. Chin, T. Ong, A. Teoh, K. Goh, Integrated biometrics template protection technique based on fingerprint and palmprint feature-level fusion. Inf. Fusion **18**, 161–174 (2014)
158. A. Kumar, A. Kumar, A cell-array based multibiometric cryptosystem. IEEE Access **4**, 15–25 (2016)
159. M. Sandhya, M.V.N.K. Prasad, Cancelable fingerprint cryptosystem based on convolution coding, in *Advances in Signal Processing and Intelligent Recognition Systems: Proceedings of Second International Symposium on Signal Processing and Intelligent Recognition Systems (SIRS)* (2016), pp. 145–157
160. Z. Jin, A.B.J. Teoh, B.M. Goi, Y.H. Tay, Biometric cryptosystems: a new biometric key binding and its implementation for fingerprint minutiae-based representation. Pattern Recogn. **56**, 50–62 (2016)
161. Z. Erkin, M. Franz, J. Guajardo, S. Katzenbeisser, I. Lagendijk, T. Toft, Privacy-preserving face recognition, in *Privacy Enhancing Technologies* (Springer, Berlin, 2009), pp. 235–253
162. A.R. Sadeghi, T. Schneider, I. Wehrenberg, Efficient privacy-preserving face recognition, in *Information, Security and Cryptology* (Springer, Berlin, 2009), pp. 229–244
163. S.D. Rane, W. Sun, A. Vetro, Secure distortion computation among untrusting parties using homomorphic encryption, in *16th International Conference on Image Processing (ICIP)* (2009), pp. 1485–1488
164. M. Barni, T. Bianchi, D. Catalano, M.D. Raimondo, R.D. Labati, P. Failla, D. Fiore, R. Lazzeretti, V. Piuri, A. Piva, F. Scotti, A privacy-compliant fingerprint recognition system based on homomorphic encryption and fingercode templates, in *International Conference on Biometrics: Theory Applications and Systems (BTAS)* (2010), pp. 1–7
165. M. Barni, T. Bianchi, D. Catalano, M. Di Raimondo, R. Donida Labati, P. Failla, D. Fiore, R. Lazzeretti, V. Piuri, F. Scotti et al., Privacy-preserving fingercode authentication, in *Proceedings of the 12th ACM Workshop on Multimedia and Security* (2010), pp. 231–240
166. M. Osadchy, B. Pinkas, A. Jarrous, B. Moskovich, SCiFi-a system for secure face identification, in *Symposium on Security and Privacy (SP)* (2010), pp. 239–254
167. R. Kulkarni, A. Namboodiri, Secure hamming distance based biometric authentication, in *International Conference on Biometrics (ICB)* (2013), pp. 1–6
168. C. Karabat, M.S. Kiraz, H. Erdogan, E. Savas, Thrive: threshold homomorphic encryption based secure and privacy preserving biometric verification system. EURASIP J. Adv. Signal Process. **2015**(1), 1–18 (2015)
169. M. Gomez-Barrero, J. Fierrez, J. Galbally, Variable-length template protection based on homomorphic encryption with application to signature biometrics, in *4th International Conference on Biometrics and Forensics (IWBF)* (2016), pp. 1–6
170. A. Nagar, A. Jain, On the security of non-invertible fingerprint template transforms, in *First International Workshop on Information Forensics and Security (WIFS)* (2009), pp. 81–85
171. A. Nagar, K. Nandakumar, A.K. Jain, Biometric template transformation: a security analysis, in *IS&T/SPIE Electronic Imaging*, vol. 7541 (2010), p. 15
172. K. Takahashi, T. Matsuda, T. Murakami, G. Hanaoka, M. Nishigaki, A signature scheme with a fuzzy private key, in *13th International Conference on Applied Cryptography and Network Security (ACNS)* (2015), pp. 105–126
173. K. Takahashi, T. Murakami, Optimal sequential fusion for multi-biometric cryptosystems. Inf. Fusion **32**, 93–108 (2016)
174. C. Li, J. Hu, J. Pieprzyk, W. Susilo, A new biocryptosystem-oriented security analysis framework and implementation of multibiometric cryptosystems based on decision level fusion. IEEE Trans. Inf. Forensics Secur. **10**(6), 1193–1206 (2015)
175. A. Adler, R. Cappelli, Template security. *Encyclopedia of Biometrics* (Springer, New York, 2009), pp. 1322–1327

176. M. Savvides, B. Kumar, P. Khosla, Cancelable biometric filters for face recognition, in *17th International Conference on Pattern Recognition (ICPR)*, vol. 3 (2004), pp. 922–925
177. J. Ortega-Garcia, J. Fierrez-Aguilar, D. Simon, J. Gonzalez, M. Faundez-Zanuy, V. Espinosa, A. Satue, I. Hernaez, J.J. Igarza, C. Vivaracho, D. Escudero, Q.I. Moro, MCYT baseline corpus: a bimodal biometric database. IEE Proc. Vision Image Signal Process. **150**(6), 395–401 (2003)
178. M. Sandhya, M.V. Prasad, Multi-algorithmic cancelable fingerprint template generation based on weighted sum rule and T-operators. Pattern. Anal. Appl. (2016). doi:10.1007/s10044-016-0584-5
179. M. Sandhya, M.V. Prasad, Cancelable fingerprint cryptosystem using multiple spiral curves and fuzzy commitment scheme. Int. J. Pattern Recognit. Artif. Intell. (2016). doi:10.1142/S0218001417560043

Chapter 15
A Survey on Cyber Security Evolution and Threats: Biometric Authentication Solutions

Leila Benarous, Benamar Kadri, and Ahmed Bouridane

15.1 Introduction

It is the age of information; almost everything is digitalized and connected, thanks to the internet. What took hours a few decades ago now takes seconds, what was considered a mere science fiction is now a reality and most of what was impossible is now possible.

The massive use of internet and electronic devices greatly requires and depends on the security and privacy as people are becoming more active in the cyberworld, sharing massive information especially via social networks which are the newest type of addiction that people cannot live without, our privacy is being threatened.

Ordinary users believe that the internet is safe and that they are immune from threats and are not targeted by attackers maybe it is the human nature to be trusty and optimistic to feel safety and security but the truth is that this is just an illusion; the cybersecurity is being threatened every day, at least one million of new viruses and malware is released daily, and more than 100,000 cyberattacks every hour which costs over 100 billion dollar yearly [1].

Ensuring security is mandatory and is rather a difficult task because of many reasons:

L. Benarous (✉) • B. Kadri (✉)
Department of telecommunications, Faculty of Technology, University of Abou Bekr Belkaid, Tlemcen, Algeria
e-mail: leila.benarous@mail.univ-tlemcen.dz; benamarkadri@yahoo.fr

A. Bouridane
Department of Computer Science and Digital Technologies, Northumbria University, Newcastle upon Tyne, UK
e-mail: ahmed.bouridane@northumbria.ac.uk

R. Jiang et al. (eds.), *Biometric Security and Privacy*, Signal Processing for Security Technologies, DOI 10.1007/978-3-319-47301-7_15

- Firstly, the number of new malwares released everyday makes it difficult to keep the track. In addition, the new viruses rely on new techniques to avoid detection such as self-changing code on each infection or the use of cryptography.
- Secondly, the attacks may come from the inside of the networks making it more effective as the attacker might have higher access rights.
- Thirdly, the attacker is unknown (his/her level, techniques and motivation).
- Finally, attacks and malwares exploit the system and software vulnerabilities which may include the vulnerabilities of the security system itself.

Regardless of all the previously mentioned challenges, the security protectors keep developing new techniques and search for new solutions to keep the networks, systems and data safe.

The main focus of this chapter is to survey the cyberworld security and its evolution as the technologies of networks, systems and data saving are developing. The survey includes a historical overview, the security objectives, its issues and their classification, its solutions including the current solutions (authentication, steganography, ..., etc.) and future tendencies such as biometric solutions for authentication, its attackers and threat creators and their different classifications. Finally, as the cybersecurity crimes are spreading and governments are taking legal action against attackers, a brief description of forensic investigation process is included.

15.2 The CyberWorld

The cyberworld and virtual world are other words referring to the internet. The simplified definition of internet would be: "a set of computers or machines linked to huge servers that provide different services for the machines to use". This is not the technical definition, rather it is an easy way to define the concept and technology.

We, humans, are the users of the machine which can be a computer, smartphone, tablets, smart TVs or any other device that can connect us to the internet.

The services are the websites we use such as social media (Facebook, twitter, YouTube, ..., etc.), mailing websites (Yahoo, Gmail, Hotmail, ..., etc.), e-commerce websites (eBay, PayPal, Alibaba, Amazon, ..., etc.) or any other websites that we frequently use.

The server for a normal network user means nothing but a black box and is totally invisible, all what is clear for a normal user is that he/she contacts an internet service provider, pays a bill and he/she is connected to the cyberworld and enjoys its services.

15.3 Darknet or Deep Internet

The internet defined above is what most of the users know but there are other terms such as the darknet or deep internet; it is a part of the internet that is accessible via special tools such as Tor [2] or Riffle from MIT [3], and it contains websites that cannot be accessed directly from normal search engines like Google. One of the definitions of darknet is: "a class of networks that aim to guarantee anonymous and untraceable access to web content and anonymity for a site" [4]. This dark side of the internet is used by hackers and dangerous criminals because it ensures intractability.

Although cybercrimes occur in both the clearnet and the darknet, the crimes of darknet are out of this chapter's scope, instead more focus is forwarded to security threats to clearnet, the systems and information.

15.4 Internet Usage

This section includes some frequently used online services that make us so dependent on the internet in general with all its components and that put our privacy at stake. The mostly used internet services are regrouped by category from Alexa websites ranking.

- *Search engines and mailing services*, such as Google, Yahoo and MSN
- *Social media networks;* include video sharing websites (YouTube and Daily motion), online TV, social services (Facebook, Twitter, Instagram, Weibo, ..., etc.) and blogging websites (Tumbler, Blogger, ..., etc.)
- *E-commerce services*: all websites allowing transactions or establishing business, banking or simply buy-sell websites such as Alibaba, Amazon, eBay, PayPal
- *E-learning*: includes websites used for learning that present lectures, webinars or training, also encyclopaedias such as Wikipedia, ask/answer forums like Stack OverFlow or Ask, and it includes also scientific journals and academic websites
- *News and entertainment*: journals, magazines, health, sports, map and weather services are few examples
- *Cloud services, storage and file sharing*: include Google drive, drop box, media fire, ..., etc.
- *Gaming*: online gaming websites such as Twitch and Battle, ..., etc.
- *Advertising*: such as onclickads
- *Personal/organizations websites*: websites of corporation and individual personalities
- *Industrial monitoring and controlling*, which includes remote access services; controlling and monitoring industrial equipment such as robots [5]

15.5 History of Cybersecurity

> "The cyber issue is not new, but rather has taken a half-century to develop. Indeed, it was already decades old before the general public and many senior leaders recognized its salience in the mid-1990s." *Michael Warner*

Historians think that security issues appeared as early as 1960s. However, in theory it appeared much earlier, more precisely in 1949 when the mathematician John von Neumann defined the virus and worm as self-replicating automata [6].

Although the theoretical concept has existed for so long, it was only after the successful security attacks have occurred that the importance of cybersecurity was emphasized. The attacks cost fortunes, exposed secrets and endangered lives.

Historians observed that the cybersecurity concerns took different trends through decades and that it was influenced by political and sociological events. For example, during 1980s, the cybersecurity concern was foreign espionage. During the 1990s, the threat has been directed towards civilian critical infrastructures and later on, the terms cyberterrorism, organized crime and hacking were used to describe the attacks [7].

Based on their chronological order, the major security issues that occurred in the past five decades can be described as follows:

- In 1964, AT&T started monitoring Phone Freaks also known as Phreaks who were able to make free phone calls.
- In 1968, potentially the world's first case of computer espionage was carried out by a German spy in German subsidiary of IBM.
- In 1971, hackers made the viral *Creeper* worm which infected ARPANET.
- In 1981, *Elk Cloner* was made by 15 years old Richard Skrenta.
- In 1982, a group of teenagers hacked high-profile computers in the USA; the operation was called *414 s break-ins*. A similar operation named *Cuckoo's eggs* occurred from 1986 to 1988.
- In 1988, *Morris* worm was released and ARPANET recorded its first major network incident.
- In 1990, the first self-modifying viruses were created.
- In 1992, *Michelangelo* virus was released overwriting the first 100 sectors of the hard disk causing what was called the first digital mass hysteria.
- In 1994, two major hacking operations were recorded, the *RomeLab* incident and the *Citibank* incident.
- In 1995, the first Microsoft word-based virus was created.
- In 1998, *Back Orifice* was released to give remote system administration (Trojan Horse). Also the hacking incidents of *Solar Sunrise* and *Moonlight Maze* were recorded that year.
- In 1999, *Melissa* spread by infecting emails.
- In 2000, *I Love You* spread by sending itself to the first fifty people in windows address book. In the same year, the computers of California University were used by hackers to crash the websites of Amazon, eBay and Yahoo.

- In 2001, *Code Red* and *Nimda* worms were used to take over the control of computers and use them for DDoS. In the same year, the first cyber world war was recorded caused by a conflict between USA and China, where hackers from many countries in the world participated.
- In 2003, *Slammer* and *Blaster* were used to launch DDoS attacks. Also Titan Rain aimed to access high-profile computers in the USA.
- In 2004, *Sasser* caused systems to crush and internet to slow down.
- In 2007, *Zeus* was used to steal banking and other information.
- In 2008, *Conficker* worm was used to form botnets. Also K*oobface* virus spreads through email or social media service such as Facebook, causing fake buying operations and thus money theft.
- In 2009, *GhostNet* and *Operation Aurora* incidents were launched, the first was for cyberspying and the second was against Google and other high-tech companies to access and modify their codes.
- In 2010, *Stuxnet* was used for industrial espionage, and *WikiLeaks* launched an attack called *wikileaks* cable gate where confidential diplomatic cables were leaked.
- In 2011, *Duqu* was released (a modified copy of *Stuxnet*) also the *Ramnit* virus which steals Facebook accounts and passwords. The year recorded highly publicized hacking attacks such as: the attacks against Sony and other corporations, against governments and the theft of CO_2 emission allowance.
- In 2013 and 2014, many hacking operations were done, and most of the attacks focused on stealing Facebook and email credentials [6, 8–10].

15.6 Security Objectives

The objectives presented here are the requirements of the security solutions. Originally, there were three principles which are the availability, the confidentiality and the integrity but three other principles were added as the security needs required which are the authenticity, non-repudiation and auditability. The six objectives are explained in this section as follows:

- *Availability*: allows the authorized parties to access to the systems and required data resources, i.e. the system can always provide its services to the authorized requester and the resources can be used when needed by the authorized requester as well. One of the most dangerous threats to availability is the Denial of Service attack which prevents the authorized users to access a system (resource) and thus preventing them from using its services [6].
- *Confidentiality (privacy)*: ensures that an asset (personal data or resource) is accessed by the authorized person only [11, 12]. One of the major threats to privacy is the exposure of data.

- *Integrity*: ensures that the data is modified by the authorized person only. This means that the content is coherent and authentic and has not been altered by a third party [11].
- *Authenticity*: the ability of a system to confirm the identity of a sender where he is first identified then authorized (or denied) to access the system [11].
- *Non-repudiation*: it ensures that the sender (originator of the message) cannot deny sending the message [11]. This principle was added because the users would deny having sent the message if they are legally prosecuted especially in cases where the message contains illegal content, threats or in e-commerce the user's denial of purchase.
- *Auditability*: it was added by the US Department of Defence and it is "the ability of a system to trace all actions related to a given asset" [11]. The system records all important events such as sending and receiving messages, the IP addresses and any other relevant information to help detect security issues and bugs, trace the intruders, and resolve and document the issues.

15.7 Security Threats

After emphasizing above how dependable we have become on the cyberworld and the importance of security by stating some of the major security threats recorded, this section discusses the security threats which can be classified into two main types: *physical* and *logical*.

The *physical attacks* require the physical presence of the attacker and his/her direct interaction with the system of network components. This usually results in the loss of the device or equipment, its destruction or generating noise and interfering with wireless signals.

On the other hand, the *logical threats* can be divided into three classes depending on the target: the *information*, *networks* and *systems*. The attacks on the networks are depicted in the table below where different kind of networks (infrastructure-based and ad hoc) are presented along with major threats related to each type. In Fig. 15.1 and Table 15.1 the previously mentioned problems and their solutions are presented which will be further explained in the next section.

The following describes some of the attacks and security threats previously mentioned:

- *Denial of Service (DoS)*: it is an attack that aims to overtake a network or a machine and prevent it from providing services. This kind of attack can be done via flooding attacks or exploiting technical vulnerabilities of a system or a protocol. The Distributed Denial of Service (DDoS) attack is more dangerous since it uses multiple machines to launch the attack in order for the attacker(s) to overtake the machines (often these attacks are also called zombie or bot machine and they form a botnet to launch the DDoS attack). DoS attacks can be addressed by filtering the IP addresses and dropping the packets coming from the attacker's

IP. This can become very hard in DDoS since the attacker uses multiple machines each sending a certain amount of packets to cause the denial of service of the victim sever (it is a synchronized attack, sending multiple packets from different machines); in this case, filtering the IP addresses and detecting the attacking machine become hard [13].

- *Spoofing* or *Impersonation*: it is an attack where the attacker pretends to be another node which in most of the times is a trusted node. Among spoofing attacks there are:
  - *The IP spoofing attack* where a node changes its IP address to another address that is different from its real source address.
  - *ARP spoofing*, it aims to send fake replies to ARP queries, which results in mapping the IP address to a fake MAC address and thus poisoning the ARP cache.
  - *The DNS spoofing* which is similar to ARP spoofing and results in directing the requester to the wrong webpage or mailing service by mapping the pages to the wrong addresses in DNS cache, also known as Kaminsky attack where the attacker tries to poison the DNS cache by fabricating a reply to a specific request; the attacker mainly succeeds in the attack because he can guess the sequence number of the DNS reply known as the Query ID [14, 15].
- *Flooding*: it is an attack that aims to saturate the network or resources of targeted node by sending enormous amount of packets such as TCP SYN packets; or PING packets.
  - In the *SYN flooding*, the attacker sends SYN messages from different IP addresses to the victim server, the server reserves a memory space for the information related to the half opened connection, replies with ACK and waits for the ACK of the attacker or the expiry of the session, with many half opened requests the memory space allocated for storing connection information will reach its limit causing the denial of service.
  - In *Ping Flooding attack* also known as the *ping of death*, the attacker sends multiple ping requests to cause the absence of response of the victim.
  - There is another type of flooding which is the *UDP flooding*; it relies on sending a large number of UDP packets to the victim resulting in the Denial of Service [13, 16].
- *Jamming*: a physical layer attack which interferes with the radio frequency used for communication between nodes of the network. The noise generated disrupts the communication between the nodes causing collision and failure [17].
- *Man in the middle* (active interference): it is an attack where the eavesdropper generates, alters or drops packets. The attacker first intercepts the traffic, breaks the authentication chain and then impersonates the hacked endpoints seamlessly [18].
- *Node isolation*: partitioning the network by preventing a node or a set of nodes from communicating with the rest of the network [19].

- *Route disruption:* one of the routing problems where the routes are modified by faking the route replies so that routing loops are created, or packets are forwarded along erroneous, non-optimal or non-existing routes [19].
- *Resource consumption:* decreasing the network performance by consuming network bandwidth or node resources such as the memory or the energy [19].
- *Eavesdropping* also known as disclosure attacks, where the attacker intercepts and analyses the network traffic (broadcasted messages); it is a passive attack since the user does not alter, create or drop the packets [19].
- *Wormhole (Tunnelling)*: the wormhole attack is possible even if the attacker has not compromised any host and even if all communication provides authenticity and confidentiality. In the wormhole attack, an attacker records the packets (or bits) at one location in the network, tunnels them (selectively) to another location and retransmits them from there into the network [20].
- *Timing attacks*: the attacker pretends to be closer to the victim nodes than it really is. This can be done either by sending a fake route reply (rush attack) or by flooding the network with hello messages (hello flood attack) that are transmitted with high power, enough to make the victim node believe that the attacker is a neighbour node [19].
- *Stealthy attack*: it is similar to DoS attack but the attacker does not control the node instead he/she manipulates the routing table of the victims so as the routed packets cause disruption to the network. Another form of stealthy attack is also based on the manipulation of routing table and allows the attacker to eavesdrop nodes that are out of his range by using the victim node [21].
- *Black hole attack:* it is an attack launched by an insider which fails to propagate messages or packets and drops it instead [22].
- *Sinkhole attack*: the attacker or malicious node attracts network traffic and then does a selective forwarding [23].
- *Spamming*: hackers often use this technique to deliver the recipient unwanted emails such as advertisements, to spoof the user credentials (phishing) or to crash the mailing server by sending millions of spams, an attack known as mail bombing [24].
- *Position faking attack and GPS spoofing*: in this attack, the attacker generates a stronger localization signal than that of the GPS signal with a fake position so as the receiver node gets fake information [25].
- *Sybil attack*: the ability for the attacker node to have multiple identities which is dangerous because the node can use the multiple identities to generate additional votes in election algorithms, to route multiple paths through it (the malicious node) or to avoid traceability [25, 26].
- *Masquerading attack*: the attacker node poses as a legitimate node to conduct other attacks such as injecting false messages [22].
- *Replay attack*: attacker node records exchanged packets to retransmit them later [27].
- *Session hijacking*: the attacker forges unprotected session after it has been initialized by forging the IP address and computing the sequence number to launch DoS attacks [27].

- *Phishing*: an attack against computer users to convince them into performing an action that can cause them harm. It can be done by sending emails to the victim where they have to download an attachment or click a link. The attachment could contain malwares such as viruses, spyware or key logger that either steal the information of the user, slow the performance of the system or destroy it. The link would direct the user to a fake webpage where he could be asked to provide sensitive credentials [28].
- *Password cracking*: it can be done either by guessing the password, stealing the file containing the password (if stored as clear text), brute force or dictionary attacks or known through social engineering techniques. Noting that a password is generally not saved as clear text but as hash values and sometimes even salted hash values (a salt is a random value added to a password before it is hashed, and it ensures that two passwords would not have the same hash) [29].
- *Teardrop*: an attack that can be used to cause denial of service. It exploits the features of IP datagram which is that a single data unit can be of a variable length and can be fragmented into smaller pieces and transmitted; each piece will indicate its position in the original data unit and length so as the receiver reassembles the pieces to get the complete data sent. The attacker would use those features, but the position of the pieces will overlap causing the system to crash while trying to assemble them [30].
- *Social engineering*: the attacker deceives the users to give him/her confidential information such as password, system configuration or contact information. The social engineering uses psychological tricks, since human often tends to trust, sympathize and help each other. The hacker can use these characteristics to trick his/her victims. The hacker often does background research about the targeted company and collects useful information which he can use in gaining the trust of the victim. The attacker pretending to be an insider can contact the victim by phone, email or in person and request sensitive information. This attack is also known as human hacking [31].
- *Technical vulnerabilities*: it includes the system, network protocol and database vulnerabilities, for example: SQL injection, buffer overflow, ..., etc. [32].
- *Data mining*: it refers to a process of nontrivial extraction of implicit, previously unknown and potentially useful information from databases [33].
- *Dumpster diving* is also known as *trashing* and is one of social engineering techniques. The attacker collects the trash of companies to look for useful information such as contacts, password, old hard disk, printouts of code source, security system design, system configuration, ..., etc. The hacker can use the collected information to break-in the system or network [30, 34].
- *Fragmentation attacks*: fragmentation of a packet into tiny pieces sometimes overlapping pieces to create problems to the receiver if there is no minimum fragment size and offset [35].
- *Malware*: it is a portion of codes intended to cause harm. It includes:

- *Viruses*: "Fred Cohen defines the term computer virus as a program that can 'infect' other programs by modifying them to include a possibly evolved copy of itself" [36].
- *Ad ware*: tools for advertising; these software might install other tools without permission [37].
- *Spyware*: tools used to gather information about a user or organization without their knowledge [37].
- *Rootkit*: a malicious code that hides its presence in the system by intercepting and filtering commands of operating systems that might lead to its discovery, and when discovered and removed, it could re-establish itself. It is named rootkit because it operates as root [30].
- *Worm*: unlike the virus which requires the user taking action such as executing or copying the carrier program for it to propagate, worms are malicious programs that self-propagate across a network exploiting security flaws in used services [38].
- *Scripts*: also known as the cross-site scripting (XSS) vulnerabilities, allow the attacker to include malicious code typically written with JavaScript in the website sent to a victim's browser. The code can log the keyboard input, steal credentials or the session identifier and many other forms of attacks [39].
- *Trojans or Trojan horses*: a program with overt and covert functions, the overt function is what expected from the program, the covert is the unwanted and unexpected function that represents the threat to the user or a system. Examples of covert function: key logging, stealing credentials, ..., etc.

  Remarque: unlike viruses, a Trojan does not replicate [40, 41].
- *Backdoor*: it is often used by hackers to control computers to create bot networks and launch DoS attacks; it is a method to access a system by bypassing authentication and security controls [42].

– *Right alteration*: if an attacker gets to access the system as an administrator or a root, he could change the rights of legitimate users giving them the least rights.
– *Repudiation and right violation:* a user can deny being the originator of data and this action is known as repudiation, while right violation is when someone tries to steal credits of a digital work pretending to be the real owner.
– *Destruction*: can either be physical attack such as smashing the electronic devices, magnetizing it, using chemical products or putting it in fire or it can also be logical such as tampering or damaging the files and data [43].
– *Fraud and theft*: fraud is altering the data integrity for gain such as falsifying transactions. Theft includes software and hardware theft and it is the act of trading information, secrets or the stolen hardware for profit [44].

15.8 Security Solutions

This section describes some of the security solutions and measures mentioned in Table 15.1 and Fig. 15.1, such as: the authentication, steganography, cryptography, anti-malware, intrusion detection and prevention systems, firewalls, freshness protection, virtualization, backup, patches and user education.

15.8.1 Triple A (AAA)

It is the first security measure against systems, networks and databases intruders as it identifies the authorized user, grants them with specific access rights and records what occurs in the system.

- *Authentication*: it is the process of identifying a user or a person, and its mechanisms can use three qualities:
    - *Knowledge-based authentication*: "*Something the user knows*" such as text passwords, schemes or image password, secret question, ..., etc. It is the most basic and common way of securing systems, information or networks. Providing the user's ID and password is the first mandatory step to get a secured access. The major problem with using password is that we can forget them, for that we often look for ways to memorize them such as writing them in a diary, using tools to organize and save passwords (password managers or password keepers) or the use of "Cookies" which stores the user ID and password (hashed) to access websites. Another problem with this technique is that passwords can be guessed or cracked [30, 41, 50].
    - *Possession-based authentication*: *"Something the user has"* like tokens, dongles, badges, smart card, ..., etc. It is commonly used in workplaces as workers often carry badges and smart cards also in hotels with electronic door locks (smart keys). It can exploit many technologies such as RFID (Radio frequency), Bluetooth or via USB. Yet, this method also has drawbacks, such as losing the hard key or forgetting it [50].
    - *Physiology-based authentication*: "*Something the user is*" also known as biometrics and is based on the physical characteristics of a person, such as: face, eyes (retina and iris), voice, veins, fingerprint and hand geometry [30] and also the behavioural characteristics such as the keystroke rate, signature, gait and handwriting. The biometric authentication relies on taking sample of the authorized persons and storing them in a database to compare them later during the verification process (to identify and authorize the persons to access) [14].

Table 15.1 Networks security threats

Network		Security threats (non-exhaustive list)	Solutions
Ad hoc networks	Mobile	Viruses and worms -Denial of Service (SYN flooding, jamming or hello flood) -Monitoring, traffic analysis and eavesdropping -Active interference -Route disruption -Node isolation -Wormhole -Resource consumption -Fabrication, modification and dropping attacks -Timing attacks [19]	-Secure routing: authentication and cryptography -Intrusion detection techniques [19]
	Vehicular	-Malware -Spamming -Denial of service -Black hole attack -Broadcast tampering -Replay attack -Position faking -Global positioning system spoofing -Masquerading -Sybil attack -Certificate/key replication -Message tampering/manipulation -Tunnelling [45]	-Redundancy provides resilience to black holes -Authentication -Cryptography -Encrypted precise positioning system to avoid GPS spoofing -Rogue repudiation basing on reputation model [22]
	Sensor	-Spoofing -Modification -Replay attack -Eavesdropping -Replication attack -Camouflage -Jamming -Tampering -Collision -Resource exhaustion -Unfairness in location -Selective forwarding -Sinkhole -Sybil -Wormhole -Flooding (hello, SYN, ACK) -De-synchronization -Stealthy attack [26]	-Authentication -Cryptography algorithms and software -Crypto-processors (against physical tampering) -Jammed area isolation -Spread spectrum techniques to prevent jamming -Randomness multiple paths in routing, using probabilistic routing and the introduction of fake messages in the network as a solution against traffic analysis -Radio source testing, random key distribution, central registration of node and position verification as a solution again Sybil attack [26, 46]

(continued)

Table 15.1 (continued)

Network		Security threats (non-exhaustive list)	Solutions
	Body	-Denial of service -Eavesdropping -Replay attack -False data injection -Unauthorized access -Selective reporting -Data alteration -Impersonation [47, 48]	-Encryption -Authentication and authorization -Intrusion detection and prevention -Freshness protection [47, 48]
Infrastructure based	Internet	-DNS cache poisoning -Spamming -DNS spoofing -Denial of Service, Distributed Denial of Service -Session hijacking -Malwares (worms, viruses, spyware, adware, scripts and Trojans) -Hacking and social engineering -Password cracking -Exploiting technical vulnerabilities of the protocols and systems (SQL injection, buffer overflow, ..., etc.) -IP spoofing -ARP spoofing -Phishing and web spoofing -Flooding -Teardrop attack [14]	-Anti-malware (antivirus, anti-spyware, etc.) -Firewall (hardware and software) -Intrusion detection and prevention -Encryption -Authentication and authorisation -Securing internet explorers and disabling scripts -Educating the users -Backups -Applying patches -Detecting vulnerabilities in protocols and systems and disabling unnecessary services -Check configurations [14]
	Internet of things	-Denial of service -Physical attacks -Privacy attacks: eavesdrop-ping, traffic analysis and data mining [49]	-Authentication and authorization -Identity management -Trust management -Key exchange and management -End to end security: applying cryptography and hashing (IPsec, TLS) [49]
	Cloud computing	-Eavesdropping -Sabotage (denial of service) -Theft -Fraud -External attacks (scanning vulnerabilities, and malware) -Fragmentation attacks -Intrusion and session hijacking -Backdoor -Spoofing -Man in the middle -Replay attack -Dumpster diving -Social engineering and password guessing [44]	-Intrusion detection -Firewalls and layered defence mechanisms -Virtualization -Authentication and authorization -Backup -Encryption -Network security protocols (for example: VPN) [44]

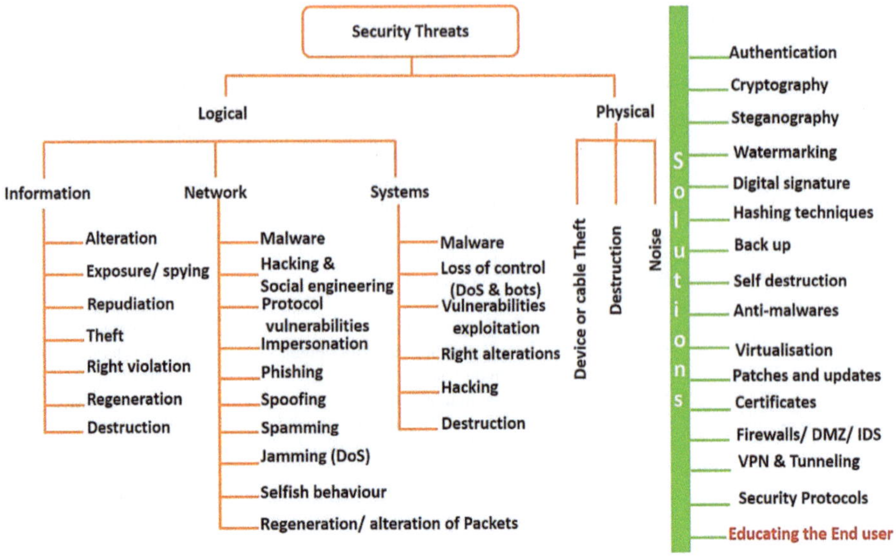

Fig. 15.1 Classification of security threats

An efficient biometric solution has the following five characteristics:

- *Robustness* which means that it remains unchanged overtime,
- *Distinctiveness* which means that it is unique for each individual,
- *Availability* which means that all individuals have it,
- *Accessibility* which means that it is easy to scan and capture its features using sensors,
- And *acceptability* which means that people do not object taking the measurement [51].

Biometrics offer a secured authentication since it is not easy to fake, steal or fabricate biometrics characteristics and it holds a very bright future. However, it is not as reliable as it was predicted to be as it became possible to fake fingerprint per example, although other physiological characteristics are still considered secure such as retina, iris, ..., etc. That is why more researches are being conducted in this field [14]. Further details of the biometric solutions are explained in the case study surveying them (see Sect. 15.10).

- *Authorization*: it enables determining the user who, once identified, is permitted to have the resources. It is implemented through the use of access control mechanisms by granting or denying the access to a resource according to a set of criteria. It differentiates the users by granting different set of rights and privileges [49].
- *Accounting or Auditing*: Monitoring and tracking the activity related to the use of a resource gives the security administrator the ability to know what actually

occurs in the system and the network. This allows him/her to have a clear view of what is supposed to happen, thus giving him/her a higher chance of detecting the vulnerabilities and correcting them [14].

15.8.2 Cryptography

Cryptography is the science of securing messages by scrambling the content so it will not be understood by a third party. One of the earliest uses of cryptography is credited to Julius Caesar. The cryptography continued to develop over the year. It has two types: the symmetric cryptography and the asymmetric cryptography. Chronologically, the secret key cryptography came first, but since the secured distribution and management of these keys were the drawbacks of this method, the asymmetric cryptography came to resolve this problem by relying on the use of a pair of keys one is public used for encryption and the other is private used for decryption [52, 53].

The cryptography is the key technique for many security solutions such as: IPsec, VPN, SSL and TLS.

15.8.3 Steganography

The concept was first given by Johannes Trithemius (1462–1516) in his book "steganographia" and is created from the combination of two word "stegano" which means hidden and "graphia" which means writing. It is the art and science of securing message exchange by hiding it with innocent object.

Digital steganography includes hiding data of digital format into another innocent cover such as images, videos, audio files, documents (word, PDF, web pages, etc.), compressed files (rar, zip, etc.), virtual machine disk file, network protocols (headers) or operating system files, ..., etc.

For each type of cover file used, the concept aims to embed the message in such a way to make the alteration undetectable visually. The alteration might cause the stego file (the resulting file with the embedded data) to be larger in size, with lower quality or include noise that can be perceptible by human eye or ear (images, videos and audio files) [54–56].

The steganography techniques can be dangerous if misused; many hackers and cybersecurity attackers are using it to launch attacks or seal deals secretly, per example your computer can be hacked as soon as you open an image in your browser [57] and data exfiltration can be successfully undetected using steganography [58].

In the following, a brief overview of some of the existing steganography techniques is given:

- *Steganography in images*: images represent a perfect steganography cover file as it is easily exchanged, can hold a fairly high size of secret data and is innocent and does not draw suspicions. Many methods are developed to embed the data such as hiding data in the least significant bit of the pixel which is one of the most used methods that hides the message by changing the least significant bit from a byte, thus making the changes undetectable especially in images since the changed byte (pixel) will give an approximately the same colour (shades of the same colour), in the Discrete Cosine Transform (DCT), Discrete Wavelet Transform (DWT), Discrete Fourier Transform (DFT), in the colour pallet or at the end of the file. Note that the JPEG images are the mostly used cover files as they have a small size (compressed) and are widely used thus non-suspicious [54].

- *Steganography in audio*: steganography can be applied to audio files by hiding the secret data within the audio signal in an imperceptible way, i.e. the resulting signal has the same characteristics and the changes made to the audio file will likely not be detected by the human ear [59].
- *Steganography in videos*: videos have a larger size than images and can store more secret data, and they are innocent and easily exchanged. Steganography techniques in videos can exploit the fact that a video is composed of a set of images where the DCT, DFT, LSB and DWT techniques can be used, other techniques embed the data in the motion vector and some techniques rely on embedding data by exploiting the encoding/decoding process or by embedding data in the audio and images of the video. Steganography techniques can be extended to work on streaming videos [54].
- *Steganography in protocols*: the steganography techniques can use protocols headers to embed data, per example the sequence number field in TCP protocol, the TTL (Time To Live) in IP protocol or the VoIP protocol which generates a great number of small packets that can be used to embed small amounts of secret data [54].
- *Text and linguistic steganography*: some steganography methods rely on the semantic of the words or its spelling to hide data, per example some methods hide data within English text by altering the spelling of the words between the US and UK [60], or by changing the dots, punctuation or the stressing of the words in Arabic language [61]. Steganography can also be applied to the spoken language and the way the words are pronounced and stressed or by replacing words with synonyms and metonymy [62].

15.8.4 Anti-Malware

Anti-malware relates to a set of tools to detect and remove malwares (viruses, worms, adware, spyware, rootkit, ..., etc.). Searching for malware can be done by two techniques:

- *Heuristic methods* which enable the detection of new threats based on the behaviour of malware and the effects left on the system.
- *Signature methods*, by searching for malware signature from the anti-malware database which is updated regularly every time a malware is detected and analysed. This method is fast but it detects known threats only.

Anti-malware detects a set of malware of different types; there are specific tools for each type of malwares such as: antivirus which detects viruses, worms and Trojans, ..., etc.; anti-spyware which removes spying software left by hackers and adware removers which remove the adware from the system. It is important to note that sometimes a malware can be an adware, spyware or a Trojan and a rootkit blended together to be more discrete and destructive [63].

Recent anti-malwares use combined methods (signature and heuristic) to provide security and efficiency in terms of scanning time. An example of such an anti-malware is presented in the patent *US 7725941 B1* [64], where an executable file is scanned for malware presence following these steps:

- Detecting, an attempt to execute a file on the computer;
- For a known file, a signature malware check is performed;
- For an unknown file (encountered for the first time), a risk analysis based on a plurality of risk factors is performed;
- Based on the risk analysis, identifying which malware detection algorithm is to be used for the file, in addition to signature detection;
- If no malware is detected using the malware detection algorithms, the execution of the file is permitted [64].

The risk analysis is based on: file source, file origin, file path, file size, whether the file is digitally signed, whether the file is a download utility, whether the file is packed, and whether the file was received from a CD, as a file downloaded from the internet is more suspicious than the one copied from a CD, also the files downloaded from the internet can be rejected by the anti-malware if its origin URL is blacklisted (even before scanning it); files that are not digitally signed, have small size, launch other processes, create copies, download other files or are saved in root directory are more suspicious than other files and potentially are (infected by) a malware [64].

The malware detection algorithms include: heuristic detection algorithms, statistical analysis (frequency and pattern of instructions), instruction-based emulation and environment-based emulation [64].

15.8.5 Intrusion Detection and Intrusion Prevention Systems

Intrusion Detection System (IDS) monitors the events occurring within a system or a network and analyses them to detect security problems. The system is passive which means that it saves the security problems details in log files and alerts the security administrator but it does not try to stop or prevent the attack. Intrusion prevention system (IPS) detects the intrusion and attempts to stop it. The key functions that an IDS/IPS (IDPS) system performs are: logging and reporting security events, alerting the security administrator and preventing the attacks [65].

There are different types of IDPS based on its used method and its spatial distribution:

- *Signature based*: this type of IDPS detects the known threats by comparing their unique pattern known as a signature to the database of known threats. It is the simplest detection method and it is ineffective at detecting unknown threats [66, 67].

- *Anomaly based* (behaviour based): the IDPS compares the normal behaviour to the observed events to detect threats. The IDPS creates normal profiles based on the typical observed activity of users, hosts, network connections, or applications, and compares it using statistical methods with the observed activities such as sudden high consumption of bandwidth or processors to detect and eliminate the threat. It is effective at detecting unknown threats [67, 68].
- *Stateful protocol analysis*: the IDPS tracks the state of network, transport and application protocols that have a notion of state, i.e. it can link requests to responses and identify unexpected sequence of commands. However this type of IDPS cannot detect threats that do not violate the proprieties of known protocols such as creating many TCP sessions to cause denial of service [67].
- *Host based*: the IDPS monitors the host which can be a machine or a server, by tracking network traffic, running processes, file/directory access and modification (edit/delete) and system logs to detect suspicious activity that can potentially be a local threat [67, 68].
- *Network based*: the IDPS monitors network traffic and analyses the network and application protocol activity to identify and eliminate suspicious activity [67, 68].
- *Hybrid IDPS* (host/network based): the IDPS has two parts, an interface that monitors the network and an interface that monitors the local host [67, 69].

Some recent researches focus at building an IDPS inspired by the human immune system known as Artificial Immune System (AIS), a new branch of artificial intelligence, basically working like our immune system which is able to distinguish self- and non-self-cells and protect the body against non-self-cells. In computer science, a non-self-cell is the intrusion. Further details can be found in Farhaoui [70].

15.8.6 Firewalls

It is very common to find a local network which enables the exchange of file and sharing resources such as the printer. The enterprise can extend its network by using the internet to provide services and information to its users. That can cause an open port that can be exploited to launch attacks and other related security problems. Firewall aims to secure the local network and isolate it from the threats, and it detects the intrusion to the system, controls the access to the resources and analyses the coming traffic. There are different categories of firewalls:

- *Stateless firewall*: the oldest and the basic method where each packet is controlled independently basing on the administrator's predefined rules. It filters the packet basing on their IP addresses and Port number

- *Stateful firewall*: a memory-based firewall; the packets are not only controlled basing on the administrator rules (IP, Port number) but also on the session state; this allows to detect and prevent some DoS attacks such as SYN flooding
- *Applicative firewall (proxy)*: this firewall must know all the protocols and rules of the application, each application, has a special process in the proxy that is responsible for the filtering
- *Authentication firewall*: the filtering is not based on the Machine addresses (IP) only but on the users as well
- *Personal firewall*: anti-malware and anti-spyware tools that are installed on the host machines [71]

15.8.7 Freshness Protection

A security measure that can be used to prevent the attacker from replaying the eavesdropped packets or data (it allows the detection of replayed data). It is a technique used in networks. It can be applied by the use of timestamp (Kerberos), nonce (Needham–Schroeder protocol) or the use of sequence numbers [72–74].

15.8.8 Virtualization

Virtualization allows having multiple operating systems on the same server or host system. It can be used for security purposes such as testing un-trusted software because virtual systems occupy specific partitions. Therefore, if the system is compromised, the infected partition(s) can be isolated and the access to resources such as the disk or networks can be severed and the system can be halted. It is recommended to back up the virtual system to restore it in case of severe security problems. Also, because the virtual system is independent from the physical architecture, virtualization is considered as one of the security solutions [75].

15.8.9 Backup, Patches and Users Education

In many cases, the security problem is due to wrong configuration, such as leaving open port or wrong firewall configuration, which is the mistake of the security administrator. Training the security administrator to do backups, monitor the network traffic and applying patches to detected vulnerabilities is mandatory to maintain the security of a system or a network. Sometimes the security system is correctly configured and relies on the combination of recent technologies such as biometry, the use of tokens, smart cards, firewalls, IDPS, and anti-malware known as the layered security system, making it hard to be attacked externally. Yet, because of the user it can be broken because they fall for social engineering attacks which rely on the authentic (insider) users to gain an authorized access. Thus, users must be aware of this attack and taught not to fall for it.

15.9 Attackers and Security Breakers

After explaining security threats and solutions, it is important to introduce and classify threat makers also known as the security attackers, breakers or hackers, which will be the content of this section.

There are many ways to classify an attacker or a security breaker. M, Raya et al. [76] classified attackers of VANETs in three dimensions which can be applied to any network attacker.

The first criterion is whether or not the attacker belongs to the networks:

- The *insider attacker* who belongs to the network
- The *outsider attacker* who does not and may be considered as an intruder

The second criterion is concerned with the intention/purpose of the attack which results in two classes as well:

- The *malicious attacker* who aims for a full destruction
- The *rational attacker* who unlike the malicious attacker has a personal reason for the attack to gain profits and thus has a specific target

The last criterion relates to the type of attack, and this also gives two classes:

- The *passive attacker* who eavesdrops the networks but does not alter or generate packets
- The *active attacker* may alter, drop or generate packets

We can consider other criteria such as the types of security threat, and the technical level of an intruder.

Considering the criterion of the technical level, we would have two categories of attackers:

- *Script kiddies* also known as *click kiddies* or the amateur hackers, who create threats by using pre-made hacking tools or copying scripts made by other skilled hackers and often available on the internet [14].
- *Expert* or professional hackers: they have high-level programming skills, write their own hacking tools, make malwares, exploit system vulnerabilities and are more dangerous.

If we consider the type of security threats, four main categories of attackers can be distinguished:

- *Hackers*: the word hacker originally was used to describe a person who is expert at writing and modifying computer programs, learning the details of a computer and stretching its capabilities. Now it is used to describe the malicious intruder. A hacker can be motivated by profit, psychological reasons such as: greed, revenge, self-satisfaction or mental illness; or governmental such as political activists (hacktivists) or spies [34].

There are three types of hackers:

- *The white hat hackers*: also known as the ethical hackers, who have the permission to access a system, and their skill level is the same as that of the black hackers but white hackers are hired by organization to evaluate their security system [77]. The ethical hackers are considered to be experts and have an advanced level of skills in network and computing technologies. Ethical hackers have their community and they organize and attend conferences and workshops to exchange knowledge. They are trained and certified hackers (for example: the EC-Council CEH certificate) [78].
- *The grey hat hackers*: they are hybrid between the black and white; sometimes they act legally and sometimes not [77].
- *The black hat hackers*: they are malicious hackers who intrude the systems illegally to cause harm [79]. Black hackers have a high level of skills, and they can detect systems and network flows and vulnerabilities and exploit them to take over the control or cause harm. They can develop their own hacking tools and methods, create malwares, use cryptography or steganography techniques or even rely on social engineering techniques. In most of the times, they succeed at launching their attacks in anonymous and untraceable ways, thus they may be considered as a real threat to corporations and states and they present a challenge to cybersecurity protectors and investigators.

– *Malware makers* are the virus, Trojan, adware, spyware, rootkit and other malware writers who are considered skilled programmers with good theory background. They develop harmful codes that can be used to exploit vulnerabilities, collect and steal data, create a backdoor and control the system or cause complete destruction. Most importantly, they develop it to avoid detection and in most of the cases it has more than one function and it uses advanced techniques such as steganography (Trojans), cryptography or rootkit to keep being hidden and undetected.
– *Phreaks* are hackers who are specialized in breaking phone networks. They are one of the earliest types or attackers, and they hack phone networks to make free calls.
– *Crackers* originally used to describe the security systems breaker (cracker), now used for describing password crackers [14].
– The above-mentioned types of attackers are not the only existing as some are specified at hacking bank cards (carders), smartphone system and applications, games and software, networks, ..., etc.

Figure 15.2 gives the classification of security attackers basing on the previously explained criteria.

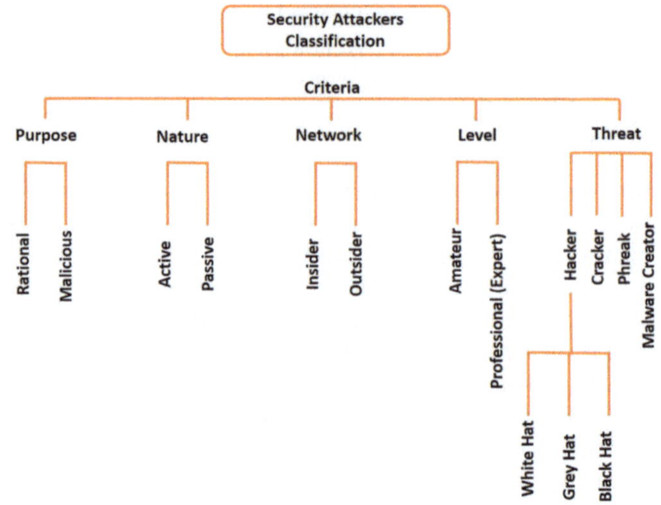

Fig. 15.2 Classification of security attackers

15.10 Forensic Investigation of Cybercrime

Cybercrime is defined as "any illegal behaviour committed by means of, or in relation to, a computer system or network, including such crimes as illegal possession and offering or distributing information by means of a computer system or network". In its general meaning, it can be any crime that considers the cyberspace (computers and internet) as a tool, or as a target.

Some cybercrime examples include:

- Damage to computer data or programs
- Computer sabotage
- Computer espionage
- Public switched telephone network (PSTN) intrusions
- Major computer network intrusions
- Network integrity violations
- Privacy violations
- Industrial/corporate espionage
- Software piracy

Cybercrime can be categorized into three categories:

- *Violent cybercrimes* (cyberterrorism, assault by threat, cyber talking, child pornography, . . . , etc.)

- *Non-violent cybercrimes* (cyber trespass, cybertheft, cyber fraud, ..., etc.)
- *Destructive cybercrimes* (hacking into a network or a system; introducing viruses, worms and malwares; mounting DoS attacks, ..., etc.)

We will discuss in this chapter the cybercrimes that aim to take down systems and networks (non-violent and destructive cybercrimes), and the investigation process done by forensic investigators.

The investigation process can be divided into three main steps:

- The first step in the investigation is to detect the cybercrime (while it is still happening or after it ends with success or failure).
- The next step is to secure the evidences.
- The last step is to analyse the evidences and extract the crime proofs.

All the three phases involve using specific tools and require skilled IT experts and investigators who are trained to preserve the evidences.

15.10.1 Detecting the Cybercrime

Cybercrimes can be detected by intrusion detection tools, firewalls or anti-malwares. It can also be detected by the network or the security administrator by checking the network state regularly, verifying the system state, studying security logs or by noticing systems slow down or non-response. In case of spams, email headers can indicate the spammers IP. Although, malicious attackers rarely leave traces and in most of the times they use bots (victims) to launch attacks such as spams and DDoS, ..., etc. Their IP address will be hidden and tracking them for their crimes would be difficult; some solutions used by forensic investigators involve using "Honey Pots" and "Honey Net" (also known as cyber stings) to lure the attackers and capture them. The honey pots and honey net are computers and networks deliberately exposed to the public. Usually, they contain data that is not harmful to be exposed, and security systems and firewalls that contain vulnerabilities that can be exploited by the attackers.

15.10.2 Securing Evidences

This phase include defining the crime scene, and it is also required to do the disk imaging either by using tools such as Encase or using a standalone disk imager so as to make a bit stream copy and create an exact image of the disk (logically and physically identical) which is important to keep the original evidence intact. Volatile data in the memory must be saved using proper commands and tools before shutting down the system where it is preferable to unplug the wires instead of using the command "shut down" to avoid self-destruction scenario. If the computer screen displays messages or executes certain programs, it is important to note that either by recording it with a camera or by memorizing the commands in a text format. It is mandatory for the IT expert to know how to save the evidence regardless of the operating system running on the machine or the type of electronic device in use (tablet, computer, smartphone, cloud, ..., etc.). Every evidence must be tagged and logged and the transfer operation must be safe to keep the evidences as found.

15.10.3 Analysing the Evidence

The IT experts in forensic labs use the images of disk to extract files and analyse them. They can extract log files, trace the security reports, check history of visited websites, temporary files, email, social media network activity and their chat log, or even the history of downloaded software to look for malicious software that can be used to cause harm. They also analyse every file that can lead them to a clue related to the crime committed.

An attacker can use cryptography and steganography to hide or encrypt the files, and thus the help of cryptanalysts and steganalysis expert could be required in the investigation.

The next part is to provide the analysed evidence as an acceptable, non-repudiated proof in the courtroom against the criminals.

It is worth noting that if the attacker physically destroys the evidence (computer used for the attack) either by smashing it, magnetizing it, putting it in fire or exposing it to chemicals, there would be no way of proving anything against him unless a witness speaks against him, other kind of evidence is found or he admits his crimes on his own [14].

15.11 Case Study: Biometric Authentication Solutions

There are many unique features that can be used to identify a person and are used as biometric security solution; in what follows we will mention some of these solutions:

15.11.1 Eyes

- *IRIS recognition*: iris (Fig. 15.3) represents a great biometrics solution for authentication; because it is small (11 mm), the identification also known as the search for a match from the database is fast, it is distinguishable, stable over time,

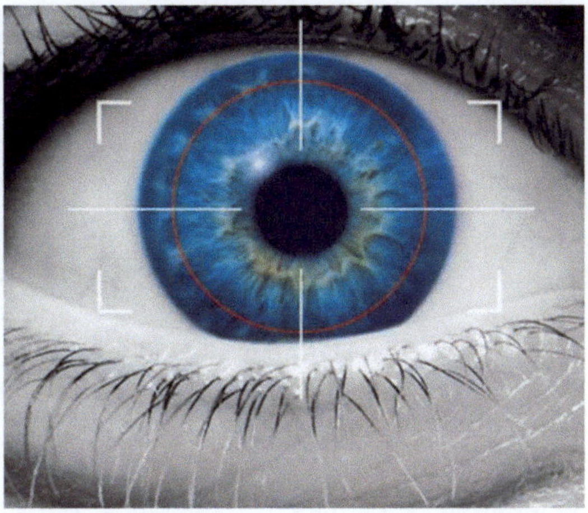

Fig. 15.3 Eye iris [80]

insensitive to angle of illumination and its distinctive annular shape facilitates reliable and precise isolation of this feature. The main steps for iris recognition system are:

1- Using a camera, the eye is captured
2- The image is then processed so as to isolate the iris by detecting its boundaries (pupil, limbus and eyelids)
3- Demodulation of iris code
4- XOR comparison of two iris codes

For more detailed explanation, we refer the reader to read this paper [81].

- *Retina authentication*: the authentication is based on the blood vessels in the retina of the eye which have unique patterns that can be used to identify a person. Since the blood vessels are at the back of the eye, the retina scan requires the use of a low intensity light before photographing and analysing it [82].

15.11.2 *Ears*

Ear authentication has more advantages than the face recognition as it degrades little with time and thus is considered more stable than the face. The authentication process in general is done in five stages:

1- *Ear detection*: it involves localizing the position of the ear in an image.
2- *Ear normalization and enhancement*: the detected segment (ear) is enhanced in terms of fidelity and may be subject to geometric or photometric correction.
3- *Feature extraction*: the segmented ear is reduced to a mathematical model (e.g. a feature vector) that summarizes the discriminatory information.

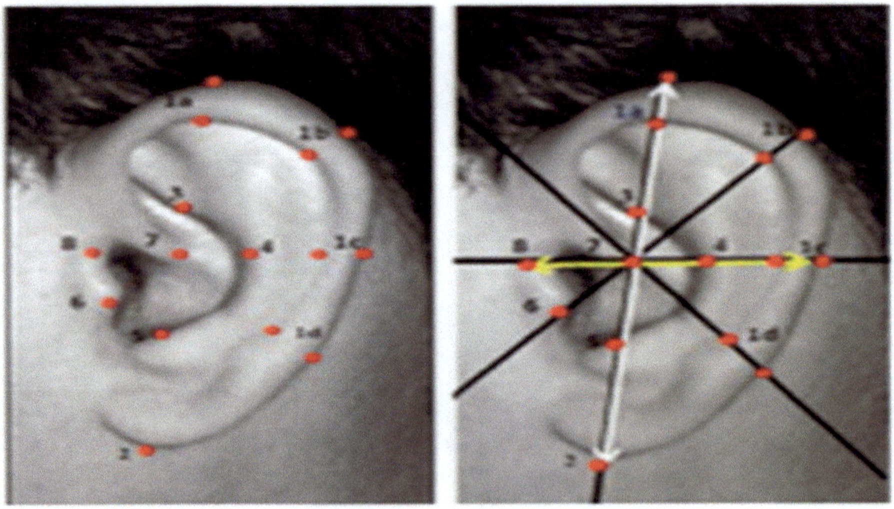

Fig. 15.4 Ear biometric, geometry-based authentication [83]

4- *Matching*: the features extracted must be compared to the features saved in the database.
5- *The decision*: after the matching, the decision is established as a (yes) or a (no).

This method of authentication relies on the geography of the ear, and the imaging can either be 2D (Fig. 15.4) or 3D [83].

Another type of ear authentication, which is based on the use of acoustic waves to distinguish the unique cavity shape of the ear, has been developed by NEC Corporation, Japan. The technology uses earphones with built-in microphone to send and receive wave sounds and extract features from the received signals which are unique to each individual basing on the unique structure of the ear (see Fig. 15.5). The earphone is used to eliminate the noise and to ensure a natural authentication even when moving or working [84].

15.11.3 Face Recognition

Face authentication is still very challenging problem. This is due to the variability of human faces under different operational scenario conditions such as illumination, rotations, expressions, camera viewpoint, aging, makeup and eyeglasses.

Face recognition method is classified into two methods:

- Feature-based methods
- Appearance-based methods

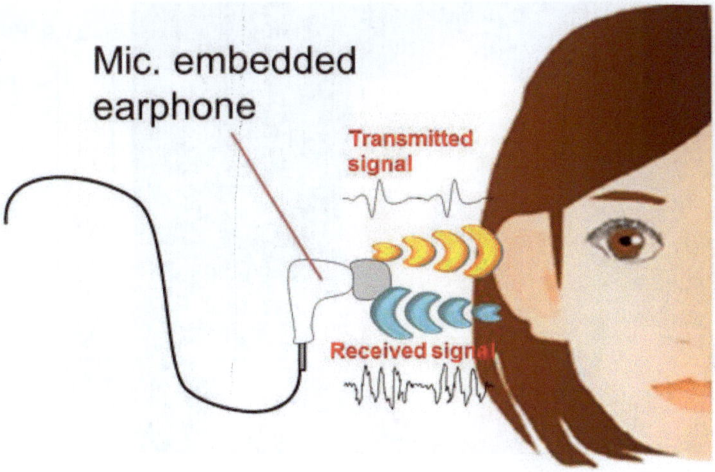

Fig. 15.5 Ear biometric, acoustic waves-based authentication [84]

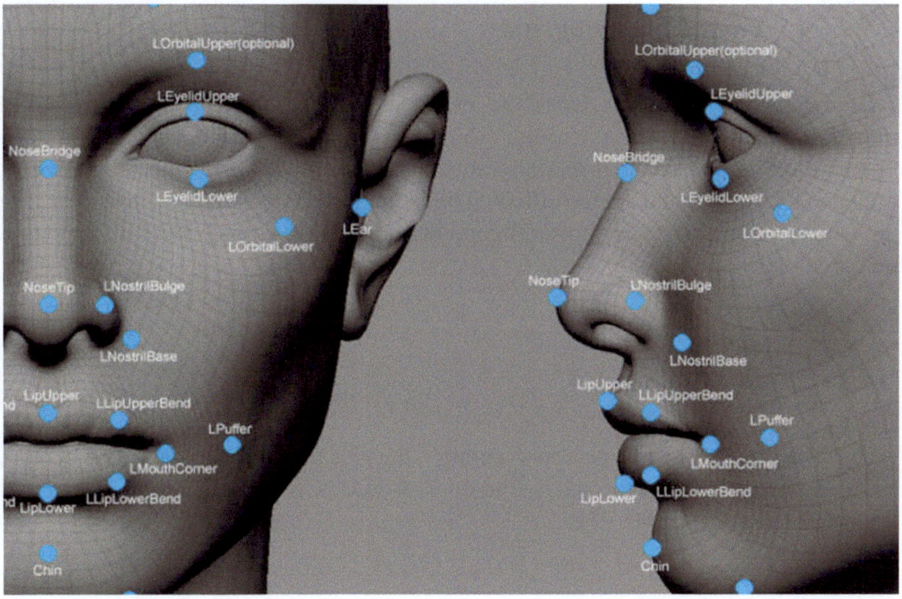

Fig. 15.6 Face biometric authentication [85]

Face recognition techniques can be either 2D (Fig. 15.6) or 3D and it uses properties and geometric relations such as the areas, distances and angles between the facial feature points such as the eyes, nose and mouth [86].

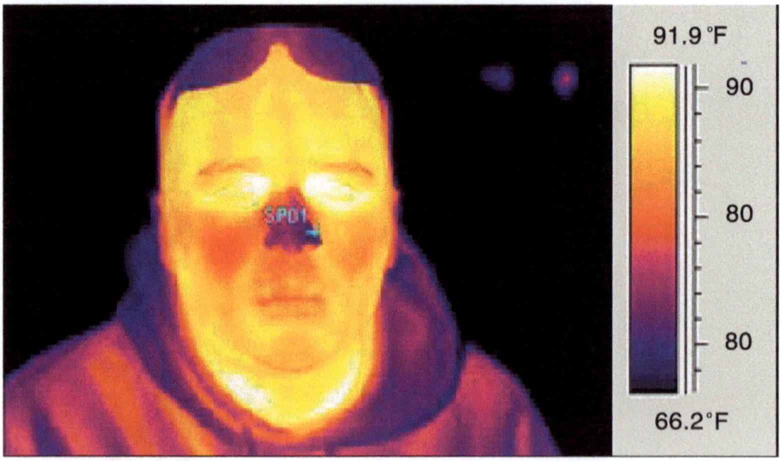

Fig. 15.7 Facial thermograms authentication [87]

15.11.4 Facial Thermograms

Visual face authentication faces the problem of lightening, changing pose and orientation. Those are the reasons why new techniques are developed for face recognition known as facial thermograms (Fig. 15.7) which are formed by the heat radiated by the face and are less affected by facial growth, pose and expressions. The images are captured using infrared camera and used to extract the authentication features [88].

15.11.5 Lip Biometrics

The lip print varies from a person to another, thus it can be used to identify a person, with its unique features formed by the lines, wrinkles, fissures, grooves, colour and shape (see Fig. 15.8) [90, 91].

15.11.6 Fingerprint

Each individual has a unique fingerprint, thus the use of fingerprint for authentication is widely adapted now. The fingerprint has three patterns loop, delta, and whorl, others features such as the valleys, the ridges, and the minutia which

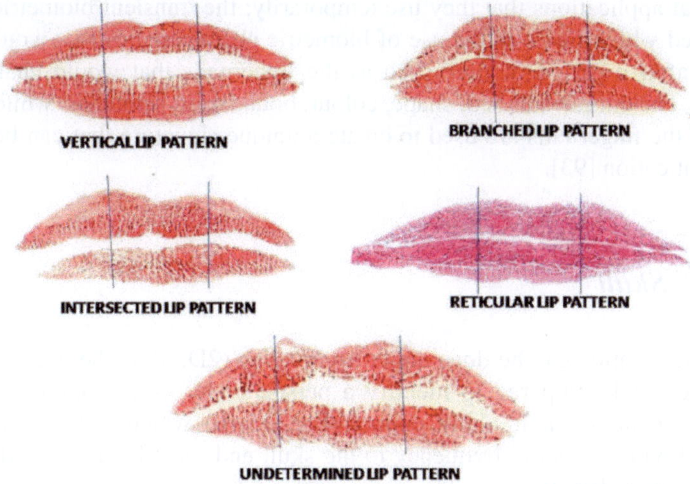

Fig. 15.8 Lip patterns [89]

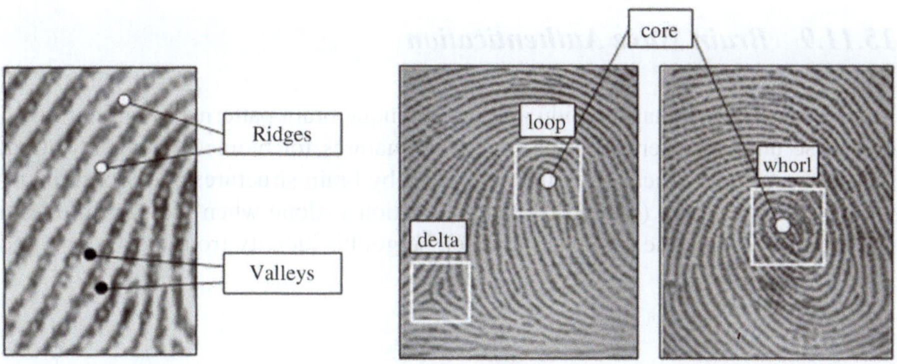

Fig. 15.9 Fingerprint [92]

refer to the various ways in which the ridges can be discontinuous (termination and bifurcation). The minutia is a key feature in identifying (authenticating) the individuals (Fig. 15.9) [92].

15.11.7 Fingernail

Some applications require secured short-term authentication and the use of biometric solutions is recommended. However, biometric characteristics such as retina, iris or fingerprint are considered as permanent characteristics of an individual, also because people would likely object having their biometric parameters saved for non-crucial applications that they use temporarily; the transient biometric solution was created which involves the use of biometric characteristics that would change naturally after a period of time such as the fingernails that would change in 3–6 months. The unique nail bed, shape, colour, boundaries, scratches, white dots and texture of the fingernails are used to create a unique signature that can be used for the authentication [93].

15.11.8 Skull

Skull authentication can be done either by imaging (2D, 3D) where a set of points are selected and compared to identify a person or by using sound waves as in the most recent research SkullConduct (Fig. 15.10) which sends sound waves (vibration) with a specific frequency to the skull and records the echoed waves to identify a person [95, 96].

15.11.9 Brain Wave Authentication

Studies have shown that individuals exhibit unique brain patterns for similar tasks, and these unique patterns can be used as signatures for biometric authentication; the patterns are electrical activities generated by brain structures and measured by electroencephalogram (EEG). The authentication is done when the person does a certain mental task (the same he did when he got his identity from the system) and

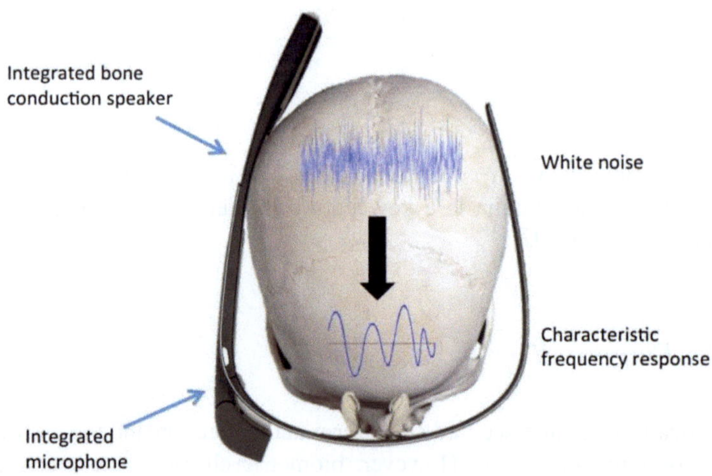

Fig. 15.10 SkullConduct, a skull authentication solution [94]

this task can be reading, multiplication, relaxing, ..., etc., and the EEG measures the electrical activity of the brain and compares it with the saved signature to accept or reject the individual.

This type of authentication is new and presents many advantages:

- Confidentiality (mental tasks)
- Difficult to mimic (each individual has a unique pattern)
- Impossible to steal (individuals emit different brain patterns when under stress, threat or mood change) [97]

15.11.10 Body Odour

Body odours can be classified into three layers or types: the first contains constituents that are stable over time regardless of diet or environmental factors, the second contains constituents that are present due to diet and environmental factors and the third contains constituents that are present because of the influence of outside sources such as perfumes. The first type of odour is unique to each individual and it is used in the authentication which uses e-noise to identify odour and individuals [98].

15.11.11 Palm Print

Palm print authentication is hand-based biometric technology. The palm is covered by the same skin as the finger tips and is used to uniquely identify a person. The palm has a set of unique features (see Fig. 15.11) that can be used to authenticate a person, such as:

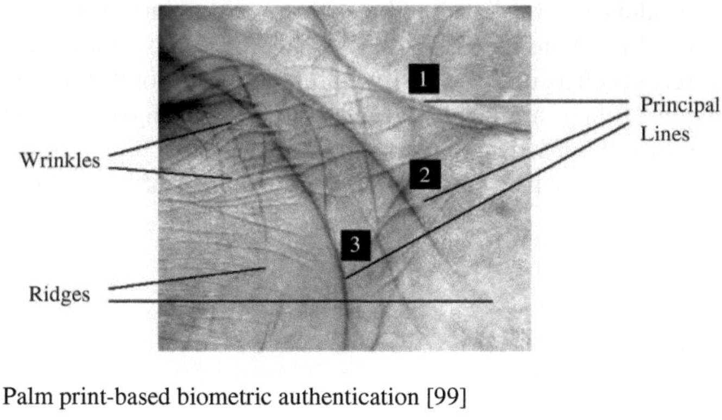

Fig. 15.11 Palm print-based biometric authentication [99]

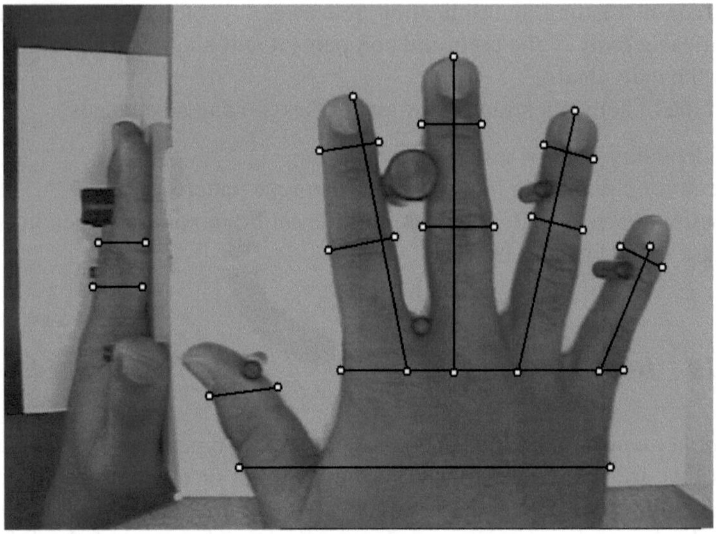

Fig. 15.12 Hand geometry-based solution [101]

- Palm shape (geometry feature)
- Form and location of principle lines (line feature)
- Wrinkles which are the thinner irregular lines (wrinkle feature)
- Delta point features at the root of fingers
- Minutiae features [100]

15.11.12 Hand Geometry

This technique measures the palm surface, length, width and shape, the fingers size and shape, and calculates the distances between a set of points in order to extract the unique features of the hand (see Fig. 15.12). This method is insensitive to changes in surface features such as tattoos, hair, cuts, scrapes, burns or dirt. However, it does not perform well if there are large bandages or casts on fingers, when the hand is deformed significantly or when a finger is missing [102].

It is also possible to use the foot geometry and print or the toes prints to authenticate a user.

15.11.13 Veins

Veins or the vascular pattern (see Fig. 15.13) can be used to authenticate a person as it is unique for each individual; this type of authentication can rely either on the veins in the face of the hand or on the fingers [104, 105].

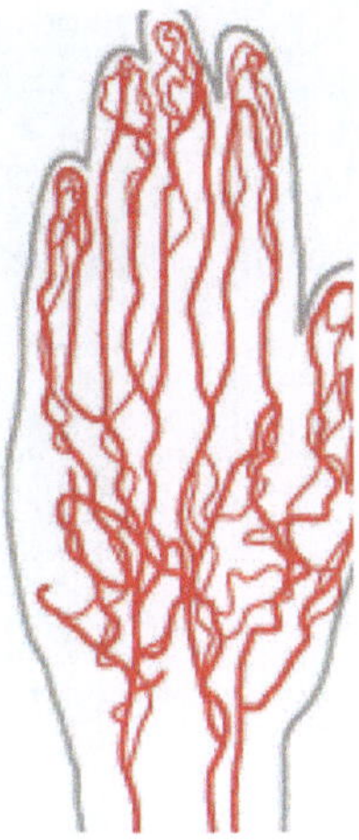

Fig. 15.13 Vein-based biometric authentication [103]

15.11.14 Keystroke and Mouse Moves

Although some researchers are against the use of keystroke as an authentication method by arguing that it is not unique all the time, other researchers are still interested in developing a keystroke based authentication systems, and they rely on some characteristics such as: the speed of typing, time between two key strokes, frequency of typing errors, the time a key is pressed, ..., etc.

Mouse dynamics is also a behaviour-based authentication method where the researchers record the mouse moves and clicks (right, middle and left buttons; drag and drop; stillness; single or double clicks) to create the user template which will be used later for the authentication [106].

15.11.15 Gait

Human way of walking is unique to each individual and it is used to authenticate living person to the system. The authentication can be done through three ways:

- *Machine vision*: relies on the use of cameras to record the person's walk and identify him. There are two approaches for machine vision, the model based (see Fig. 15.14) and the silhouette based.
- *Wearable sensors*: a set of non-intrusive wearable sensors can be attached to the person to identify his gait such as smartphones, smartwatches or shoes with sensors.
- *Floor sensor approach*: the gait of a person is saved and identified as he walks on a sensor-monitored floor as signals are generated from the stepped-on sensors [108].

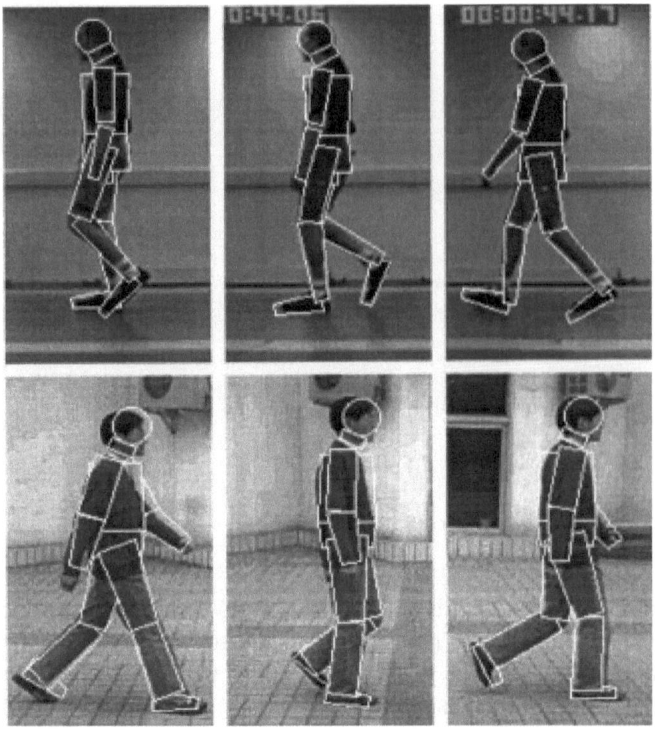

Fig. 15.14 Model-based gait authentication [107]

15.11.16 Speaker Recognition

Human voice is a unique identifier of a person and it differs from a person to another because of the different anatomy (throat shape) and the learned behaviour patterns. The user would provide a vocal passphrase to identify himself to the system. However, since the user's voice can change due to illnesses such as cold, and some individuals are expert in imitating voices, this system might not perform well in these cases, and it would be advisable to use it in a layered biometric security system instead of using it alone [91].

15.11.17 Heart beat

The heartbeat pattern (Fig. 15.15) is unique for each individual, and it represents a robust biometric security solution. One of the authentication approaches uses the electrocardiogram (ECG) which describes the electrical activity of the heart over

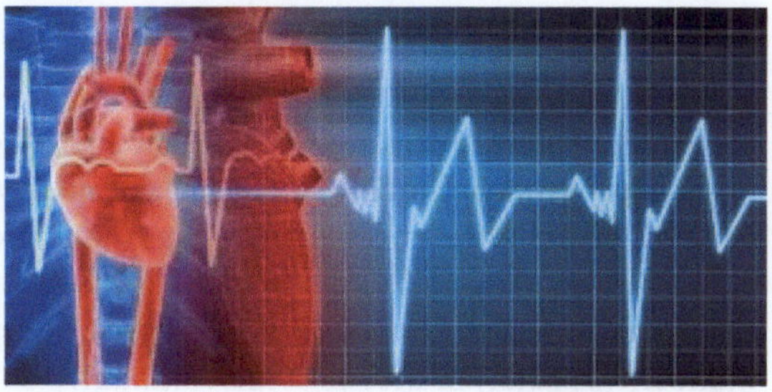

Fig. 15.15 Heartbeat authentication [109]

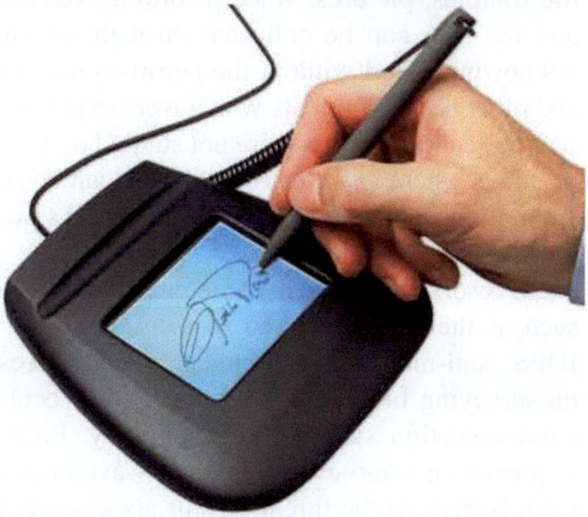

Fig. 15.16 Signature-based authentication [111]

time to identify the person. The ECG can be captured by external sensors attached to the wrist or from the fingers to provide easiness and acceptance so as this solution gets used in smartphones and other devices [110].

15.11.18 Signature and Handwriting

Handwriting and signatures (Fig. 15.16) are one of the oldest and most common behavioural biometric solutions where the idea of signing and authorizing papers is adopted as a method of authentication. This solution is done either online where the sensor compares not only the written passphrase but also the speed, acceleration and the pressure applied to the pen or offline where the passphrase image is compared to the one stored. The major problem of this approach is the forgers who can imitate the handwritings [112].

15.12 Conclusion

In the age we are living, being ignorant and illiterate does not refer to being unable to read or write but refers to the lack of ability to use computer and other digital devices. Kids since a very young age now use computers, tablets, smartphones, smartwatches, smart TVs, ..., etc. All those devices are connected to the internet almost all the time. Users are asked to provide credentials (ID, password, credit card number, ..., etc.) or they provide voluntarily other information related to their daily life routines, pictures, voice recording, videos, ..., etc., through social networks, and the data can be collected through spywares or the intrusions to the system unknowingly and without the permission of the user which creates privacy issues and other security threats which were explained in this chapter.

The security threats were not studied only for the internet, but also to the internet of things where any electronic device can be connected to the internet, in the cloud computing, and even too ad hoc networks such as sensors, vehicles, mobile and body networks.

To resolve the security problems, many solutions were proposed and developed such as the use of layered security solutions which include the use of firewalls, IDPS, anti-malware, authentication and access control where we focused more on surveying biometric solutions. Other security solutions were created to secure communication such as steganography. Each solution is a research field where improvements and evolution are always in progress and what boost this improvement is the fact the threat creator always tries to break the security; this challenge between the security protectors and breakers is what makes both sides work harder to outperform the other party.

The chapter included the presentation and the classification of the security breakers, their different targets, motivations, level and techniques as well as how cyber forensic investigator tries to catch them and what are the relevant steps in the investigation process.

In nutshell, the security issues and solutions will continue their evolution and development as new technologies, systems and protocols are invented and improved.

References

1. Cyber attack (2016) [Online], http://money.cnn.com/2015/04/14/technology/security/cyber-attack-hacks-security/
2. Tor, [Online], https://www.torproject.org/. Accessed 26 July 2016
3. A. Kwon, D. Lazar, S. Devadas, B. Ford, Riffle an efficient communication system with strong anonymity, in *Proceedings on Privacy Enhancing Technologies*, vol. 1, no. 20, 2016
4. G.I. Rathod, D.A. Nikam, Darknet: a class of networks to share anonymous digital content. Int. J. Innov. Res. Comput. Commun. Eng. **3**(7), 8 (2015)
5. Top 500 websites, ALEXA [Online], http://www.alexa.com/topsites. Accessed 25 Mar 2016
6. K. Geers, *Strategic Cyber Security* (CCD COE, Tallinn, 2011)

7. S. Lawson, *Beyond Cyber-Doom: Cyberattack Scenarios and the Evidence of History* (Mercatus Center at George Mason University, 2011), p. 38
8. R. Brown, Computer security threats: a brief history (2014) [Online], https://powermore.dell.com/technology/computer-security-threats-brief-history/. Accessed 01 Mar 2016
9. M. Warner, Cybersecurity: a pre-history. Intell. Natl. Secur. **27**(5), 19 (2012)
10. M.D. Cavelty, *Cyber Security* (Oxford University Press, Oxford, 2012)
11. C.P. Pfleeger, S.L. Pfleeger, J. Margulies, *Security in Computing*, 5th edn. (Prentice Hall, Upper Saddle River, NJ, 2015)
12. K.M. Lord, T. Sharp, *America's Cyber Future Security and Prosperity in the Information Age volume II, Washington, DC : Center of New American Security*, 2011
13. Q. Gu, P. Liu, *Denial of Service Attacks*. Handbook of Computer Networks: Distributed Networks, Network Planning, Control, Management, and New Trends and Applications, 3, 454–468, 2007
14. D.L. Shinder, E. Tittel, Tittel, *Cybercrime Scene of the Computer Forensics Handbook* (Syngress, Rockland, USA, 2002)
15. S. Friedl's, An Illustrated Guide to the Kaminsky DNS Vulnerability, Unixwiz.net Tech Tips [Online], http://unixwiz.net/techtips/iguide-kaminsky-dns-vuln.html. Accessed 27 July 2016
16. S. Fontaine, Attaque DOS (Deny of service), authsecu [Online], http://www.authsecu.com/dos-attaque-deny-of-service/dos-attaque-deny-of-service.php#Attaque_SynFlood. Accessed 19 Mar 2016
17. Y. Wang, G. Attebury, B. Ramamurthy, A survey of security issues in wireless sensor networks. IEEE Commun. Surv. Tutorials **8**(2), 23 (2006)
18. "Man in the middle attack," valency networks (2008) [Online], http://www.valencynetworks.com/articles/cyber-attacks-explained-man-in-the-middle-attack.html. Accessed 31 July 2016
19. S. Şen, J.A. Clark, J.E. Tapiador, Security threats in mobile ad hoc networks, in *Security of Self-Organizing Networks: MANET, WSN, WMN, VANET*, Auerbach Publications, 2011
20. Y.-C. Hu, A. Perrig, D.B. Johnson, Wormhole attacks in wireless networks. IEEE J. Sel. Areas Commun. **24**(2), 11 (2006)
21. M. Jakobsson, S. Wetzel, B. Yener, Stealth attacks on ad-hoc wireless networks, in *Vehicular Technology Conference, IEEE 58th*, p. 9, 2003
22. C. Laurendeau, M. Barbeau, *Threats to Security in DSRC/WAVE* (Springer, Heidelberg, 2006), pp. 266–279
23. S.M.K.R. Raazi, Z. Pervez, S. Lee, Key management schemes of wireless sensor networks a survey, in *Security of Self-Organizing Networks MANET, WSN, WMN, VANET*, Auerbach Publications, 2011
24. J. Chirillo, *Hack Attacks Revealed, a Complete Reference with Custom Security Hacking Toolkit* (John Wiley & Sons, New York, NY, 2001)
25. M.N. Mejri, J. Ben-Othman, M. Hamdi, Survey on VANET security challenges and possible cryptographic solutions. Veh. Commun. **1**(2), 53–66 (2014)
26. J. Sen, *A Survey on Wireless Sensor Network Security*, 2010
27. Z.M. Fadlullah, T. Taleb, M. Schöller, Combating against security attacks against mobile ad hoc networks (MANETs), in *Security of Self-Organizing Networks MANET, WSN, WMN, VANE*, Auerbach Publications, 2011
28. S. Piper, *Intrusion Prevention Systems for Dummies* (Wiley Publishing, Inc., 2011)
29. S. Martin, M. Tokutomi, *Password Cracking*, researchers report, Arizona University, USA (2012)
30. C.P. Pfleeger, S.L.R. Pfleeger, *Security in Computing*, 4th edn. (Prentice Hall, 2006)
31. K.D. Mitnick, W.L. Simon, *The Art of Deception Controlling the Human Element of Security* (Wiley Publishing, Indiana, USA and simultaneously in Canada 2002)
32. P.A.H. Peterson, P. Reiher, Exploits: buffer overflows, pathname attacks, and SQL injections, mathcs [Online], http://mathcs.slu.edu/~chambers/spring11/security/assignments/lab05.html. Accessed 26 Mar 2016
33. S. Noel, D. Wijesekera, C. Youman, *Modern Intrusion Detection, Data Mining, and Degrees of Attack Guilt* (Springer, 2002)

34. J.M. Kizza, *A Guide to Computer Network Security* (Springer, 2009)
35. R.L. Krutz, R.D. *Vines, Cloud Security A Comprehensive Guide to Secure Cloud Computing* (Wiley Publishing, 2010)
36. W. Apolinarski, *Cohen and the First Computer Virus, Seminar "Malware" Prof. Dr. Joachim von zur Gathen, Daniel Loebenberger WS* (2007–2008)
37. G. White, S. Black, *Malware, Spyware, Adware, Viruses*, Information Technology Services (Clark College), 2011
38. N. Weaver, V. Paxson, S. Staniford, R. Cunningham, A taxonomy of computer worms, in *ACM Workshop on Rapid Malcode*, 2003
39. M. Madou, E. Lee, J. West, B. Chess, Watch what you write: preventing cross-site scripting by observing program output, in *OWASP AppSec Conference*, 2008
40. M. Bishop, *Introduction to Computer Viruses* (Pearson Education India, 2006)
41. M. Egan, T. Mather, *The Executive Guide to Information Security Threats, Challenges, and Solutions* (Addison Wesley Professional, 2004)
42. C. Wysopal, C. Eng, T. Shields, *Static Detection of Application Backdoors* (Black Hat, USA, 2007)
43. E. Casey, *Digital Evidence and Computer Crime* 3rd edn. (Academic Press, 2011)
44. T. Roosta, S. Shieh, S. Sastry, Taxonomy of security attacks in sensor networks and countermeasures, in *The First IEEE International Conference on System Integration and Reliability Improvements*, 2006, p. 25
45. P. Seuwou, D. Patel, G. Ubakanma, Vehicular ad hoc network applications and security: a study into the economic and the legal implications. Int. J. Electron. Secur. Digit. Forensics **6**(2), 115–129 (2014)
46. T. Roosta, Taxonomy of security attacks in sensor networks and countermeasures, in *The first IEEE International Conference on System Integration and Reliability Improvements,* vol. 25, 2006
47. M. Mana, M. Feham, B.A. Bensaber, SEKEBAN (secure and efficient key exchange for wireless body area network). Int. J. Adv. Sci. Technol. **12**, 15 (2009)
48. M.A. Ameen, J. Liu, K. Kwak, Security and privacy issues in wireless sensor networks for healthcare applications. J. Med. Syst. **36**(1), 8 (2012)
49. M. Abomhara, G.M. Koien, Security and privacy in the internet of things: current status and open issues, in *Privacy and Security in Mobile Systems (PRISMS), International Conference on. IEEE*, 2014
50. S.J. Stolfo, S.M. Bellovin, S. Hershkop, A.D. Keromytis, *Insider Attack and Cyber Security Beyond the Hacker* (Springer, 2008)
51. J. Wayman, A. Jain, D. Maltoni, D. Maio, *An Introduction to Biometric Authentication Systems* (Springer, London, 2005)
52. I. Curry, *An Introduction to Cryptography and Digital Signatures*, Version 2.0 (Entrust 2001)
53. S. Singh, *The Code Book: The Science of Secrecy from Ancient Egypt to Quantum Cryptography* (Anchor, 2011)
54. L. Benarous, M. Djoudi, A. Bouridane, *Etudes Comparatives d'outils de stéganographie et d'outils de stéganalyse: Application aux images et aux vidéos* (Amar Telidji University, Laghouat, Algeria, 2015)
55. Y. Miche, Developing fast machine learning techniques with applications to steganalysis problems, Thèse de doctorat., Institut National Polytechnique de Grenoble-INPG, 2010
56. T. Sereyvathana, Discriminative algorithms for large-scale image steganalysis and their limitation, Electronic Theses, Treatises and Dissertations, The Florida State University, Florida, 2012
57. Stegosploit, the hack news [Online], http://thehackernews.com/2015/06/Stegosploit-malware.html. Accessed 25 July 2016
58. Hackers exfiltrating data with video steganography, tripwire [Online], http://www.tripwire.com/state-of-security/incident-detection/hackers-exfiltrating-data-with-video-steganography-via-cloud-video-services/. Accessed 25 July 2016
59. S.K.J. Pooja, P. Balgurgi, Audio steganography used for secure data transmission, in *Proceedings of International Conference on Advances in Computing*, Springer, India, 2012

60. M. Shirali-Shahreza, Text Steganography by changing words spelling, in *Advanced Communication Technology, 2008. ICACT 2008. 10th International Conference on, Volume: 3*, Gangwon-do, 2008
61. M.L. Bensaad, Steganography and digital watermarking, Ph.D. Thesis, University of Laghouat (Amar Telidji), Laghouat, Algeria, 2014
62. R. Bergmair, *Towards Linguistic Steganography: A Systematic Investigation of Approaches Systems, and Issues*, Final year thesis, B. Sc.(Hons.) in Computer Studies, The University of Derby, UK, 2004
63. A. Harper, S. Harris, J. Ness, C. Eagle, G. Lenkey, T. Williams, *Gray Hat Hacking the Ethical Hacker's Handbook*, 3rd edn. (McGraw-Hill, 2011)
64. M.A. Pavlyushchik, Method and system for antimalware scanning with variable scan settings. Patent U.S. 7725941 B1, 25 May 2010
65. J. Alexander, *Intrusion Detection and Prevention Systems (IDS/IPS) Good Practice Guide Jason Alexander* (NHS Connecting for Health, 2009)
66. S. Dinesh, Intrusion Prevention Systems: security's silver bullet? Bus. Commun. Rev. **33**(3), 36–41 (2003)
67. K. Scarfone, P. Mell, *Guide to Intrusion Detection and Prevention Systems (IDPS)*, NIST special publication, 800, 94, 2007
68. C. Martin, Intrusion detection and prevention systems in the industrial automation and control systems environment, in *Process Control Systems Industry Conference*, Industrial Defender Inc. 2008.
69. I. Mukhopadhyay, M. Chakraborty, S. Chakrabarti, A comparative study of related technologies of intrusion detection & prevention systems. J. Inf. Secur. **2**, 11 (2011)
70. Y. Farhaoui, Intrusion prevention system inspired immune systems. Indones. J. Electr. Eng. Comput. Sci. **2**(1), 168–179 (2016)
71. Masquelier, Mottier, Pronzato, Les Firewalls, Institut d'électronique et d'informatique Gaspard-Monge (IGM), France, 2000
72. M.A. Ameen, J. Liu, K. Kwak, *Security and Privacy Issues in Wireless Sensor Networks for Healthcare Applications* (Springer, 2010), p. 9
73. Y. Yan, Y. Qian, H. Sharif, D. Tipper, A survey on cyber security for smart grid communications. IEEE Commun. Surv. Tutorials **14**(4), 998–1010 (2012)
74. S. Goldwasser, M. Bellare, *Lecture Notes on Cryptography* (MIT, 2008)
75. J. Hoopes, *Virtualization for Security Including Sandboxing, Disaster Recovery, High Availability, Forensic Analysis, and Honeypotting, Syngress*, Burlington, USA, 2008
76. M. Raya, J. P. Hubaux, The security of vehicular ad hoc networks. J. Comput. Secur. **15**(1), 39–68 (2007)
77. A. Cecil, A summary of hacking organizations, conferences, publications, and effects on society. [online], http://www.cse.wustl.edu/~jain/cse571-07/ftp/hacking_orgs/. Accessed 25 July 2016
78. Certified Ethical Hacker, Eccouncil [Online], https://www.eccouncil.org/Certification/certified-ethical-hacker. Accessed 09 June 2016
79. L.A. Long, *Profiling Hackers* (The SANS Institut, 2012)
80. Eye Scaning 2012, [Online], http://www.messagetoeagle.com/images/eyescanning.jpg. Accessed 25 July 2016
81. D. John, How iris recognition works. IEEE Trans. Circuits Syst. Video Technol. **14**(1), 21–30 (2004)
82. D. Bhattacharyya, R. Ranjan, F. Alisherov, M. Choi, Biometric authentication: a review. Int. J. u–e-Serv. Sci. Technol. **2**(3), 13–28 (2009)
83. A. Abaza, A. Ross, C. Hebert, M.A.F. Harrison, M.S. Nixon, A survey on ear biometrics. ACM Comput. Surv. **45**(2), 35 (2013)
84. NEC biometrics technology uses sound to distinguish individually unique ear shape. NEC, 7 Mar 2016 [Online], http://www.nec.com/en/press/201603/global_20160307_01.html. Accessed 30 Apr 2016

85. Face recognition, 2016 [Online], http://reconocimientofacial.site/wp-content/uploads/2016/01/reconocimiento-facial-orna-innovations.jpg Accessed 07 October 2016
86. M. Savvides, J. Heo, S.W. Park, Face Recognition, in *Handbook of Biometrics*, ed. by A.K. Jain, P. Flynn, A.A. Ross (Springer Science & Business Media, New York, 2007), p. 43
87. human facial recognition, 2002, [Online]. http://www.nationalinfrared.com/images/Human_facial_imaging_recognition.jpg. Accessed 25 July 2016
88. S. Vasikarla, H. Madasu, Online biometric authentication using facial thermograms, in *Applied Imagery Pattern Recognition Workshop (AIPR), IEEE*, 2012
89. Lip print 2013, [Online], http://www.jfds.org/articles/2013/5/2/images/JForensicDentSci_2013_5_2_110_ 119777_f7.jpg. Accessed 25 July 2016
90. M. Chora, *The Lip as a Biometric* (Springer, 2009)
91. O.S. Adeoye, A survey of emerging biometric technologies. Int. J. Comput. Appl. **9**(10), 0975–8887 (2010)
92. D. Maltoni, R. Cappelli, Handbook of biometrics, in *Fingerprint Recognition*, ed. by A.K. Jain, P. Flynn, A.A. Ross (Springer, New York, 2008), pp. 23–42
93. I.B. Barbosa, T. Theoharis, A.E. Abdallah, On the use of fingernail images as transient biometric identifiers Biometric recognition using fingernail images. Mach. Vis. Appl. **27**(1), 65–76 (2016)
94. SkullConduct, 2016, [Online], http://s.newsweek.com/sites/www.newsweek.com/files/styles/embed-lg/public/2016/04/25/biometrics-skull-skullconduct-password-security.jpg. Accessed 25 July 2016
95. S. Schneegass, Y. Oualil, A. Bulling, SkullConduct: biometric user identification on eyewear computers using bone conduction through the skull, in *Proceedings of the 34th ACM SIGCHI Conference on Human Factors in Computing Systems (CHI 2016)*, 2016
96. H.O. Alanazi, B.B Zaidan, A.A Zaidan, 3D Skull recognition using 3D matching technique. J. Comput. **2**(1) (2010), p121–126
97. C.R. Hema, M.P. Paulraj, H. Kaur, Brain signatures: a modality for biometric authentication, in *International Conference on Electronic Design*, Penang, Malaysia, 2008
98. P. Inbavalli, G. Nandhini, Body odor as a biometric authentication. Int. J. Comput. Sci. Inform. Technol. **5**(5), 6270–6274 (2014)
99. Intech, 2011, [Online], http://www.intechopen.com/source/html/17745/media/image2.png. Accessed 01 Aug 2016
100. G. Lu, D. Zhang, W.K. Kong, M. Wong, A palmprint authentication system, in *Handbook of Biometrics*, ed. by A.K. Jain, P. Flynn, A.A. Ross (Springer, New York, 2008), p. 171–187
101. [Online], http://www.360biometrics.com/img/hand_features.gif. Accessed 25 July 2016
102. S.T. David, P. Sidlauskas, in Hand Geometry Recognition. *Handbook of Biometrics* (Springer, 2008), p. 91–107
103. Palm Veins, 2012, [Online], https://crisisboom.files.wordpress.com/2012/01/how-palm-vein-works.gif. Accessed 25 July 2016
104. D. Mulyono, H.S. Jinn, A study of finger vein biometric for personal identification, in *Biometrics and Security Technologies, IEEE*, pp. 1–8, 2008
105. K. Wang, Z. Yuan, D. Zhuang, Hand vein recognition based on multi supplemental features of multi-classifier fusion decision, in *Mechatronics and Automation, Proceedings of the 2006 IEEE International Conference* (Luoyang, Henan: IEEE, 2006)
106. G. Ioan Buciu, Biometrics systems and technologies: a survey. Int. J. Comput. Commun. Control **11**(3), 315–330 (2016)
107. L. Wang, H. Ning, T. Tan, W. Hu, Fusion of static and dynamic body biometrics for gait recognition. IEEE Trans. Circuits Syst. Video Technol. **14**(2), 149–158 (2004)
108. J.E. Mason, I. Traoré, I. Woungang, Gait (Canada) biometric recognition, in *Machine Learning Techniques for Gait Biometric Recognition*, ed. by J.E. Mason, I. Traoré, I. Woungang (Springer International Publishing, Switzerland), 9–35, 2016.
109. Pulse,2012, [Online], http://www.homelandsecuritynewswire.com/sites/default/files/imagecache/stand ard/pulse_biometrics-1.jpg. Accessed 25 July 2016

110. F. Agrafioti, D. Hatzinakos, J. Gao, *Heart Biometrics: Theory, Methods and Applications* (INTECH Open Access Publisher, 2011)
111. Hand signature, [Online], http://www.b2bedocuments.com/images/signaturepad08.jpg. Accessed 25 July 2016
112. R. Das, S. Dhar, S. Das, S. Dutta, S. Mukherjee, A comparative study of biometric authentication based on handwritten signature. Int. J. Res. Eng. Technol. **02**(12), 2321–7308 (2013)

Chapter 16
Data Protection and Biometric Data: European Union Legislation

Pedro Miguel Freitas, Teresa Coelho Moreira, and Francisco Andrade

16.1 Introductory Remarks

The protection of personal data is enshrined in several important legal texts. In 1981, under the auspices of the Council of Europe, the first binding international instrument concerning the protection of an individual against misuses in the collection and processing of personal data opened for signature. The Convention for the Protection of Individuals with regard to Automatic Processing of Personal Data set as a purpose the protection of personal data in a world of growing automatic data processing on a large scale, although it did not foresee biometric data as a special category of data deserving particular protection.

On a different level, the European Union's Treaty on European Union, on article 39, states that "the Council shall adopt a decision laying down the rules relating to the protection of individuals with regard to the processing of personal data by the Member States when carrying out activities which fall within the scope of this chapter, and the rules relating to the free movement of such data. Compliance with these rules shall be subject to the control of independent authorities". Meanwhile, the Treaty on the Functioning of the European Union concedes everyone the right to protection of personal data concerning them (article 16). The protection of personal data is also found on the Charter of Fundamental Rights of the European Union. This Charter embodies the core fundamental values of the Member States of the European Union, as important as human dignity (article 1), life (article 2) and the right to physical and mental integrity (article 3). Amongst them emerges the protection of personal data as a fundamental freedom that everyone is entitled to (article 8). This freedom sits right at the heart of the fundamental values that the European Union considers necessary to be promoted and strengthened, especially

P.M. Freitas (✉) • T.C. Moreira • F. Andrade
Law School, University of Minho, Braga, Portugal
e-mail: pfreitas@direito.uminho.pt; tmoreira@direito.uminho.pt; fandrade@direito.uminho.pt

R. Jiang et al. (eds.), *Biometric Security and Privacy*, Signal Processing for Security Technologies, DOI 10.1007/978-3-319-47301-7_16

"in light of changes in society, social progress and scientific and technological developments" (Preamble). In this sense, the protection of personal data should be safeguarded by all those to whom this charter is addressed to, including not only the institutions, bodies and agencies of the European Union, but also the Member States themselves when implementing European Union legislation. It might appear, in consequence, that, when it comes to the Member States, the Charter has a somewhat limited field of application: it is limited by the powers and competences that have been conferred by the Member States to the European Union. This idea that the European Union has competences that are subjected to the principle of subsidiarity, meaning that, on the one hand, the European Union only acts within the limits of the conferred competences and to meet the objectives set out in the Treaties and, on the other hand, if there is not exclusive competence regarding a certain area, can only act if the objectives should be better achieved at the European level, is an important limitation that should not be overlooked. Yet, it is clear that the protection of personal data is paramount to the objectives of the European Union and, as a fundamental right, portrays an important role in society.

16.2 Data Protection and Personal Data

The concept of data protection as a fundamental right has been further developed in the General Data Protection Regulation (Regulation (EU) 2016/679 of the European Parliament and of the Council on the protection of natural persons with regard to the processing of personal data and on the free movement of such data).[1] The reasoning behind the Regulation is the promotion of a consistent level of protection concerning personal data throughout the European Union while allowing a free flow of personal data between Member States. The free flow of personal data is necessary to a deeper economic and social integration of European stakeholders. Without the exchange of personal data, the public and private sector would be severely limited when performing their activities, the European internal market would cave in and public authorities would not be able to perform their tasks. Since 1995, with the directive 95/46/EC of the European Parliament and of the Council of 24 October 1995, that the European Union has a common set of rules on data protection of natural persons. However, globalization and rapid technological developments felt since then made this Directive completely obsolete and incapable of striking a balance between the protection of personal data and the fulfilling of society modern needs.

A conceptual description of personal data is given by the European legislator in the beginning provisions of the Regulation. According to article 4/1 of the

[1]This regulation repealed the Directive 95/46/EC of the European Parliament and of the Council of 24 October 1995 on the protection of individuals with regard to the processing of personal data and on the free movement of such data. Further developments on the concept of personal data are found on opinion 4/2007 of the Article 29 Working Party (2007).

Regulation, personal data means any information relating to a data subject,[2] including:

- Name;
- Identification number;
- Location data;
- Online identifier; and
- Specific factors such as physical, physiological, genetic, mental, economic, cultural or social identity of that natural person.

16.3 Biometric Data as Personal Data

Biometrics involves techniques used to identify[3] individuals based on a particular trait or physical characteristic unique to that individual or on a behavioural characteristic of an individual.[4] Any human physiological and/or behavioural characteristic can be used as a biometric characteristic as long as it satisfies some requirements like universality, distinctiveness, permanence and collectability. Biometric systems record personal information[5] about identifiable individuals.[6]

[2]Only identified or identifiable natural persons are data subjects. The regulation does not afford protection to legal persons, but only natural living persons. The rules concerning the processing of personal data of deceased persons should be established by the Member States and this regulation does not apply to this type of personal data, even when personal data is archived for historical research and genealogical purposes.

[3]Biometric systems are "applications that use biometric technologies, which allow the automatic identification, and/or authentication/verification of a person. Authentication/verification applications are often used for various tasks in completely different areas, for different purposes and under the responsibility of a wide range of different entities", Article 29 Data Protection Working Party "Opinion 3/2012 on developments in biometric technologies", adopted on 27th April 2013.

[4]"Biometric systems are tightly linked to a person because they can use a certain unique property of an individual for identification and or/authentication. While a person's biometric data can be deleted or altered the source from which they have been extracted can in general neither be altered nor deleted", Article 29 Data Protection Working Party "Opinion 3/2012 on developments in biometric technologies", adopted on 27th April 2013.

[5]Although "Biometric systems are not 100 % accurate. Biometric systems accuracy during the template comparison process of authentication depends on external variables, namely, temperature, training level of the enrollment process technicians, physical condition of the individual to be authenticated, etc. Biometric systems accuracy is also dependent on internal variables such as quality of the equipment and the proprietary algorithms being used." Chinchilla, Rigoberto "Ethical and Social consequences of Biometric Technologies".

[6]"While other new technologies that target large populations and have recently raised data protection concerns do not necessarily focus on establishing a direct link to a specific individual—or creating this link requires considerable efforts—biometric data, by their very nature, are directly linked to an individual", Article 29 Data Protection Working Party "Opinion 3/2012 on developments in biometric technologies" adopted on 27th April 2012.

Biometric data is thus implicitly included in the definition of personal data because it implies retrieving and processing unique identification characteristics[7] of an individual.[8] It does not matter the nature of the specific identification processing technique used, e.g., facial recognition,[9] iris scan or dactyloscopic data,[10] insofar this technique allows the identification of a natural living person. Biometric data should be understood, for the purposes of the Regulation, as any technical means of retrieval or confirmation of the identity of a natural living person using their physical, physiological or behavioural characteristics.[11]

The same conclusion—biometric data being conceived as personal data—was already possible with the Directive 95/46/EC. In Article 2 a), "personal data" was defined as "any information relating to an identified or identifiable natural person (...); an identifiable person is one who can be identified, directly or indirectly, in particular by reference to an identification number or to one or more factors specific to his physical, physiological, mental (...) identity". In accordance with this definition, measures of biometric identification or their digital translation in a template form in most cases are personal data.

Some biometric data could also be considered sensitive data (a special category of personal data) in the meaning of Article 8 of Directive 95/46/EC, because biometric data could reveal the racial or ethnic origin of the data subject or concern

[7]Physical biometrics are faced with the unequivocal fact that people get older and the body changes over time, cfr. Anton Alterman, "A piece of yourself: Ethical issues in biometric identification". See also Rebera, Andrew P. and Mordini, Emilio "Biometrics and ageing: social and ethical considerations": "That the ageing process poses a problem for biometrics is well-understood. No biometric is 100 % permanent: people's biometrics change over time. Hence the technical challenge is to develop techniques whereby an individual may be identifiable by his or her biometrics, throughout his or her lifetime, despite this mutability".

[8]We should be wary when an author writes that "increasingly, the way to keep information secure is to offer up a piece of yourself... to be recorded and used to verify your identity", Anton Alterman "A piece of yourself: Ethical issues in biometric identification".

[9]On facial recognition in online and mobile services, see Article 29 Working Party (2012).

[10]Article 29 Working Party (2007) gives as examples of biometric data: "fingerprints, retinal patterns, facial structure, voices, but also hand geometry, vein patterns or even some deeply ingrained skill or other behavioural characteristic (such as handwritten signature, keystrokes, particular way to walk or to speak, etc ...)".

[11]His concept of biometric data, that is found on article 4/14 of the Regulation is substantially aligned with the opinion of the Article 29 Working Party (2007) that biometric data should be defined "as biological properties, physiological characteristics, living traits or repeatable actions where those features and/or actions are both unique to that individual and measurable, even if the patterns used in practice to technically measure them involve a certain degree of probability".

their health.[12] Biometric data was not explicitly categorized as sensitive data but, taking in consideration the type of data it usually involves, implicitly ought to be regarded as sensitive data.

16.4 The New Data Protection Regulation: Requirements and Challenges for the Treatment of Personal Data

This state of affairs has substantially changed with the new data protection regulation. The European legislator makes now absolutely clear that biometric data is a species of personal data, alongside sensitive personal data revealing racial or ethnic origin, political opinions, religious or philosophical beliefs, or trade-union membership, genetic data, data concerning health or data concerning a natural person's sex life or sexual orientation (article 9). The immediate consequence of categorizing certain data as a special category of personal data is the general prohibition of its processing.[13] Simply put, processing biometric data is forbidden. There are, however, exceptions: if the data subject consents (article 9/a) or, if he or she is—physically or legally—incapable of giving consent, and there is the need of protection of vital interests (article 9/c); for employment, social security or social protection law purposes (article 9/b); data concerning members of a legal entity related to political, philosophical, religious or trade-union aims and the data is not disclosed outside the legal entity (article 9/d); data made public by the data subject (article 9/e); processing of data necessary to the exercise of judicial activity (article 9/f); reasons of substantial public interest (article 9/g); purposes of preventive or occupational medicine, for the assessment of the working capacity of the employee, medical diagnosis, the provision of health or social care or treatment or the management of health or social care systems and services (article 9/h); reasons of public health (article 9/i); archiving purposes in the public interest, scientific or historical research purposes or statistical purposes (article 9/j). Each of these exceptions to the referred prohibition is herein only superficially mentioned.

---

[12]Article 29 Working Party (2003): "Some biometric data could be considered sensitive in the meaning of Article 8 of Directive 95/46/EC and in particular, data revealing racial or ethnic origin or data concerning health. For example, DNA data of a person often include health data or can reveal the racial or ethnic origin. In this case DNA data are sensitive data and the special safeguards provided by article 8 must apply in addition to the general data protection principles of the Directive. In order to assess the sensitivity of data processed by a biometric system the context of the processing should also be taken into account", Article 29 Data Protection Working Party "Opinion 3/2012 on developments in biometric technologies", referred.

[13]"'Processing' means any operation or set of operations which is performed on personal data or on sets of personal data, whether or not by automated means, such as collection, recording, organisation, structuring, storage, adaptation or alteration, retrieval, consultation, use, disclosure by transmission, dissemination or otherwise making available, alignment or combination, restriction, erasure or destruction" (article 4/2 of the Regulation).

A thorough analysis would be necessary in order to fully understand the scope and limitation of each and every one of the exceptions. To perform a full-depth analysis of each and every exception to the general rule that prohibits processing of biometric data would be inopportune and even bothersome given the objective of this chapter but these exceptions will be of utmost importance to those whose field of activity is connected, to some extent, to biometric data.

The sensitive nature of biometric data means that it can only be used in very restrictive and specific situations.[14] The misuse of biometric information in various domains, e.g., medical, commercial and security profiling, is a matter of concern to a vast number of individuals. According to survey done in the United Kingdom up to 45 % of respondents considered biometric data to be extremely sensitive [1], which makes a strong argument in favour of conceding a strong legal protection to this type of personal data. This major concern over one's privacy means, in turn, that the use cases of biometric data are severely limited, especially when free, informed and explicit consent[15] is not given by the individual, and even if there this consent exists some legal restrictions may withdraw its validity.[16] In the online technology sector consent might be given by ticking a box when visiting an internet website that processes personal data (recital 32 of the Regulation). On mobile phones that rely on security features such as fingerprint[17] or facial recognition the consent requisite implies explicit agreement of the user with the processing of biometric data during the enrolment stage. The data controller, in other words, the natural or legal person responsible for determining the purposes and means of the processing of personal data, must be able to demonstrate that the data subject has given consent to the processing after being made aware of the fact that he or she was consenting and the extent of the consent. Irrespective of being located in European territory, this consent must be obtained for the purposes of the Regulation, as long as the data subjects are

[14]Even as a means of identification of the individual, according to the new regulation on electronic identification and trust services for electronic transactions (Regulation EU Nr 910/2014) electronic identification must comply with the principles relating to the protection of personal data provided for in Directive 95/46/EC (still in force but that will be replaced in 2018 by the new European Regulation 2016/679 of the European Parliament and Council of the 27th April 2016) and authentication for online service should concern processing of only those identification data that are adequate, relevant and not excessive to grant access to that service online.

[15]A definition of consent is found on article 4/11 of the Regulation.

[16]The European Union and national laws may foresee situations where the general prohibition of processing biometric data or other special personal data is not lifted by consent of the data subject (article 2/a, *in fine* of the Regulation).

[17]Even this kind of biometric data may encompass difficulties of application: "How will population with disabilities (or lacking physical traits) be enrolled or authenticated in biometric databases? People with just one hand, no iris or retina, no fingers, and in general people lacking physicals characteristics in need of using a biometric facility, may suffer discrimination and unnecessary delays in biometric systems. A well-developed, well-designed biometric system should allow these persons alternative ways to enroll and authenticate, yet delays and processes of bypassing the biometric systems may give them hardships each time they want to access a resource or use a facility which may be an ethical violation of their rights", Chinchilla, Rigoberto "Ethical and Social consequences of Biometric Technologies".

in the European Union and the processing activities relate to the offering of goods or services to such data subjects in the Union, or to the monitoring of their behaviour as far as their behaviour takes place within the Union.

Apart from the consent requisite, the data controller is also responsible for complying with additional requisites. The collection of personal data must be done with a specified, explicit and legitimate purpose. The so-called purpose limitation means that the data controller must process the data in accordance with the purpose that initially motivated the processing. If the data controller desires to process data for a purpose that is incompatible[18] with the one that the data was initially collected for then either the data subject gives consent for further processing or the further processing falls within the scope of public interest, scientific or historical research purposes or statistical purposes.

Another concern has to do with data accuracy. The data controller must show that has taken all reasonable steps to ensure that personal data, including biometric data, under his control, is accurate and up to date. The Regulation states that personal data shall be "accurate and, where necessary, kept up to date; every reasonable step must be taken to ensure that personal data that are inaccurate, having regard to the purposes for which they are processed, are erased or rectified without delay". Having in regard that the accuracy of biometric data is challenged [2] and that biometric data may need updating by force of natural factors such as ageing [3], this requisite demands from the data controller an ongoing effort of compliance.

One final remark concerning integrity and confidentiality of personal data. Processing and storage of personal data, especially biometric, must be carried out with strong security measures.[19] Reality displays numerous examples of how insecure information systems are. Hospitals, for instance, have become a prime target of cyber-attacks. Old and insecure computer systems coupled with poor security policies seem to create the perfect prey for hackers after easy money. The increasing economic value of health records in the black market, often surpassing credit card information, make the health industry an attractive victim of cyber-attacks such as data theft or crypto-ransomware. High security standards are not usually implemented and the exponential computerization of health care records creates new opportunities for security breaches. Although the regulation does not describe which security measures should be put in place, it requires proper security of the personal data that is object of processing, including protection against unauthorized or unlawful processing and against accidental loss, destruction or damage, using appropriate technical or organizational measures (article 5/f of the Regulation). The legislative choice of not determining exactly the security measures

[18]Incompatibility of purposes arises when there is not any link between the initial purpose and the further purpose, for instance. The data controller should also evaluate the context in which the personal data have been collected, in particular regarding the relationship between data subjects and the controller; the nature of the personal data; the possible consequences of the intended further processing for data subjects and the existence of appropriate safeguards, which may include encryption or pseudonymization.

[19]"If biometric databases are not protected properly and information is stolen, the consequences can be permanently devastating", Chinchilla, Rigoberto "Ethical and Social consequences of Biometric Technologies".

is an understandable one. Security measures as well as attack techniques change over time in their nature and complexity. They are a product of the technological state of the art. Given this, a neutral approach to wording is wiser, meaning that the European legislator opted for establishing certain objectives, e.g., confidentiality, integrity, availability and resilience, while stating the factors that should be taken into account when deciding which security measures to deploy. In this sense, the required technical and organizational security measures depend on the state of the art, the costs of implementation and the nature, scope, context and purposes of processing as well as the risk of varying likelihood and severity for the rights and freedoms of natural persons.

16.5 Conclusions

Since 1995, with the Directive 95/46/EC of the European Parliament and the Council, the European Union has established minimum rules concerning the protection of personal data which is thus enshrined in several important legal texts.

The concept of data protection as a fundamental right has been further developed in the General Data Protection Regulation (Regulation (EU) 2016/679 of the European Parliament and of the Council on the protection of natural persons with regard to the processing of personal data and on the free movement of such data).

Human physiological and/or behavioural characteristic can be used as a biometric characteristic as long as it satisfies some requirements like universality, distinctiveness, permanence and collectability. Biometric systems record personal information about identifiable individuals.

Biometric data should be understood, for the purposes of the Regulation, as any technical means of retrieval or confirmation of the identity of a natural living person using their physical, physiological or behavioural characteristics.

Some biometric data must also be considered as sensitive data and the processing sensitive biometric data is generally forbidden, but there are exceptions such as situations whenever there is a free informed consent or if someone is—physically or legally—incapable of giving consent, and there is the need of protection of his/her vital interests, among many others. Anyway, the sensitive nature of biometric data means that it can only be used in very restrictive and specific situations.

The collection of personal data must be done with a specified, explicit and legitimate purpose. Besides that, biometric data may need updating by force of natural factors such as ageing.

Processing and storage of personal data, especially biometric, must be carried out with strong security measures.

References

1. Bustard, J. (2015). The Impact of EU Privacy Legislation on Biometric System Deployment: Protecting citizens but constraining applications. IEEE Signal Processing Magazine, 32(5), 101–108.
2. Liu, Y. (2008). Identifying Legal Concerns in the Biometric Context, Journal of the International Commercial Law and Technology, 3(1), 45–54
3. Mordini E. & Rebera A. (2013). Social Factors in Ageing and Relevance to Biometrics. in M. Fairhurst (ed.), Age Factors in Biometric Processing, Springer: Berlin, 37–62

Correction to: Face Anti-spoofing in Biometric Systems

Zinelabidine Boulkenafet, Zahid Akhtar, Xiaoyi Feng, and Abdenour Hadid

Correction to:
Chapter 13 in: R. Jiang et al. (eds.), *Biometric Security and Privacy*, Signal Processing for Security Technologies, https://doi.org/10.1007/978-3-319-47301-7_13

The book was inadvertently published without clear indication of the provenance of some of the images in the figures of this chapter. This has now been corrected, and the images published without permission were removed.

The updated version of the chapter can be found at
https://doi.org/10.1007/978-3-319-47301-7_13

R. Jiang et al. (eds.), *Biometric Security and Privacy*, Signal Processing for Security Technologies, DOI 10.1007/978-3-319-47301-7_17

Index

R. Jiang et al. (eds.), *Biometric Security and Privacy*, Signal Processing for Security Technologies, DOI 10.1007/978-3-319-47301-7

MIX
Papier aus verantwortungsvollen Quellen
Paper from responsible sources
FSC® C105338

If you have any concerns about our products,
you can contact us on
ProductSafety@springernature.com

In case Publisher is established outside the EU,
the EU authorized representative is:
Springer Nature Customer Service Center GmbH
Europaplatz 3, 69115 Heidelberg, Germany

Printed by Libri Plureos GmbH
in Hamburg, Germany